Control Engineering

Control Engineering publishes research monographs and advanced graduate texts dealing with areas of current research in all areas of control engineering and its applications.

We encourage the preparation of manuscripts in TeX—LaTeX is also acceptable—for delivery as camera-ready hard copy, which leads to rapid publication, or on a diskette.

Proposals should be sent directly to the editor or to Birkhäuser Boston, Computational Science and Engineering Program, 675 Massachusetts Avenue, Cambridge, MA 02139, USA.

Robust Kalman Filtering for Signals and Systems with Large Uncertainties
I.R. Petersen, A.V. Savkin

Qualitative Theory of Hybrid Dynamical Systems
A.S. Matveev, A.V. Savkin

Lyapunov-Based Control of Mechanical Systems
M.S. de Queiroz, D.M. Dawson, S.P. Nagarkatti, F. Zhang

Nonlinear Control and Analytical Mechanics
H.G. Kwatny, G.L. Blankenship

Control Systems Theory with Engineering Applications
S.E. Lyshevski

Control Systems with Actuator Saturation
T. Hu, Z. Lin

Deterministic and Stochastic Time Delay Systems
E.-K. Boukas, Z.-K. Liu

Deterministic and Stochastic Time Delay Systems

El-Kébir Boukas
Zi-Kuan Liu

Birkhäuser
Boston • Basel • Berlin

El-Kébir Boukas
Zi-Kuan Liu
Mechanical Engineering Department
École Polytechnique de Montreal
P.O. Box 6079, Station "Centre-Ville"
Montreal, Quebec H3C 3A7
Canada

Library of Congress Cataloging-in-Publication Data
Boukas, El-Kébir.
    Deterministic and stochastic time delay systems / El-Kébir Boukas, Zi-Kuan Liu.
      p. cm.—(Control engineering)
    Includes bibliographical references and index.
    ISBN 0-8176-4245-5 (alk. paper)
      1. Feedback control systems.   2. Time delay systems.   I. Liu, Zi-Kuan.   II. Title.
    III. Control engineering (Birkhäuser)
    T216 .B65 2002
    629.8′3—dc21   2001043506

Printed on acid-free paper.
© 2002 Birkhäuser Boston   ***Birkhäuser*** ®

ISBN 0-8176-4245-5
ISBN 3-7643-4245-5   SPIN 10838829

Production managed by Louise Farkas; manufacturing supervised by Jacqui Ashri.
Typeset by the authors in LaTeX2e.
Printed and bound by Maple-Vail Book Manufacturing Group, York, PA.
Printed in the United States of America.

9 8 7 6 5 4 3 2 1

Birkhäuser   Boston   Basel   Berlin
A member of BertelsmannSpringer Science+Business Media GmbH

To the memory of my father
E.K. Boukas

Contents

II   Stochastic Control   175

Preface

Most practical processes such as chemical reactor, industrial furnace, heat exchanger, etc., are nonlinear stochastic systems, which makes their control in general a hard problem. Currently, there is no successful design method for this class of systems in the literature. One common alternative consists of linearizing the nonlinear dynamical stochastic system in the neighborhood of an operating point and then using the techniques for linear systems to design the controller. The resulting model is in general an approximation of the real behavior of a dynamical system. The inclusion of the uncertainties in the model is therefore necessary and will certainly improve the performance of the dynamical system we want to control.

The control of uncertain systems has attracted a lot of researchers from the control community. This topic has in fact dominated the research effort of the control community during the last two decades, and many contributions have been reported in the literature.

Some practical dynamical systems have time delay in their dynamics, which makes their control a complicated task even in the deterministic case. Recently, the class of uncertain dynamical deterministic systems with time delay has attracted some researchers, and some interesting results have been reported in both deterministic and stochastic cases. But we can't claim that the control problem of this class of systems is completely solved; more work must be done for this class of systems.

The presence of time delay in dynamical systems is a well-known cause of instability and performance degradation. The results on uncertain dynamical systems with time delay are few. Recently, Mahmoud [135] published a monograph on this subject. He dealt with the problems of stability, stabilizability, $\mathcal{H}_\infty$ control, filtering, and their robustness for deterministic linear systems with time delay.

Our goal in this book is to complement the work done by Mahmoud by developing new general results in the deterministic case and treating at the same time the control of uncertain dynamical stochastic systems with time delay. The results we are presenting in the area of uncertain dynamical systems with time delay are new. All the results we will report in this book except for the nonlinear case are stated using the linear matrix inequality (LMI) formalism, which is becoming a standard in control theory.

This book is divided into two parts and an introduction. Part I deals with the class of linear deterministic systems with time delay. It is composed of five chapters. Part I defines this class of systems. Mainly, we define the different concepts presented in the subsequent chapters of Part I, such as stability, stabilizability, $\mathcal{H}_\infty$ control, and filtering. Their robustness is also stated. The guaranteed cost control problem is also presented. Part II covers the class of dynamical stochastic systems with time delay. It is composed of six chapters.

Part II describes the class of dynamical systems with Markovian jumps and time delay and defines the concepts of stochastic stability, stochastic stabilizability, $\mathcal{H}_\infty$ control, filtering and their robustness. The guaranteed cost control of this class of systems is also defined.

The introduction gives, after presenting some dynamical time delay systems, the notation we will use throughout the book, states the problems we will deal with, and clarifies the difference between the two classes of systems under discussion. The terminology used in this book is defined. The first class of systems is restricted to the deterministic dynamical systems with time delay and the second one covers the stochastic dynamical systems with time delay.

Chapter 2 defines the different concepts we are dealing with for the class of deterministic time delay systems. Mainly, we define the stability and stabilizability concepts, the $\mathcal{H}_\infty$ control problem, the filtering problem and their robustness.

Chapter 3 contains the major results on stability and stabilizability of dynamical deterministic systems with time delay. Delay-independent and delay-dependent sufficient conditions are developed to check the stability or stabilizability of dynamical systems with time delay. The output feedback stabilization is also considered. Algorithms are developed to solve the different problems considered in this chapter.

Chapter 4 contains the results on robust stability and robust stabilizability of uncertain dynamical systems with time delay. Only norm-bounded uncertainty is considered in this chapter. Delay-independent and delay-dependent sufficient conditions are established to check the robust stability and robust stabilizability of the class of uncertain dynamical systems with time delay. Different design algorithms for state feedback are proposed. The robust output feedback stabilization is also considered.

Chapter 5 deals with the $\mathcal{H}_\infty$ control problem, the filtering problem, and the output feedback guaranteed cost control problem. Memory and memoryless controllers are used to stabilize the class of systems we are considering. Delay-independent and delay-dependent algorithms are developed for these problems. The LMI framework is used in all the problems.

Chapter 6 deals with the uncertain class of dynamical linear systems with time delay. Under the assumption that the uncertainties are of the norm-bounded form, the robust $\mathcal{H}_\infty$ control problem and the filtering problems are solved. Delay-independent and delay-dependent algorithms are developed to solve these problems using the LMI framework. For the $\mathcal{H}_\infty$ control problem different types of controllers (memoryless feedback, memory feedback, and output feedback) are used to stabilize the class of systems under consideration. The guaranteed cost control is also considered. Both delay-independent and delay-dependent conditions are established.

Chapter 7 defines the different concepts treated in Part II. Mainly, we define the stochastic stability and stochastic stabilizability concepts, the $\mathcal{H}_\infty$ control problem, the filtering problem and their robustness.

Chapter 8 covers the stochastic stability and stochastic stabilizability of the class of linear systems with Markovian jumps and time delay. Delay-independent and delay-dependent conditions are established to check the stochastic stability and the stochastic stabilizability of the class of systems under study. The output feedback stabilization of this class of systems is also treated. Finally, the stability of the class of systems with Markovian jumps and mode-dependent time delay is considered.

Chapter 9 considers the class of uncertain linear systems with Markovian jumps and time delay. The norm-bounded uncertainty type is covered in this chapter. The same problems covered in Chapter 8 are treated here and sufficient conditions that guarantee the robustness of the problem considered. For the stabilization problem different types of controllers are considered.

Chapter 10 considers the class of dynamical time delay systems with Markovian jumps and deals with the $\mathcal{H}_\infty$ control problem, filtering problems, and the output feedback guaranteed cost control problem. Memory and memoryless controllers are used to stochastically stabilize the class of systems we are considering. Delay-independent and delay-dependent algorithms are developed in the discussion of these problems. The LMI framework is used in all the problems.

Chapter 11 considers the uncertain class of dynamical time delay linear systems with Markovian jumps. Under the assumption that the uncertainties are of the norm-bounded form, the robust $\mathcal{H}_\infty$ control problem is solved. Delay-independent and delay-dependent algorithms are developed to solve these problems using the LMI framework. For the $\mathcal{H}_\infty$ control problem different types of controllers (memoryless feedback, memory feedback, and output feedback) are used to stochastically stabilize the class of systems under consideration. Guaranteed cost control is also considered. Both delay-independent and delay-dependent conditions are established.

Chapter 12 considers the class of nonlinear dynamical time delay systems with Markovian jumps. The stability problem and the stabilization problem are studied using the LMI formalism. For the stabilization problem linear state-feedback and nonlinear state-feedback are used to stochastically stabilize the class of systems we are considering.

Finally, several appendixes are included to make the book self-contained.

Most of the results presented in the different chapters are illustrated by numerical examples to show their usefulness. The LMI Matlab toolbox is used to solve the different required sufficient conditions. The Matlab code of all the examples we developed in this book can be obtained from http://www.meca.polymtl.ca/boukas/.

This book can be used as a textbook for graduate-level courses in engineering or as a reference for practicing control engineers, graduate students, and researchers in control engineering. Prerequisites to reading this book are elementary courses on mathematics, matrix theory, probability, optimization techniques, and control system theory.

We are deeply indebted to W. S. Levine, V. Dragan, A. Benzaouia, and P. Shi for reading the manuscript, in full or in part, and making corrections and suggestions to improve the book. We would like also to thank Louise Farkas for her comments on how to improve the readability of the book and Nicole Paradis from our Groupe d'Études et de Recherche en Analyse des Décisions (GERAD) for making the last changes and preparing the final version of the book.

We would like also to thank Mathworks for allowing us to reproduce some helpful LMI toolbox functions.

The draft of this book was completed in April 2001. We have tried to add new results of our own and others that are related to the topics covered in this book as we have become aware of them through journals and proceedings of conferences. However, because of the rapid developments in the field, it is possible that we have inadvertently omitted some results and references. We apologize to any author or reader who feels that we have not given credit where it is due.

Montréal, Québec, Canada   El-Kébir Boukas and Zi-Kuan Liu

1
Introduction

The purpose of this chapter is to define the class of dynamical systems with time delay that we are treating in this book. This class of systems can be defined as a group of systems in which there exists a time delay between the instant the input or the control to the system is applied and the instant or moment effect is observed. In most cases, the time delay cannot be neglected; it should be taken into account during the design phase of an experiment if we want to assure good performance.

Time delays occur in different industrial systems such as:

- mechanical systems

- electrical systems

- metallurgical systems

- chemical systems

- economic systems

- biological systems

For examples in these fields, we refer the reader to Malek-Zavarei and Jamshidi [141], Gorecki et al. [91], Kolmanovskii and Myshkis [117], Mahmoud [135], and the references therein.

A lot of attention has been paid to the class of dynamical systems with time delay. Among the problems that have been tackled we quote:

- stability

- stabilizability with different types of controllers like memoryless state feedback controller, memory state feedback controller, output feedback controller, etc.

- the $\mathcal{H}_\infty$ control problem

- the filtering problem

The robustness of these problems has also been addressed. Interesting results have been reported in the literature. The problem of stability has been extensively studied and delay-independent and delay-dependent conditions have been developed. For more information on this direction of research, we refer the reader to [44, 45, 94, 95, 96, 101, 114, 129, 158, 173, 208] and the references therein. In the case of uncertain linear time delay, robust stability has been also considered by many researchers. The reader can find more information in [42, 103, 104, 114, 115, 125, 128, 130, 131, 156, 204, 213], and the references therein.

For the stabilizability problem we refer the reader to [100, 132, 133, 140, 146, 153, 160, 188, 216] and the references therein. Robust stabilizability has also attracted many researchers from the control community. Some interesting results in this direction can be found in [46, 102, 123, 124, 134, 137, 138, 157, 198, 202, 219], and the references therein. Notice that different types of controllers have been considered.

For systems with time delay and exogenous disturbances, $\mathcal{H}_\infty$ control is one solution for eliminating the effect of the external disturbance. This type of problem has been extensively studied. For more information we refer the reader to [87, 106, 113, 130, 139, 154, 170, 174]. For the robust case, the reader can consult [130, 178, 184, 187, 196, 217, 221], and the references therein.

Filtering is an important control problem that has attracted many researchers. The following references summarize the mainstream in this direction of research: [12, 13, 89, 176, 199, 200, 203, 218, 210]. For robust filtering we refer the reader to [46, 76, 167, 166] and the references therein. Guaranteed cost control has also attracted many researchers. For more information on this topic, we refer the reader to [72, 126, 136, 144, 145, 220] and the references therein.

For stochastic systems, some interesting results have been published, mainly on the class of systems with Markovian jumps and time delay. In 1996 Benjelloun, Boukas, and Yang studied stochastic stability and stochastic stabilizability and their robustness. Boukas and his coauthors considered other problems like stochastic stability, stochastic stabilizability, $\mathcal{H}_\infty$ control, filtering, output feedback control and their robustness. More recently, Boukas and Liu have considered the class of Markovian jump linear systems with mode-dependent time delay and have extended their results to this class of systems. See for example [9] and the references therein for more information. Other contributions have been made by other researchers. Among them we mention [41].

To fix our idea on the topic we are dealing with, let us consider the following example from classical control theory. In 1942, Ziegler and Nichols proposed a model that can be used to describe a general class of single input/single output systems that arises in process control. The transfer function of this model is

$$G(s) = K \frac{e^{-\tau s}}{Ts + 1}$$

where K is the system gain, τ is the time delay in the system, and T is the system constant time.

Let $y(t)$ and $u(t)$ denote respectively the system output and input at time t respectively. Based on the classical control theory (see Boukas [16] or any equivalent textbook on classical control theory), we have

$$T\dot{y}(t) + y(t) = Ku(t - \tau).$$

If we assume that the initial condition is zero and letting $x(t) = y(t)$, the corresponding state space representation is given by

$$\dot{x}(t) = Ax(t) + Bu(t - \tau), \ x(\sigma) = \phi(\sigma) = 0, \ \sigma \in [-\tau, 0]$$
$$y(t) = Cx(t),$$

where $\phi(.)$ is the initial function, $A = \frac{-1}{T}$, $B = \frac{K}{T}$ and $C = 1$.

Remark 1.1 *Notice that the time delay is only on the control variable* $u(t)$. *Such a situation arises in many systems like fluid systems where for instance long pipes are used to transport fluids.*

As a second example, let us consider a dynamical system with the following state space representation:

$$\dot{x}(t) = Ax(t) + Bu(t), \ x(0) = x_o$$

where $x(t) \in \mathbb{R}^n$ is the state vector at time t, $u(t) \in \mathbb{R}^m$ is the input vector at time t, and A and B are constant matrices with appropriate dimensions.

Let us assume that the system is stabilizable and we have complete access to the state vector at time t, or the system allows us to get an estimate of the state vector at time t. Then, a state feedback controller can be used to control our system and get the desired performances. However, since this requires the measurement or the estimation of the state vector at time t, this will entail a delay τ in order to get the state vector and the controller to have the following form:

$$u(t) = Kx(t - \tau)$$

where K is the controller gain to be designed.

Plugging this controller into the previous state space representation we get

$$\dot{x}(t) = Ax(t) + A_d x(t - \tau), \ x(s) = x_o, \ s \in [-\tau, 0]$$

with $A_d = BK$.

Remark 1.2 *Notice that we have considered only constant time delay in the two examples. In general, time delay can be time-varying or stochastic. In some parts of the book, time delay will be considered time-varying, sometimes it will be stochastic, and in some situations both.*

These two examples show that dynamical systems with time delay can be easily described by a state space representation; and this will be our choice for the rest of the book. The representation by transfer function also can be used to study this class of systems. This representation will not be used through this manuscript.

The rest of this chapter is organized as follows. In Section 1.1 we develop some engineering examples to show how the time delay intervenes in the system dynamics. Deterministic and stochastic examples are provided. Section 1.2 gives the notations we will use in this book. In Section 1.3, we give the general dynamics we will use through the book and formulate the problems we will cover in this book. Section 1.4 presents the organization of the book.

1.1   Examples

In this section we give examples of some engineering examples that belong to the class of systems we are studying in this book. The examples are divided into two categories: the deterministic engineering systems and the stochastic engineering systems.

1.1.1   Deterministic Systems

In this subsection, we will give some deterministic engineering systems with time delay. Three examples from different areas are developed. The emphasis is made on the establishment of the state representation.

Example 1.1 *As a first example, let us consider a system for metal rolling. Its goal is to produce metal sheets of desired thicknesses. Figure 1.1 illustrates the different components of a metal rolling system. In principle, ribbons of hot metal are drawn through rollers at high speed. The lower roller is fixed and the upper one is used to adjust the spacing between the two rollers, which controls the thickness of the metal sheet. This can be done by controlling the position of the upper roller. For obvious reasons, the thickness sensor cannot be placed at the rollers, so it is placed downstream,*

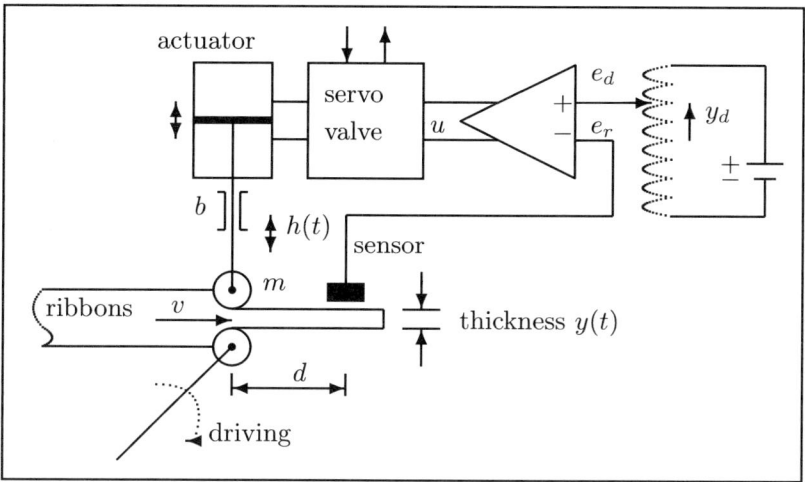

Figure 1.1. Roller system.

which results in delay in the measurement. This delay will depend on the distance d between the rollers and the sensor and the speed v at which the ribbons are fed into the system. If we denote by y(t) the real thickness of the produced sheet, the measured one will be $y(t - \tau)$, where $\tau = \frac{d}{v}$. To control the position of the upper roller, we can use a drive motor. Based on classical control theory (see Boukas [16]) we can get the block diagram of Figure 1.2 of the corresponding system with G(s) the transfer function between the position of the upper roller and the input voltage of the drive motor; H(s) the transfer function of the sensor ($H(s) = e^{-\tau s}$), and C(s) the transfer function of the used controller. The input, P(s), is introduced to represent the effect of the exogenous disturbance on the system. In the rest of this example, we assume that the controller is a proportional one with a gain k_p, and we consider two cases of modeling the group drive motor and upper roller when the system is disturbance free, i.e., p(t) for all $t \geq 0$.

- *Let us assume that the transfer function G(s) is given by*

$$G(s) = \frac{k_m}{s}.$$

Based on the block diagram and classical control theory, we get

$$\dot{y}(t) + k_p k_m y(t - \tau) = k_p k_m y_d(t)$$

with $y_d(t)$ the desired thickness. This gives the following state space representation:

$$\dot{x}(t) = A_d x(t - \tau) + Br(t), \ x(0) = x_0 \ initial \ thikness$$
$$y(t) = Cx(t)$$

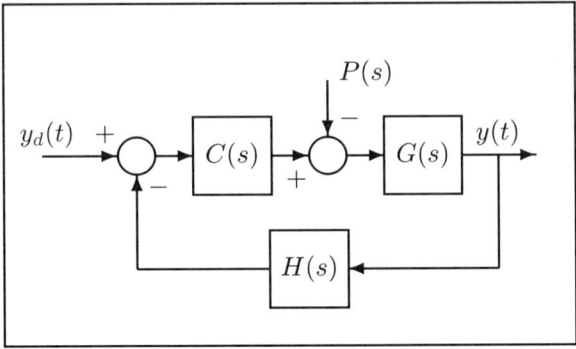

Figure 1.2. Block diagram of rolling system.

where $y(t) = x(t)$, $A_d = -k_p k_m$, $B = k_p k_m$ and $C = 1$.

- *Let us assume that the transfer function $G(s)$ is given by*

$$G(s) = \frac{k_m}{s(Ts+1)}.$$

Following the same steps as before, we get

$$\ddot{y}(t) + \dot{y}(t) + k_p k_m y(t - \tau) = k_p k_m y_d(t).$$

If we let $y(t) = x_1(t)$, $x_2(t) = \dot{y}(t)$, we get the following state space representation:

$$\dot{x}(t) = Ax(t) + A_d x(t - \tau) + Bu(t), \quad x(0) = x_0$$
$$y(t) = Cx(t)$$

where

$$A = \begin{pmatrix} 0 & 1 \\ 0 & -\frac{1}{T} \end{pmatrix}$$

$$A_d = \begin{pmatrix} 0 & 0 \\ 0 & -\frac{k_p k_m}{T} \end{pmatrix}$$

$$B = \begin{pmatrix} 0 \\ \frac{k_p k_m}{T} \end{pmatrix}$$

$$C = \begin{pmatrix} 1 & 0 \end{pmatrix}$$

$$u(t) = y_d(t).$$

In both cases, we have seen that we can put our system in the state space representation with delayed state.

Example 1.2 *As a second engineering example, let us consider a heating system (Figure 1.3). This system uses hot air to control the temperature of a given room that is located a little bit far from the heater. Due to this long distance that separates the heater and the space that we would like to* control its temperature, a time delay will result. If the distance is d and the velocity of the hot air in the pipe is v, the time delay is then given by

$$\tau = \frac{d}{v}$$

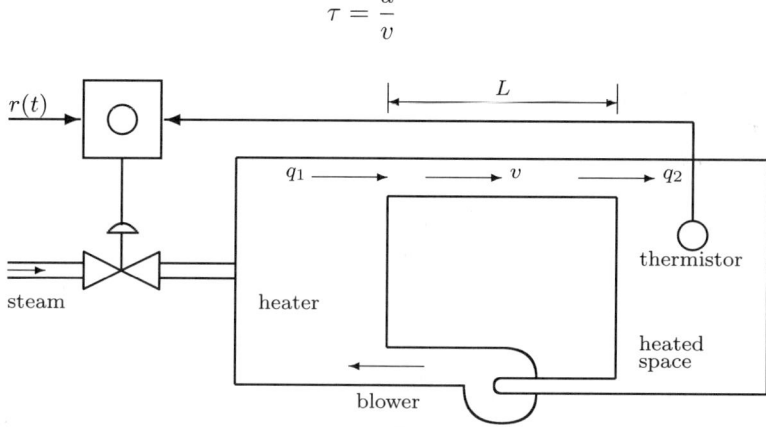

Figure 1.3. Heating system.

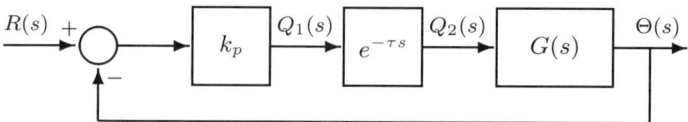

Figure 1.4. Block diagram of the heating system.

Let us denote by $q_1(t)$ and $q_2(t)$ the heat flow rate from the heater and the one acting on the room, respectively. Due to the long pipe, we have the following relationship between these two flow rates

$$q_2(t) = q_1(t - \tau).$$

From classical control theory (see Boukas [16]), we can show that the transfer function between the temperature $\theta(t)$ of the room and the flow rate $q_2(t)$ is approximated by

$$G(s) = \frac{\Theta(s)}{Q_2(s)} = \frac{k}{Ts + 1}$$

where T and k are given constants that depend on the system.

A block diagram that represents the control system is illustrated by Figure 1.4.

If we define $y(t) = \theta(t) = x(t)$ and $u(t) = r(t)$ where $r(t)$ is the desired temperature, we get

$$\dot{x}(t) = Ax(t) + A_d x(t - \tau) + Bu(t - \tau), \quad x(0) \text{ given}$$

$$y(t) = Cx(t)$$

where $A = -\frac{1}{T}$, $A_d = -\frac{k_p k}{T}$, $B = \frac{k_p k}{T}$ and $C = 1$.

Here again, we were able to put the dynamics of the system in the state space representation with delayed state and control.

Example 1.3 *As a third example, let us consider the control of a feeder of a production system (see Figure 1.5). The goal consists of feeding the downstream machine with a constant flow of mass. The feed is brought to the control of the opening of a mechanism that allows the mass flow. The sensor (for the same reasons as for the rolling system) is placed at a distance, d, from the opening. If the material is flowing with a speed v the time delay is $\tau = \frac{d}{v}$, and the corresponding block diagram is similar to the rolling system.*

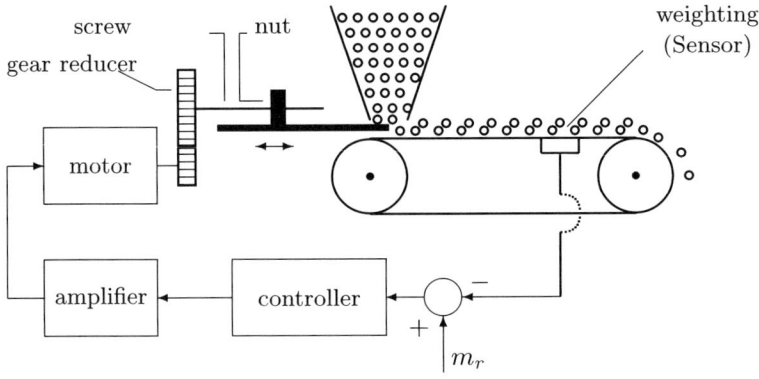

Figure 1.5. Control feeding of production system.

If we choose a proportional controller and we assume that the transfer function of the group amplifier-motor and the mechanical part that controls the opening is given by:

$$G(s) = \frac{L(s)}{Z(s)} = \frac{k}{s(Ts + 1)}$$

where $l(t)$ is the position of the mechanical part, $z(t)$ is the applied voltage to the group amplifier, motor, and mechanical part, k and T are given constants, then the state space representation of this system becomes

$$\dot{x}(t) = Ax(t) + A_d x(t - \tau) + Bu(t), \quad x(0) = x_0$$
$$y(t) = Cx(t)$$

where

$$A = \begin{pmatrix} 0 & 0 \\ 0 & -\frac{1}{T} \end{pmatrix}$$

$$A_d = \begin{pmatrix} 0 & 0 \\ -\frac{k_p k_m}{T} & 0 \end{pmatrix}$$

$$B = \begin{pmatrix} 0 \\ \frac{k_p k_m}{T} \end{pmatrix}$$

$$C = \begin{pmatrix} 1 & 0 \end{pmatrix}$$

$$u(t) = r(t)$$

with $y(t) = x_1(t) = l(t)$, $x_2(t) = \dot{l}(t)$.

Remark 1.3 *These engineering examples show us how we can model dynamical systems with time delay with a linear state space representation, where the time delay affects the state vector or the control vector or both.*

So far, we have not touched on the main subject of our book, which is the stochastic behavior of dynamical systems with time delay. This will be covered in the next section.

1.1.2   Stochastic Systems

In this subsection we will consider the class of systems with Markovian jumping parameters and show how this class of systems can model different types of practical systems.

Example 1.4 *Let us consider a production system with failure-prone machines. For simplicity of presentation, let us assume that the production system is composed of one machine that produces one part type. Let us also assume that the machine state is described by a continuous-time Markov process $\{r(t), t \geq 0\}$ with finite state space $\mathcal{S} = \{0, 1\}$. Equation $r(t) = 0$ means that the machine is under repair and no part can be produced, and $r(t) = 1$ means that the machine is operational and it can produce parts. Switching between these two states is described by the following probability transitions:*

$$\mathrm{P}[r(t + \Delta t) = j | r(t) = i] = \begin{cases} \lambda_{ij} \Delta t + o(\Delta t) & \text{if } i \neq j \\ 1 + \lambda_{ii} \Delta t + o(\Delta t) & \text{otherwise} \end{cases}$$

with $\lambda_{ij} \geq 0$ when $i \neq j$, $\lambda_{ii} = -\sum_{j \neq i} \lambda_{ij}$ and $\lim_{\Delta t \to 0} \frac{o(\Delta t)}{\Delta t} = 0$.

Let $x(t)$ denote the stock level of the production system at time t. When $x(t) \geq 0$, it represents a surplus, and when $x(t) < 0$, it represents the backlog. Let us assume that the produced parts deteriorate with time at a constant rate, ρ, and there exists a transport time delay $\tau(t)$, which can model the processing time of the part. Let $u(t)$ and d represent respectively the production rate at time t and the constant demand rate.

The stock level is then described by the dynamic

$$\dot{x}(t) = -\rho x(t) + b(r_t) u(t - \tau(t)) - d$$

with

$$b(r_t) = \begin{cases} 1 & \text{if } r_t = 1, \\ 0 & \text{otherwise.} \end{cases}$$

The production rate $u(t)$ is assumed to satisfy the constraints

$$0 \le u(t) \le \bar{u}$$

with $\bar{u}$ a given positive constant.

When the production system is producing p parts, the state space representation becomes

$$\dot{x}(t) = Ax(t) + B(r_t)u(t - \tau(t)) - Id$$

where

$$A = \begin{pmatrix} -\rho_1 & & \\ & \ddots & \\ & & -\rho_p \end{pmatrix}$$

$$B = \begin{pmatrix} b_1(r_t) & \cdots & \\ & \ddots & \\ & & b_p(r_t) \end{pmatrix}$$

$$I = \begin{pmatrix} 1 & & \\ & \ddots & \\ & & 1 \end{pmatrix}$$

with ρ_i, $i = 1, \ldots, p$ is the deterioration rate of the part i and $b_j(r_t)$ is defined as previously.

Remark 1.4 *Notice that we can make the time delay in the previous example mode-dependent. This can be done by assuming that the Markov process $\{r(t), t \ge 0\}$ modeling the machine state takes values in a state space $\mathcal{S} = \{0, 1, 2, \ldots, s - 1\}$. Equation $r(t) = 0$ means that the machine is under repair; and $r(t) = 1$ means that the machine is good and the produced parts are of good quality. $r(t) = s - 1$ means that the machine is in bad condition. Notice that when the process takes values 1, 2, 3, etc. the machine state degrades. In this case, the processing time will be affected by the process $r(t) \in \mathcal{S}$ and therefore the time delay will be $\tau(r(t))$.*

Example 1.5 *Let us assume that we have a dynamical system that is described by the dynamic*

$$\dot{x}(t) = Ax(t) + Bu(t),$$
$$x(0) = x_o,$$

where $x(t)$ is the state vector of the system, $u(t)$ is the control input vector, A and B are constant matrices with appropriate dimensions.

Let us assume that the system is stabilizable and we have complete access to the state vector $x(t)$ for feedback, but assume also that the measurement of the state vector is available with a certain time delay that depends on the state of the network used to control our system. Let us assume that the state of the network is described by a continuous-time Markov process $\xi(t)$ with finite state space $\mathcal{S}_1 = \{1, 2, \ldots, N_1\}$, i.e.,

$$u(t) = Kx(t - \tau(\xi(t))$$

where K is a constant gain to be designed.

Plugging this control law into the previous dynamic gives

$$\dot{x}(t) = Ax(t) + A_d x(t - \tau(\xi(t))), \; x(0) = x_0$$

with $A_d = BK$.

This system belongs to the class of systems we are considering in this book.

Example 1.6 *We assume now, besides what we have supposed for the previous example, that even the dynamic of the system is described by a continuous-time Markov process $\zeta(t)$ taking values in a finite state space $\mathcal{S}_2 = \{1, \ldots, N_2\}$, i.e., the matrices $A(\zeta(t))$ and $B(\zeta(t))$ depend on the process $\{\zeta(t) : t \geq 0\}$. In this case, the dynamic becomes*

$$\dot{x}(t) = A(\zeta(t))x(t) + B(\zeta(t))u(t),$$
$$x(0) = x_o,$$

where $x(t)$ is the state vector of the system, $u(t)$ is the control input vector, $A(\zeta(t))$ and $B(\zeta(t))$ are constant matrices with appropriate dimensions for each mode $\zeta(t)$ in $\mathcal{S}_2$.

Let us assume that the system is stochastically stabilizable and we have complete access to the state vector $x(t)$ for feedback, but assume also that the measurement of the state vector is available with a certain time delay that depends on the state of the network used to control our system. Let us assume that the state of the network is described by a continuous-time Markov process $\xi(t)$ with finite state space $\mathcal{S}_1 = \{1, 2, \ldots, N_1\}$, i.e.,

$$u(t) = K(\zeta(t))x(t - \tau(\xi(t)))$$

where $K(\zeta(t))$ is a constant gain to be designed for each $\zeta(t) \in \mathcal{S}_2$.

If we define the resulting Markov process by $r(t) = (\zeta(t), \xi(t))$, which values in $\mathcal{S} = \mathcal{S}_1 \times \mathcal{S}_2 = \{1, 2, \ldots, N\}$, with $N = N_1 \times N_2$, the resulting dynamic becomes

$$\dot{x}(t) = A(r(t))x(t) + A_d(r(t))x(t - \tau(r(t))),$$

with $A_d(r(t)) = B(r(t))K(r(t))$.

In this section we have presented some engineering systems that can be described with the class of systems we are considering in this book. In the subsequent chapters we will deal with different problems for the dynamical time delay systems. The LMI test and algorithms are developed to facilitate the resolution of problems we will deal with. In the next section, we will define the notation we will use in the rest of this book.

1.2 Notation and Abbreviations

The notation we are using in this book is quite standard in control theory. We will study dynamical time delay systems with state space representation, and therefore, we will extensively use vectors and matrices in all our developments. Vectors and matrices are represented by lower case and upper case letters, respectively. The following table summarizes most of the symbols in this book.

Symbol	*Meaning*
$\mathbb{R}^n$	set of real n-dimensional vectors
$\mathbb{R}^{n \times m}$	set of real $n \times m$ matrices
$\mathbb{N}$	set of natural numbers
$\mathcal{S}$	mode state space, $\mathcal{S} = \{1, 2, \ldots, N\}$
	where N is a positive natural number
$x(t)$	state vector at time t, $x(t) \in \mathbb{R}^n$
$y(t)$	measured output vector at time t, $y(t) \in \mathbb{R}^q$
$z(t)$	controlled output vector at time t, $z(t) \in \mathbb{R}^p$
τ	constant time delay in the system, $\tau \geq 0$
$\tau(r(t))$	mode-dependent time delay, $\tau(r(t)) \geq 0$ for every $r(t) \in \mathcal{S}$
$x(t - \tau)$	delayed state vector at time t
$\bar{\tau}$	maximum time delay in the system,
	$\bar{\tau} = \max(\tau_1, \ldots, \tau_m)$ for the deterministic case
	or $\bar{\tau} = \max(\tau(1), \ldots, \tau(N))$ for the stochastic case
$\underline{\tau}$	minimum time delay in the system,
	$\underline{\tau} = \min(\tau_1, \ldots, \tau_N)$ for the deterministic case
	$\underline{\tau} = \min(\tau_{r_1}, \ldots, \tau_{r_N})$ for the stochastic case
$\bar{\lambda}$	$\bar{\lambda} = \max(|\lambda_{11}|, \ldots, |\lambda_{NN}|)$
$\mathbb{C}[-\tau, 0]$	space of continuous functions defined on $[-\tau, 0]$
$\mathbf{x}(t)$	$\mathbf{x}(t) \triangleq (x(s), t - \bar{\tau} \leq s \leq t) \in \mathbb{C}[-\tau, 0]$
$u(t)$	control input vector at time t, $u(t) \in \mathbb{R}^m$
$u(t - \tau)$	delayed control input vector at time t
$w(t)$	external disturbance vector at time t, $w(t) \in \mathbb{R}^l$
I	identity matrix (the size can be obtained from the context)
A	state matrix, $A \in \mathbb{R}^{n \times n}$
A_d	state matrix of the delayed state vector, $B \in \mathbb{R}^{n \times m}$
B	input matrix
$r(t)$	Markov process describing the system mode at time t

Symbol	Meaning
$P[.]$	probability measure
λ_{ij}	jump rate from mode i to mode j, $(i,j \in \mathcal{S})$
$\Delta(t)$	norm-bounded uncertainty for deterministic system
$\Delta(r_t, t)$	norm-bounded uncertainty for stochastic system
$\|\cdot\|_2$	2-norm (vector, signal, or system)
$\|\cdot\|_\infty$	∞-norm (vector, signal or system)
$\mathbb{E}\{.\}$	mathematical expectation
$\mathcal{L}_2\left([t_1, t_2], \mathbb{R}^l\right)$	set of integrable functions on $[t_1, t_2]$, taking values in $\mathbb{R}^l$: $\mathcal{L}_2\left[[t_1, t_2], \mathbb{R}^l\right] \triangleq \{f(t)\| \int_{t_1}^{t_2} f^\top(s)f(s)ds < \infty\}$
$\operatorname{diag}(A_1, \ldots, A_n)$	real matrix with diagonal elements $A_1, \ldots, A_n$
$\operatorname{tr}(A)$	trace of the square matrix A, $A \in \mathbb{R}^{n \times n}$
$A^\top$	transpose of the real matrix A, $A \in \mathbb{R}^{n \times m}$
$\mathcal{X}_i$	$\mathcal{X}_i = \operatorname{diag}\{X_1, \ldots, X_{i-1}, X_{i+1}, \ldots, X_N\}$
$\mathcal{Y}_i$	$\mathcal{Y}_i = \operatorname{diag}\{Y_1, \ldots, Y_{i-1}, Y_{i+1}, \ldots, Y_N\}$
$\mathcal{B}_\perp$	orthogonal complement of $\mathcal{B}$
$\lambda_{\max}(A)$	maximum eigenvalue of the square matrix A
$\lambda_{\min}(A)$	minimum eigenvalue of the square matrix A
$\dot{V}(.)$	derivative of the Lyapunov functional candidate with respect to time t
$*$	denotes a symmetric term in a matrix
$P > 0$	symmetric, positive-definite matrix
MJLS	Markov jump linear system
SS	stochastically stable
RSS	robustly stochastically stable
MSQS	mean square quadratically stable
LMI	linear matrix inequality
GEVP	generalized eigenvalue problem
s.t.	subject to

1.3   Dynamical Systems with Time Delay

In this section, we will first try to give the dynamics of the class of systems with time delay we consider in this book. Then we define what we will mean by free (unforced) system, nominal systems, deterministic systems and stochastic systems.

In the previous engineering examples, we made assumptions to get a linear model representing the behavior of the real system under consideration.

But in practice, this model never describes the real dynamics of the system; a discrepancy always exits between the linear model and the real behavior. This discrepancy is referred to as the *system uncertainty*, which can be the consequence of things including neglected dynamics, and changes in systems parameters. Consequently, it is a must that system uncertainties be considered in the analysis and design phases in order to guarantee the required performance of the system. The model that takes care of these uncertainties and which will describe correctly the system dynamics is referred to as uncertain.

In practice, besides uncertainties we also can have systems that have different state space representation, and the switching between these different representation, referred to as the modes, is random and described by a continuous-time Markov process with finite state space $\mathcal{S} = \{1, 2, \ldots, N\}$. The class of systems with such behavior is referred to as the class of systems with Markovian jump parameters. It is a hybrid class of systems with two components in the state vector, i.e., $(x(t), r_t)$. The first component, $x(t)$, taking values in $\mathbb{R}^n$, is referred to as the continuous state vector and the second component, r_t, taking values in $\mathcal{S}$, is referred to as the mode of the system. A mathematical representation for the linear case of the class of systems we described is governed by the following dynamics:

$$\begin{cases} \dot{x}(t) = A(r_t, t)x(t) + A_d(r_t, t)x(t - \tau) + B_1(r_t, t)w(t) + B(r_t, t)u(t) \\ z(t) = C_1(r_t, t)x(t) + D_{11}(r_t, t)w(t) + D_{12}(r_t, t)u(t) \\ y(t) = C_2(r_t, t)x(t) + D_{21}(r_t, t)w(t) \\ x(t) = \phi(t), t \in [-\tau, 0], \ r_0 \text{ given} \end{cases} \tag{1.1}$$

where $x(t) \in \mathbb{R}^n$ is the state vector at time t, $u(t) \in \mathbb{R}^m$ is the control input vector at time t, $w(t) \in \mathbb{R}^l$ is the square-integrable disturbance input vector at time t, $z(t) \in \mathbb{R}^p$ is the controlled output vector at time t, $y(t) \in \mathbb{R}^q$ is the measured output vector at time t, τ is the time delay, $\phi(.)$ is the initial function, and $A(r_t, t)$, $A_d(r_t, t)$, $B_1(r_t, t)$, $B(r_t, t)$, $C_1(r_t, t)$, $C_2(r_t, t)$, $D_{11}(r_t, t)$, $D_{12}(r_t, t)$, and $D_{21}(r_t, t)$ are as follows:

$$A(r_t, t) = A(r_t) + D_A(r_t)\Delta_1(r_t, t)E_A(r_t)$$
$$A_d(r_t, t) = A_d(r_t) + D_{A_d}(r_t)\Delta_2(r_t, t)E_{A_d}(r_t)$$
$$B_1(r_t, t) = B_1(r_t) + D_{B_1}(r_t)\Delta_1(r_t, t)E_{B_1}(r_t)$$
$$B(r_t, t) = B(r_t) + D_B(r_t)\Delta_1(r_t, t)E_B(r_t)$$
$$C_1(r_t, t) = C_1(r_t) + D_{C_1}(r_t)\Delta_1(r_t, t)E_{C_1}(r_t)$$
$$C_2(r_t, t) = C_2(r_t) + D_{C_2}(r_t)\Delta_1(t)(r_t, t)E_{C_2}(r_t)$$
$$D_{11}(r_t, t) = D_{11}(r_t) + D_{D_{11}}(r_t)\Delta_1(r_t, t)E_{D_{11}}(r_t)$$
$$D_{12}(r_t, t) = D_{12}(r_t) + D_{D_{12}}(r_t)\Delta_1(r_t, t)E_{D_{12}}(r_t)$$
$$D_{21}(r_t, t) = D_{21}(r_t) + D_{D_{21}}(r_t)\Delta_1(r_t, t)E_{D_{21}}(r_t)$$

with $A(r_t)$, $A_d(r_t)$, $B_1(r_t)$, $B(r_t)$, $C_1(r_t)$, $C_2(r_t)$, $D_{11}(r_t)$, $D_{12}(r_t)$, $D_{21}(r_t)$, $D_A(r_t)$, $E_A(r_t)$, $D_{A_d}(r_t)$, $E_{A_d}(r_t)$, $D_{B_1}(r_t)$, $E_{B_1}(r_t)$, $D_B(r_t)$,

$E_B(r_t)$, $D_{C_1}(r_t)$, $E_{C_1}(r_t)$, $D_{C_2}(r_t)$, $E_{C_2}(r_t)$, $D_{D_{11}}(r_t)$, $E_{D_{11}}(r_t)$, $D_{D_{12}}(r_t)$, $E_{D_{12}}(r_t)$, $D_{D_{21}}(r_t)$, and $E_{D_{21}}(r_t)$ constant matrices with appropriate dimensions for each mode $r(t)$ in $\mathcal{S}$. $\Delta_1(r_t,t)$ and $\Delta_2(r_t,t)$ unknown time-varying matrices of appropriate dimensions representing the parameter uncertainties in the system.

Remark 1.5 *Notice that the dimension of the different matrices can easily be obtained from the dynamics. We leave this for the reader.*

The switching between the different modes of the stochastic systems is described by the following probability transitions (see Appendix D and the references cited there):

$$P[r_{t+\Delta t} = j | r_t = i] = \begin{cases} \lambda_{ij}\Delta t + o(\Delta t) & \text{if } i \neq j \\ 1 + \lambda_{ii}\Delta t + o(\Delta t) & \text{otherwise} \end{cases} \quad (1.2)$$

with $\lambda_{ij} \geq 0$ for all $i \neq j$, $\lambda_{ii} = -\sum_{j \neq i} \lambda_{ij}$, and $\lim_{\Delta t \to 0} \frac{o(\Delta t)}{\Delta t} = 0$.

The uncertainties $\Delta_1(r_t,t)$ and $\Delta_2(r_t,t)$ will be admissible if they satisfy the following conditions for each $r(t)$ in $\mathcal{S}$:

$$\Delta_1^\top(r_t,t)\Delta_1(r_t,t) \leq I$$
$$\Delta_2^\top(r_t,t)\Delta_2(r_t,t) \leq I.$$

The mathematical model for the nominal system with Markovian jumps and time delay is described by the following dynamics:

$$\begin{cases} \dot{x}(t) = A(r_t)x(t) + A_d(r_t)x(t-\tau) + B_1(r_t)w(t) + B(r_t)u(t) \\ z(t) = C_1(r_t)x(t) + D_{11}(r_t)w(t) + D_{12}(r_t)u(t) \\ y(t) = C_2(r_t)x(t) + D_{21}(r_t)w(t) \\ x(t) = \phi(t), \ t \in [-\tau, 0]. \end{cases} \quad (1.3)$$

Remark 1.6 *When the finite state space $\mathcal{S}$ is reduced to one element, the corresponding system (1.1) or (1.3) has only one mode and is referred to as a deterministic uncertain system with time delay. In this case we will drop $r(t)$ from the dynamics. This will be the case for the systems in all chapters of Part I.*

Remark 1.7 *When the uncertainties $\Delta_1(r_t,t) = \Delta_2(r_t,t) \equiv 0$, the system is referred to as the nominal system. System (1.3) with $u(t) \equiv 0$ is called an unforced or free system.*

For the class of stochastic dynamical systems with time delay (1.1) or (1.2) many questions have to be solved. Among them, we quote the following:

- Which conditions can we use to check whether the unforced system ($u(t) \equiv 0$ and $w(t) \equiv 0$) is stochastically stable or not? robustly stochastically stable or not?

- Which type of controller can we use to stochastically stabilize or to robustly stabilize this class of systems, and which conditions can we use to check whether the systems under the chosen controller are stochastically stable or not? robust stochastically stable or not?

- In case of nonaccessibility to the state vector at time t, which type of filter can we use to estimate the state vector, and how do we design the filter parameters?

- In most cases, the time delay is not known and we are interested in establishing a method that allows us to determine what will be the maximum time delay that a given system can have to remain stochastically stable or stochastically unstable under a given control law.

- In practice, we are interested in rejecting the effect of the exogenous disturbance, therefore, it is helpful to establish a condition that allows us to check if a given system will have or not have the desired disturbance rejection level and how to determine what will be the minimum disturbance rejection level. How do we determine the minimum disturbance rejection level that we can get?

Remark 1.8 *Different extensions of the previous dynamics can be obtained. Among these extensions we quote the following:*

- *The time delay in the system can be considered to be mode-dependent, that is, $\tau(r(t))$.*

- *The time delay in the dynamics can be chosen time-varying, that is, $\tau(t)$, and in this case some assumptions are required. Commonly, the following assumptions are used:*

$$0 \leq \tau(t) \leq \bar{\tau}$$
$$0 \leq \dot{\tau}(t) \leq 1$$

  where $\bar{\tau} > 0$ is a given positive number.

- *The last extension consists of considering systems with multiple time delays that can be deterministic or stochastic.*

In this book our first goal is to develop delay-independent and delay-dependent conditions that the reader can use to solve its appropriate control problem. We would like to advise the reader that delay-independent conditions are easy to get, but conservative since they don't depend on the system time delay. On the other hand, the delay-dependent conditions depend on the time delay, and therefore are less conservative and valid only on a finite interval in which the time delay system belongs. This guarantees that the studied problem remains valid.

1.4   Outline of the Book

The rest of this monograph is divided in two parts. Part I deals with the class of linear deterministic systems with time delay. It is composed of five chapters. Part II covers the class of dynamical stochastic systems with time delay. It is composed of six chapters.

In Part I, we will deal with the following problems:

- stability and stabilizability and their robustness

- $\mathcal{H}_\infty$ control, filtering and their robustness

- guaranteed cost.

Part II will treat the stochastic version of the problems studied in Part I. In both parts, we are interested in delay-dependent and delay-independent conditions.

1.5   Notes

Time delay is inherent to some practical systems whose control is considered the key to the success of industrialized societies. The presence of this time delay renders the control of systems more complicated. This was the challenge for the control community during the last three decades. Progress has been made on this class of systems, and we have seen the development of results that can we can use presently to control such systems. Malek-Zavarei [141], Gorecki et al. [91], Kolmanovskii and Myshkis [117], Mahmoud [135], and the references therein summarize most of the current results on this class of systems.

Part I

Deterministic Control

2
Deterministic Time Delay Systems

It is clear from the examples given previously that the the investigation of the class of dynamical time delay systems is of great importance in dealing with the world economy. It is why much attention has been paid to this class of systems by many researchers from different communities. During recent decades, the control community has contributed to many problems of this class of systems. Part I focuses on this class of systems and gives a summary of what has been done in the area. It addresses mainly the problems of stability and stabilizability of the class of linear systems with time delay. It also deals with the robustness of stability and stabilizability when uncertainties are assumed to be of the norm-bounded type. Memoryless and memory state feedback controllers and output feedback controllers are used to stabilize dynamical time delay systems and to ensure the desired performance.

$\mathcal{H}_\infty$ control and estimation problems for the class of linear systems with time delay are also treated. The robustness of these problems is also considered. Finally, the guaranteed cost control problem is investigated. In all the problems delay-independent and delay-dependent conditions are developed. The LMI framework is used to facilitate computation and to take advantage of existing tools including the Matlab LMI toolbox and Scilab.

Part I is organized as follows. Section 2.1 presents the dynamics of the class of deterministic dynamical systems with time delay and enumerates the different problems we will cover in this part. In Section 2.2, the different concepts are defined and the problems are stated. Section 2.3 presents the objectives of Part I.

2.1   Class of Dynamical Linear Systems with Time Delay

Based on the examples in the introduction, a general mathematical representation of the class of dynamical linear systems with time delay can be described by the following dynamics:

$$\begin{cases} \dot{x}(t) = A(t)x(t) + A_d(t)x(t-\tau) + B_1(t)w(t) + B(t)u(t) \\ z(t) = C_1(t)x(t) + D_{11}(t)w(t) + D_{12}(t)u(t) \\ y(t) = C_2(t)x(t) + D_{21}(t)w(t) \\ x(t) = \phi(t), t \in [-\tau, 0] \end{cases} \tag{2.1}$$

where $x(t) \in \mathbb{R}^n$ is the state vector at time t, $u(t) \in \mathbb{R}^m$ is the control input vector at time t, $w(t) \in \mathbb{R}^l$ is the square-integrable disturbance input vector at time t (unknown but not necessarily random), $z(t) \in \mathbb{R}^p$ is the controlled output vector at time t, $y(t) \in \mathbb{R}^q$ is the measured output vector at time t, τ is the time delay, $\phi(.)$ is the initial function, and $A(t)$, $A_d(t)$, $B_1(t)$, $B(t)$, $C_1(t)$, $C_2(t)$, $D_{11}(t)$, $D_{12}(t)$, and $D_{21}(t)$ are as follows:

$$A(t) = A + D_A \Delta_1(t) E_A$$
$$A_d(t) = A_d + D_{A_d} \Delta_2(t) E_{A_d}$$
$$B_1(t) = B_1 + D_{B_1} \Delta_1(t) E_{B_1}$$
$$B(t) = B + D_B \Delta_1(t) E_B$$
$$C_1(t) = C_1 + D_{C_1} \Delta_1(t) E_{C_1}$$
$$C_2(t) = C_2 + D_{C_2} \Delta_1(t) E_{C_2}$$
$$D_{11}(t) = D_{11} + D_{D_{11}} \Delta_1(t) E_{D_{11}}$$
$$D_{12}(t) = D_{12} + D_{D_{12}} \Delta_1(t) E_{D_{12}}$$
$$D_{21}(t) = D_{21} + D_{D_{21}} \Delta_1(t) E_{D_{21}}$$

with A, A_d, B_1, B, C_1, C_2, D_{11}, D_{12}, D_{21}, D_A, E_A, D_{A_d}, E_{A_d}, D_{B_1}, E_{B_1}, D_B, E_B, D_{C_1}, E_{C_1}, D_{C_2}, E_{C_2}, $D_{D_{11}}$, $E_{D_{11}}$, $D_{D_{12}}$, $E_{D_{12}}$, $D_{D_{21}}$, and $E_{D_{21}}$ are constant matrices with appropriate dimensions and $\Delta_1(t)$ and $\Delta_2(t)$ are unknown time-varying matrices of appropriate dimensions representing the parameter uncertainties in the system and satisfying the following:

$$\Delta_1^\top(t)\Delta_1(t) \le I$$
$$\Delta_2^\top(t)\Delta_2(t) \le I$$

which are called admissible uncertainties.

Part I is composed of five chapters that deal with the following problems and their robustness:

- stability;

- stabilizability using different types of controllers like memoryless and memory state feedback controllers and output feedback controller;

- $\mathcal{H}_\infty$ control;

- the filtering problem.

Besides this we will tackle the guaranteed cost control problem. In the rest of this part we will define the basic concepts involving linear systems with time delay, that is, stability, stabilizability, $\mathcal{H}_\infty$ control, filtering, and their robustness.

2.2   Definitions

Let the initial time be t_0 and assume $w(t), u(t)$ are defined on $[0, \infty)$. Notice that for given initial conditions of the form

$$x(t_0 + s) = \phi(s), \ \forall s \in [-\tau, 0]$$

system (2.1) admits the unique solution $x(t, t_0, \phi(.))$ defined on $[t_0 - \tau, \infty)$. In the rest of this book, we will refer to this solution as $x(t)$.

Let us now define the concepts we will use throughout this book. The definitions we will give will be brief. For details on these concepts, we refer the reader to the appropriate chapter or to the listed references at the end of this book.

2.2.1   Stability and Stabilizability

Stability is the basic property of any system. The stability problem of a system with time delay is much more complex than the one of a non-delayed system. For time delay system (2.1), the definitions of stability can be given as follows:

Definition 2.1 *System (2.1) with* $u(t) \equiv 0$ *and* $w(t) \equiv 0$ *is said to be stable if for every positive* ε *there exists a positive* δ, *which may depend on the initial time and* ε, *such that if* $\|\phi(.)\| < \delta$, *then* $\|x(t, t_0, \phi(.))\| < \varepsilon$ $\forall t \geq t_0$.

Remark 2.1 *The norm we are using in the previous definition is the standard uniform norm, that is*

$$\|\phi(.)\| = \max_{s \in [-\bar{\tau}, 0]} \|\phi(s)\|.$$

Remark 2.2 *Notice that* δ *may be chosen such that in addition the solution* $x(t, t_0, \phi(.))$ *goes to zero as time* t *goes to infinity, in this case we said that the nominal system (2.1) is asymptotically stable.*

Remark 2.3 *System (2.1) with* $u(t) \equiv 0$ *is also referred to as unforced system or free system.*

Definition 2.2 *A state $x_0 \in \mathbb{R}^n$ is called an equilibrium state of system (2.1) if for some $t_0 \geq 0$*

$$x(t_0, \phi(\cdot)) = x_0 \implies x(t, \phi) = x_0, \ \forall t \geq t_0 \tag{2.2}$$

The relation (2.2) means that once the state trajectory reaches the point x_0 it will stay there forever. Obviously, the equilibrium of system (2.1) is zero.

Definition 2.3 *The zero state equilibrium is said to be*

(a) *stable as $t \to \infty$, if for any given positive numbers t_0 and ε, there exists $\delta > 0$ that may depend on t_0 and ε such that, if*

$$\max_{t_0 \leq t \leq t_0 + \bar{\tau}} \|x(t)\| \leq \delta, \tag{2.3}$$

then

$$\max_{t_0 \leq t < \infty} \|x(t)\| \leq \varepsilon \tag{2.4}$$

holds;

(b) *uniformly stable if, for any given $\varepsilon > 0$, there exists a $\delta > 0$ (dependent on ε, but not on to) such that if $x(t)$ satisfies (2.3) for any $t_0 \geq 0$, then $x(t)$ satisfies (2.4);*

(c) *asymptotically stable if*

    (i) *it is stable;*
    (ii) *for each $t_0 > 0$ there is a $\delta(t_0)$ such that*

$$\lim_{t \to \infty} x(t) = 0, \tag{2.5}$$

for any initial $\phi(\cdot)$ in $\{g \in C[-\bar{\tau}, 0] | \|g\| \leq \delta(t_0)\}$.

To develop stability conditions for the class of systems we are considering, two approaches can be used. The known results can be divided into two groups. The first one is mainly based on the Lyapunov–Krasovskii approach, which consists basically in constructing a Lyapunov functional candidate, which can be used to determine the required condition for stability. The second group of results is based on the Lyapunov–Razimukhin approach. For more information on these two approaches, we refer the reader to [93, 127, 128, 156, 158, 171, 172] and the references therein for more information.

In case of stabilization or improvement of system performances, we can design a control law (memory, memoryless, output feedback) that solves the posed problem.

Remark 2.4 *It is required that the stability test depend on the system time delay.*

Definition 2.4 *A piecewise continuous function* $u : [t_0, \infty) \to \mathbb{R}^m$ *is called an admissible control, or control, for simplicity, if for an arbitrary* $(t_1, \psi) \in [t_0, \infty) \times C[-\bar{\tau}, 0]$, *(2.1) has a solution on* $[t_1, \infty)$ *with initial condition* ψ *under this control law* $u(\cdot)$.

Definition 2.5 *A dynamical time delay system (2.1) with* $w(t) \equiv 0$ *and* $\Delta_1(t) = \Delta_2(t) \equiv 0$ *is said to be stabilizable if there exists a control law* $u(t)$ *such that the closed-loop system is stable in the sense of Definition 2.1.*

As was pointed out in the introduction, the dynamical class of systems we are considering can't be described precisely using nominal dynamics. The inclusion of uncertainties in the dynamics is a must. In this case, the concepts of robust stability and robust stabilizability are defined by the following definitions.

Definition 2.6 *The uncertain time delay system (2.1) is said to be robustly stable if the trivial solution* $x(t) = 0$ *of the functional differential equation associated to (2.1) with* $u(t) \equiv 0$ *and* $w(t) \equiv 0$ *is asymptotically stable for all admissible uncertainties* $\Delta A_1(t)$ *and* $\Delta A_2(t)$.

Definition 2.7 *The uncertain time delay system (2.1) with* $w(t) \equiv 0$ *is said to be robustly stabilizable if there exists a control law* $u(t)$ *such that the closed-loop system is robustly stable in the sense of Definition 2.6.*

More details on the concepts of stability and stabilizability and their robustness can be found in Chapter 3.

2.2.2 $\mathcal{H}_\infty$ Control Problem

Practical systems are always affected by external disturbances that can degrade system performance. To avoid this, we should develop a technique that allows us to eliminate the effect of disturbance on system performance. One way of doing this is to use the $\mathcal{H}_\infty$-optimal controller which minimizes the worst-case gain of the system. This optimization problem can be stated as a game optimization problem with two players: the designer, who is seeking a controller that minimizes the output and nature that seeks an external disturbance that maximizes the output.

$\mathcal{H}_\infty$-optimization seeks a state-feedback controller that minimizes the $\mathcal{H}_\infty$-norm of the system's closed-loop transfer function between the controlled output $z(t)$ and the external disturbance $w(t)$, which belongs to $\mathcal{L}_2[0, T]$, with $\Delta_1(t) = \Delta_2(t) \equiv 0$; i.e.,

$$\|G_{zw}\|_\infty = \sup_{\|w(t)\|_{2,[0,T]} \neq 0} \frac{\|z(t)\|_{2,[0,T]}}{\|w(t)\|_{2,[0,T]}}. \tag{2.6}$$

The $\mathcal{H}_\infty$ control problem can be defined on either finite or infinite horizon $(T \to \infty)$. In the rest of this section, we will develop the finite horizon case.

To get the infinite horizon case, we let T go to infinity with the appropriate assumptions.

The $\mathcal{H}_\infty$-norm cost function (2.6) is not acceptable as an objective function since this cost depends on the controller, that is, the supremum makes this function independent of a particular disturbance input.

A quadratic objective function that yields tractable solutions of the differential game(which is referred to as a suboptimal solution to the $\mathcal{H}_\infty$-optimization control problem) can be obtained by considering the following bound on the closed-loop $\mathcal{H}_\infty$-norm:

$$\|G_{zw}\|_\infty = \sup_{\|w(t)\|_{2,[0,T]} \neq 0} \frac{\|z(t)\|_{2,[0,T]}}{\|w(t)\|_{2,[0,T]}} < \gamma,$$

where γ is referred to as the performance bound.

This suboptimal controller must also satisfy the following bound:

$$\|G_{zw}\|_\infty^2 = \sup_{\|w(t)\|_{2,[0,T]} \neq 0} \frac{\|z(t)\|_{2,[0,T]}^2}{\|w(t)\|_{2,[0,T]}^2} < \gamma^2. \tag{2.7}$$

To make the supremum satisfy this inequality, the following must hold:

$$\frac{\|z(t)\|_{2,[0,T]}^2}{\|w(t)\|_{2,[0,T]}^2} \leq \gamma^2 - \varepsilon^2, \tag{2.8}$$

which gives

$$\|z(t)\|_{2,[0,T]}^2 - \gamma^2 \|w(t)\|_{2,[0,T]}^2 \leq -\varepsilon^2 \|w(t)\|_{2,[0,T]}^2. \tag{2.9}$$

Note that the satisfaction of this inequality for all disturbance inputs and some ε is equivalent to the bound on the closed-loop $\mathcal{H}_\infty$-norm (2.8). Therefore, the left expression of (2.9) can be used as an objective function of our $\mathcal{H}_\infty$-optimization control problem. Therefore, the optimization problem we must solve is given by

$$\min_{u(.)} \int_0^T \left[z^\top(t)z(t) - \gamma^2 w^\top(t)w(t) \right] dt$$

subject to the autonomous dynamics of system (2.1).

When uncertainties are present in the dynamics, in this case the robust $\mathcal{H}_\infty$ control consists of making the gain from the exogenous input $w(t)$ to the controlled output $z(t)$ less or equal to $\gamma > 0$; that is,

$$\int_0^T \|z(t)\|^2 dt \leq \gamma^2 \int_0^T \|w(t)\|^2 dt$$

for all $T > 0$ and for all admissible uncertainties. Note that T can be chosen to be infinite.

Mathematically the robust $\mathcal{H}_\infty$ control problem can be stated as follows: Given a positive γ, find a controller that robustly stabilizes the system and guarantees the following:

$$\sup_{w(.)\in\mathcal{L}_2[0,\infty]} \frac{\|z_t\|_2^2}{\|w_t\|_2^2} \leq \gamma^2$$

for all admissible uncertainties.

More details on this subject can be found in Chapter 4 and Chapter 5.

2.2.3 Filtering

The filtering problem consists of determining an estimate $\hat{z}(t)$ of the controlled output $z(t)$ using the measurement of the output $y(t)$.

this filtering problem consists of choosing the filter that minimizes the $\mathcal{H}_\infty$-norm from the disturbance inputs to the filter error output. Our goal in this subsection is to show how the filtering problem can be solved using $\mathcal{H}_\infty$-control theory. For this purpose, let the control $u(t) \equiv 0$ for all $t \geq 0$. Notice that this is not a restriction, since, if the control is *not* equal to zero, the way to handle this case is similar to the one we develop here. $\mathcal{H}_\infty$-filtering is one of the approaches we can use. This problem can be stated as follows: Given a nominal dynamical time delay system with exogenous input that can be deterministic but not known, and a measured output, design a filter to estimate an unmeasured output such that the mapping from the exogenous input to the estimation error is minimized or no larger than some prescribed level in terms of the $\mathcal{H}_\infty$-norm. Mathematically, the $\mathcal{H}_\infty$-filtering problem is stated as follows: Given $\gamma > 0$, find a filter such that

$$\sup_{w\in\mathcal{L}_2[0,\infty]} \frac{\|z(t) - \hat{z}(t)\|_2^2}{\|w(t)\|_2^2} < \gamma^2$$

holds for all $w(t) \in \mathcal{L}_2[0, \infty]$.

Therefore, the filtering problem can be regarded as a special $\mathcal{H}_\infty$ control problem that keeps the $\mathcal{H}_\infty$-norm of the system transfer function between the estimation error and the exogenous disturbance less than a given positive constant γ.

The design of a linear time-invariant filter of order n for system (2.1) with the form

$$\begin{cases} \dot{\xi}_t = K_A\xi_t + K_B y(t) \\ \xi_0 = x_0, \ \xi_s = 0, s \in [-\tau, 0) \\ \hat{z}(t) = K_C\xi_t \end{cases} \tag{2.10}$$

is brought to an $\mathcal{H}_\infty$ control problem that can make the extended system $\{(x(t), \xi_t), t \geq 0\}$ asymptotically stable when t goes to infinity and the estimation error $e_t = z(t) - \hat{z}(t)$ satisfies the condition

$$\|e(t)\|_2 \leq \gamma\|w(t)\|_2. \tag{2.11}$$

Notice that the design of this filter requires the observability (or detectability) of the class of systems we are treating.

When the uncertainties are acting on the system dynamics, the robust $\mathcal{H}_\infty$-filtering can be treated in the same way as the robust $\mathcal{H}_\infty$ control problem and it consists of making the extended system $\{(x(t), \xi_t), t \geq 0\}$ asymptotically stable when t goes to infinity and the estimation error $e_t = z(t) - \hat{z}(t)$ satisfies (2.11) for all admissible uncertainties.

The last problem that will be treated in this part is the guaranteed cost control problem. It consists of designing a controller that robustly stabilizes system (2.1) and at the same time guarantees a given bound for the cost function. Mathematically, it can be formulated as follows: Given a dynamical system with the appropriate assumptions, find a controller $u(.)$ that robustly stabilizes the systems and at the same time guarantees that the following cost is bounded for all admissible uncertainties $\Delta_1(t)$ and $\Delta_2(t)$:

$$ J = \int_0^\infty \left[x^\top(t) R_1 x(t) + u^\top(t) R_2 u(t) \right] dt $$

with R_1 and R_2 being symmetric and positive-definite matrices.

2.3   Objective of Part I

Part I deals with different problems like stability, stabilizability, $\mathcal{H}_\infty$ control, filtering, and their robustness. We also cover the guaranteed cost control problem.

Our goal in Chapter 3 is to establish delay-dependent and delay-independent sufficient conditions that we can use to check whether a given system with time delay is stable or not when the uncertainties are null. The stabilizability problem is also considered. Different types of control law are considered, mainly state feedback and output feedback.

In Chapter 4, the norm-bounded type of uncertainty is considered and delay-independent and delay-independent sufficient conditions for stability and stabilizability are developed. Here we will also consider different types of controllers.

In Chapter 5, the $\mathcal{H}_\infty$ control problem and the filtering problem are both considered and delay-dependent and delay-independent sufficient conditions are established to design a controller that rejects the disturbance acting of the system at a level γ. We also develop a design algorithm to estimate the state vector of the class of systems with time delay.

Chapter 6 treats the robust part of Chapter 5. The uncertainties are of norm-bounded type. The guaranteed cost control problem is also investigated.

2.4 Notes

Linear system with constant time delay are addressed in [77, 118, 147, 148, 149] and references therein. Systems with time-varying delay are investigated in [77, 105, 109]. Other variants of time delay system including parameter uncertainty are considered in [49, 109, 137, 175] and references therein.

3
Stability and Stabilizability

This chapter is devoted to the study of the stability and the stabilizability of linear time invariant systems with time delay using the Lyapunov method and linear matrix inequality (see Appendix A). For this purpose, let us consider the following linear continuous-time system with time delay:

$$
\begin{cases}
\dot{x}(t) = Ax(t) + \sum_{j=1}^{m} A_j x(t - \tau_j) + Bu(t) \\
y(t) = Cx(t) \\
x(t) = \phi(t), \ t \in [-\bar{\tau}, 0),
\end{cases}
\tag{3.1}
$$

where $x(t) \in \mathbb{R}^n$ is the state vector, $u(t) \in \mathbb{R}^k$ is the control input, $y(t) \in \mathbb{R}^r$ is the output vector of the system, $\phi(\cdot)$ is the initial condition, $A \in \mathbb{R}^{n \times n}$, $A_j \in \mathbb{R}^{n \times n}, j = 1, \ldots, m, \ B \in \mathbb{R}^{n \times k}, \ C \in \mathbb{R}^{r \times n}$ are constant matrices, $\tau_j > 0, j = 1, \cdots, m$ are constant delays in the system and $\bar{\tau} = \max\{\tau_j, 1 \leq j \leq m\}$.

Our goal is to establish the conditions under which system (3.1) will be stable when $u(t) \equiv 0$ and when the system is stabilizable. Two types of conditions are of interest: the delay-independent and delay-dependent conditions. The first one is the condition that guarantees the stability of the system for all values of the delay. In general, this condition is conservative and a condition that depends on of the value of the delay is needed in order to obtain less conservative results.

Different type of controllers will be addressed in this chapter. Among the controllers we will use to stabilize our class of time delay linear systems we will consider the state feedback controller (memoryless and memory), the output feedback controller, and the observer-based output feedback controller.

In the present literature there exist two techniques that can be used to study stability and stabilizability and the related topics. The first one is based on the Lyapunov–Razimukhin technique. It consists of considering a Lyapunov function of the form, $V(x(t)) = x^\top(t)Px(t)$, with P a symmetric, positive-definite matrix of appropriate dimension to create the conditions that can be used to check if the system with time delay under study is stable or not and/or stabilizable or not. This technique gives a condition that depends on the maximum value of the delay. The reader can consult the work of [158, 171] and the references therein for more information.

The second technique is based on the Lyapunov–Krasovskii approach. It entails considering a more complicated Lyapunov functional to determine the appropriate condition that can be delay-independent or delay-independent. This technique has been used extensively as the large number of references using it confirms. See for example the works of [93, 127, 128, 156, 158, 172] and the references therein for more information. In this book, we will use mainly the second approach.

The chapter is organized as follows. Section 3.1 develops delay-independent sufficient conditions for stability and stabilizability. Different types of controllers are used to stabilize the class of systems we are considering in this chapter. In Section 3.2, the conservatism of Section 3.1 is overcome by developing delay-dependent sufficient conditions for stability and stabilizability. Along the same lines, different types of state feedback controllers are designed to stabilized our class of systems. Section 3.3 covers the output feedback control and develops design algorithms. In Section 3.4, we use a more general Lyapunov functional and develop more general results. The power of these results is shown by a numerical example.

3.1 Delay-Independent Stability and Stabilizability

The goal of this section is to establish sufficient conditions for system (3.1) with $u(t) \equiv 0$ to be stable for any $\tau_j \in [0, \infty)$, $j = 1, \ldots, m$ and to design a control law that stabilizes system (3.1) for any $\tau_j \in [0, \infty)$, $j = 1, \ldots, m$.

Remark 3.1 *The results of this chapter can be extended to the case of the following class of linear time invariant systems with time varying delays without difficulty:*

$$\begin{cases} \dot{x}(t) = Ax(t) + \sum_{j=1}^{m} A_j x(t - h_j(t)) + Bu(t) \\ y(t) = Cx(t) \end{cases} \tag{3.2}$$

where $h_j(t)$ denotes the time-varying time delay in the system satisfying $\dot{h}_j(t) \leq \mu < 1$, with μ a given constant, and $u(t) \in \mathbb{R}^k$ the control input.

The study of the stability problem of system (3.2) can be done using the direct Lyapunov method. The generalized quadratic Lyapunov functional

$$V(\mathbf{x}_t) = x^\top(t) P x(t) + \sum_{j=1}^{m} x^\top(t) \int_{-h_j(t)}^{0} P_1(s) x(t+s) ds$$

$$+ \sum_{j=1}^{m} \int_{-h_j(t)}^{0} x^\top(t+s) P_1^\top(s) x(t) ds$$

$$+ \sum_{j=1}^{m} \int_{-h_j(t)}^{0} \int_{-h_j(t)}^{0} x^\top(t+s) P_2(s,v) x(t+v) ds dv$$

can be used for this purpose. This technique involves solving algebraic, ordinary, and partial differential equations with appropriate boundary conditions to get P, P_1, P_2, which is obviously unpromising. Thus, the authors in this field prefer to invent concrete functionals $V(.)$ for special classes of delayed equations. By using special Lyapunov functionals, this section will establish delay-independent and delay-dependent stability conditions for system (3.1) using LMI techniques.

3.1.1   Delay-Independent Stability

Let us now consider the unforced system (3.2) and try to develop sufficient conditions for stability. The following theorem gives our first delay-independent sufficient condition.

Theorem 3.1 *(Delay-independent stability) If there exists a set of symmetric, positive-definite matrices $P > 0, Q_j > 0, j = 1, \cdots, m$ satisfying the LMI*

$$\Theta = \begin{pmatrix} A^\top P + PA + \sum_{j=1}^{m} Q_j & PA_1 & \cdots & PA_m \\ A_1^\top P & -Q_1 & & \\ \vdots & & \ddots & \\ A_m^\top P & & & -Q_m \end{pmatrix} < 0, \qquad (3.3)$$

then system (3.1) with $u(t) \equiv 0$ is stable.

Proof: Consider a Lyapunov functional candidate as

$$V(\mathbf{x}(t)) = x^\top(t) P x(t) + \sum_{j=1}^{m} \int_{t-\tau_j}^{t} x^\top(s) Q_j x(s) ds$$

where $\mathbf{x}(t) = (x(s), t - \bar{\tau} \le s \le t)$. Then, the derivative of $V(\mathbf{x}(t))$ with respect to time t is given by

$$\dot{V}(\mathbf{x}(t)) = \dot{x}^\top(t) P x(t) + x^\top(t) P \dot{x}(t)$$

$$+ x^\top(t) \left(\sum_{j=1}^{m} Q_j \right) x(t) - \sum_{j=1}^{m} x^\top(t - \tau_j) Q_j x(t - \tau_j).$$

Using now (3.1) with $u(t) \equiv 0$, we get

$$\dot{V}(\mathbf{x}(t)) = x^\top \left(A^\top P + PA + \sum_{j=1}^{m} Q_j \right) x_t + 2 \sum_{j=1}^{m} x^\top(t - \tau_j) A_j^\top P x(t)$$

$$- \sum_{j=1}^{m} x^\top(t - \tau_j) Q_j x(t - \tau_j),$$

which after using (3.3) can be rewritten as follows:

$$\dot{V}(\mathbf{x}(t)) = \eta_t^\top \Theta \eta_t,$$

where $\eta_t^\top = (x^\top(t), x^\top(t - \tau_1), \cdots, x^\top(t - \tau_m))$, which implies that $\dot{V}(\mathbf{x}(t))$ is negative by (3.3). This proves that the unforced (3.1) system is stable. □

Remark 3.2 *Notice that this condition doesn't depend on the delays. It is clear that the condition is restrictive because it is hard to find a system that will be stable for all the ranges of the delays.*

The condition of this theorem is a sufficient one. This means that if the system doesn't meet this condition; it does not imply that the system is not stable.

To establish a simple test that system (3.1) with $u(t) \equiv 0$ is asymptotically stable, Theorem 3.1 utilizes a very special Lyapunov functional. Generally speaking, the more general the Lyapunov functional used, the more powerful the stability test will be. To get a more powerful test, one has to consider a more complex Lyapunov functional. However, the use of the generalized quadratic form will require the solution of complex differential equations with boundary conditions. For instance, in the next theorem we will use a Lyapunov functional that has time-varying matrix $Q(.)$ in its structure. We will consider $Q(s)$ to be a piecewise-constant matrix-valued function.

For notational simplicity, let us consider system (3.1) with $u(t) \equiv 0$ and $m = 1$. In this case, let $\tau_1 = \tau$. The following theorem summarizes the results for this case.

Theorem 3.2 *If there exist symmetric, positive-definite matrices $P > 0$ and $Q = (Q_0, Q_1, \ldots, Q_r) > 0$ such that the following holds:*

$$\begin{pmatrix} A^\top P + PA + Q_0 & \mathbf{0} & PA_1 \\ \mathbf{0} & \bar{Q} & \mathbf{0} \\ A_1^\top P & \mathbf{0} & -Q_r \end{pmatrix} < 0, \tag{3.4}$$

where $\bar{Q} = \text{diag}\{Q_1 - Q_0, \cdots, Q_r - Q_{r-1}\}$ and r is the step size of the partition of the interval $[0, \tau]$ (it gives $r + 1$ sub-intervals), then system (3.1) with $u(t) \equiv 0$ is asymptotically stable.

Proof: Consider the following Lyapunov functional candidate:

$$V(\mathbf{x}(t)) = x^\top(t)Px(t) + \int_{t-\tau}^{t} x^\top(s)Q(s)x(s)ds.$$

Let us discretize the interval $[0, \tau]$ by step $\tau_i, 0 \leq i \leq r$ with $\tau_i = (i/r)\tau$ and suppose that the matrix $Q(.)$ is constant in the interval $[\tau_i, \tau_{i+1})$. Let us denote this matrix by $Q(\tau_i) = Q_i$. Then, $V(\mathbf{x}(t))$ becomes

$$\bar{V}(\mathbf{x}(t)) = x^\top(t)Px(t) + \sum_{i=0}^{r-1} \int_{t-\tau_{i+1}}^{t-\tau_i} x^\top(s)Q_i x(s)ds.$$

The remaining part of the proof is the same as in Theorem 3.1.   □

Let us work out a numerical example to show the usefulness of the previous results.

Example 3.1 *Consider the following dynamical time delay system*

$$\dot{x}(t) = \begin{pmatrix} -2.1 & 0 \\ 0 & -0.91 \end{pmatrix} x(t) + \begin{pmatrix} -1.0 & 0 \\ -1 & -1.1 \end{pmatrix} x(t-1).$$

For this system, we can't find a feasible solution to LMI (3.3). This means that the delay-independent stability condition given in Theorem 3.1 is not satisfied and therefore we can't conclude whether the system is stable or not. However, if we use the results of Theorem 3.2 by choosing $r = 3$, we get the following symmetric, positive-definite feasible solutions to (3.4):

$$P = \begin{pmatrix} 134.4424 & 12.8127 \\ 12.8127 & 84.8292 \end{pmatrix} \quad Q_0 = \begin{pmatrix} 357.962 & 23.4175 \\ 23.4175 & 75.9710 \end{pmatrix}$$

$$Q_1 = \begin{pmatrix} 267.9032 & 33.5210 \\ 33.5210 & 70.1691 \end{pmatrix} \quad Q_2 = \begin{pmatrix} 187.6826 & 44.1502 \\ 44.1502 & 64.6126 \end{pmatrix}$$

$$Q_3 = \begin{pmatrix} 279.0897 & 102.7875 \\ 12.7875 & 118.3276 \end{pmatrix}.$$

Therefore, based on Theorem 3.2, this system is asymptotically stable.

3.1.2 Delay-Independent State Feedback Stabilization

In this subsection, we will assume that the state of the system is perfectly available for state feedback and restrict our study to the case of linear state feedback control, that is,

$$u(t) = Kx(t) + \sum_{j=1}^{m} K_j x(t - \tau_j), \tag{3.5}$$

where $K, K_j, 1 \leq j \leq m$ are constant matrices of appropriate dimension.

First, let us consider the stabilizability problem using memoryless controller $K_j = 0$ for all $j = 1, \cdots, m$. The stabilizability of system (3.1) using a linear memoryless controller consists of finding a gain matrix K such that the closed-loop system

$$\begin{cases} \dot{x}(t) = [A + BK]\,x(t) + \sum_{j=1}^{m} A_j x(t - \tau_j) \\ x(t) = \phi(t), \ t \in [-\bar{\tau}, 0] \end{cases} \qquad (3.6)$$

is asymptotically stable. The following theorem provides an LMI-based design technique for finding a stabilizing gain matrix K.

Theorem 3.3 *If there exist symmetric, positive-definite matrices $X > 0$, $U = (U_1, \ldots, U_m) > 0$ and a matrix Y such that the following LMI holds:*

$$\begin{pmatrix} AX + XA^\top + BY + Y^\top B^\top + \sum_{i=1}^{m} A_i^\top U_i A_i & AX \\ X\mathcal{A} & -\bar{U} \end{pmatrix} < 0, \qquad (3.7)$$

where $\mathcal{A} = \begin{pmatrix} A_1, \cdots, A_m \end{pmatrix}$ and $\bar{U} = \text{diag}\{U_1, \cdots, U_m\}$, then system (3.1) is asymptotically stable under control $u(t) = Kx(t)$ with $K = YX^{-1}$.

Proof: In view of Theorem 3.1, to prove that the closed-loop system is asymptotically stable it suffices to show that

$$\begin{pmatrix} [A + BK]^\top P + P[A + BK] + \sum_{j=1}^{m} Q_j & P\mathcal{A} \\ \mathcal{A}^\top P & -\bar{Q} \end{pmatrix} < 0 \qquad (3.8)$$

has a set of feasible solutions $P > 0, Q_i > 0, 1 \leq i \leq m$, and $\bar{Q} = \text{diag}\{Q_1, \cdots, Q_m\}$. Pre- and post-multiplying (3.8) by

$$\text{diag}\{P^{-1}, P^{-1}, \cdots, P^{-1}\}$$

and letting $X = P^{-1}, Y = KP^{-1}, U_i = P^{-1}Q_iP^{-1}$ yields (3.7). Therefore, if $X > 0, U_i > 0, 1 \leq i \leq m$, and Y satisfy (3.7), then $P = X^{-1}, Q_i = X^{-1}U_iX^{-1}$, and $K = YX^{-1}$ satisfies (3.8). This concludes the proof of Theorem 3.3. $\qquad \square$

Theorem 3.3 provides a memoryless state feedback controller design technique. When the time delays $\tau_j, 1 \leq j \leq m$, are known constants, a controller (3.5) that contains a delayed state can be constructed in the same way. Using the control (3.5) the closed-loop system becomes

$$\dot{x}(t) = [A + BK]x(t) + \sum_{j=1}^{m} [A_j + BK_j]x(t - \tau_j).$$

The following theorem develops a delay-independent stabilizability condition and provides a method to design a stabilizing controller.

Theorem 3.4 *If there exist symmetric, positive-definite matrices $X > 0, U_i > 0, 1 \leq i \leq m$, and the matrices $Y, Y_i, 1 \leq i \leq m$ such that the following holds:*

$$\begin{pmatrix} \vartheta(X,Y) & A_1 X + BY_1 & \cdots & A_m X + BY_m \\ * & -U_1 & & 0 \\ \vdots & & \ddots & \\ * & 0 & & -U_m \end{pmatrix} < 0, \qquad (3.9)$$

where $\vartheta(X,Y) = AX + XA^\top + BY + Y^\top B^\top + \sum_{j=1}^{m} U_j$, then system (3.1) is asymptotically stable under the control (3.5) with gains $K = YX^{-1}$ and $K_j = Y_j X^{-1}$, $j = 1, \ldots, m$.

Proof: Applying controller (3.5), the closed-loop system dynamic becomes

$$\dot{x}(t) = [A + BK]x(t) + \sum_{j=1}^{m} [A_j + BK_j]x(t - \tau_j). \qquad (3.10)$$

In view of Theorem 3.1, to prove the asymptotic stability of system (3.10), it suffices to show that

$$\begin{pmatrix} \bar{A}^\top P + P\bar{A} + \sum_{j=1}^{m} Q_j & P\bar{A}_1 & \cdots & P\bar{A}_m \\ \bar{A}_1^\top P & -Q_1 & & 0 \\ \vdots & & \ddots & \\ \bar{A}_m^\top P & 0 & & -Q_m \end{pmatrix} < 0 \qquad (3.11)$$

has a set of feasible solutions $P > 0, Q_i > 0, 1 \le i \le m$, where

$$\bar{A} = A + BK, \bar{A}_i = A_i + BK_i, \ 1 \le i \le m.$$

Pre- and post-multiplying (3.11) with $\mathrm{diag}\{P^{-1}, \cdots, P^{-1}\}$ and letting $X = P^{-1}, Y = KX, U_i = P^{-1}Q_i P^{-1}, Y_i = K_i P^{-1}$ yields (3.9). This ends the proof. □

Theorem 3.4 develops a sufficient delay-independent condition for system (3.1) to be asymptotically stable under control (3.5) and provides a control design algorithm. Obviously, LMI (3.9) contains variables $X, Y, Y_1, \cdots, Y_m$, which are to be determined. If the variables $Y, Y_1, \cdots, Y_m$ can be eliminated, the computation burden can be reduced greatly. Lemma A.3 can be used for this purpose.

Let $U \in \mathbb{R}^{n \times k}$ be a given matrix. $U_\perp$ is said to be the orthogonal complement of U if $U^\top U_\perp = 0$ and $[UU_\perp]$ is of maximum rank (which means that $[UU_\perp]$ is nonsingular).

Using Lemma A.3 to eliminate $Y, Y_i, 1 \le i \le m$ in Theorem 3.4 leads to the following theorem.

Theorem 3.5 *System (3.1) is asymptotically stabilizable with controller (3.5) if one of the following conditions is satisfied:*

(i) *There exist symmetric, positive-definite matrices $X > 0, U_i > 0, 1 \leq i \leq m$ such that*

$$\begin{pmatrix} \vartheta(X) + \sigma BB^\top & A_1 X & \cdots & A_m X \\ XA_1^\top & -U_1 & & \mathbf{0} \\ \vdots & & \ddots & \\ XA_m^\top & \mathbf{0} & & -U_m \end{pmatrix} < 0 \qquad (3.12)$$

holds for some scalar $\sigma > 0$, where $\vartheta(X) = AX + XA^\top + \sum_{j=1}^m U_j$;

(ii) *There exist symmetric, positive-definite matrices $X > 0, U_i > 0, 1 \leq i \leq m$ such that*

$$\mathcal{B}^\top \begin{pmatrix} \vartheta(X) & A_1 X & \cdots & A_m X \\ XA_1^\top & -U_1 & & \mathbf{0} \\ \vdots & & \ddots & \\ XA_m^\top & \mathbf{0} & & -U_m \end{pmatrix} \mathcal{B} < 0, \qquad (3.13)$$

holds, where $\mathcal{B} = \mathrm{diag}\{B_\perp, I, \cdots, I\}$.

Proof: Note that

$$\begin{pmatrix} \vartheta(X,Y) & A_1 X + BY_1 & \cdots & A_m X + BY_m \\ * & -U_1 & & \mathbf{0} \\ \vdots & & \ddots & \\ * & \mathbf{0} & & -U_m \end{pmatrix} =$$

$$\begin{pmatrix} \vartheta(X) & A_1 X & \cdots & A_m X \\ * & -U_1 & & \mathbf{0} \\ \vdots & & \ddots & \\ * & \mathbf{0} & & -U_m \end{pmatrix} + \begin{pmatrix} B \\ 0 \\ \vdots \\ 0 \end{pmatrix} (Y \ Y_1 \ \cdots \ Y_m)$$

$$+ \left[\begin{pmatrix} B \\ 0 \\ \vdots \\ 0 \end{pmatrix} (Y \ Y_1 \ \cdots \ Y_m) \right]^\top \qquad (3.14)$$

which combined with results (ii) of Lemma A.3 yields the proof of (i).

Note that the orthogonal complement of $(B^\top \ \mathbf{0} \ \cdots \ \mathbf{0})^\top$ is $\mathcal{B}$, which combined with results (i) of Lemma A.3 terminates the proof of (ii) of this theorem.   □

Remark 3.3 *All the results we developed so far are delay-independent and therefore conservative. The next section overcomes this conservatism by providing delay-dependent sufficient conditions.*

3.2 Delay-Dependent Stability and Stabilizability

This section develops delay-dependent stability and stabilizability conditions. In the stabilization case, we are seeking a state feedback control law $u(t)$ that asymptotically stabilizes the closed-loop system.

3.2.1 Delay-Dependent Stability

Theorem 3.1 provides a delay-independent sufficient condition for system (3.1) with $u(t) \equiv 0$ to be asymptotically stable. The fact that (3.3) doesn't depend on time delay means that (3.3) assures that system (3.1) with $u(t) \equiv 0$ is asymptotically stable for any time delay. This condition is too restrictive. To overcome this drawback, we proceed to the development of a delay-dependent stability condition for system (3.1) with $u(t) \equiv 0$.

For notational simplicity, we let $A = A_0$ and $0 = \tau_0$. With this notation, we have the following delay-dependent stability condition for system (3.1) with $u(t) \equiv 0$.

Theorem 3.6 *If there exist symmetric, positive-definite matrices $P > 0, Q_l > 0, 0 \leq l \leq m$, such that the following holds:*

$$\begin{pmatrix} \bar{A}^\top P + P\bar{A} + \sum_{j=1}^{m}\sum_{l=0}^{m} \tau_j Q_l & M_1^\top \\ M_1 & -\bar{Q} \end{pmatrix} < 0, \tag{3.15}$$

where

$$\bar{A} = \sum_{i=0}^{m} A_i$$

$$M_1^\top = (\tau_1 PA_1 A_0 \quad \cdots \quad \tau_1 PA_1 A_m \quad \cdots \quad \tau_m PA_m A_0 \quad \cdots \quad \tau_m PA_m A_m)$$

$$\bar{Q} = \text{diag}\{\tau_1 Q_0, \cdots, \tau_1 Q_m, \cdots, \tau_m Q_0, \cdots, \tau_m Q_m\}$$

then system (3.1) with $u(t) \equiv 0$ is asymptotically stable.

Proof: Noting that

$$x(t) - x(t - \tau_j) = \int_{t-\tau_j}^{t} \dot{x}(s)ds,$$

we have

$$x(t - \tau_j) = x(t) - \int_{t-\tau_j}^{t} \dot{x}(s)ds$$

$$= x(t) - \int_{t-\tau_j}^{t} \sum_{l=0}^{m} A_l x(s - \tau_l)ds. \tag{3.16}$$

Substituting (3.16) into (3.1) with $u(t) \equiv 0$ yields

$$\dot{x}(t) = Ax(t) + \sum_{j=1}^{m} A_j \left[x(t) - \int_{t-\tau_j}^{t} \sum_{l=0}^{m} A_l x(s - \tau_l)ds \right]$$

$$= \bar{A}x(t) - \sum_{j=1}^{m}\sum_{l=0}^{m} A_j A_l \int_{t-\tau_j}^{t} x(s-\tau_l)ds$$

$$= \bar{A}x(t) - \sum_{j=1}^{m}\sum_{l=0}^{m} A_j A_l \int_{t-\tau_j-\tau_l}^{t-\tau_l} x(s)ds. \tag{3.17}$$

Let us now consider the following dynamical time delay system:

$$\begin{cases} \dot{x}(t) = \bar{A}x(t) - \sum_{j=1}^{m}\sum_{l=0}^{m} A_j A_l \int_{t-\tau_j-\tau_l}^{t-\tau_l} x(s)ds \\ x(t) = \phi(t), \ t \in [-2\bar{\tau}, 0] \end{cases} \tag{3.18}$$

where $\bar{\tau} = \max(\tau_1, \ldots, \tau_m)$.

Note that (3.1) with $u(t) \equiv 0$ is a special case of (3.18), and thus any solution of (3.1) with $u(t) \equiv 0$ is also a solution of (3.18) (see, e.g., [98], p. 156). Hence, the asymptotic stability of (3.18) ensures the asymptotic stability of (3.1) with $u(t) \equiv 0$. Thus, to study the asymptotic stability of system (3.1) with $u(t) \equiv 0$ it suffices to study the asymptotic stability of (3.18). For this purpose, let us define the Lyapunov functional as follows:

$$V(\mathbf{x}(t)) = x^{\top}(t)Px(t) + \sum_{j=1}^{m}\sum_{l=0}^{m} \int_{\tau_l}^{\tau_j+\tau_l} \left[\int_{t-s}^{t} x^{\top}(z)Q_l x(z)dz \right] ds \tag{3.19}$$

where $\mathbf{x}_t = (x(s), t - 2\bar{\tau} \leq s \leq t)$. Then, it follows from (3.18) that

$$\dot{V}(\mathbf{x}(t)) = 2x^{\top}(t)P\dot{x}(t) + \sum_{j=1}^{m}\sum_{l=0}^{m} \int_{\tau_l}^{\tau_j+\tau_l} [x^{\top}(t)Q_l x(t)$$

$$- x^{\top}(t-s)Q_l x(t-s)]ds$$

$$= 2x^{\top}(t)P\dot{x}(t) + \sum_{j=1}^{m}\sum_{l=0}^{m} \tau_j x^{\top}(t)Q_l x(t)$$

$$- \sum_{j=1}^{m}\sum_{l=0}^{m} \int_{\tau_l}^{\tau_j+\tau_l} x^{\top}(t-s)Q_l x(t-s)ds.$$

Adding and subtracting

$$x^{\top}(t) \left[\sum_{j=1}^{m}\sum_{l=0}^{m} \tau_j P(A_j A_l)Q_l^{-1}(A_j A_l)^{\top} P \right] x(t)$$

on the right-hand side of the above equality yield

$$\dot{V}(\mathbf{x}(t)) = x^{\top}(t) \left[\bar{A}^{\top}P + P\bar{A} + \sum_{j=1}^{m}\sum_{l=0}^{m} \tau_j Q_l \right.$$

$$\left. + \sum_{j=1}^{m}\sum_{l=0}^{m} \tau_j P(A_j A_l)Q_l^{-1}(A_j A_l)^{\top} P \right] x(t)$$

$$-\sum_{j=1}^{m}\sum_{l=0}^{m}\int_{t-\tau_j-\tau_l}^{t-\tau_l}[x^{\top}(t)P(A_jA_l)Q_l^{-1}(A_jA_l)^{\top}Px(t)$$

$$+2x^{\top}(t)PA_jA_lx(s)+x^{\top}(s)Q_lx(s)]ds$$

$$=x^{\top}(t)\left[\bar{A}^{\top}P+P\bar{A}+\sum_{j=1}^{m}\sum_{l=0}^{m}\tau_jQ_l\right.$$

$$\left.+\sum_{j=1}^{m}\sum_{l=0}^{m}\tau_jP(A_jA_l)Q_l^{-1}(A_jA_l)^{\top}P\right]x(t)$$

$$-\sum_{j=1}^{m}\sum_{l=0}^{m}\int_{t-\tau_j-\tau_l}^{t-\tau_j}\left[Q_lx(s)+(A_jA_l)^{\top}Px(t)\right]^{\top}$$

$$Q_l^{-1}\left[Q_lx(s)+(A_jA_l)^{\top}Px(t)\right]ds$$

$$\leq x^{\top}(t)\left[\bar{A}P+P\bar{A}+\sum_{j=1}^{m}\sum_{l=0}^{m}\tau_jQ_l\right.$$

$$\left.+\sum_{j=1}^{m}\sum_{l=0}^{m}\tau_jP(A_jA_l)Q_l^{-1}(A_jA_l)^{\top}P\right]x(t).$$

The last inequality is obtained from (3.15) and Lemma A.2 and therefore the system is asymptotically stable. This completes the proof of Theorem 3.6. □

Remark 3.4 *When the delay in the dynamics is known, the condition (3.15) can be used to check whether the system under study is stable or not. In a real system, it is almost impossible to know exactly the value of the delay and therefore an upper bound on the system delay for which the system will remain stable is of interest.*

When $m = 1$, the determination of the upper bound of the delay for which system (3.18) will remain asymptotically stable can be cast into a (GEVP). In this case, (3.15) becomes

$$\left(\begin{bmatrix}(A_0+A_1)^{\top}P+P(A_0+A_1)\\+\tau[Q_0+Q_1]\end{bmatrix}\quad \tau PA_1A_0\quad \tau PA_1A_1\\ \tau A_0^{\top}A_1^{\top}P\qquad -\tau Q_0\qquad 0\\ \tau A_1^{\top}A_1^{\top}P\qquad 0\qquad -\tau Q_1\right)<0 \quad (3.20)$$

which is equivalent to

$$\begin{pmatrix}Q_0+Q_1 & PA_1A_0 & PA_1A_1\\ A_1^{\top}A_1^{\top}P & -Q_0 & 0\\ A_1^{\top}A_1^{\top}P & 0 & -Q_1\end{pmatrix}<\frac{1}{\tau}\begin{pmatrix}J & 0 & 0\\ 0 & 0 & 0\\ 0 & 0 & 0\end{pmatrix}, \quad (3.21)$$

where $J = -(A_0 + A_1)^\top P - P(A_0 + A_1)$. Thus, our optimization problem can be stated as

$$\max_{\tau>0,P>0,Q_0>0,Q_1>0} \tau \tag{3.22}$$

$$\text{s.t. } (3.21)$$

Let $\eta = 1/\tau$. Then, (3.21) becomes

$$\begin{pmatrix} Q_0 + Q_1 & PA_1A_0 & PA_1A_1 \\ A_0^\top A_1^\top P & -Q_0 & 0 \\ A_1^\top A_1^\top P & 0 & -Q_1 \end{pmatrix} < \eta \begin{pmatrix} J & 0 & 0 \\ 0 & 0 & 0 \\ 0 & 0 & 0 \end{pmatrix} \tag{3.23}$$

and then, the optimization problem (3.22) becomes

$$\min_{\eta>0,P>0,Q_0>0,Q_1>0} \eta \tag{3.24}$$

$$\text{s.t. } (3.23)$$

where s.t. denotes "subject to".

Due to Remark A.1, (3.24) can not be solved directly using the GEVP. To cast it into the framework of the GEVP, let us introduce an auxiliary matrix $\Gamma > 0$ and rewrite (3.23) as

$$\begin{pmatrix} Q_0 + Q_1 & PA_1A_1 & PA_0A_1 \\ A_0^\top A_1^\top P & -Q_0 & \\ A_1^\top A_1^\top P & & -Q_1 \end{pmatrix} < \begin{pmatrix} \Gamma & 0 & 0 \\ 0 & 0 & 0 \\ 0 & 0 & 0 \end{pmatrix} \tag{3.25}$$

$$\Gamma < \eta \left[-(A_0 + A_1)^\top P - P(A_0 + A_1) \right]. \tag{3.26}$$

Since $Q_0 > 0, Q_1 > 0$, to make Q_0, Q_1, and P satisfy (3.20), we need $(A_0 + A_1)^\top P + P(A_0 + A_1) < 0$. Therefore, we can require the right-hand side of (3.26) to be positive-definite. In this way, our optimization problem becomes a standard generalized eigenvalue problem, then which can be solved using the GEVP technique. From this discussion, we have the following theorem.

Theorem 3.7 *Let η_0 be the optimal solution of the following GEVP*

$$\min_{\eta>0,P>0,Q_0>0,Q_1>0,\Gamma>0} \eta$$

$$\text{s.t. } (3.25) \text{ and } (3.26)$$

Then for any $\tau \in [0, 1/\eta_0]$, system (3.1) ($m = 1$) with $u(t) \equiv 0$ is asymptotically stable.

Let us now show the applicability of these results by working out the following numerical example.

Example 3.2 *Consider a linear system with constant time delay and $m = 1$. The system parameters are as follows:*

$$A = \begin{pmatrix} -2 & 0 \\ 0 & -0.91 \end{pmatrix}, \quad A_1 = \begin{pmatrix} -1 & 0 \\ -1 & -1.1 \end{pmatrix}.$$

Let us assume that the delay τ is an unknown constant. Using the LMI tool-box of Matlab, it is easy to check that there doesn't exist a feasible solution to the LMI (3.3). However, by solving the GEVP problem with constraints (3.25) and (3.26), we get an upper bound for the time delay τ, which is 0.8979. Based on the results of Theorem 3.7, for any delay in the interval $[0, 0.8979)$, the system remains asymptotically stable.

Let us now return to the design of a delay-dependent stabilizing controller for our class of systems. This will be the subject of the next subsection.

3.2.2 Delay-Dependent State Feedback Stabilization

In this subsection we study the stabilizability of system (3.1). A delay-dependent condition will be developed to guarantee that system (3.1) is stable under state-feedback control.

We first consider the stabilizability of system (3.1) using a linear memoryless state feedback controller, that is,

$$u(t) = Kx(t). \tag{3.27}$$

Using Theorem 3.6 we get the following theorem.

Theorem 3.8 *If there exist symmetric, positive-definite matrices $X > 0, U_l > 0, 0 \leq l \leq m$, and a matrix Y with appropriate dimensions such that the following holds:*

$$\left(\begin{array}{cc} \left[\begin{array}{c} X\bar{A}^\top + \bar{A}X + \sum_{j=1}^m \sum_{l=0}^m \tau_j U_l \\ +BY + Y^\top B^\top \end{array} \right] & M^\top(X,Y) \\ M(X,Y) & -\bar{U} \end{array} \right) < 0 \tag{3.28}$$

where

$$\bar{A} = \sum_{i=0}^m A_i$$

$$M^\top(X,Y) = \Big(\tau_1 A_1(A_0 X + BY), \cdots, \tau_1 A_1 A_m, \cdots,$$

$$\tau_m A_m(A_0 X + BY), \cdots, \tau_m A_m A_m \Big)$$

$$\bar{U} = \text{diag}\{\tau_1 U_0, \cdots, \tau_1 U_m, \cdots, \tau_m U_0, \cdots, \tau_m U_m\},$$

then system (3.1) is asymptotically stable under the state feedback control law and $K = YX^{-1}$ is a stabilizing gain.

Proof: In view of Theorem 3.6, to prove the asymptotic stability of the closed-loop system with control $u(t) = Kx(t)$, it suffices to show that there exist symmetric, positive-definite matrices $P > 0, Q_l > 0, 0 \leq l \leq m$ such that (3.15) remains valid with A_0 replaced by $A_0 + BK$. Pre- and post-multiplying both sides of (3.15) by $\text{diag}\{P^{-1}, \cdots, P^{-1}\}$ and letting $X =$

$P^{-1}, Y = KX, P^{-1}Q_l P^{-1} = U_l$ leads to (3.28). Thus, if $X > 0, U > 0$, and Y are a set of feasible solutions to LMI (3.28), then $P = X^{-1}, Q_l = PU_l P$, and $K = YX^{-1}$ satisfy (3.15) with A_0 replaced by $A_0 + BK$. This ends the proof. □

Theorem 3.3 provides an LMI-based technique to design a state feedback control law that stabilizes system (3.1). Theorem 3.6 gives us a method to find an upper bound $1/\eta_0$ such that for any $\tau \in [0, 1/\eta_0]$ system (3.1) is asymptotically stable. An immediate idea is to apply the delay-dependent stability condition to develop an LMI-based control design method for computing a control law that stabilizes the system when the delay is maximized.

When $m = 1$, the equivalent condition to (3.28) becomes

$$\begin{pmatrix} J & \tau A_1(AX + BY) & \tau A_1 A_1 X \\ * & -\tau U_0 & 0 \\ * & 0 & -\tau U_1 \end{pmatrix} < 0 \qquad (3.29)$$

where $J = (A + A_1)X + X(A + A_1)^\top + BY + Y^\top B^\top + \tau[U_0 + U_1]$.

To get the upper bound of the time delay, τ, for which system (3.1) remains asymptotically stable, one needs only to maximize τ in (3.29). To this end, letting $\eta = 1/\tau$ and multiplying both sides of (3.29) by η yields

$$\begin{pmatrix} U_0 + U_1 & A_1(AX + BY) & A_1 A_1 X \\ * & -U_0 & 0 \\ * & 0 & -U_1 \end{pmatrix} < \eta \begin{pmatrix} \# & 0 & 0 \\ 0 & 0 & 0 \\ 0 & 0 & 0 \end{pmatrix} \qquad (3.30)$$

where $\# = -(A + A_1)X - X(A + A_1)^\top - BY - Y^\top B^\top$.

Thus, the optimization problem that determines this upper bound can be stated as follows:

$$\min_{\eta > 0, X > 0, Y > 0} \eta \qquad (3.31)$$
$$\text{s.t. (3.30)}.$$

To solve this optimization problem, let us introduce an auxiliary matrix $\Gamma > 0$ and cast (3.30) into the following equivalent form:

$$\begin{pmatrix} U_0 + U_1 & A_1(AX + BY) & A_1 A_1 X \\ * & -U_0 & 0 \\ * & 0 & -U_1 \end{pmatrix} < \begin{pmatrix} \Gamma & 0 & 0 \\ 0 & 0 & 0 \\ 0 & 0 & 0 \end{pmatrix} \qquad (3.32)$$

$$\Gamma < \eta[-(A + A_1)X - X(A + A_1)^\top - BY - Y^\top B^\top]. \qquad (3.33)$$

From the above discussion and Theorem 3.8, we have the following theorem.

Theorem 3.9 *Let $X > 0, Y, \Gamma > 0$ and η_0 be the solution to the following GEVP problem*

$$\min_{\eta > 0, X > 0, Y, \Gamma > 0} \eta$$

s.t. (3.32), (3.33).

Then, the control law $u(t) = Kx(t)$ with $K = YX^{-1}$ stabilizes system (3.1) with $m = 1$ for any $\tau \in [0, 1/\eta_0]$, and η_0 is the upper bound of the time delay in system (3.1).

To show the usefulness of this result, let us consider the following numerical example.

Example 3.3 *Consider a linear time delay system with the following dynamics*

$$\dot{x}(t) = \begin{pmatrix} -1.6 & 0.2 \\ 0.2 & -1.9 \end{pmatrix} x(t) + \begin{pmatrix} -2.1 & -1 \\ -1 & -0.6 \end{pmatrix} x(t - \tau)$$
$$+ \begin{pmatrix} 0.3 & 0 \\ -0.1 & 0.5 \end{pmatrix} u(t),$$

where τ is an unknown constant delay. Using Theorem 3.9, we get the suboptimal gain matrix

$$K = \begin{pmatrix} -100.1588 & 127.1173 \\ 59.1486 & -154.894 \end{pmatrix},$$

which assures an upper delay bound $\bar{\tau} = 0.3215$.

Now we are in a position to consider the delay-dependent stabilizability using a controller containing time delay. Unlike the delay-independent case, the delay-dependent stability of Theorem 3.6 cannot provide the delay-dependent stabilizability by replacing A_i by $A_i + BK_i, 0 \leq i \leq m$ in Theorem 3.6 and using the technique of variable change. To establish the delay-dependent stabilizability condition for system (3.1), we need another delay-dependent stability condition. For this purpose, let us introduce the following lemma.

Lemma 3.1 *If there exists a symmetric, positive-definite matrix $P > 0$ such that*

$$\bar{A}^\top P + P\bar{A} + \left[\sum_{j=1}^m \tau_j \right] \sum_{l=0}^m A_l^\top P A_l$$
$$+ (m+1) \sum_{j=1}^m \tau_j P A_j P^{-1} A_j^\top P < 0 \tag{3.34}$$

holds, then system (3.1) is asymptotically stable.

Proof: As in Theorem 3.6, let us consider the Lyapunov functional candidate as follows:

$$V(\mathbf{x}(t)) = x^\top(t)Px(t) + V_1(\mathbf{x}(t)) \tag{3.35}$$

where

$$V_1(\mathbf{x}(t)) = \sum_{j=1}^{m}\sum_{l=0}^{m}\int_{\tau_l}^{\tau_j+\tau_l}\left[\int_{t-s}^{t-\tau_l}x^\top(v)Q_l x(v)dv\right]ds$$

with $Q_l = A_l^\top P A_l, 0 \le l \le m$. Then

$$\dot{V}_1(\mathbf{x}(t)) = \sum_{j=1}^{m}\sum_{l=0}^{m}\left[\tau_j x^\top(t)Q_l x(t) - \int_{t-\tau_j-\tau_l}^{\tau_l}x^\top(s)Q_l x(s)ds\right]. \tag{3.36}$$

In view of (3.18), we have

$$\frac{d}{dt}(x^\top(t)Px(t)) = x^\top(t)\left[\bar{A}^\top P + P\bar{A}\right]x(t)$$
$$- 2\sum_{j=1}^{m}\sum_{l=0}^{m}\int_{t-\tau_j-\tau_l}^{t-\tau_l}x^\top(t)PA_j A_l x(s)ds.$$

Using Lemma A.1, we obtain

$$\frac{d}{dt}(x^\top(t)Px(t)) \le x^\top(t)\left(\bar{A}^\top P + P\bar{A}\right)x(t)$$
$$+ \sum_{j=1}^{m}\sum_{l=0}^{m}\int_{t-\tau_j-\tau_l}^{t-\tau_l}\left[x^\top(t)PA_j P^{-1}A_j^\top Px(t)\right.$$
$$\left. + x^\top(s)A_l^\top PA_l x(s)\right]ds. \tag{3.37}$$

Combining now (3.35)-(3.37) leads to

$$\dot{V}(\mathbf{x}(t)) \le x^\top(t)\left[\bar{A}^\top P + P\bar{A} + \left(\sum_{j=1}^{m}\tau_j\right)\sum_{l=0}^{m}A_l^\top PA_l\right.$$
$$\left. + (m+1)\sum_{j=1}^{m}\tau_j PA_j P^{-1}A_j^\top P\right]x(t).$$

Therefore, if (3.34) holds, then $\dot{V}(\mathbf{x}(t)) < 0$, which proves that system (3.1) is asymptotically stable.   □

Note that (3.34) can be rewritten as

$$\bar{A}^\top P + P\bar{A} + \left(\sum_{j=1}^{m}\tau_j\right)\sum_{l=0}^{m}A_l^\top PA_l$$

$$+ (m + 1) \sum_{j=1}^{m} \tau_j P A_j P^{-1} P P^{-1} A_j^\top P < 0. \tag{3.38}$$

Let $X = P^{-1}$. Pre- and post-multiplying (3.34) by X yields

$$X \bar{A}^\top + \bar{A} X + \left(\sum_{j=1}^{m} \tau_j \right) \sum_{l=0}^{m} X A_l^\top X^{-1} A_l X$$

$$+ (m + 1) \sum_{j=1}^{m} \tau_j A_j X X^{-1} X A_j^\top < 0 \tag{3.39}$$

which is equivalent to

$$\mathcal{J}_d = \begin{pmatrix} X \bar{A}^\top + \bar{A} X & J_{12} & J_{13} \\ J_{12}^\top & -J_{22} & \mathbf{0} \\ J_{13}^\top & \mathbf{0} & -J_{33} \end{pmatrix} < 0,$$

where

$$J_{12} = \begin{pmatrix} X A_0^\top & \cdots & X A_m^\top \end{pmatrix}$$
$$J_{13} = \begin{pmatrix} A_1 X & \cdots & A_m X \end{pmatrix}$$
$$J_{22} = \frac{1}{\sum_{j=1}^{m} \tau_j} \mathrm{diag}\{X, \cdots, X\}$$
$$J_{33} = \frac{1}{m+1} \left\{ \frac{1}{\tau_1} X, \cdots, \frac{1}{\tau_m} X \right\}.$$

As a result, we get the following theorem.

Theorem 3.10 *If there exists a symmetric, positive-definite matrix $X > 0$ such that*

$$\mathcal{J}_d = \begin{pmatrix} X \bar{A}^\top + \bar{A} X & J_{12} & J_{13} \\ J_{12}^\top & -J_{22} & \mathbf{0} \\ J_{13}^\top & \mathbf{0} & -J_{33} \end{pmatrix} < 0, \tag{3.40}$$

where

$$J_{12} = \begin{pmatrix} X A_0^\top & \cdots & X A_m^\top \end{pmatrix}$$
$$J_{13} = \begin{pmatrix} A_1 X & \cdots & A_m X \end{pmatrix}$$
$$J_{22} = \frac{1}{\sum_{j=1}^{m} \tau_j} \mathrm{diag}\{X, \cdots, X\}$$
$$J_{33} = \frac{1}{m+1} \left\{ \frac{1}{\tau_1} X, \cdots, \frac{1}{\tau_m} X \right\}.$$

hold, then system (3.1) is asymptotically stable.

From Theorem 3.10, we can establish the following asymptotic stabilizability condition for system (3.1).

Theorem 3.11 *If there exist a symmetric, positive-definite matrix $X > 0$ and matrices $Y, Y_1, \cdots, Y_m$ satisfying*

$$\begin{pmatrix} M_{11} + M_{11}^\top & M_{12} & M_{13} \\ M_{12}^\top & -J_{22} & \mathbf{0} \\ M_{13}^\top & \mathbf{0} & -J_{33} \end{pmatrix} < 0, \tag{3.41}$$

where

$$M_{11} = \bar{A}X + BY + \sum_{j=1}^{m} BY_m$$

$$M_{12} = \left(XA_0^\top + Y^\top B^\top, \; XA_1^\top + Y_m^\top B^\top, \; \cdots, XA_m^\top + Y^\top B^\top \right)$$

$$M_{13} = \left(A_1 X + BY_1, \; \cdots, \; A_m X + BY_m \right),$$

then system (3.1) is asymptotically stable under control (3.5) and $K = YX^{-1}, K_i = Y_iX^{-1}, 1 \le i \le m$ are a set of stabilizing gain matrices.

Proof: Replace A_0 and $A_i, 1 \le i \le m$ with $A_0 + BK$, and $A_i + BK_i$, respectively, in $\mathcal{J}_d$ and let $Y = KX, Y_i = K_iX$, then the proof of Theorem 3.11 follows from Theorem 3.10.   □

To show the efficiency of the stabilization design algorithm let us consider the following numerical algorithm.

Example 3.4 *Consider the dynamical time delay system*

$$\dot{x}(t) = \begin{pmatrix} 0 & 1 \\ -1 & -2 \end{pmatrix} x(t) + \begin{pmatrix} 0.5 & 0 \\ 0.2 & 0.1 \end{pmatrix} x(t-\tau) + \begin{pmatrix} 0.4 \\ 0.1 \end{pmatrix} u(t).$$

Using Theorem 3.11, we get the following gains for different time delays ($\tau = 1$ and $\tau = 2$):

τ	K	K_1
1	$\begin{pmatrix} 0.2055 & -0.4683 \end{pmatrix}$	$\begin{pmatrix} -2.0377 & -0.8619 \end{pmatrix}$
2	$\begin{pmatrix} 1.9825 & -2.3857 \end{pmatrix}$	$\begin{pmatrix} -1.7443 & -0.7324 \end{pmatrix}$

By the same argument as in Theorem 3.9 the upper bound for the time delay can be obtained by solving a GEVP. The maximum delay is then equal to 24.1899 and the corresponding gain matrices are

$$K = \begin{pmatrix} 3.4653 & 5.3012 \end{pmatrix} \quad K_1 = \begin{pmatrix} -1.5362 & -0.3998 \end{pmatrix}.$$

3.3 Output Feedback Stabilization

In the previous sections we used the state feedback controller to stabilize the class of systems we are dealing with. To do this, we required the complete access to the state vector of the system at any time t. But more often, this is not the case and the state vector cannot be accessed for many reasons well-known by the control community. In case of nonaccessibility of the state vector, an alternate approach consists of designing an output feedback controller that stabilizes the class of systems under study. In this section, we consider the output feedback stabilizability of system (3.1). Two kinds of control laws are considered, delay-independent memoryless output feedback control and observer-based output feedback control.

3.3.1 Delay-Independent Dynamic Output Feedback Stabilization

Let us consider the following output feedback law:

$$\begin{cases} \dot{\xi}(t) = K_A \xi(t) + K_B y(t) \\ u(t) = K_C \xi(t) + K_D y(t) \end{cases} \tag{3.42}$$

where $\xi(t) \in \mathbb{R}^n$ is the state of the controller, $y(t) \in \mathbb{R}^r$ is the output vector, and K_A, K_B, K_C, K_D are constant matrices of appropriate dimensions. Applying controller (3.42) to system (3.1) and letting $\eta^\top(t) = \begin{pmatrix} x^\top(t) & \xi^\top(t) \end{pmatrix}$, we get the closed-loop system

$$\dot{\eta}(t) = \bar{A}\eta(t) + \sum_{j=1}^{m} \bar{A}_j I_0 \eta(t - \tau_j), \tag{3.43}$$

where

$$I_0 = \begin{pmatrix} I & \mathbf{0} \end{pmatrix} \quad \bar{A} = \begin{pmatrix} A + BK_D C & BK_C \\ K_B C & K_A \end{pmatrix} \quad \bar{A}_j = \begin{pmatrix} A_j \\ \mathbf{0} \end{pmatrix}.$$

By the same argument as in Theorem 3.1 one gets the following theorem.

Theorem 3.12 For given matrices K_A, K_B, K_C, K_D, if there exist symmetric and positive-definite matrices $Q_i > 0, i = 1, \cdots, m$ such that the LMI

$$\begin{pmatrix} \bar{A}^\top P + P\bar{A} + \sum_{j=1}^{m} I_0^\top Q_j I_0 & P I_0^\top A_1 & \cdots & P I_0^\top A_m \\ A_1^\top I_0 P & -Q_1 & & \\ \vdots & & \ddots & \\ A_m^\top I_0 P & & & -Q_m \end{pmatrix} < 0 \tag{3.44}$$

has a feasible solution $P > 0$, then the closed-loop system is asymptotically stable.

Proof: Let us consider the following Lyapunov functional candidate

$$V(\eta(t)) = \eta^\top(t) P\eta(t) + \sum_{j=1}^{m} \int_{t-\tau_j}^{t} x^\top(s) Q_j x(s) ds$$

$$= \eta^\top(t) P\eta(t) + \sum_{j=1}^{m} \int_{t-\tau_j}^{t} \eta^\top(s) I_0^\top Q_j I_0 \eta(s) ds,$$

where $\eta(t) = (\eta(s), t - \bar{\tau} \le s \le t)$. The remaining part of the proof is the same as in Theorem 3.1. $\qquad\square$

Obviously, (3.44) is nonlinear with respect to P and K_A, K_B, K_C, K_D. To cast the controller design problem into the LMI framework, let us use the following notations

$$K = \begin{pmatrix} K_D & K_C \\ K_B & K_A \end{pmatrix} \quad \bar{A}_0 = \begin{pmatrix} A & 0 \\ 0 & 0 \end{pmatrix},$$

$$B_0 = \begin{pmatrix} B & 0 \\ 0 & I \end{pmatrix} \quad C_0 = \begin{pmatrix} C & 0 \\ 0 & I \end{pmatrix}.$$

With this notation, we have

$$\bar{A} = \bar{A}_0 + B_0 K C_0.$$

In view of Lemma A.2, P and $Q_i, i = 1, \cdots, m$, satisfy (3.44) if and only if

$$P\bar{A} + \bar{A}^\top P + \sum_{j=1}^{m} I_0^\top Q_j I_0 + \sum_{j=1}^{m} PI_0^\top A_j Q_j^{-1} A_j^\top I_0 P < 0$$

holds, that is,

$$\Xi + PB_0 K C_0 + C_0^\top K^\top B_0^\top P < 0, \tag{3.45}$$

where $\Xi \overset{\triangle}{=} P\bar{A}_0 + \bar{A}_0^\top P + \sum_{j=1}^{m} PI_0^\top A_j Q_j^{-1} A_j^\top I_0 P + \sum_{j=1}^{m} I_0^\top Q_j I_0$. Note that $(PB_0)_\perp = P^{-1} B_{0\perp}$. Then, from Lemma A.3 it follows that there exists a matrix K satisfying (3.45) if and only if

$$\begin{cases} B_{0\perp}^\top P^{-1} \Xi P^{-1} B_{0\perp} < 0 \\ C_{0\perp} \Xi C_{0\perp}^\top < 0 \end{cases} \tag{3.46}$$

that is,

$$\begin{cases} B_{0\perp}^\top \left[\bar{A}_0 P^{-1} + P^{-1} \bar{A}_0^\top + \sum_{j=1}^{m} I_0^\top A_j Q_j^{-1} A_j^\top I_0 \right. \\ \qquad\qquad \left. + \sum_{j=1}^{m} P^{-1} I_0^\top Q_j I_0 P^{-1} \right] B_{0\perp} < 0 \\ C_{0\perp} \Xi C_{0\perp}^\top < 0. \end{cases}$$

Let us now suppose that the matrices P and P^{-1} are partitioned as follows:

$$P = \begin{pmatrix} Y & N \\ N^\top & W_1 \end{pmatrix} \quad P^{-1} = \begin{pmatrix} X & M \\ M^\top & W_2 \end{pmatrix}, \tag{3.47}$$

where $X, Y \in \mathbb{R}^{n \times n}$; $W_1, W_2 \in \mathbb{R}^{n \times n}$; $M, N \in \mathbb{R}^{n \times n}$. From the fact that $PP^{-1} = I$, it follows that

$$\begin{cases} YX + NM^\top = I \\ YM + NW_2 = 0 \\ N^\top X + W_1 M^\top = 0. \end{cases} \tag{3.48}$$

Then, we get

$$P^{-1} \Xi P^{-1} = \bar{A}_0 P^{-1} + P^{-1} \bar{A}_0^\top + \begin{pmatrix} \sum_{j=1}^m A_j Q_j^{-1} A_j^\top & 0 \\ 0 & 0 \end{pmatrix}$$
$$+ \begin{pmatrix} X \\ M^\top \end{pmatrix} \left[\sum_{j=1}^m Q_j \right] \begin{pmatrix} X & M \end{pmatrix} \tag{3.49}$$

and

$$\Xi = \begin{pmatrix} YA_0 + A_0^\top Y & 0 \\ 0 & 0 \end{pmatrix} + \begin{pmatrix} Y \\ N^\top \end{pmatrix} \left[\sum_{j=1}^m A_j Q_j^{-1} A_j^\top \right] \begin{pmatrix} Y & N \end{pmatrix}$$
$$+ \begin{pmatrix} \sum_{j=1}^m Q_j & 0 \\ 0 & 0 \end{pmatrix}. \tag{3.50}$$

Note that

$$B_{0\perp} = \begin{pmatrix} B_\perp \\ 0 \end{pmatrix} \quad \text{and} \quad C_{0\perp} = \begin{pmatrix} C_\perp^\top & 0 \end{pmatrix},$$

which, combined with (3.49) and (3.50), implies that (3.46) is equivalent to

$$\begin{cases} B_\perp^\top \left[AX + XA^\top + \sum_{j=1}^m A_j Q_j^{-1} A_j^\top + \sum_{j=1}^m X Q_j X \right] B_\perp < 0 \\ C_\perp^\top \left[YA + A^\top Y + \sum_{j=1}^m Y A_j Q_j^{-1} A_j^\top Y + \sum_{j=1}^m Q_j \right] C_\perp < 0. \end{cases} \tag{3.51}$$

Using now Lemma A.2, (3.51) is equivalent to

$$\mathcal{B}^\top \begin{pmatrix} \beta_1 & X & \cdots & X \\ X & -Q_1^{-1} & & \\ \vdots & & \ddots & \\ X & & & -Q_m^{-1} \end{pmatrix} \mathcal{B} < 0 \tag{3.52}$$

$$\mathcal{C} \begin{pmatrix} \beta_2 & YA_1 & \cdots & YA_m \\ A_1^\top Y & -Q_1 & & \\ \vdots & & \ddots & \\ A_m^\top Y & & & -Q_m \end{pmatrix} \mathcal{C}^\top < 0, \tag{3.53}$$

where

$$\beta_1 = AX + XA^\top + \sum_{j=1}^{m} A_j Q_j^{-1} A_j^\top$$

$$\beta_2 = YA + A^\top Y + \sum_{j=1}^{m} Q_j$$

$$\mathcal{B} = \mathrm{diag}\{B_\perp, I, \cdots, I\}$$

$$\mathcal{C} = \mathrm{diag}\{C_\perp^\top, I, \cdots, I\}.$$

From the above discussion, we have the following theorem.

Theorem 3.13 *(i) If there exist symmetric, positive-definite matrices $K_A, K_B, K_C, K_D, P > 0$, and $Q > 0$ satisfying (3.44), which means that the closed-loop system (3.43) is asymptotically stable, then LMIs (3.52), (3.53), and*

$$\begin{pmatrix} X & I \\ I & Y \end{pmatrix} \geq 0 \tag{3.54}$$

have a set of feasible solutions $X > 0, Y > 0$.

(ii) On the other hand, if LMIs (3.52), (3.53), and (3.54) have a set of feasible solutions $X > 0, Y > 0$ for given positive-definite matrices $Q = (Q_1, \ldots, Q_m)$, then there exists an output feedback controller of form of (3.42) that stabilizes system (3.1).

Proof: *Proof of (i).* Suppose K_A, K_B, K_C, K_D, and P satisfy (3.44) and partition P and P^{-1} as in (3.47). Since (3.44) is strictly negative, by the same argument as in Lemma 4.2 of [48], we can assume with no loss of generality that M, N are of full rank. Then, the above derivation shows that $X > 0, Y > 0$ satisfy (3.52) and (3.53). So, to complete the proof of (i) it suffices to prove (3.54). For this purpose, let us define the following two matrices

$$\Phi_1 = \begin{pmatrix} X & I \\ M^\top & 0 \end{pmatrix} \quad \Phi_2 = \begin{pmatrix} I & Y \\ 0 & N^\top \end{pmatrix}. \tag{3.55}$$

Since M is of full rank, Φ_1 is invertible. Moreover, direct manipulation gives

$$\Phi_1^\top P \Phi_1 = \begin{pmatrix} X & I \\ I & Y \end{pmatrix} > 0, \tag{3.56}$$

establishing (3.54). This completes the proof of (i).

Proof of (ii). Let X, Y satisfy (3.52), (3.53), and (3.54). Then, it follows from Lemma A.4 that there exists a symmetric, positive-definite matrix $P > 0$ satisfying

$$P^{-1} = \begin{pmatrix} X & \# \\ \# & \# \end{pmatrix} \quad P = \begin{pmatrix} Y & \# \\ \# & \# \end{pmatrix},$$

which can be constructed as follows. With X and Y given, using singular value decomposition of $I - XY$, we can obtain M, N satisfying $MN^\top = I - XY$. Using (3.48), if we let $W_1 = -N^\top X[M^{-1}]^\top$, $W_2 = -N^{-1}YM$, and we define P by (3.47), then it is easy to check that P satisfies (3.45). Moreover, from the fact that (3.46) is feasible if and only if (3.45) is feasible. Therefore, there exists matrix $K = \begin{pmatrix} K_D & K_C \\ K_B & K_A \end{pmatrix}$ satisfying (3.45), which means that there exists an output feedback controller (3.42) that stabilizes system (3.1). This completes the proof of (ii) and concludes the proof of Theorem 3.13. □

Eliminating the design parameters K from (3.45), Theorem 3.12 provides a sufficient condition for system (3.1) to be stabilizable instead of giving a design method for the output feedback controller. However, a stabilizing controller can obtained by the following algorithm.

Algorithm 3.1 *(Design algorithm for output feedback controller)*

Step 1. Solve the feasible problem (3.52), (3.54) to get matrices X, Y.

Step 2. Compute two full rank matrices $M, N \in \mathbb{R}^{n \times n}$ such that $MN^\top = I - XY$.

Step 3. Solve the matrix equation

$$\begin{pmatrix} I & Y \\ 0 & N^\top \end{pmatrix} = P \begin{pmatrix} X & I \\ M^\top & 0 \end{pmatrix}. \tag{3.57}$$

to get P.

Step 4. With P obtained in Step 3, solve the feasible problem (3.44) and a set of control parameters K can be obtained.

To illustrate the usefulness of Algorithm 3.1, let us work out a numerical example.

Example 3.5 *Consider a linear time delay system with dynamics described by*

$$\begin{cases} \dot{x}(t) = \begin{pmatrix} -2 & 1 \\ -1 & 1 \end{pmatrix} x(t) + \begin{pmatrix} 0.2 & 0.1 \\ 0.3 & 0.1 \end{pmatrix} x(t-1) + \begin{pmatrix} 1 \\ 0 \end{pmatrix} u(t) \\ y(t) = \begin{pmatrix} 1 & 0 \end{pmatrix} x(t). \end{cases}$$

Letting $Q = I$ and using Algorithm 3.1, we get the following results:

$$X = \begin{pmatrix} 2.0367 & 0.18712 \\ 0.1871 & 5.3330 \end{pmatrix} \qquad Y = \begin{pmatrix} 11.8185 & -4.6526 \\ -4.6526 & 2.7491 \end{pmatrix}$$

$$M = \begin{pmatrix} -23.8867 & 1.6989 \\ 25.9218 & 1.5651 \end{pmatrix} \qquad N = \begin{pmatrix} 0.8984 & -0.4392 \\ -0.4392 & -0.8984 \end{pmatrix}$$

$$K_A = \begin{pmatrix} -41.5886 & -2.0227 \\ -2.0227 & -4.4046 \end{pmatrix} \qquad K_B = \begin{pmatrix} -316.2049 \\ -13.8697 \end{pmatrix}$$

$$K_C = \begin{pmatrix} -7,2781 & -1.6916 \end{pmatrix} \qquad K_D = -59.6668.$$

The gains of the output feedback controller that stabilize our example are then given by K_A, K_B, K_C, and K_D.

3.3.2   Observer-Based Output Feedback Stabilization

Theorem 3.13 provides a sufficient condition for the stabilizability of system (3.1) using a memoryless output feedback controller. In this subsection, we will consider a kind of observer-based output feedback controller, which contains time delays in its dynamics, that is,

$$\begin{cases} \dot{\xi}(t) = (A + BK - LC)\xi(t) + \sum_{j=1}^{m} A_j \xi(t - \tau_j) + Ly(t) \\ u(t) = K\xi(t), \end{cases} \qquad (3.58)$$

where $\xi(t)$ is the state controller, $K \in \mathbb{R}^{k \times n}$ is the feedback gain, and $L \in \mathbb{R}^{n \times r}$ is the observer gain.

Let $\eta^\top(t) = \begin{pmatrix} x^\top(t) & \xi^\top(t) \end{pmatrix}$. Then applying (3.58) to system (3.1) gives a closed-loop system of order $2n$ (extended system) described by

$$\dot{\eta}(t) = \bar{A}\eta(t) + \sum_{j=1}^{m} \bar{A}_j \eta(t - \tau_j) \qquad (3.59)$$

where

$$\bar{A} = \begin{pmatrix} A & BK \\ LC & A + BK - LC \end{pmatrix} \qquad \bar{A}_j = \begin{pmatrix} A_j & 0 \\ 0 & A_j \end{pmatrix}.$$

The following theorem provides a method to design a controller given by (3.58) that stabilizes system (3.1).

Theorem 3.14 *If there exist symmetric, positive-definite matrices $X, Y \in \mathbb{R}^{n \times n}, U_i \in \mathbb{R}^{2n \times 2n}, 1 \leq i \leq m$, and $\mathcal{A}, \mathcal{B} \in R^{k \times n}$, such that the LMI*

$$\begin{pmatrix} J_{11} + \sum_{j=1}^{m} U_j & M_1 & \cdots & M_m \\ M_1^\top & -U_1 & & 0 \\ \vdots & & \ddots & \\ M_m^\top & & & -U_m \end{pmatrix} < 0 \qquad (3.60)$$

is feasible and the matrix equation

$$\begin{cases} \mathcal{A} = BKY + XC^\top L^\top \\ \mathcal{B} = BKY - LCY \end{cases} \tag{3.61}$$

has a set of solutions K, L, where

$$J_{11} = \begin{pmatrix} AX + XA^\top & \mathcal{A} \\ \mathcal{A}^\top & AY + YA^\top + \mathcal{B} + \mathcal{B}^\top \end{pmatrix}$$

$$M_i = \begin{pmatrix} A_i X & 0 \\ 0 & A_i Y \end{pmatrix},$$

then system (3.1) is asymptotically stable under controller (3.58) with gains K, L.

Proof: Based on Theorem 3.1, to establish the asymptotic stability of system (3.59) it suffices to show that there exist $P > 0, Q_i > 0, 1 \leq i \leq m$, such that

$$\begin{pmatrix} \bar{A}^\top P + P\bar{A} + \sum_{j=1}^m Q_i & P\bar{A}_1 & \cdots & P\bar{A}_m \\ \bar{A}_1^\top P & -Q_1 & & \\ \vdots & & \ddots & \\ \bar{A}_m^\top P & & & -Q_m \end{pmatrix} < 0. \tag{3.62}$$

Pre- and post-multiplying both sides of (3.62) by $\text{diag}\{P^{-1}, P^{-1}, \cdots, P^{-1}\}$ yields

$$\begin{pmatrix} \tilde{W} & \bar{A}_1 P^{-1} & \cdots & \bar{A}_m P^{-1} \\ P^{-1}\bar{A}_1^\top & -P^{-1}Q_1 P^{-1} & & \\ \vdots & & \ddots & \\ P^{-1}\bar{A}_m^\top & & & -P^{-1}Q_m P^{-1} \end{pmatrix} < 0 \tag{3.63}$$

with $\tilde{W} = P^{-1}\bar{A}^\top + \bar{A}P^{-1} + \sum_{j=1}^m P^{-1}Q_i P^{-1}$. If we assume that the matrix P^{-1} is partitioned as

$$P^{-1} = \begin{pmatrix} X & 0 \\ 0 & Y \end{pmatrix}$$

with $X > 0, Y > 0$, then noting that

$$P^{-1}\bar{A}^\top + \bar{A}P^{-1} = \begin{pmatrix} AX + XA^\top & \mathcal{A} \\ \mathcal{A}^\top & AY + YA^\top + \mathcal{B} + \mathcal{B}^\top \end{pmatrix}$$

$$\bar{A}_i P^{-1} = \begin{pmatrix} A_i X & 0 \\ 0 & A_i Y \end{pmatrix}$$

and letting $P^{-1}Q_i P^{-1} = U_i$, $\mathcal{A} = BKY + XC^\top L^\top$, $\mathcal{B} = BKY - LCY$ we get (3.60). Therefore, if (3.60) has a set of feasible solutions $X > 0, Y > 0, U_i > 0, \mathcal{A}, \mathcal{B}$, and (K, L) is a set of solutions of (3.61), then

$$P = \begin{pmatrix} X^{-1} & \mathbf{0} \\ \mathbf{0} & Y^{-1} \end{pmatrix}, Q_i = PU_iP \text{ will satisfy (3.62). Thus, the system is}$$

asymptotically stable. This completes the proof of Theorem 3.14.   □

Remark 3.5 *With $\mathcal{A}, \mathcal{B}, X, Y$ at hand, the matrix equation (3.61) can be solved using the Kronecker product (see Appendix B) as follows: First of all, the gain L can be obtained by solving*

$$\left[(I \otimes (XC^\top))\mathbf{U} + ((Y^\top C^\top) \otimes I)\right] \text{vec}(L) = \text{vec}(\mathcal{A} - \mathcal{B}) \qquad (3.64)$$

where $\mathbf{U}$ is the permutation matrix as defined in Appendix B. Then, solve the following to get the gain K

$$(Y \otimes B)\text{vec}(K) = \text{vec}(\mathcal{A} - XC^\top L^\top). \qquad (3.65)$$

In fact, from (3.61) it follows that

$$\mathcal{A} - \mathcal{B} = XC^\top L^\top + LCY.$$

Thus, using Property 1 of Appendix B, we have

$$\text{vec}(\mathcal{A} - \mathcal{B}) = (I \otimes (XC^\top))\text{vec}(L^\top) + ((CY)^\top \otimes I)\text{vec}(L)$$

which combined with Property 2 of Appendix B yields (3.64). With L at hand, (3.65) follows from the first equation of (3.61) and Property 2 of Appendix B.

To show the validity of Theorem 3.14, we give the following numerical example.

Example 3.6 *Consider a time delay system with the following dynamics:*

$$\begin{cases} \dot{x}(t) = \begin{pmatrix} 0 & 1 \\ -1 & -2 \end{pmatrix} x(t) + \begin{pmatrix} 0 & 0 \\ 0.2 & 0.1 \end{pmatrix} x(t-\tau) + \begin{pmatrix} 0 & 0.1 \\ 1 & 0 \end{pmatrix} u(t) \\ y(t) = \begin{pmatrix} 0 & 1 \\ 1 & 0 \end{pmatrix} x(t). \end{cases}$$

For this system, if our choice is fixed on an observer-based output feedback controller, the solution is given by Theorem 3.14. Solving LMI (3.60) gives

$$X = \begin{pmatrix} 151.9746 & -53.0935 \\ -53.0935 & 59.9963 \end{pmatrix}$$

$$Y = \begin{pmatrix} -78.1362 & 51.9118 \\ 51.9118 & 7.4424 \end{pmatrix}$$

$$U = \begin{pmatrix} 53.0935 & -7.1043 & 0 & 0 \\ -7.1043 & 66.8991 & 0 & 0 \\ 0 & 0 & 69.664 & 0 \\ 0 & 0 & 0 & 69.664 \end{pmatrix}$$

$$\mathcal{A} = 0$$

$$\mathcal{B} = \begin{pmatrix} 173.4875 & -18.2450 \\ -18.2450 & -63.9290 \end{pmatrix}.$$

Solving matrix equation (3.61) gives the gain matrices

$$K = \begin{pmatrix} 0.3404 & -0.8427 \\ -30.4744 & 13.9397 \end{pmatrix} \quad L = \begin{pmatrix} -0.9197 & -2.3643 \\ 0.9709 & 1.3119 \end{pmatrix}.$$

3.4 Discretization Scheme of Lyapunov Functional

Section 3.2 shows that the more general the Lyapunov functional used, the more powerful the stability test. This section addresses the asymptotic stability problem by considering a general Lyapunov functional which has matrix-valued time-varying functions $Q(\cdot), R(\cdot, \cdot)$, and $S(\cdot)$. By discretizing the delay τ into subintervals in which the matrices $Q(\cdot), S(\cdot), R(\cdot, \cdot)$ are piecewise linear continuous functions, a new sufficient condition guaranteeing the asymptotic stability of the system is developed. For notational simplicity, this subsection will consider the simplest case, that is,

$$\dot{x}(t) = Ax(t) + A_1 x(t - \tau). \tag{3.66}$$

To establish the stability of system (3.66), let us consider a Lyapunov functional of the following generalized quadratic form:

$$V(\mathbf{x}(t)) = \frac{1}{2} x^\top(t) P x(t) + x^\top(t) \int_{-\tau}^{0} Q(s) x(t + s) ds$$

$$+ \frac{1}{2} \int_{-\tau}^{0} \int_{-\tau}^{0} x^\top(t + z) R(z, v) x(t + v) dz dv$$

$$+ \frac{1}{2} \int_{-\tau}^{0} x^\top(t + s) S(s) x(t + s) ds \tag{3.67}$$

where

$$\begin{cases} P \in \mathbb{R}^{n \times n}, P > 0 \\ Q(s) : [-\tau, 0] \to \mathbb{R}^{n \times n} \\ S(s) : [-\tau, 0] \to \mathbb{R}^{n \times n}, \ S^\top(s) = S(s) > 0 \\ R : [-\tau, 0] \times [-\tau, 0] \to \mathbb{R}^{n \times n}, \ R(u, v) = R^\top(v, u) > 0. \end{cases} \tag{3.68}$$

Then, we can conclude (see Hale [98]) that system (3.66) is asymptotically stable if there exist symmetric, positive-definite matrix $P > 0$, matrix-valued functions $Q(s), S(s), R(s, t)$ satisfying (3.68), and a scalar $\varepsilon > 0$ such that

$$V(\mathbf{x}_t) \geq \varepsilon x^\top(t) x(t) \tag{3.69}$$

and its derivative along the trajectory of system (3.66) satisfies

$$\dot{V}(\mathbf{x}_t) \leq -\varepsilon x^\top(t) x(t). \tag{3.70}$$

Obviously the Lyapunov functional is more general than the functionals used in previous theorems, thus to get $V(\cdot)$ one has to handle complex partial differential and algebraic equations with boundary conditions. In the following, we will restrict $S(s), R(z, v), Q(s)$ to the subspace of piecewise linear continuous functions and thus develop an LMI-based stability condition for system (3.66). For this purpose, let us introduce some notations as follows:

$$\varsigma = [-\tau, 0], \quad \varsigma_i = [-\tau_{i-1}, -\tau_i],$$

where

$$\tau_i = -\tau + ih, \quad h = \frac{\tau}{N}$$

with N being a fixed integer. Similarly,

$$\mathcal{S} = [-\tau, 0] \times [-\tau, 0],$$
$$\mathcal{S}_{ij} = [\tau_{i-1}, \tau_i] \times [\tau_{j-1}, \tau_j], \quad i, j = 1, \cdots, N.$$

Each $\mathcal{S}_{ij}$ is further divided into two triangular regions

$$\mathcal{T}_{ij}^1 = \left\{ \tau_{i-1} + \alpha h, \tau_{j-1} + \beta h \;\middle|\; \begin{array}{l} 0 \le \beta \le 1 \\ 0 \le \alpha \le \beta \end{array} \right\}$$

$$\mathcal{T}_{ij}^2 = \left\{ \tau_{i-1} + \alpha h, \tau_{j-1} + \beta h \;\middle|\; \begin{array}{l} 0 \le \alpha \le 1 \\ 0 \le \beta \le \alpha \end{array} \right\}.$$

Let $R_{ij}, W_i, Q_i, S_i, i, j = 1, \cdots, N$ be symmetric, positive-definite matrices. Then, define Q and S to be continuous in ς and linear with each ς_i, and R to be continuous in $\mathcal{S}$ and linear within each $\mathcal{T}_{ij}^1, \mathcal{T}_{ij}^2$, that is,

$$Q(\tau_{i-1} + \alpha h) = Q^i(\alpha)$$
$$= (1 - \alpha)Q_{i-1} + \alpha Q_i, 0 \le \alpha \le 1 \qquad (3.71)$$

$$S(\tau_{i-1} + \alpha h) = S^i(\alpha)$$
$$= (1 - \alpha)S_{i-1} + \alpha S_i, 0 \le \alpha \le 1 \qquad (3.72)$$

and

$$R(\tau_{i-1} + \alpha h, \tau_{j-1} + \beta h) = R_{ij}(\alpha, \beta)$$
$$= \begin{cases} (1 - \alpha)R_{i-1,j-1} + \beta R_{i,j} + (\alpha - \beta)R_{i,j-1}, & \alpha \ge \beta \\ (1 - \beta)R_{i-1,j-1} + \alpha R_{i,j} + (\beta - \alpha)R_{i,j-1}, & \alpha < \beta \end{cases}$$
$$0 \le \alpha \le 1, 0 \le \beta \le 1. \qquad (3.73)$$

Evidently, $S_i = S_i^\top, R_{ij} = R_{ji}^\top$.

To state the stability condition, let us introduce the following notations

$$\tilde{R} = \begin{pmatrix} R_{00} & R_{01} & \cdots & R_{0N} \\ R_{10} & R_{11} & \cdots & R_{1N} \\ \vdots & \vdots & \vdots & \vdots \\ R_{N0} & R_{N1} & \cdots & R_{NN} \end{pmatrix}$$

$$\tilde{Q} = \begin{pmatrix} Q_0 & Q_1 & \cdots & Q_N \end{pmatrix}$$

$$\hat{S} = \operatorname{diag}\{2S_0, S_1, \cdots, S_{N-1}, 2S_N\}$$

and

$$S_{di} = \frac{1}{h}(S_i - S_{i-1})$$
$$S_d = \text{diag}\{S_{d1}, S_{d2}, \cdots, S_{dN}\}$$
$$R_{dij} = \frac{1}{h}(R_{ij} - R_{i-1,j-1})$$
$$\tilde{R}_d = \begin{pmatrix} R_{d11} & R_{d12} & \cdots & R_{d1N} \\ R_{d21} & R_{d22} & \cdots & R_{d2N} \\ \vdots & \vdots & \vdots & \vdots \\ R_{dN1} & R_{dN2} & \cdots & R_{dNN} \end{pmatrix}$$
$$D_{1i}^k = A^\top Q_{i-1+k} - \frac{1}{h}(Q_i - Q_{i-1}) + R_{i-1+k,N}^\top$$
$$D_{2i}^k = B_1^\top Q_{i-1+k} - R_{i-1+k,0}^\top$$
$$D_j^k = (D_{j1}^k, D_{j2}^k, \cdots, D_{jN}^k).$$

Theorem 3.15 *Let a given set of symmetric, positive-definite matrices* $R_{ij}, Q_i, S_i, i, j = 0, 1, \cdots, N$, *and* S, Q, R *be defined by (3.71), (3.72), and (3.73), respectively.*

(i) *If there exist* $W_i, 0 \leq i \leq N$, *and* $U > 0$ *satisfying*

$$\begin{pmatrix} U & \tilde{W} \\ \tilde{W}^\top & \tilde{R} \end{pmatrix} \geq 0 \qquad (3.74)$$

$$\begin{pmatrix} \frac{P-U}{h} & \tilde{Q} - \tilde{W} \\ (\tilde{Q} - \tilde{W})^\top & \hat{S} \end{pmatrix} > 0, \qquad (3.75)$$

where $\tilde{W} = (W_0, W_1, \cdots, W_N)$, *then* V *defined by (3.67) satisfies (3.69).*

(ii) *If there exist matrices* $X_{ij} \in \mathbb{R}^{n \times n}, i = 1, 2; j = 1, 2, \cdots, N$, *and* $T_{ij}^\top = T_{ji}, Y_{ij}^\top = Y_{ji}, i, j = 1, 2$ *such that*

$$\begin{pmatrix} T_{11} & T_{12} & X_1 \\ * & T_{22} & X_2 \\ * & * & R_d \end{pmatrix} \geq 0 \qquad (3.76)$$

$$\begin{pmatrix} \Delta_{11} - T_{11} + Y_{11} & -\Delta_{12} - T_{12} + Y_{12} & -D_1^0 - X_1 \\ * & \Delta_{22} - T_{22} + Y_{22} & -D_2^0 - X_2 \\ * & * & \frac{1}{h}S_d \end{pmatrix} \geq 0 \quad (3.77)$$

$$\begin{pmatrix} \Delta_{11} - T_{11} - Y_{11} & -\Delta_{12} - T_{12} - Y_{12} & -D_1^1 - X_1 \\ * & \Delta_{22} - T_{22} - Y_{22} & -D_2^1 - X_2 \\ * & * & \frac{1}{h}S_d \end{pmatrix} \geq 0 \quad (3.78)$$

where

$$X_i = (X_{i1}, X_{i2}, \cdots, X_{iN})$$

$$\Delta_{11} = -PA - A^\top P - Q_N - Q_N^\top - S_N$$
$$\Delta_{12} = PB - Q_0$$
$$\Delta_{22} = S_0,$$

then $V(.)$ satisfies (3.70).

(iii) If (i) and (ii) are satisfied, then system (3.66) is asymptotically stable.

Proof: For notational brevity, let us write

$$\phi(t + \cdot) = \{x(t + s), \ -\tau \le s \le 0\}, \ \forall t \in (0, \infty)$$
$$\phi^i(\alpha) = \phi(\tau_{i-1} + \alpha h) \tag{3.79}$$

and

$$W(\tau_{i-1} + \alpha h) = W^i(\alpha) = (1 - \alpha)W_{i-1} + \alpha W_i. \tag{3.80}$$

Proof of (i). Put $V(\cdot)$ as follows:

$$V(\phi) = V_1(\phi) + V_2(\phi)$$

where

$$V_1(\phi) = \frac{1}{2}\phi^\top(0)U\phi(0) + \phi^\top(0)\int_{-\tau}^0 W(s)\phi(s)ds$$
$$+ \frac{1}{2}\int_{-\tau}^0\int_{-\tau}^0 \phi^\top(s)R(s, u)\phi(u)duds$$
$$V_2(\phi) = \frac{1}{2}\phi^\top(0)(P - U)\phi(0) + \phi^\top(0)\int_{-\tau}^0 [Q(s) - W(s)]\phi(s)ds$$
$$+ \frac{1}{2}\int_{-\tau}^0 \phi^\top(s)S(s)\phi(s)ds.$$

To prove (i), it suffices to show that

(a) (3.74) guarantees

$$V_1(\phi) \ge 0, \tag{3.81}$$

(b) (3.75) implies

$$V_2(\phi) \ge \varepsilon\phi^\top(0)\phi(0) \tag{3.82}$$

for some sufficiently small $\varepsilon > 0$.

We proceed to prove (a). Using (3.80), (3.73), and (3.79) with a change of order of integration between α and β leads to

$$V_1(\phi) = \frac{1}{2}\phi^\top(0)U\phi(0) + h\phi^\top(0)\sum_{i=1}^N \int_0^1 W^i(\alpha)\phi^i(\alpha)d\alpha$$
$$+ h^2\sum_{i=1}^N\sum_{j=1}^N \int_0^1\int_0^\alpha \phi^{i\top}(\alpha)R_{ij}(\alpha, \beta)\phi^j(\beta)d\beta d\alpha.$$

Introducing a change of variable

$$\psi^i(\alpha) = \int_0^\alpha \phi^i(\tau) d\tau$$

or

$$\phi^i(\alpha) = \dot{\psi}^i(\alpha), \psi^i(0) = 0$$

and using integration by parts, we have

$$V_1(\phi) = \frac{1}{2}\phi^\top(0)U\phi(0)$$

$$+ h\sum_{i=1}^N \left[\phi^\top(0)W_i\psi^i(1) - \phi(0)\int_0^1 (W_i - W_{i-1})\psi^i(\alpha)d\alpha \right]$$

$$+ h^2\sum_{i=1}^N\sum_{j=1}^N \left[\psi^{i\top}(1)R_{ij}\psi^j(1) - \psi^{i\top}(1)(R_{ij} - R_{i,j-1})\int_0^1 \psi^j(\beta)d\beta \right]$$

$$+ \frac{h^2}{2}\sum_{i=1}^N\sum_{j=1}^N \int_0^1 \psi^{i\top}(\alpha)[R_{ij} - R_{i,j-1} - R_{i-1,j} + R_{i-1,j-1}]\psi^j(\alpha)d\alpha.$$

Let

$$\tilde{\psi}(\alpha) = \left(\ \psi^{1\top}(\alpha), \psi^{2\top}(\alpha), \cdots, \psi^{N\top}(\alpha) \ \right)^\top.$$

Then $V_1(\phi)$ can be rewritten as

$$V_1(\phi) = \frac{1}{2}\phi^\top(0)U\phi(0) + h\phi^\top(0)\tilde{W}I_r^\top\tilde{\psi}(1)$$

$$- h\phi^\top(0)\tilde{W}I_d^\top \int_0^1 \tilde{\psi}(\alpha)d\alpha + \frac{h^2}{2}\tilde{\psi}^\top(1)I_r\tilde{R}I_r^\top\tilde{\psi}(1)$$

$$- h^2\tilde{\psi}^\top(1)I_r\tilde{R}I_d^\top \int_0^1 \psi^j(\beta)d\beta + \frac{h^2}{2}\int_0^1 \tilde{\psi}^\top(\alpha)I_d\tilde{R}I_d^\top\psi^j(\alpha)d\alpha,$$

where

$$I_r = (0_{N\times 1}\ I), I_l = (I\ 0_{N\times 1}), \ I_d = I_l - I_r$$

with $\mathbf{0}_{k\times m}$ being a $kn \times mn$-dimensional zero matrix, that is,

$$V_1(\phi) = \frac{1}{2}\int_0^1 \left(\phi^\top(0)\ \ h[\psi^\top(1)I_r - \psi^\top(\alpha)I_d] \right)$$

$$\times \begin{pmatrix} \tilde{U} & \tilde{W} \\ \tilde{W}^\top & \tilde{R} \end{pmatrix} \begin{pmatrix} \phi(0) \\ h[I_r^\top\psi(1) - I_d^\top\psi(\alpha)] \end{pmatrix} d\alpha$$

which is nonnegative. This proves (a).

We are now in a position to prove (b). Noting that

$$V_2(\phi) = \frac{1}{2}\int_{-\tau}^0 \left(\ \phi^\top(0)\ \ \phi^\top(s) \ \right)$$

$$\times \begin{pmatrix} \frac{1}{\tau}[P - U] + Z(s) & Q(s) - W(s) \\ [Q(s) - W(s)]^\top & S(s) \end{pmatrix} \begin{pmatrix} \phi(0) \\ \phi(s) \end{pmatrix} ds$$

for any $Z(s)$ satisfying

$$\int_{-\tau}^{0} Z(s) ds = 0, \tag{3.83}$$

thus, to prove (b) we need only to find a Z satisfying (3.83) such that

$$\begin{pmatrix} \frac{1}{\tau}[P - U] + Z(s) & Q(s) - W(s) \\ [Q(s) - W(s)]^\top & S(s) \end{pmatrix} > 0, \quad -\tau \le s \le 0. \tag{3.84}$$

Choose $Z(s)$ to be piecewise linear:

$$Z(\tau_{i-1} + \alpha h) = (1 - \alpha)Z_{i-1} + \alpha Z_i, \ 0 \le \alpha \le 1. \tag{3.85}$$

Then inequality (3.84) is satisfied for all $-\tau \le s \le 0$ if it is satisfied for $s = \tau_i, i = 0, 1, \cdots, N$; in other words,

$$\begin{pmatrix} \frac{1}{\tau}[P - U] + Z_i & Q_i - W_i \\ [Q_i - W_i]^\top & S_i \end{pmatrix} > 0, \ i = 0, 1, \cdots, N, \tag{3.86}$$

and (3.83) is equivalent to

$$\frac{1}{2}(Z_0 + Z_N) + \sum_{i=1}^{N-1} Z_i = 0. \tag{3.87}$$

Therefore, we need only to prove that (3.75) implies the existence of Z_i such that (3.87) and (3.86) are satisfied. Let

$$\Gamma_i = \frac{1}{\tau}(P - U) - (Q_i - V_i)S_i^{-1}(Q_i - V_i)^\top$$

and

$$\Gamma = \frac{1}{2}(\Gamma_0 + \Gamma_N) + \sum_{i=1}^{N-1} \Gamma_i.$$

Then, in view of Schur's complement, (3.75) implies that

$$\Gamma > 0.$$

Choose sufficiently small $\delta_i > 0$, and let

$$Z_i = \delta_i I - \Gamma_i, \ i = 1, 2, \cdots, N$$

$$Z_0 = 2 \left[\Gamma - \left(\frac{1}{2}\delta_N + \sum_{i=1}^{N-1} \delta_i \right) I \right] - \Gamma_0.$$

It follows that

$$\Gamma_i + Z_i > 0, \ i = 0, 1, \cdots, N,$$

which implies (3.86) using Schur's complement, and (3.87) can be verified by direct calculation. This completes the proof of (i).

The proof of (ii). Using (3.67) and (3.66), direct manipulation gives that

$$\dot{V}(\phi) = \phi^\top(0)P[A\phi(0) + A_1\phi(-\tau)]$$
$$+ [A\phi(0) + A_1\phi(-\tau)]^\top \int_{-\tau}^0 Q(s)\phi(s)ds + \phi^\top(0)\int_{-\tau}^0 Q(s)\dot{\phi}(s)ds$$
$$+ \int_{-\tau}^0 ds \int_{-\tau}^0 \phi^\top(s)R(s,u)\dot{\phi}(u)du + \int_{-\tau}^0 \phi^\top(s)S(s)\dot{\phi}(s)ds.$$

Using integration by parts and noting the fact that $R(s,u) = R^\top(u,s)$, it is easy to check that

$$\dot{V}(\phi) = -\frac{1}{2}\phi^\top(0)\left[-PA - A^\top P - Q(0) - Q^\top(0) - S(0)\right]\phi(0)$$
$$- \frac{1}{2}\phi^\top(-\tau)S(-\tau)\phi(-\tau) - \frac{1}{2}\int_{-\tau}^0 \phi^\top(s)\dot{S}(s)\phi(s)ds$$
$$- \frac{1}{2}\int_{-\tau}^0 ds \int_{-\tau}^0 \phi^\top(s)\left[\left(\frac{\partial W}{\partial s} + \frac{\partial W}{\partial u}\right)R(s,u)\phi(u)\right]du$$
$$+ \phi^\top(0)[PA_1 - Q(-\tau)]\phi(-\tau) + \phi^\top(0)\int_{-\tau}^0 [A^\top Q(s) - \dot{Q}(s)$$
$$+ R^\top(s,0)]\phi(s)ds + \phi^\top(0)\int_{-\tau}^0 [B^\top Q(s) - R(s,-\tau)]\phi(s)ds. \quad (3.88)$$

Substituting (3.72), (3.71), and (3.73) into (3.88) leads to

$$\dot{V}(\phi) = -\frac{1}{2}\phi^\top(0)\Delta_{11}\phi(0) - \frac{1}{2}\phi^\top(-\tau)\Delta_{22}\phi(-\tau)$$
$$+ \phi^\top(0)\Delta_{12}\phi(-\tau) - \frac{h}{2}\sum_{i=1}^N \int_0^1 \phi^{i\top}(\alpha)S_{di}\phi^i(\alpha)d\alpha$$
$$- \frac{h^2}{2}\sum_{i=1}^N \sum_{j=1}^N \left(\int_0^1 \phi^i(\alpha)d\alpha\right)^\top R_{dij}\left(\int_0^1 \phi^i(\alpha)d\alpha\right)$$
$$+ h\phi^\top(0)\sum_{i=1}^N \int_0^1 [(1-\alpha)D_{1i}^0 + \alpha D_{2i}^1]\phi^i(\alpha)d\alpha$$
$$+ h\phi^\top(-\tau)\sum_{i=1}^N \int_0^1 [(1-\alpha)D_{2i}^0 + \alpha D_{2i}^1]\phi^i(\alpha)d\alpha$$

that is,

$$\dot{V}(\phi) = -\frac{h^2}{2}\left(\int_0^1 \tilde{\phi}(\alpha)d\alpha\right)^\top R_d \left(\int_0^1 \tilde{\phi}(\alpha)d\alpha\right)$$

$$- \frac{1}{2} \int_0^1 \rho^\top(\alpha)\Theta_0(\alpha)\rho(\alpha)d\alpha \qquad (3.89)$$

where

$$\rho^\top(\alpha) = \begin{pmatrix} \phi^\top(0) & \phi^\top(-\tau) & h\tilde{\phi}^\top(\alpha) \end{pmatrix}$$

$$\Theta_0(\alpha) = \begin{pmatrix} -\Delta_{11} & -\Delta_{12} & -(1-\alpha)D_1^0 - \alpha D_1^1 \\ * & \Delta_{22} & -(1-\alpha)D_2^0 - \alpha D_2^1 \\ * & * & -\frac{1}{h}S_d \end{pmatrix}.$$

(3.89) can be further written as

$$\dot{V}(\phi) = -\frac{1}{2} \int_0^1 \rho^\top(\alpha)\Theta_1(\alpha)\rho(\alpha)d\alpha - \frac{1}{2}\rho_1^\top\Theta_2\rho,$$

where

$$\Theta_1(\alpha) = \begin{pmatrix} \vartheta_{11} & -\Delta_{12} - T_{12} + (1-2\alpha)Y_{12} & -(1-\alpha)D_1^0 - \alpha D_1^1 \\ * & \Delta_{22} - T_{22} + (1-2\alpha)Y_{22} & -(1-\alpha)D_2^0 - \alpha D_2^1 \\ * & * & \frac{1}{h}S_d \end{pmatrix}$$

$$\Theta_2 = \begin{pmatrix} T_{11} & T_{12} & X_1 \\ * & T_{22} & X_2 \\ * & * & R_d \end{pmatrix},$$

with $\vartheta_{11} = \Delta_{11} - T_{11} + (1-2\alpha)Y_{11}$, and

$$\rho_1^\top(\alpha) = \begin{pmatrix} \phi^\top(0) & \phi^\top(-\tau) & h\int_0^1 \tilde{\phi}^\top(\alpha)d\alpha \end{pmatrix}.$$

Therefore, to prove (3.70) it suffices to prove

$$\Theta_1(\alpha) > 0 \forall\ 0 \le \alpha \le 1$$
$$\Theta_2 \ge 0,$$

which are equivalent to (3.76) and (3.78) since $\Theta_1(\alpha)$ depends on α linearly and therefore needs only to be satisfied for $\alpha = 0$ and $\alpha = 1$. This proves (ii).

The proof of (iii) follows directly from (i) and (ii).   □

Obviously, Theorem 3.15 has a large number of parameters. Some reduction of parameters is possible by introducing constraints. For example, we can set

$$W_i = Q_i, \ R_{ij} = R_{i-j}, \ T_{ij} = 0, \ X_{ij} = 0.$$

Then we have the following corollary.

Corollary 3.1 *System (3.66) is asymptotically stable if there exist $n \times n$ matrices $P = P^\top$, $Q_i, i = 0, 1, \cdots, N$, $R_i = R_i^\top, i = 0, 1, \cdots, N$, $S_i =$*

$S_i^\top, i = 0, 1, \cdots, N$ $U = U^\top$, *and* $Y_{ij} = Y_{ji}^\top, i, j = 1, 2$, *such that*

$$\begin{pmatrix} U & \tilde{Q} \\ \tilde{Q}^\top & \tilde{R} \end{pmatrix} \geq 0 \qquad (3.90)$$

$$\begin{cases} P > U \\ S_i > 0 \end{cases} \qquad (3.91)$$

and

$$\begin{pmatrix} \Delta_{11} + Y_{11} & -\Delta_{12} + Y_{12} & -D_1^0 \\ * & \Delta_{22} + Y_{22} & -D_2^0 \\ * & * & \frac{1}{h} S_d \end{pmatrix} > 0$$

$$\begin{pmatrix} \Delta_{11} - Y_{11} & -\Delta_{12} - Y_{12} & -D_1^1 \\ * & \Delta_{22} - Y_{22} & -D_2^1 \\ * & * & \frac{1}{h} S_d \end{pmatrix} > 0$$

where

$$\Delta_{11} = -PA - A^\top P - Q_N - Q_N^\top - S_N$$
$$\Delta_{12} = PB - Q_0$$
$$\Delta_{22} = S_0$$
$$S_d = \mathrm{diag}\{S_{d1}, S_{d2}, \cdots, S_{dN}\}$$
$$S_{di} = \frac{1}{h}(S_i - S_{i-1})$$
$$D_j^k = \left(D_{j1}^i, D_{j2}^k, \cdots, D_{jN}^k\right)$$
$$D_{1i}^k = A^\top Q_{i-1+k} - \frac{1}{h}(Q_i - Q_{i-1}) + R_{i-1+k-N}^\top$$
$$D_{2i}^k = B^\top Q_{i-1+k} - R_{i-1+k}^\top.$$

By dividing the time delay interval into N segments and choosing the kernel as piecewise linear, a complex partial differential equation problem is converted into an LMI problem. When N is large enough, piecewise linear matrix-valued functions $Q(s), S(s), R(s, u)$ constructed by (3.71), (3.72), (3.73) can approximate general continuous differential matrix-valued functions Q, S, R with symmetric, positive-definite properties to any accuracy. However, as N increases, the number of parameters in Theorem 3.15 increases rapidly and thus the computation burden increases greatly. Numerical examples show that for small N this approach requires rather modest computational requirements, and results in significant improvements over the existing results.

The following numerical example illustrates the usefulness of Theorem 3.15 and Corollary 3.1.

Example 3.7 *Consider the linear system with time delay where the dynamic is described by*

$$\dot{x}(t) = \begin{pmatrix} -2 & 0 \\ 0 & -0.9 \end{pmatrix} x(t) + \begin{pmatrix} -1 & 0 \\ -1 & -1 \end{pmatrix} x(t - \tau).$$

The upper bound, $\bar{\tau}$, can be estimated by using Theorem 3.15 and corollary 3.1 with different N. The results are listed as follows:

N	1	2	5	10	20
$\bar{\tau}$ *(Theorem 3.15)*	5.30	5.74	6.07	6.14	6.16
$\bar{\tau}$ *(Corollary 3.1)*	5.30	5.74	5.99	6.06	6.10

From these computations, it is clear that increasing N results in a greater upper bound $\bar{\tau}$, and the system will be stable for a larger delay.

3.5   Notes

The stability and stabilizability problems for the class of systems with time delay in the state have attracted many researchers from the control community. Both delay-independent and delay-dependent stability and stabilizability have been considered and some sufficient conditions have been reported to the literature. For instance, the delay-independent stability conditions based on the algebraic Riccati equation for the class of systems (3.1) can be found in [70, 118, 141, 205] and the references therein. The delay-dependent stability conditions are provided by [38, 70, 98, 105, 108, 109, 118, 175] and references therein. In the literature on systems with time-varying time delay, the delay-dependent conditions depend on the maximum value of the time derivative of the time-varying delay. The stability problem of a neutral system was addressed by [60, 97, 159, 186].

The output feedback control design technique presented in this chapter is adapted from [109], which developed the H_∞-control problem of linear systems with time-varying delay.

The observer-based output feedback stabilization problem is considered by Ge et al. [88]. They studied H_∞-control using the algebraic Riccati equation technique.

The discretization scheme for general Lyapunov functional in Section 3.4 was adapted from Gu [93, 94].

The results of this chapter can be extended to the following time delay linear system with polytopic uncertain parameters:

$$\begin{cases} \dot{x}(t) = Ax(t) + \sum_{i=1}^{m} A_i^h x(t - \tau_i) + Bu(t) \\ y(t) = Cx(t) \\ x(t) = \phi(t), \ t \in [-\bar{\tau}, 0) \end{cases}$$

where $x(t) \in \mathbb{R}^n, A, A_j \in \mathbb{R}^{n \times n}, B \in \mathbb{R}^{n \times k}, C \in \mathbb{R}^{r \times n}$ are uncertain parameters taking values in a polytope set

$$(A, A_1^h, \cdots, A_m^h, B, C) \in \left\{ \sum_{j=1}^N \alpha_j (A_j, A_{1j}^h, \cdots, A_{mj}^h, B_j, C_j), \right.$$

$$\left. \alpha_j \geq 0, \ \sum_{j=1}^N \alpha_j = 1 \right\}$$

with $A_j, A_{1j}^h, \cdots, A_{mj}^h, B_j, C_j, 1 \leq j \leq N$, being given matrices with appropriate dimensions. The polytope models may result from convex interpolation of a set of time delay systems with parameters $A_j, A_{1j}^h, \cdots, A_{mj}^h$, $B_j, C_j, 1 \leq j \leq N$. Such uncertain systems have been studied by [38, 48, 89, 90, 164].

4

Robust Stability and Robust Stabilizability

Most real systems cannot be represented by linear dynamics, but sometimes, under some assumptions, it is possible to model the dynamical behavior of practical systems with a linear model having some uncertainties. The presence of these uncertainties in the dynamics requires the establishment of robust conditions that can guarantee the stability and/or the stabilizability of the practical system under study. This topic has in fact dominated the research effort of the control community during the last two decades. Among the contribution on this area of research we quote the work of [77, 105, 114] on robust stability and the work of [49, 108, 110, 118, 128, 155, 189, 190, 206, 207, 215] on robust stabilizability. This chapter will deal with the robust stability and robust stabilizability of the class of uncertain continuous-time linear time delay systems. Our results are mainly based on the Lyapunov second method. In some sense, given an uncertain dynamical system with time delay, we answer the following questions:

- How can we check whether the unforced nominal system with time delay is stable or not?

- How can we check if the unforced uncertain dynamical system with time delay is robust stable for all admissible uncertainties or not?

- When the unforced uncertain dynamical system with time delay is unstable, how can we design a memoryless state feedback, or memory state feedback or output feedback controller to stabilize the system and guarantee that it will remain stable for all admissible uncertainties?

Toward this end, let us consider uncertain continuous-time linear time delay systems where the dynamic is described by the following differential delay equation

$$\begin{cases} \dot{x}(t) = A(t)x(t) + \sum_{j=1}^{m} A_j(t)x(t - \tau_j) + B(t)u(t) \\ x(t) = \phi(t), \ \forall t \in [-\bar{\tau}, 0], \end{cases} \tag{4.1}$$

where $x(t) \in \mathbb{R}^n$ is the state, $u(t) \in \mathbb{R}^k$ is the control input $\tau_j, 1 \le j \le m$, is the time delay in the system $\bar{\tau} = \max\{\tau_1, \cdots, \tau_m\}$; $\phi(\cdot)$ is the initial condition $A(t) = A + \Delta A(t) \in \mathbb{R}^{n \times n}$, $A_j(t) = A_j + \Delta A_j(t) \in \mathbb{R}^{n \times n}$, $j = 1, \ldots, m$, $B(t) = B + \Delta B(t) \in \mathbb{R}^{n \times k}$ with $A, A_j, \ j = 1, \ldots, m$ and B being known real constant matrices of appropriate dimensions which describe the nominal system of (4.1); and $\Delta A(t), \Delta A_j(t), \ j = 1, \ldots, m$, $\Delta B(t)$ are unknown real norm-bounded matrix functions, which represent time-varying parameter uncertainties.

The admissible uncertainties are assumed to be of the form

$$\begin{cases} \begin{bmatrix} \Delta A(t) & \Delta B(t) \end{bmatrix} = D\Delta(t) \begin{bmatrix} E_A & E_B \end{bmatrix} \\ \begin{bmatrix} \Delta A_1(t) & \cdots & \Delta A_m(t) \end{bmatrix} = \begin{bmatrix} D_1 \Delta_1(t) E_1 & \cdots & D_m \Delta_m(t) E_m \end{bmatrix} \end{cases} \tag{4.2}$$

where $\Delta(t), \Delta_i(t), \ i = 1, \ldots, m$ are unknown real time-varying matrices with Lebesgue measurable elements satisfying

$$\begin{cases} \Delta^\top(t)\Delta(t) \le I \\ \Delta_j^\top(t)\Delta_j(t) \le I, \ \forall t \ge 0, 1 \le j \le m, \end{cases} \tag{4.3}$$

and $D, D_j, E_A, E_B, E_j, 1 \le j \le m$ are known real constant matrices that characterize how the uncertain parameters in $\Delta(t), \Delta_j(t)$ enter the nominal matrices A, A_j, and B.

Remark 4.1 *In the rest of this chapter, we will assume complete access to the state vector for feedback control when it is required.*

In the remainder of this chapter, we will use the following concepts of robust stability and robust stabilization of system (4.1)-(4.3).

Definition 4.1 *The uncertain time delay system (4.1)-(4.3) is said to be robustly stable if the trivial solution $x(t) = 0$ of the functional differential equation associated to (4.1) with $u(t) \equiv 0$ is asymptotically stable for all admissible uncertainties $\Delta A(t), \Delta A_j(t), 1 \le j \le m$.*

Definition 4.2 *The uncertain time delay system (4.1)-(4.3) is said to be robustly stabilizable if there exists a control law $u(t)$ such that the closed-loop system is robustly stable in the sense of Definition 4.1.*

In the rest of this chapter, we will deal with the robust stability of the class of time delay systems such as the one given by (4.1)-(4.3). We will establish LMI-based delay-independent and LMI-based delay-dependent sufficient conditions that ensure the time delay system to be robustly stable for all the admissible uncertainties.

We will also deal with the robust stabilization of the class of time delay systems (4.1)-(4.3) by designing memoryless state feedback controller, memory state feedback controller, and output feedback controller. LMI-based delay-independent and LMI-based delay-dependent sufficient conditions are developed in each case to guarantee that the closed-loop system will remain stable for all admissible uncertainties. Some numerical examples are provided to show the usefulness of the applicability of the established results.

4.1   Delay-Independent Stability and Stabilizability

Let us consider the class of time delay system described by (4.1)-(4.3) and try first to establish under which conditions the system will be stable when $u(t) \equiv 0$ for all admissible uncertainties. Delay-independent sufficient conditions are first developed, but since they are conservative, we also develop delay-dependent sufficient conditions to overcome this conservatism.

4.1.1   Delay-Independent Stability

In this subsection, we will study the stability of system (4.1) with $u(t) \equiv 0$. For this purpose, let us introduce the following definition.

Definition 4.3 *System (4.1)-(4.3) is said to be quadratically stable if there exist symmetric and positive-definite matrices $P > 0, Q_j > 0, 1 \le j \le m$, such that*

$$
\begin{pmatrix}
\# & P(A_1 + \Delta A_1(t)) & \cdots & P(A_m + \Delta A_m(t)) \\
(A_1 + \Delta A_1(t))^\top P & -Q_1 & & 0 \\
\vdots & \vdots & \ddots & \vdots \\
(A_m + \Delta A_m(t))^\top P & 0 & & -Q_m
\end{pmatrix}
$$
$$
\triangleq \hat{\Theta}(t) < 0 \tag{4.4}
$$

holds for all admissible uncertainties, where $\# = (A + \Delta A(t))^\top P + P(A + \Delta A(t)) + \sum_{j=1}^{m} Q_j$.

The following theorem gives the relation between the robust quadratic stability and the robust stability of the system.

Theorem 4.1 *If system (4.1) is quadratically stable, then it is robustly stable.*

Proof: Consider a Lyapunov functional candidate as

$$V(\mathbf{x}_t) = x^\top(t)Px(t) + \sum_{j=1}^{m} \int_{t-\tau_j}^{t} x^\top(s)Q_j x(s)ds$$

where $\mathbf{x}_t \in \mathbb{C}[-\bar{\tau}, 0]$ defined by $\mathbf{x}_t(s) = x(t+s), -\bar{\tau} \leq s \leq 0$. To prove Theorem 4.1, it suffices to show that

$$\dot{V}(\mathbf{x}_t) < 0$$

holds for all admissible uncertainties. Evidently,

$$\dot{V}(\mathbf{x}_t) = 2x^\top(t)P\dot{x}(t) + x^\top(t)\left(\sum_{j=1}^{m} Q_j\right)x(t)$$

$$- \sum_{j=1}^{m} x^\top(t-\tau_j)Q_j x(t-\tau_j)$$

$$= x^\top(t)\left(A^\top(t)P + PA(t) + \sum_{j=1}^{m} Q_j\right)x(t)$$

$$+ 2\sum_{j=1}^{m} x^\top(t)A_j^\top(t)Px(t-\tau_j) - \sum_{j=1}^{m} x^\top(t-\tau_j)Q_j x(t-\tau_j)$$

$$= \eta_t^\top \hat{\Theta}(t)\eta_t$$

where $\eta_t^\top = (x^\top(t), x^\top(t-\tau_1), \cdots, x^\top(t-\tau_m))$, which implies that $\dot{V}(\mathbf{x}_t)$ is negative since $\hat{\Theta}(t) < 0$. This concludes the proof of Theorem 4.1. $\quad\square$

Theorem 4.1 tells us that the quadratic stability provides a sufficient condition for the uncertain system to be robustly stable. However, $\hat{\Theta}(t)$ contains the system uncertainties $\Delta(t), \Delta_j(t), j = 1, \ldots, m$, which are unknown and thus, it is not convenient to verify the quadratic stability using (4.4). The development of a condition that can be used to test the quadratic stability of system (4.1) is then necessary. The following theorem gives such a condition.

Theorem 4.2 *System (4.1) with $u(t) \equiv 0$ is quadratically stable if there exist symmetric, positive-definite matrices $P > 0, Q_j > 0, 1 \leq j \leq m$, and a positive constant $\varepsilon > 0$ such that*

$$\begin{pmatrix} A^\top P + PA + \sum_{j=1}^{m} Q_j + \varepsilon E_A^\top E_A & P\tilde{A} & P\tilde{D} \\ \tilde{A}^\top P & \tilde{Q}_E & 0 \\ \tilde{D}^\top P & 0 & -\varepsilon I \end{pmatrix} < 0 \qquad (4.5)$$

where $\tilde{A} = (A_1 \cdots A_m)$, $\tilde{D} = (D \; D_1 \cdots D_m)$, and

$$\tilde{Q}_E = \begin{pmatrix} \varepsilon E_1^\top E_1 - Q_1 & & \mathbf{0} \\ & \ddots & \\ \mathbf{0} & & \varepsilon E_m^\top E_m - Q_m \end{pmatrix}.$$

Proof: To prove this theorem, first of all note that

$$\hat{\Theta}(t) = \begin{pmatrix} A^\top P + PA + \sum_{j=1}^m Q_j & PA_1 & \cdots & PA_m \\ A_1^\top P & -Q_1 & & \\ \vdots & & \ddots & \\ A_m^\top P & & & -Q_m \end{pmatrix}$$

$$+ \begin{pmatrix} \Delta A^\top(t) P + P\Delta A(t) & P\Delta A_1(t) & \cdots & P\Delta A_m(t) \\ \Delta A_1^\top(t) P & & & \\ \vdots & & \mathbf{0} & \\ \Delta A_1^\top(t) P & & & \end{pmatrix}$$

$$= \begin{pmatrix} A^\top P + PA + \sum_{j=1}^m Q_j & PA_1 & \cdots & PA_m \\ A_1^\top P & -Q_1 & & \\ \vdots & & \ddots & \\ A_m^\top P & & & -Q_m \end{pmatrix}$$

$$+ \hat{D}\tilde{\Delta}(t)\tilde{E} + \tilde{E}^\top \tilde{\Delta}^\top(t)\hat{D}^\top,$$

where

$$\hat{D} = \begin{pmatrix} PD & PD_1 & \cdots & PD_m \\ \mathbf{0} & \mathbf{0} & \mathbf{0} & \mathbf{0} \\ \vdots & \vdots & \vdots & \vdots \\ \mathbf{0} & \mathbf{0} & \mathbf{0} & \mathbf{0} \end{pmatrix}$$

$$\tilde{\Delta}(t) = \text{diag}\{\Delta(t), \Delta_1(t), \cdots, \Delta_m(t)\}$$

$$\tilde{E} = \text{diag}\{E_A, E_1, \cdots, E_m\}.$$

In view of Lemma A.5, $\hat{\Theta}(t) < 0$ if there exists $\varepsilon > 0$ such that

$$\begin{pmatrix} A^\top P + PA + \sum_{j=1}^m Q_j & PA_1 & \cdots & PA_m \\ A_1^\top P & -Q_1 & & \\ \vdots & & \ddots & \\ A_m^\top P & & & -Q_m \end{pmatrix}$$

$$+ \frac{1}{\varepsilon} \begin{pmatrix} PDD^\top P + \sum_{i=1}^m PD_i D_i^\top P & \mathbf{0} & \cdots & \mathbf{0} \\ \mathbf{0} & \mathbf{0} & \cdots & \mathbf{0} \\ \vdots & \vdots & \ddots & \vdots \\ \mathbf{0} & \mathbf{0} & \cdots & \mathbf{0} \end{pmatrix}$$

$$+\,\varepsilon\begin{pmatrix} E_A^\top E_A & & & 0 \\ & E_1^\top E_1 & & \\ & & \ddots & \\ 0 & & & E_m^\top E_m \end{pmatrix} < 0$$

which is equivalent to (4.5) using the Schur complement. This concludes the proof of Theorem 4.2. □

Remark 4.2 *As can be seen, (4.5) is linear in* P, $Q_1, \ldots, Q_m$, *and* ε, *therefore tools like Matlab LMI toolbox, Scilab, or any equivalent software can be used to solve the problem of robust stability of the class of systems we are addressing. In practice, it can happen that a system by construction can be unstable, so we must find a way to stabilize it before we can use it. The next subsection deals with the stabilization problem.*

4.1.2 Delay-Independent Stabilization with Memoryless Control

Theorem 4.2 provides a sufficient condition for system (4.1) to be robust quadratically stable, which can be easily checked using Matlab LMI toolbox or any equivalent software. Now we come to address the robust quadratic stabilizability of system (4.1).

Definition 4.4 *System (4.1) is said to be robust quadratically stabilizable if there exists a state feedback controller,* $u(t)$, *such that the closed-loop system is quadratically stable.*

We first consider the memoryless state feedback control, that is, a controller of the form of (3.27).

Applying this control law to system (4.1), the closed-loop system becomes

$$\dot{x}(t) = [A + BK + D\Delta(t)(E_A + E_B K)]x(t)$$
$$+ \sum_{j=1}^{m} [A_j + D_j(t)\Delta_j(t)E_j(t)]x(t - \tau_j). \qquad (4.6)$$

The following theorem gives an LMI-based method for the design of a memoryless controller that quadratically stabilizes system (4.1).

Theorem 4.3 *If there exist a positive constant* $\varepsilon > 0$ *and matrices* Y, X, U_j, $1 \leq j \leq m$ *with* X, U_j *being symmetric, positive-definite such that the* *following LMI is feasible:*

$$\begin{pmatrix} M_{11} & A_1X & \cdots & A_mX & XE_A^\top + Y^\top E_B^\top & \mathbf{0} \\ * & -U_1 & & \mathbf{0} & \mathbf{0} & \mathcal{E}_X \\ & & \ddots & & & \\ * & \mathbf{0} & & -U_m & \mathbf{0} & \mathbf{0} \\ * & \mathbf{0} & \cdots & \mathbf{0} & -\varepsilon I & \mathbf{0} \\ \mathbf{0} & \mathcal{E}_X^\top & & \mathbf{0} & \mathbf{0} & -\varepsilon I \end{pmatrix} < 0 \qquad (4.7)$$

where

$$M_{11} = AX + BY + XA^\top + Y^\top B^\top + \sum_{j=1}^m U_j + \varepsilon DD^\top + \varepsilon \sum_{j=1}^m D_j D_j^\top$$

$$\mathcal{E}_X = \mathrm{diag}(XE_1^\top, \ldots, XE_m^\top),$$

then system (4.1) is robust quadratically stabilizable, and a memoryless stabilizing controller is $u(t) = Kx(t)$ *with* $K = YX^{-1}$.

Proof: In view of Theorem 4.1, to prove Theorem 4.3, it suffices to show that the closed-loop system is quadratically stable. By definition, we need only to verify that there exist symmetric, positive-definite $P > 0, Q_i > 0, 1 \le i \le m$, such that

$$\begin{pmatrix} \begin{bmatrix} (\bar{A} + D\Delta(t)\bar{E}_A)^\top P \\ + P(\bar{A} + D\Delta(t)\bar{E}_A) \\ + \sum_{j=1}^m Q_j \end{bmatrix} & P\tilde{A}(t) \\ \tilde{A}^\top(t)P & -\bar{Q} \end{pmatrix} \triangleq \tilde{\Theta}(t) < 0$$

holds for all admissible uncertainties, where

$$\bar{A} = A + BK$$
$$\bar{E}_A = E_A + E_B K$$
$$\tilde{A}(t) = (A_1 + D_1\Delta_1(t)E_1, \cdots, A_m + D_m\Delta_m(t)E_m)$$
$$\bar{Q} = \mathrm{diag}\{Q_1, \cdots, Q_m\}.$$

By the same argument as in Theorem 4.2, we can conclude that $\tilde{\Theta}(t) < 0$ if there exists a positive constant $\varepsilon > 0$ such that

$$\begin{pmatrix} J_1 & P\tilde{A} \\ \tilde{A}^\top P & -\bar{Q} \end{pmatrix} + \frac{1}{\varepsilon} \begin{pmatrix} \bar{E}_A^\top \bar{E}_A & & & \\ & E_1^\top E_1 & & \\ & & \ddots & \\ & & & E_m^\top E_m \end{pmatrix} < 0 \qquad (4.8)$$

where $\tilde{A} = (A_1, \ldots, A_m)$ and $J_1 = \bar{A}^\top P + P\bar{A} + \sum_{j=1}^m Q_j + \varepsilon PDD^\top P + \sum_{j=1}^m \varepsilon PD_j D_j^\top P$. Using the Schur complement, we conclude that (4.8) is equivalent to

$$
\begin{pmatrix}
J_1 & P\tilde{A} & \bar{E}_A^\top & 0 \\
\tilde{A}^\top P & -\bar{Q} & 0 & \mathcal{E} \\
\bar{E}_A & 0 & -\varepsilon I & 0 \\
0 & \mathcal{E}^\top & 0 & -\varepsilon I
\end{pmatrix} < 0
\tag{4.9}
$$

where $\mathcal{E} = \text{diag}\{E_1^\top, \cdots, E_m^\top\}$. Let $X = P^{-1}$. Pre- and post-multiplying (4.9) by $\text{diag}\{X, \mathcal{X}, I, I\}$ with $\mathcal{X} = \text{diag}\{X, \cdots, X\}$ and letting $Y = KX, U_i = XQ_iX, 1 \leq i \leq m$ we get (4.7). Thus from above argument, we find that if $X > 0, Y$ and $U_i, 1 \leq i \leq m$ satisfy (4.7) then $P = X^{-1}, Q_i = PU_i^{-1}P, 1 \leq i \leq m$ satisfy (4.7). This completes the proof of Theorem 4.3. $\qquad\square$

To show the usefulness of these results, let us consider the following numerical example.

Example 4.1 *Consider a continuous-time linear system with time delay and let its dynamic be described by (4.1) with $m = 1$ and its system parameters as*

$$
A = \begin{pmatrix} 5 & 1 \\ 0 & -2 \end{pmatrix} \quad A_1 = \begin{pmatrix} -1 & 0 \\ -1 & -1 \end{pmatrix}
$$

$$
B = \begin{pmatrix} 1 \\ 0.1 \end{pmatrix} \quad D = \begin{pmatrix} 1 \\ 0.3 \end{pmatrix}
$$

$$
D_1 = \begin{pmatrix} 0.2 \\ 1 \end{pmatrix} \quad E_1 = \begin{pmatrix} 0.1 & 0.2 \end{pmatrix}
$$

$$
E_B = 0.3 \quad E_d = \begin{pmatrix} 0 & 1 \end{pmatrix}.
$$

Solving (4.7) yields

$$
X = \begin{pmatrix} 0.238 & -0.1640 \\ -0.1640 & 0.9265 \end{pmatrix} \quad Y = (-2.9990 \ -0.6419)
$$

$$
U = \begin{pmatrix} 1.006 & -0.1044 \\ -0.1044 & 1.7297 \end{pmatrix} \quad \varepsilon = 1.1987.
$$

Therefore, Theorem 4.3 gives the following gain controller

$$
K = YX^{-1} = \begin{pmatrix} -14.8968 & -3.3297 \end{pmatrix}
$$

which stabilizes system (4.1).

4.1.3 Delay-Independent Stabilization under Memory Controller

Theorem 4.3 provides an algorithm that can be used to find a memoryless controller that robustly stabilizes system (4.1). Evidently, if the delay effect is taken into account during the design phase, the controller will give better performance. For this purpose, let us consider now a state feedback control of the form of (3.5).

In the remainder of this subsection, we will assume the following:

$$\begin{cases} D_1 = \cdots = D_m = D \\ \Delta_1(t) = \cdots = \Delta_m(t) = \Delta(t). \end{cases}$$

Substituting (3.5) into (4.1) gives the dynamics of the closed-loop system as follows:

$$\dot{x}(t) = [A + BK + D\Delta(t)(E_A + E_B K)]x(t)$$
$$+ \sum_{j=1}^{m} [A_j + BK_j + D\Delta(t)(E_j + E_B K_j)]x(t - \tau_j) \qquad (4.10)$$

The following theorem establishes a design method for such controller.

Theorem 4.4 *If there exists a positive constant $\varepsilon > 0$, symmetric and positive-definite matrices $X > 0, V_i > 0$, and matrices $Y_i, 1 \leq i \leq m$, such that the following holds:*

$$\begin{pmatrix} M_{11} & M_1 & \cdots & M_m & \beta \\ M_1^\top & -V_1 & & 0 & \beta_1 \\ \vdots & \vdots & \ddots & & \vdots \\ M_m^\top & 0 & & -V_m & \beta_m \\ \beta^\top & \beta_1^\top & \cdots & \beta_m^\top & -\varepsilon I \end{pmatrix} < 0 \qquad (4.11)$$

where $M_{11} = AX + BY + XA^\top + Y^\top B^\top + \varepsilon DD^\top + \sum_{j=1}^{m} V_j$; $M_i = A_i X + BY_i, 1 \leq i \leq m$; $\beta = XE_A^\top + Y^\top E_B^\top$, $\beta_i = XE_i^\top + Y_i^\top E_B^\top, 1 \leq i \leq m$, then controller (3.5) with $K = YX^{-1}, K_i = Y_i X^{-1}, 1 \leq i \leq m$, robustly stabilizes system (4.1).

Proof: Applying control (3.5) to system (4.1) leads to the closed-loop dynamical system

$$\dot{x}(t) = \left[\bar{A} + D\Delta(t)\bar{E}_A\right] x(t) + \sum_{j=1}^{m} \left[\bar{A}_j + D\Delta(t)\bar{E}_j\right] x(t - \tau_j),$$

where $\bar{A} = A + BK$, $\bar{A}_j = A_j + BK_j$, $\bar{E}_j = E_j + E_B K_j, 1 \leq j \leq m$. By Definition 4.4, to guarantee that the system is robust quadratically stable, it suffices to show that there exist positive constant $\varepsilon > 0$ and symmetric, positive-definite matrices $P > 0, Q_i > 0, 1 \leq i \leq m$, such that

$$\Theta_d(t) = \begin{pmatrix} \# & P(\bar{A}_1 + D\Delta(t)\bar{E}_1) & \cdots & P(\bar{A}_m + D\Delta(t)\bar{E}_m) \\ * & -Q_1 & & \\ \vdots & & \ddots & \\ * & & & -Q_m \end{pmatrix} < 0$$

holds for all admissible uncertainties, where $\# = (\bar{A} + D\Delta(t)\bar{E}_A)^\top P + P(\bar{A} + D\Delta(t)\bar{E}_A) + \sum_{j=1}^{m} Q_j$.

Note that

$$
\Theta_d(t) = \begin{pmatrix} \bar{A}^\top P + P\bar{A} + \sum_{j=1}^m Q_j & P\bar{A}_1 & \cdots & P\bar{A}_m \\ \bar{A}_1^\top P & -Q_1 & & \\ \vdots & & \ddots & \\ \bar{A}_m^\top P & & & -Q_m \end{pmatrix}
$$

$$
+ \begin{pmatrix} PD \\ 0 \\ \vdots \\ 0 \end{pmatrix} \Delta(t) \begin{pmatrix} \bar{E}_A & \bar{E}_1 & \cdots & \bar{E}_m \end{pmatrix}
$$

$$
+ \left[\begin{pmatrix} PD \\ 0 \\ \vdots \\ 0 \end{pmatrix} \Delta(t) \begin{pmatrix} \bar{E}_A & \bar{E}_1 & \cdots & \bar{E}_m \end{pmatrix} \right]^\top .
$$

Thus, it follows from Lemma A.5 that $\Theta_d(t) < 0$ if and only if

$$
\begin{pmatrix} \# & P\bar{A}_1 & \cdots & P\bar{A}_m \\ \bar{A}_1^\top P & -Q_1 & & \\ \vdots & & \ddots & \\ \bar{A}_m^\top P & & & -Q_m \end{pmatrix}
$$

$$
+ \frac{1}{\varepsilon} \begin{pmatrix} \bar{E}_A^\top \\ \bar{E}_1^\top \\ \vdots \\ \bar{E}_m^\top \end{pmatrix} \begin{pmatrix} \bar{E}_A & \bar{E}_1 & \cdots & \bar{E}_m \end{pmatrix} < 0
$$

holds, where $\# = \varepsilon PDD^\top P + \bar{A}^\top P + P\bar{A} + \sum_{j=1}^m Q_j$.

Letting $X = P^{-1}, Y_i = K_i X$ and Pre- and post-multiplying both sides of above inequality by $\mathrm{diag}\{X, \cdots, X\}$ yields

$$
\begin{pmatrix} \# & M_1 & \cdots & M_m \\ M_1^\top & -XQ_1X & & \\ \vdots & & \ddots & \\ M_m^\top & & & -XQ_mX \end{pmatrix}
$$

$$
+ \frac{1}{\varepsilon} \begin{pmatrix} \beta \\ \beta_1 \\ \vdots \\ \beta_m \end{pmatrix} \begin{pmatrix} \beta^\top & \beta_1^\top & \cdots & \beta_m^\top \end{pmatrix}
$$

with $\# = \varepsilon DD^\top + X\bar{A}^\top + \bar{A}X + \sum_{j=1}^m XQ_jX$. From this we get (4.11) by letting $Y = KX$, $V_i = XQ_iX$ and using the Schur complement. Therefore, suppose that a constant positive scalar ε and matrices $X > 0, V_i > 0$ and

Y_i satisfy (4.11), and let $K = YX^{-1}, K_i = Y_iX^{-1}, P = X^{-1}, Q_i = PV_iP$ then we can conclude that $P, Q_i, 1 \leq i \leq m$, satisfy (4.11) and thus the closed-loop system is quadratically stable. This completes the proof of Theorem 4.4. $\qquad\square$

Let us now consider the following numerical example that shows how the design algorithm works.

Example 4.2 *To illustrate the results of Theorem 4.4, let us consider a system of the form of the one described by (4.1) with $m = 1$ with the following parameters*

$$A = \begin{pmatrix} 5 & 1 \\ 0 & -2 \end{pmatrix} \quad A_1 = \begin{pmatrix} -1 & 0 \\ -1 & -1 \end{pmatrix}$$

$$B = \begin{pmatrix} 1 \\ 0.1 \end{pmatrix} \quad D = \begin{pmatrix} 1 \\ 0.3 \end{pmatrix}$$

$$E_1 = \begin{pmatrix} 0.1 & 0.2 \end{pmatrix} \quad E_B = \begin{pmatrix} 0.3 \end{pmatrix}$$

$$E_D = \begin{pmatrix} 0 & 1 \end{pmatrix}.$$

For this set of parameters, we get the following feasible solution to LMI (4.11)

$$X = \begin{pmatrix} 19.8571 & -10.4726 \\ -10.4726 & 35.2969 \end{pmatrix} \quad V_1 = \begin{pmatrix} 60.7259 & 0.3028 \\ 0.3028 & 73.2009 \end{pmatrix}$$

$$Y = \begin{pmatrix} -220.9053 & -10.9895 \end{pmatrix} \quad Y_1 = \begin{pmatrix} 16.1332 & 10.7613 \end{pmatrix}$$

$$\varepsilon = 108.3199.$$

According to Theorem 4.4, system (4.1) is stable under control (3.5) with gains K and K_1 given by:

$$K = YX^{-1} = \begin{pmatrix} -13.3831 & -4.2821 \end{pmatrix}$$
$$K_1 = Y_1X^{-1} = \begin{pmatrix} 1.1538 & 0.6472 \end{pmatrix}.$$

4.2   Delay-Dependent Stability and Stabilizability

As seen in Chapter 3, delay-independent stability condition is generally conservative, especially in situations where delays are small. This is represented in practice by the impossibility of finding a feasible solution of the corresponding LMI. However, the delay-dependent stability condition can provide less conservative results which makes it more attractive.

4.2.1   Delay-Dependent Stability

In this subsection, we will consider the robustness of the delay-dependent stability condition. The following theorem provides an LMI-based sufficient condition for system (4.1) to be asymptotically stable. Since the time delays are contained in the LMI, such stability is called delay-dependent stability.

In the sequel of this subsection, $A(t), A$ will be denoted by $A_0(t), A_0$, respectively, for notational brevity.

Theorem 4.5 *If there exist symmetric, positive-definite matrices $X > 0, Q_{ij} > 0, P_{ij} > 0, i = 1, \cdots, m, j = 0, 1, \cdots, m$, and positive scalars $\varepsilon_i, \eta_i, \rho_{ij}, 1 \leq i \leq m, 0 \leq j \leq m$, such that the LMI*

$$\begin{pmatrix} M_{11} & H_1 & H_2 & H_3 \\ H_1^\top & -J_1 & 0 & 0 \\ H_2^\top & 0 & -J_2 & 0 \\ H_3^\top & 0 & 0 & -J_3 \end{pmatrix} < 0 \qquad (4.12)$$

holds, where

$$M_{11} = \left(\sum_{i=0}^m A_i \right) X + X \left(\sum_{i=0}^m A_i \right)^\top + \sum_{i=0}^m \eta_i D_i D_i^\top + \sum_{i=1}^m \varepsilon_i \tau_i D_i D_i^\top$$
$$+ \sum_{i=1}^m \tau_i A_i W_i A_i^\top$$

with

$$W_i = \sum_{j=1}^m P_{ij}$$
$$H_1 = \begin{pmatrix} X E_A^\top & X E_1^\top & \cdots & X E_m^\top \end{pmatrix}$$
$$H_2 = \begin{pmatrix} \tau_1 A_1 W_1 E_1^\top & \cdots & \tau_m A_m W_m E_m^\top \end{pmatrix}$$
$$J_1 = \begin{pmatrix} \eta_1 I & & 0 \\ & \ddots & \\ 0 & & \eta_m I \end{pmatrix}$$
$$J_2 = \begin{pmatrix} \varepsilon_1 \tau_1 I - \tau_1 E_1 W_1 E_1^\top & & 0 \\ & \ddots & \\ 0 & & \varepsilon_m \tau_m I - \tau_m E_m W_m E_m^\top \end{pmatrix}$$
$$C_i = \tau_i \begin{pmatrix} X A_0^\top & X E_A & X A_1^\top & X E_1^\top \cdots X A_m^\top & X E_m^\top \end{pmatrix}$$
$$U_{ij} = \begin{pmatrix} -\tau_i P_{ij} + \rho_{ij} \tau_i D_j D_j^\top & \\ & \rho_{ij} \tau_i I \end{pmatrix}$$
$$U_i = \begin{pmatrix} U_{i0} & & & 0 \\ & U_{i1} & & \\ & & \ddots & \\ 0 & & & U_{im} \end{pmatrix}$$
$$H_3 = \begin{pmatrix} \tau_1 C_1 & \cdots & \tau_m C_m \end{pmatrix}$$

$$J_3 = \begin{pmatrix} U_1 & & \mathbf{0} \\ & \ddots & \\ \mathbf{0} & & U_m \end{pmatrix},$$

then system (4.1) is asymptotically stable.

Proof: Let $x(t), t \geq 0$ be the solution of linear time delay system (4.1) when the initial time and state are 0 and $\phi(\cdot)$, respectively. Since $x(t)$ is continuously differentiable for $t \geq 0$, following the Newton-Leibniz formula, one can write

$$x(t - \tau_i) = x(t) - \int_{-\tau_i}^0 \dot{x}(t + s)ds$$

$$= x(t) - \int_{-\tau_i}^0 \left(\sum_{j=0}^m A_j(t + s)x(t - \tau_j + s) \right) ds \qquad (4.13)$$

for $t \geq \tau_j$. Substituting (4.13) into (4.1) yields

$$\dot{x}(t) = \sum_{j=0}^m A_j(t)x(t) - \sum_{i=1}^m \sum_{j=0}^m \int_{-\tau_i}^0 A_i(t)A_j(t + s)x(t - \tau_j + s)ds$$

$$x(s) = \phi(s), \ s \in [-2\bar{\tau}, 0] \qquad (4.14)$$

where $\phi(s)$ is a smooth vector-valued initial function.

Thus, the asymptotic stability of system (4.14) will ensure the asymptotic stability of system (4.1). Let us consider the following Lyapunov functional candidate

$$V(\mathbf{x}_t) = x^\top(t)Px(t) + W(\mathbf{x}_t, t) \qquad (4.15)$$

where $\mathbf{x}_t \in \mathbb{C}[-\bar{\tau}, 0]$ defined by $\mathbf{x}_t(s) = x(t + s), -2\bar{\tau} \leq s \leq 0$, P is a symmetric, positive-definite matrix, and

$$W(\mathbf{x}_t, t) = \sum_{i=1}^m \sum_{j=0}^m \int_{-\tau_i}^0 \int_{t-\tau_j+\theta}^t x^\top(s)Q_{ij}x(s)dsd\theta$$

where $Q_{ij} > 0$ are to be determined. Simple manipulations give

$$\dot{W}(\mathbf{x}_t, t) = \sum_{i=1}^m \sum_{j=0}^m \left[\tau_i x^\top(t)Q_{ij}x(t) - \int_{-\tau_i}^0 x^\top(t - \tau_j + \theta) \right.$$

$$\left. Q_{ij}x(t - \tau_j + \theta)d\theta \right].$$

Then, the derivative of $V(\mathbf{x}_t)$ with respect to time t along the trajectory of system (4.14) is given by

$$\dot{V}(\mathbf{x}_t, t) = x^\top(t) \left[P\left(\sum_{j=0}^m A_j(t) \right) + \left(\sum_{j=0}^m A_j(t) \right)^\top P \right] x(t)$$

$$+ h(\mathbf{x}_t, t) + \dot{W}(\mathbf{x}_t, t) \tag{4.16}$$

where

$$h(\mathbf{x}_t, t) = -\sum_{i=1}^{m}\sum_{j=0}^{m} \int_{-\tau_i}^{0} 2x^{\top}(t)PA_i(t)A_j(t+\theta)x(t-\tau_j+\theta)d\theta.$$

Using Lemma A.1 and Lemma A.6, we have

$$\dot{V}(\mathbf{x}_t, t) \leq x^{\top}(t)\Theta_1 x(t) + h(\mathbf{x}_t, t) + \dot{W}(\mathbf{x}_t, t)$$

for any $\eta_i > 0$, where

$$\Theta_1 = P\sum_{j=0}^{m} A_j + \sum_{j=0}^{m} A_j^{\top} P + \sum_{i=0}^{m}\left(\eta_i PD_i D_i^{\top} P + \frac{1}{\eta_i} E_i^{\top} E_i\right).$$

Using now Lemma A.1, it follows that

$$-\int_{-\tau_i}^{0} 2x^{\top}(t)PA_i(t)A_j(t+\theta)x(t-\tau_j+\theta)d\theta$$

$$\leq \tau_i x^{\top}(t)PA_i(t)P_{ij}A_i^{\top}(t)Px(t)$$

$$+ \int_{-\tau_i}^{0} x^{\top}(t-\tau_j+\theta)A_j^{\top}(t+\theta)P_{ij}^{-1}A_j(t+\theta)x(t-\tau_j+\theta)d\theta$$

for any $P_{ij} > 0$.

Let $W_i = \sum_{j=0}^{m} P_{ij}$. Using (ii) of Lemma A.6, we get

$$\sum_{i=1}^{m} \left[\tau_i A_i(t)W_i A_i^{\top}(t)\right] \leq \Theta_2$$

where

$$\Theta_2 = \sum_{i=1}^{m} \tau_i \left[A_i W_i A_i^{\top} + \varepsilon_i D_i D_i^{\top} \right.$$

$$\left. + A_i W_i E_i^{\top}\left(\varepsilon_i I - E_i W_i E_i^{\top}\right)^{-1} E_i W_i A_i^{\top}\right]$$

for any $\varepsilon_i > 0$ with $\varepsilon_i I - E_i W_i E_i^{\top} > 0$.

Using (iii) of Lemma A.6, we have the following inequality:

$$A_j^{\top}(t+\theta)P_{ij}^{-1}A_j(t+\theta) \leq A_j^{\top}(P_{ij}-\rho_{ij}D_j D_j^{\top})^{-1}A_j + \rho_{ij}^{-1}E_j^{\top} E_j, \quad \forall t \geq 0$$

where $\rho_{ij} > 0$ satisfying $P_{ij} - \rho_{ij}D_j D_j^{\top} > 0$.

By choosing now

$$Q_{ij} = A_j^{\top}(P_{ij}-\rho_{ij}D_j D_j^{\top})^{-1}A_j + \frac{1}{\rho_{ij}}E_j^{\top} E_j,$$

we have

$$h(\mathbf{x}_t, t) < x^{\top}(t)P\Theta_2 Px(t)$$

$$+ \sum_{i=1}^{m} \sum_{j=0}^{m} \int_{-\tau_i}^{0} x^\top(t - \tau_j + \theta) Q_{ij} x(t - \tau_j + \theta) d\theta$$

which, combined with (4.15) and (4.16), yields

$$\dot{V}(\mathbf{x}_t, t) \leq x^\top(t) \Theta_3 x(t)$$

where

$$\Theta_3 = P \sum_{i=0}^{m} A_i + \sum_{i=0}^{m} A_i^\top P + \sum_{i=0}^{m} \left(\eta_i P D_i D_i^\top P + \frac{1}{\eta_i} E_i^\top E_i \right)$$

$$+ \sum_{i=1}^{m} \sum_{j=0}^{m} \tau_i Q_{ij} + \sum_{i=1}^{m} \tau_i P A_i W_i E_i^\top \left(\varepsilon I - E_i W_i E_i^\top \right)^{-1} E_i W_i A_i^\top P$$

$$+ \sum_{i=1}^{m} \tau_i P (A_i W_i A_i^\top + \varepsilon_i D_i D_i^\top) P.$$

Let $X = P^{-1}$. Pre- and post-multiplying Θ_3 by X leads to

$$X \Theta_3 X = \sum_{i=0}^{m} A_i X + X \sum_{i=0}^{m} A_i^\top + \sum_{i=0}^{m} \eta_i D_i D_i^\top + \sum_{i=0}^{m} \frac{1}{\eta_i} X E_i^\top E_i X$$

$$+ \sum_{i=1}^{m} \sum_{j=0}^{m} \tau_i X A_j^\top (P_{ij} - \rho_{ij} D_j D_j^\top)^{-1} A_j X$$

$$+ \sum_{i=1}^{m} \sum_{j=0}^{m} \frac{\tau_i}{\rho_{ij}} X E_j^\top E_j X$$

$$+ \sum_{i=1}^{m} \tau_i A_i W_i E_i^\top (\varepsilon_i I - E_i W_i E_i^\top)^{-1} E_i W_i A_i^\top$$

$$+ \sum_{i=1}^{m} \tau_i (A_i W_i A_i^\top + \varepsilon_i D_i D_i^\top). \tag{4.17}$$

Using the Schur complement, we have that $X \Theta_3 X < 0$ if and only if (4.12) holds. So, if there exist a set of symmetric, positive-definite matrices $X > 0, P_{ij} > 0, 1 \leq i \leq m, 0 \leq j \leq m$, and positive scalars $\varepsilon_i, \eta_i, \rho_{ij}, 1 \leq i \leq m, 0 \leq j \leq m$, satisfying (4.12), then $X \Theta_3(t) X < 0$, which means that $\dot{V}(\mathbf{x}_t) < 0$. This completes the proof of Theorem 4.5. $\square$

Remark 4.3 *Notice that the previous LMIs are nonlinear in* τ_i, $i = 1, \ldots, m$. *A search method can be used to determine the maximum time delay for which the system will remain stable.*

Let us now show the applicability of the results of this theorem with the following numerical example.

Example 4.3 *Consider an uncertain continuous-time linear system with time delay as described by (4.1) with $m = 1$ and the following parameters:*

$$A = \begin{pmatrix} -5 & 0.2 \\ 0 & -1 \end{pmatrix} \quad A_1 = \begin{pmatrix} -0.6 & 0.4 \\ 0 & 0 \end{pmatrix},$$

$$D = \begin{pmatrix} 0.4 & 0 \\ 0 & 0.4 \end{pmatrix} \quad D_1 = \begin{pmatrix} 0 & 0.2 \\ 0 & 0.3 \end{pmatrix}$$

$$E_A = \begin{pmatrix} 0.4 & 0 \\ 0 & 0.4 \end{pmatrix}, \quad E_B = \begin{pmatrix} 0.2 & 0 \\ 0.1 & 0.1 \end{pmatrix}.$$

With these data and by a one-dimensional search we get that the LMI (4.12) remains feasible for any $\tau \in [0, 1.45]$. In case of $\tau = 1.45$, solving LMI (4.12) gives

$$X = \begin{pmatrix} 363.2075 & 119.5673 \\ 119.5673 & 336.0476 \end{pmatrix} \quad P_{01} = \begin{pmatrix} 2863.1770 & 823.6382 \\ 823.6382 & 1594.3764 \end{pmatrix}$$

$$P_{11} = \begin{pmatrix} 481.3088 & 289.5405 \\ 289.5405 & 477.5186 \end{pmatrix}$$

$$\rho_{10} = 7.0, \rho_{11} = 8.0, \qquad\qquad \eta_0 = 6.1826, \eta_1 = 147.8947, \varepsilon = 795.596.$$

Therefore, using Theorem 4.5 we conclude that the system under study is stable for all $\tau \in [0, 1.45]$.

4.2.2 Delay-Dependent Stabilizability using Memoryless Control

In this subsection, we consider the design of a memoryless controller (3.27), such that the closed-loop system is asymptotically stable. Substituting now $u(t) = Kx(t)$ into (4.1) gives the dynamics of the closed-loop system

$$\begin{cases} \dot{x}(t) = \bar{A}(t)x(t) + \sum_{j=1}^{m} A_j(t)x(t - \tau_j) \\ x(t) = \phi(t), \ \forall t \in [-\bar{\tau}, 0], \end{cases}$$

where $\bar{A}(t) = A + BK + D\Delta(t)(E_A + E_B K)$. Using Theorem 4.5 we get the following theorem.

Theorem 4.6 *If there exist symmetric, positive-definite matrices $X > 0, Q_{ij} > 0, P_{ij} > 0, i = 1, \cdots, m, j = 0, 1, \cdots, m,$ and $Y,$ and positive scalars $\varepsilon_i, \eta_i, \varepsilon_{ij}, 1 \leq i \leq m, 0 \leq j \leq m,$ such that the following holds:*

$$\begin{pmatrix} \tilde{M}_{11} & \tilde{H}_1 & \tilde{H}_2 & \tilde{H}_3 \\ \tilde{H}_1^\top & -J_1 & 0 & 0 \\ \tilde{H}_2^\top & 0 & -J_2 & 0 \\ \tilde{H}_3^\top & 0 & 0 & -J_3 \end{pmatrix} < 0 \qquad (4.18)$$

where

$$\tilde{M}_{11} = \left(\sum_{i=0}^{m} A_i\right) X + X \left(\sum_{i=0}^{m} A_i\right)^\top + BY + Y^\top B^\top + \sum_{i=0}^{m} \eta_i D_i D_i^\top$$

$$+ \sum_{i=1}^{m} \varepsilon_i \tau_i D_i D_i^\top + \sum_{i=1}^{m} \tau_i A_i W_i A_i^\top$$

$$\tilde{H}_1 = \left(XE_A^\top + Y^\top E_B^\top, \ XE_1^\top, \ \cdots, \ XE_m^\top \right)$$

$$\tilde{C}_i = \tau_i \left(XA_0^\top + Y^\top B^\top, \ XE_A^\top + Y^\top E_B^\top \right.$$

$$\left. XA_1^\top, \ XE_1^\top, \ \cdots, \ XA_m^\top, \ XE_m^\top \right)$$

$$\tilde{H}_3 = \left[\tau_1 \tilde{C}_1 \ \cdots \ \tau_m \tilde{C}_m \right]$$

H_2, J_1, J_2, J_3 are defined as in Theorem 4.5, then controller (3.27) with $K = YX^{-1}$ robustly stabilizes system (4.1).

Proof: Replacing A_0 and E_A in (4.12) by $A + BK$ and $E_A + E_B K$, respectively, and letting $Y = KX$ leads to (4.18). The proof of Theorem 4.6 follows from steps of the proof of Theorem 4.5. □

Let us now consider the following numerical example to show the validity of the results of the previous theorem.

Example 4.4 *Consider an uncertain continuous-time linear system with time delay as described by (4.1) with the parameters $m = 1$, and*

$$A = \begin{pmatrix} 5 & 0.2 \\ 0 & -1 \end{pmatrix} \quad A_1 = \begin{pmatrix} -0.6 & 0.4 \\ 0 & 0 \end{pmatrix}$$

$$B = \begin{pmatrix} 1 \\ 1 \end{pmatrix} \quad D = \begin{pmatrix} 0.4 \\ 0.4 \end{pmatrix}$$

$$E_B = \begin{pmatrix} 0.2 \end{pmatrix} \quad D_1 = \begin{pmatrix} 0 \\ 0.3 \end{pmatrix}$$

$$E_A = \begin{pmatrix} 1 & 0 \end{pmatrix} \quad E_D = \begin{pmatrix} 0.1 & 0.1 \end{pmatrix}.$$

With these data, by a one-dimensional search we get that for any $\tau \in [0, 1.94]$ LMI (4.18) is feasible, and with $\tau = 1.94$ we have the following set of feasible solutions

$$X = \begin{pmatrix} 0.4918 & -0.9110 \\ -0.9110 & 6.63791 \end{pmatrix} \quad Y = \begin{pmatrix} -7.3786 & -1.1119 \end{pmatrix}$$

$$P_{10} = \begin{pmatrix} 10.0802 & 17.9488 \\ 17.9488 & 32.8918 \end{pmatrix} \quad P_{11} = \begin{pmatrix} 6.4338 & 0.8897 \\ 0.8897 & 1.6892 \end{pmatrix}$$

$\rho_{10} = 8.0, \rho_{11} = 9.0, \eta_0 = 1.12, \quad \eta_1 = 2.126, \varepsilon = 0.9322.$

According to Theorem 4.6, system (4.1) under control (3.27) with $K = YX^{-1} = \begin{pmatrix} -20.5316 & -2.9853 \end{pmatrix}$ is robustly stable for all admissible uncertainties.

Let us now see how we can establish a delay-dependent sufficient condition for robust stabilizability using a memory state feedback controller.

4.2.3 Delay-Dependent Stabilization with Memory Controller

In this subsection, we study the stabilization of system (4.1) using memory controller (3.5). For this purpose, in the remainder of this subsection, we will assume that

$$\begin{cases} D_1 = \cdots = D_m = D \\ \Delta_1(t) = \cdots = \Delta_m(t) = \Delta(t). \end{cases}$$

In the first step let us give the following theorem, which will be used later to design the memory controller.

Theorem 4.7 *System (4.1) with $u(t) \equiv 0$ is robustly stable if there exist symmetric, positive-definite matrices $X > 0$, $U_j > 0$, $0 \le j \le m$, scalars $\varepsilon_j > 0, 0 \le j \le m$, and $\rho > 0$ such that the following LMIs hold for all $j = 1, \ldots, m$:*

$$T_1(j) = \begin{pmatrix} \Gamma_{11} & \Gamma_{12} & \sum_{j=0}^m X E_j^\top \\ \Gamma_{12}^\top & -\Gamma_{22} & \Gamma_{23} \\ \sum_{j=0}^m E_j X & \Gamma_{23}^\top & -\rho I \end{pmatrix} < 0 \qquad (4.19)$$

$$T_2(j) = \begin{pmatrix} -U_j & X A_j^\top & X E_j^\top \\ A_j X & -X + \varepsilon DD^\top & 0 \\ E_j X & 0 & -\varepsilon_j I \end{pmatrix} < 0 \qquad (4.20)$$

where

$$\Gamma_{11} = \sum_{j=0}^m A_j X + X A_j^\top + \sum_{i=1}^m \sum_{j=0}^m \tau_i U_j + \rho DD^\top$$

$$\Gamma_{12} = (A_1 X \quad \cdots \quad A_m X)$$

$$\Gamma_{22} = \text{diag} \left\{ \frac{1}{(m+1)\tau_1} X, \cdots, \frac{1}{(m+1)\tau_m} X \right\}$$

$$\Gamma_{23}^\top = (E_1 X \quad \cdots \quad E_m X).$$

Proof: Suppose that the matrices P and $Q_j, 0 \le j \le m$ are a set of positive-definite matrices. Let us consider the Lyapunov functional candidate

$$V(\mathbf{x}_t) = x^\top(t) P x(t) + W(\mathbf{x}_t, t) \qquad (4.21)$$

where

$$W(\mathbf{x}_t, t) = \sum_{i=1}^m \sum_{j=0}^m \int_{-\tau_i}^0 \int_{t-\tau_j+\theta}^t x^\top(s) Q_j x(s) \, ds \, d\theta.$$

Simple manipulations give

$$\dot{W}(\mathbf{x}_t, t) = \sum_{i=1}^m \sum_{j=0}^m \left[\tau_i x^\top(t) Q_j x(t) \right.$$

$$\left. - \int_{-\tau_i}^0 x^\top(t - \tau_j + \theta) Q_j x(t - \tau_j + \theta) \, d\theta \right].$$

Using (4.21) we get

$$\dot{V}(\mathbf{x}_t, t) = x^\top(t) \left[\sum_{j=0}^{m} PA_j(t) + \sum_{j=0}^{m} A_j^\top(t)P \right] x(t)$$
$$+ h(\mathbf{x}_t, t) + \dot{W}(\mathbf{x}_t, t) \tag{4.22}$$

where

$$h(\mathbf{x}_t, t) = -\sum_{i=1}^{m}\sum_{j=0}^{m}\int_{-\tau_i}^{0} 2x^\top(t)PA_i(t)A_j(t+\theta)x(t-\tau_j+\theta)d\theta.$$

Using now Lemma A.1, we have

$$h(\mathbf{x}_t, t) \le x^\top(t)\left[\sum_{i=1}^{m}\sum_{j=0}^{m}\tau_i PA_i(t)P^{-1}A_i^\top(t)P\right]x(t)$$
$$+\sum_{i=1}^{m}\sum_{j=0}^{m}\int_{-\tau_i}^{0} x^\top(t-\tau_j+\theta)A_j^\top(t)PA_j(t)x(t-\tau_j+\theta)d\theta.$$

$$\tag{4.23}$$

Combining (4.22) and (4.22), to prove Theorem 4.7 it suffices to prove that there exist symmetric, positive-definite matrices P and $Q_j, 0 \le j \le m$ such that

$$\sum_{j=0}^{m} PA_j(t) + \sum_{j=0}^{m} A_j^\top(t)P + \sum_{i=1}^{m}\sum_{j=0}^{m}\tau_i Q_j$$
$$+\sum_{i=1}^{m}\sum_{j=0}^{m}\tau_i PA_i(t)P^{-1}A_i^\top(t)P < 0 \tag{4.24}$$

and

$$Q_j > A_j^\top(t)PA_j(t) \tag{4.25}$$

hold for all admissible uncertainties.

Let $X = P^{-1}$. Multiplying both sides of (4.24) by X gives

$$\sum_{j=0}^{m} A_j(t)X + \sum_{j=0}^{m} XA_j^\top(t) + \sum_{i=1}^{m}\sum_{j=0}^{m}\tau_i XQ_jX$$
$$+\sum_{i=1}^{m}\sum_{j=0}^{m}\tau_i A_i(t)XX^{-1}XA_i^\top(t) < 0$$

which is equivalent to

$$
\begin{pmatrix}
\#_0 & A_1(t)X & \cdots & A_m(t)X \\
XA_1^\top(t) & -\frac{1}{(m+1)\tau_1}X & & 0 \\
\vdots & & \ddots & 0 \\
XA_m^\top(t) & 0 & 0 & -\frac{1}{(m+1)\tau_m}X
\end{pmatrix} < 0 \qquad (4.26)
$$

where

$$
\#_0 = \sum_{j=0}^{m} A_j(t)X + \sum_{j=0}^{m} XA_j^\top(t) + \sum_{i=1}^{m}\sum_{j=0}^{m} \tau_i XQ_j X.
$$

Note that the left-hand side of (4.26) can be rewritten as

$$
\begin{pmatrix}
\#_1 & A_1 X & \cdots & A_m X \\
XA_1^\top & -\frac{1}{(m+1)\tau_1}X & & 0 \\
\vdots & & \ddots & 0 \\
XA_m^\top & 0 & 0 & -\frac{1}{(m+1)\tau_m}X
\end{pmatrix}
$$

$$
+ \begin{pmatrix} D \\ 0 \end{pmatrix} \Delta(t) \begin{pmatrix} \sum_{j=0}^{m} E_j X & E_1 X & \cdots & E_m X \end{pmatrix}
$$

$$
+ \begin{pmatrix} \sum_{j=0}^{m} XE_j^\top \\ XE_1^\top \\ \vdots \\ XE_m^\top \end{pmatrix} \Delta^\top(t) \begin{pmatrix} D^\top & 0 \end{pmatrix} \qquad (4.27)
$$

where

$$
\#_1 = \sum_{j=0}^{m} A_j X + \sum_{j=0}^{m} XA_j^\top + \sum_{i=1}^{m}\sum_{j=0}^{m} \tau_i XQ_j X.
$$

In view of Lemma A.5, (4.27) holds if and only if there exists $\rho > 0$ such that

$$
\begin{pmatrix}
\#_1 + \rho DD^\top & A_1 X & \cdots & A_m X \\
XA_1^\top & -\frac{1}{(m+1)\tau_1}X & & 0 \\
\vdots & & \ddots & 0 \\
XA_m^\top & & & -\frac{1}{(m+1)\tau_m}X
\end{pmatrix}
$$

$$
+ \frac{1}{\rho} \begin{pmatrix} \sum_{j=0}^{m} XE_j^\top \\ XE_1^\top \\ \vdots \\ XE_m^\top \end{pmatrix} \begin{pmatrix} \sum_{j=0}^{m} E_j X, & E_1 X, \cdots, & E_m X \end{pmatrix}. \qquad (4.28)
$$

Let $U_j = XQ_j X$. Then, combining (4.28) and the Schur complement lead to (4.19).

Using now the Schur complement the following $Q_j - A_j^\top(t)PA_j(t) > 0$ holds if and only if

$$\begin{pmatrix} -Q_j & A_j^\top(t) \\ A_j(t) & -P^{-1} \end{pmatrix} < 0. \tag{4.29}$$

Note that

$$\begin{pmatrix} -Q_j & A_j^\top(t) \\ A_j(t) & -P^{-1} \end{pmatrix} = \begin{pmatrix} -Q_j & A_j^\top \\ A_j & -P^{-1} \end{pmatrix} + \begin{pmatrix} \mathbf{0} \\ D \end{pmatrix} \Delta(t) \, (E_j \ \mathbf{0})$$

$$+ \begin{pmatrix} E_j^\top \\ \mathbf{0} \end{pmatrix} \Delta^\top(t) \, (\mathbf{0} \ D).$$

Thus, (4.25) is equivalent to

$$\begin{pmatrix} -Q_j + \frac{1}{\varepsilon_j} E_j^\top E_j & A_j^\top \\ A_j & -X + \varepsilon_j DD^\top \end{pmatrix} < 0 \tag{4.30}$$

for some $\varepsilon_j > 0$. Pre- and post-multiplying (4.30) by diag$\{X, I\}$ and using the Schur complement yields (4.20).

If X, U_j satisfy (4.19) and (4.20), from the above argument it follows that $Q_j = X^{-1}U_j X^{-1}, P = X^{-1}$ satisfy (4.24) and (4.25). This completes the proof of Theorem 4.7. □

Now, we come to consider the design of a delayed controller (3.5) that stabilizes system (4.1).

Applying (3.5) to system (4.1), we get the closed-loop system

$$\dot{x}(t) = [A_0 + BK + D\Delta(E_0 + E_B K)]x(t)$$

$$+ \sum_{j=1}^m [A_j + BK_j + D\Delta(E_j + E_B K_j)]x(t - \tau_j). \tag{4.31}$$

From Theorem 4.7, it follows that for given matrices $K_j, 0 \le j \le m$, if

$$\bar{\mathcal{T}}_1 < 0 \tag{4.32}$$

$$\bar{\mathcal{T}}_2 < 0 \tag{4.33}$$

then the closed-loop system is robustly stable, where $\bar{\mathcal{T}}_1$ and $\bar{\mathcal{T}}_2$ are obtained from $\mathcal{T}_1$ and $\mathcal{T}_2$, by replacing $A_j, E_j, 0 \le j \le m$ by $A_j + BK_j, E_j + E_B K_j$, respectively. Letting $Y_j = K_j X, 0 \le j \le m$, in (4.32) and (4.33), we get the following theorem.

Theorem 4.8 *If there exist symmetric, positive-definite matrices $X > 0, U_j > 0, 0 \le j \le m$, $Y_j, 0 \le j \le m$, and positive scalars $\varepsilon, \varepsilon_j, 0 \le j \le m$ and ρ such that the following LMIs hold:*

$$\begin{pmatrix} \tilde{\Gamma}_{11} & \tilde{\Gamma}_{12} & \sum_{j=0}^m [XE_j^\top + Y_j^\top E_B^\top] \\ \tilde{\Gamma}_{12}^\top & -\Gamma_{22} & \Gamma_{23} \\ \sum_{j=0}^m [E_j X + E_B Y_j] & \tilde{\Gamma}_{23}^\top & -\rho I \end{pmatrix} < 0 \tag{4.34}$$

$$\begin{pmatrix} -U_j & XA_j^\top + Y_j^\top B^\top & XE_j^\top + Y_j^\top E_B^\top \\ A_jX + BY_j & -\dot{X} + \varepsilon DD^\top & \mathbf{0} \\ E_jX + E_BY_j & \mathbf{0} & -\varepsilon_j I \end{pmatrix} < 0 \quad (4.35)$$

$j = 0, \cdots, m,$ where

$$\tilde{\Gamma}_{11} = \sum_{j=0}^{m} A_jX + BY_j + XA_j^\top + Y_j^\top B^\top + \sum_{i=1}^{m}\sum_{j=0}^{m} \tau_i U_j + \rho DD^\top$$

$$\tilde{\Gamma}_{12} = (A_1X + BY_1, \cdots, A_mX + BY_m)$$

$$\tilde{\Gamma}_{23}^\top = (E_1X + E_BY_1, \cdots, E_mX + E_BY_m)$$

then system (4.1) is stable under control (3.5) with $K = YX^{-1}, K_j = Y_jX^{-1}.$

Remark 4.4 *Theorem 4.8 gives conditions that are nonlinear in τ_i and U_i, $i = 1, \ldots, m$, and therefore, we can't use the LMI toolbox to determine the maximum time delay for which the system will remain stable for all admissible uncertainties as we did previously. A search method can be used to overcome this, which consists in fixing the time delays and searching for U_i, $i = 1, \ldots, m$.*

The following numerical example is used to show the usefulness of the results of this theorem.

Example 4.5 *Let us consider the same system studied in Example 4.4. With the same data and by a one-dimensional search we get that (4.34) and (4.35) are feasible for any $\tau \in [0, 2.75]$. In case of $\tau = 2.75$, solving (4.34) and (4.35) gives*

$$X = \begin{pmatrix} 8.4903 & -51.9254 \\ -51.9254 & 355.2551 \end{pmatrix} \quad Y = (-34.8785 \quad 210.3070)$$

$$Y_1 = (22.3135 \quad -150.7578) \quad U_0 = \begin{pmatrix} 0.9444 & -7.2300 \\ -7.2300 & 60.0519 \end{pmatrix}$$

$$U_1 = \begin{pmatrix} 1.5045 & -9.6773 \\ -9.6773 & 64.6216 \end{pmatrix}$$

$$\varepsilon = 0.0245, \varepsilon_0 = 220.0904, \qquad \varepsilon_1 = 133.1744, \rho = 1.0181.$$

From Theorem 4.8, the closed-loop system (4.1) is stable for all admissible uncertainties under control (3.5) with

$$K = YX^{-1} = (-4.5958 \ - 0.079)$$

$$K_1 = Y_1X^{-1} = (0.3089 \ - 0.3792).$$

Example 4.5 shows that by using memory controller the system can be stabilized for any $\tau \in [0, 2.75]$, while Example 4.4 shows that a memoryless control can stabilize the system for $[0, 1.94]$. This implies that a memory controller is more powerful than memoryless one.

Remark 4.5 *In what we did previously regarding stabilizability we assumed the access of the state vector at each time t for feedback. This* *assumption is not valid all the time for many reasons well known by the control community. In this case an alternate method to stabilize dynamical systems with time delay is to use output feedback control, which will be the next subject.*

4.3 Output Feedback Stabilization

In this section, we consider an uncertain time delay system described by the state equation of the form

$$
\begin{cases}
\dot{x}(t) = [A + D_x \Delta(t) E_A]\, x(t) + [B + D_x \Delta(t) E_B] u(t) \\
\qquad + \sum_{j=1}^{m} [A_j + D_j \Delta_j(t) E_j] x(t - \tau_j) \\
x(t) = \phi(t), t \in [-\bar{\tau}, 0] \\
y(t) = [C + D_y \Delta(t) E_A] x(t) + [F + D_y \Delta(t) E_B] u(t) \\
\qquad + \sum_{j=1}^{m} [C_j + D_{Cj} \Delta_j(t) E_j] x(t - \tau_j)
\end{cases}
\tag{4.36}
$$

where $x(t) \in \mathbb{R}^n$ is the system state, $y(t) \in \mathbb{R}^k$ is the measured output; $u(t) \in \mathbb{R}^k$ is the control input; $\tau_j > 0, 1 \leq j \leq m$, are the time delays in the system; $\bar{\tau} = \max\{\tau_j, 1 \leq j \leq m\}$, and $\phi(.)$ is the initial function; A, D_x, E_A, E_B, A_j, D_j, E_j, $j = 1, \ldots m$, C, D_y, F, C_j, D_{Cj}, $j = 1, \ldots, m$ are constant known matrices; and $\Delta(t), \Delta_j(t), 1 \leq j \leq m$, are time-varying matrices representing the uncertainty on system parameters and satisfying the condition

$$
\begin{cases}
\Delta^\top(t) \Delta(t) \leq I \\
\Delta_j^\top \Delta_j(t) \leq I, \ 1 \leq j \leq m.
\end{cases}
$$

In the rest of this chapter, we use $(x(0), x(s)) = (x_0, \phi(s)), -\bar{\tau} \leq s \leq 0$, to denote the initial conditions.

For this system, we will consider the full order memoryless controller of the form

$$
\begin{cases}
\dot{\xi}(t) = K_A \xi(t) + K_B y(t) \\
u(t) = K_C \xi(t)
\end{cases}
\tag{4.37}
$$

where $\xi(t) \in \mathbb{R}^n$ is the controller state, $y(t) \in \mathbb{R}^r$ is the output system, and $K_A \in \mathbb{R}^{n \times n}, K_B \in \mathbb{R}^{n \times k}, K_C \in \mathbb{R}^{k \times n}$ are matrices to be determined. Let $\eta^\top(t) = \left(x^\top(t)\ \xi^\top(t) \right)$. Then, applying this controller to system (4.36) yields the state space representation for our closed-loop system

$$
\dot{\eta}(t) = \bar{A}(t)\eta(t) + \sum_{j=1}^{m} \bar{A}_j(t) x(t - \tau_j)
$$

$$
= \bar{A}(t)\eta(t) + \sum_{j=1}^{m} \bar{A}_j(t) I_0 \eta(t - \tau_j)
\tag{4.38}
$$

where $\bar{A}(t) = \bar{A}_0 + \bar{D}_x\Delta(t)\bar{E}_x$, $\bar{A}_j(t) = \bar{A}_j + \bar{D}_j\Delta_j(t)\bar{E}_j$, $I_0 = (I, 0)$ with

$$\bar{A}_0 = \begin{pmatrix} A & BK_C \\ K_BC & K_A + K_BFK_C \end{pmatrix} \quad \bar{D}_x = \begin{pmatrix} D_x \\ K_BD_y \end{pmatrix}$$

$$\bar{E}_x = \begin{pmatrix} E_A & E_BK_C \end{pmatrix} \quad \bar{A}_j = \begin{pmatrix} A_j \\ K_BC_j \end{pmatrix}$$

$$\bar{D}_j = \begin{pmatrix} D_j \\ K_BD_{Cj} \end{pmatrix} \quad \bar{E}_j = E_j.$$

In view of Theorem 4.1, to establish the asymptotic stability of system (4.38), it suffices to show that the system is quadratically stable. Furthermore, by definition, system (4.38) is quadratically robustly stable if and only if there exist symmetric, positive-definite matrices $P \in \mathbb{R}^{2n \times 2n}, Q_j \in \mathbb{R}^{n \times n}, 1 \leq j \leq m$, such that

$$\Theta_o(t) = \begin{pmatrix} \# & P\bar{A}_1(t) & \cdots & P\bar{A}_m(t) \\ \bar{A}_1^\top(t)P & -Q_1 & & \\ \vdots & & \ddots & \\ \bar{A}_m^\top(t)P & & & -Q_m \end{pmatrix} < 0$$

holds for all admissible uncertainties, where $\# = \bar{A}^\top(t)P + P\bar{A}(t) + \sum_{j=1}^m I_0^\top Q_j I_0$.

By using Lemma A.5, we get the following theorem, which is independent of the uncertain term $\Delta(t), \Delta_j(t), 1 \leq j \leq m$.

Theorem 4.9 *System (4.36) is quadratically robustly stable if there exist symmetric, positive-definite matrices $P > 0$ and $Q_i > 0, 1 \leq i \leq m$, and a positive scalar ε such that the following holds:*

$$\begin{pmatrix} \bar{A}_0^\top P + P\bar{A}_0 & PT_{12} & PT_{13} & \bar{E}_x^\top & \mathcal{I}_0^\top \\ T_{12}^\top P & T_{22} & & & \\ T_{13}^\top P & & -\varepsilon I & & \\ \bar{E}_x & & & -\frac{1}{\varepsilon}I & \\ \mathcal{I}_0 & & & & -\bar{Q}^{-1} \end{pmatrix} \overset{\Delta}{=} \Theta < 0 \qquad (4.39)$$

where

$$T_{12} = \begin{pmatrix} \bar{A}_1 \cdots \bar{A}_m \end{pmatrix}$$
$$T_{13} = \begin{pmatrix} \bar{D}_x & \bar{D}_1 & \cdots & \bar{D}_m \end{pmatrix}$$
$$T_{22} = \text{diag}\{\varepsilon E_1^\top E_1 - Q_1, \cdots, \varepsilon E_m^\top E_m - Q_m\}$$
$$\mathcal{I}_0 = \begin{pmatrix} I_0^\top & \cdots & I_0^\top \end{pmatrix}$$
$$\bar{Q} = \text{diag}\{Q_1, \cdots, Q_m\}.$$

Proof: First of all note that

$$\Theta_o(t) = \begin{pmatrix} \bar{A}^\top P + P\bar{A} + \sum_{j=1}^m I_0^\top Q_j I_0 & P\bar{A}_1 & \cdots & P\bar{A}_m \\ * & & -Q_1 & \\ \vdots & & & \ddots \\ * & & & -Q_m \end{pmatrix}$$

$$+ \hat{T}_{13}\tilde{\Delta}(t) \begin{pmatrix} \bar{E}_x & & & \\ & E_1 & & \\ & & \ddots & \\ & & & E_m \end{pmatrix}$$

$$+ \begin{pmatrix} \bar{E}_x^\top & & & \\ & E_1^\top & & \\ & & \ddots & \\ & & & E_m^\top \end{pmatrix} \tilde{\Delta}^\top(t)\hat{T}_{13}^\top,$$

where

$$\hat{T}_{13} = \begin{pmatrix} P\bar{D}_x & P\bar{D}_1 & \cdots & P\bar{D}_m \\ 0 & 0 & 0 & 0 \\ \vdots & \vdots & \ddots & \vdots \\ 0 & 0 & \cdots & 0 \end{pmatrix}$$

$$\tilde{\Delta}(t) = \text{diag}\{\Delta(t), \Delta_1(t), \cdots, \Delta_m(t)\}.$$

In view of Lemma A.5, $\Theta_o(t) < 0$ if there exists $\varepsilon > 0$ such that

$$\begin{pmatrix} \bar{A}_0^\top P + P\bar{A}_0 + \sum_{j=1}^m I_0^\top Q_j I_0 & P\bar{A}_1 & \cdots & P\bar{A}_m \\ \bar{A}_1^\top P & & -Q_1 & \\ \vdots & & & \ddots \\ \bar{A}_m^\top P & & & -Q_m \end{pmatrix}$$

$$+ \frac{1}{\varepsilon} \begin{pmatrix} P\bar{D}_x \bar{D}_x^\top P + \sum_{i=1}^m P\bar{D}_i \bar{D}_i^\top P & 0 & \cdots & 0 \\ 0 & & 0 & \cdots & 0 \\ \vdots & & \vdots & \cdots & \vdots \\ 0 & & 0 & \cdots & 0 \end{pmatrix}$$

$$+ \varepsilon \begin{pmatrix} \bar{E}_x^\top \bar{E}_x & & & 0 \\ & \bar{E}_1^\top \bar{E}_1 & & \\ & & \ddots & \\ 0 & & & \bar{E}_m^\top \bar{E}_m \end{pmatrix} < 0,$$

which combined with the Schur complement leads to (4.39).   □

In (4.39), the unknown matrices $P, Q_i, 1 \le i \le m$, and the controller parameters K_A, K_B, K_C occur in nonlinear fashion and therefore (4.39) can't be considered an LMI. However, if a method of change of variables is applied, (4.39) can be cast into two LMIs for given scalar $\varepsilon > 0$ and matrices $Q_i > 0, 1 \leq i \leq m$.

To this end, let us assume that the matrices P and P^{-1} are partitioned as in (3.47) and define matrices Φ_1, Φ_2 by (3.55). Then, using (3.48), we get

$$P\Phi_1 = \begin{pmatrix} YX + NM^\top & Y \\ N^\top X + W_1 M^\top & N^\top \end{pmatrix} = \Phi_2$$

and

$$\Phi_1^\top P\Phi_1 = \Phi_1^\top \Phi_2 = \begin{pmatrix} X & I \\ I & Y \end{pmatrix}. \tag{4.40}$$

Let $X > 0, Y > 0$ and $\mathcal{A}, \mathcal{B}, \mathcal{C}$ be matrices with appropriate dimensions and consider the LMIs

$$\begin{pmatrix} X & I \\ I & Y \end{pmatrix} > 0 \tag{4.41}$$

$$\begin{pmatrix} J_{11} & J_{12} & J_{13} & J_{14} & J_{15} \\ J_{12}^\top & J_{22} & 0 & 0 & 0 \\ J_{13}^\top & 0 & -\varepsilon I & 0 & 0 \\ J_{14}^\top & 0 & 0 & -\frac{1}{\varepsilon}I & 0 \\ J_{15}^\top & 0 & 0 & 0 & -\bar{Q}^{-1} \end{pmatrix} < 0, \tag{4.42}$$

where

$$J_{11} = \begin{pmatrix} \mathcal{A}X + X\mathcal{A}^\top + \mathcal{B}\mathcal{C} + \mathcal{C}^\top \mathcal{B}^\top & \mathcal{A} + \mathcal{A}^\top \\ \mathcal{A}^\top + \mathcal{A} & Y\mathcal{A} + \mathcal{B}\mathcal{C} + \mathcal{A}^\top Y + \mathcal{C}^\top \mathcal{B}^\top \end{pmatrix}$$

$$J_{12} = \begin{pmatrix} A_1 & \cdots & A_m \\ YA_1 + \mathcal{B}C_1 & \cdots & YA_m + \mathcal{B}C_m \end{pmatrix}$$

$$J_{13} = \begin{pmatrix} D_x & D_1 & \cdots & D_m \\ YD_x + \mathcal{B}D_y & YD_1 + \mathcal{B}D_{C1} & \cdots & YD_m + \mathcal{B}D_{Cm} \end{pmatrix}$$

$$J_{14} = \begin{pmatrix} XE_x^\top + \mathcal{C}^\top E_B^\top \\ E_x^\top \end{pmatrix}$$

$$J_{15} = \begin{pmatrix} X & \cdots & X \\ I & \cdots & I \end{pmatrix}$$

$$J_{22} = \begin{pmatrix} \varepsilon E_1^\top E_1 - Q_1 & & 0 \\ & \ddots & \\ 0 & & \varepsilon E_m^\top E_m - Q_m \end{pmatrix}.$$

Before providing the main results of this section, let us give the following lemma, which will be used in the coming theorem.

Lemma 4.1 *Suppose that a symmetric, positive-definite matrix $P > 0$ and matrices K_A, K_B, K_C satisfy (4.39) and partition P as in (3.47).*

If N is singular, then there exists a nonsingular matrix $\tilde{N}$, such that $\tilde{P} = \begin{pmatrix} Y & \tilde{N} \\ \tilde{N}^\top & W_1 \end{pmatrix}$ satisfies (4.39).

Proof: Suppose that N is singular. Perturb N to $\tilde{N}$ such that $\tilde{N}$ is nonsingular. Note the restrictive negativeness of (4.39); thus when the perturbation is small enough, $\tilde{P}$ satisfies (4.39).   □

Remark 4.6 *In view of Lemma 4.1, we can assume that N is invertible in (3.47) and satisfies (4.39) without loss of generality. Moreover, note that in (3.47) M can be represented as $M^\top = -[W_1 - N^\top Y^{-1} N]^{-1} N^\top Y^{-1}$ when N is nonsingular. Thus, we can assume M, N to be nonsingular without loss of generality.*

Theorem 4.10 *There exist matrices $K_A, K_B, K_C, P > 0, Q_j > 0, 1 \leq j \leq m$ satisfying (4.39) if and only if there exists a positive scalar ε such that LMIs (4.41) and (4.42) are feasible for some given matrices $X > 0, Y > 0, \mathcal{A}, \mathcal{B}, \mathcal{C}$.*

Proof: *Necessity.* Let $K_A, K_B, K_C,$ and P satisfy (4.39) and partition P and P^{-1} as in (3.47). Using Lemma 4.1, we assume that M, N are nonsingular. Consider rewriting here Φ_1, Φ_2 from (3.55). Evidently, $\Phi_i, i = 1, 2,$ are invertible. So (4.41) follows from (4.40).

Now, define a set of new matrices as follows:

$$\begin{cases} \mathcal{A} = YAX + NK_B CX + YBK_C M^\top + N(K_A + K_B F K_C) M^\top \\ \mathcal{B} = NK_B \\ \mathcal{C} = K_C M^\top. \end{cases} \tag{4.43}$$

Let us proceed with the proof that matrices $\mathcal{A}, \mathcal{B}, \mathcal{C}, Y, X$ satisfy (4.42). Note that

$$\Phi_1^\top P \bar{A}_0 \Phi_1 = \begin{pmatrix} \begin{bmatrix} AX + BK_C M^\top \\ YAX + NK_B CX \\ +YBK_C M^\top \\ +N[K_A + K_B F K_C] M^\top \end{bmatrix} & \begin{matrix} A \\ \\ YA + NK_B C \end{matrix} \end{pmatrix}$$

$$= \begin{pmatrix} AX + B\mathcal{C} & A \\ \mathcal{A} & YA + \mathcal{B}C \end{pmatrix},$$

from which it follows that

$$\Phi_1^\top [\bar{A}_0^\top P + P \bar{A}_0] \Phi_1 = J_{11}. \tag{4.44}$$

Furthermore, direct manipulation gives

$$\Phi_1^\top P T_{12} = \begin{pmatrix} I & \mathbf{0} \\ Y & N \end{pmatrix} \begin{pmatrix} A_1 & A_2 & \cdots & A_m \\ K_B C_1 & K_B C_2 & \cdots & K_B C_m \end{pmatrix}$$

$$= \begin{pmatrix} A_1 & A_2 & \cdots & A_m \\ YA_1 + NK_B C_1 & YA_2 + NK_B C_2 & \cdots & YA_m + NK_B C_m \end{pmatrix}$$

$$= J_{12} \tag{4.45}$$

$$\Phi_1^\top P T_{13} = \begin{pmatrix} I & \mathbf{0} \\ Y & N \end{pmatrix} \begin{pmatrix} D_x & D_1 & \cdots & D_m \\ K_B D_y & K_B D_{C1} & \cdots & K_B D_{Cm} \end{pmatrix}$$

$$= \begin{pmatrix} D_x & D_1 & \cdots & D_m \\ Y D_x + N K_B D_y & Y D_1 + N K_B D_{C1} & \cdots & Y D_m + N K_B D_{Cm} \end{pmatrix}$$

$$= J_{13} \tag{4.46}$$

$$\Phi_1^\top \bar{E}_x^\top = \begin{pmatrix} X & M \\ I & \mathbf{0} \end{pmatrix} \begin{pmatrix} E_A^\top \\ K_C^\top E_B^\top \end{pmatrix} = \begin{pmatrix} X E_A^\top + M K_C^\top E_B^\top \\ E_A^\top \end{pmatrix}$$

$$= J_{14} \tag{4.47}$$

$$\Phi_1^\top \mathcal{I}_0 = J_{15}. \tag{4.48}$$

Pre- and post-multiplying both sides of (4.39) by $\Xi^\top = \mathrm{diag}\{\Phi_1^\top, I, I, I, I\}$, and Ξ and using (4.44) through (4.47) produces (4.42). This proves necessity.

Sufficiency. Let $\mathcal{A}, \mathcal{B}, \mathcal{C}, X > 0$ and $Y > 0$ be a set of feasible solutions of (4.41) and (4.42). Since $X > 0, Y > 0$ satisfy (4.41), in view of Lemma A.2 we have

$$X - Y^{-1} > 0,$$

from which it follows that $I - XY$ is invertible. Using now singular decomposition, we get invertible matrices M, N such that $MN^\top = I - XY$. With $\mathcal{A}, \mathcal{B}, \mathcal{C}, X, Y, M, N$ given, solving (4.43) yields K_A, K_B, K_C. Let Φ_1, Φ_2 be defined by (3.55) and let $P = \Phi_2 \Phi_1^{-1}$. Direct manipulation gives that $\Xi^\top \Theta \Xi < 0$ is equivalent to (4.42), which proves sufficiency. □

Combining Theorem 4.9 and Theorem 4.10 yields the following theorem.

Theorem 4.11 *If LMIs (4.41) and (4.42) have a set of feasible solutions $X > 0, Y > 0, \mathcal{A}, \mathcal{B}, \mathcal{C}$, then there exist matrices K_A, K_B, K_C such that controller (4.37) robustly stabilizes system (4.36) in the quadratic sense.*

Theorem 4.11 provides a design method of a dynamical output feedback controller that stabilizes system (4.36). The design procedure is as follows:

Algorithm 4.1 *(Control design algorithm)*

Step 1. Solve (4.41) and (4.42) to get $X > 0, Y > 0$ and $\mathcal{A}, \mathcal{B}, \mathcal{C}$.

Step 2. Using the singular decomposition method, compute the invertible matrices M, N satisfying $MN^\top = I - XY$.

Step 3. With $M, N, \mathcal{A}, \mathcal{B}, \mathcal{C}$ obtained, solve matrix equations (4.43) to get the gain matrices K_A, K_B, K_C, that is,

$$
\begin{cases}
K_A N^{-1} \left[\mathcal{A} - Y A X + \mathcal{B} C X + Y B \mathcal{C} - \mathcal{B} F \mathcal{C} \right] [M^\top]^{-1} \\
K_B = N^{-1} \mathcal{B} \\
K_C = \mathcal{C} [M^\top]^{-1}.
\end{cases}
$$

Let us now work out a numerical example to show the validity of the results of this theorem.

Example 4.6 *Consider an uncertain system described by (4.36) with parameters $m = 1$ and*

$$
A = \begin{pmatrix} 2 & 0.2 \\ 0 & -1 \end{pmatrix} \quad A_1 = \begin{pmatrix} 0.6 & 0.4 \\ 0 & 0 \end{pmatrix}
$$

$$
B = \begin{pmatrix} 1 & 0 \\ 1 & 1 \end{pmatrix} \quad D_x = \begin{pmatrix} 0.4 \\ 0.4 \end{pmatrix} \quad D_1 = \begin{pmatrix} 1 \\ 0 \end{pmatrix}
$$

$$
E_B = \begin{pmatrix} 0 & 0 \end{pmatrix} \quad C = \begin{pmatrix} 1 & 1 \\ 0 & 0 \end{pmatrix} \quad D_y = \begin{pmatrix} 1 \\ 0 \end{pmatrix}
$$

$$
F = \begin{pmatrix} 1 & 0 \\ 1 & 0 \end{pmatrix} \quad C_1 = \begin{pmatrix} 0 & 1 \\ 1 & 0 \end{pmatrix} \quad D_{c1} = \begin{pmatrix} 0 \\ 0.3 \end{pmatrix}
$$

$$
E_A = \begin{pmatrix} 1 & 0 \end{pmatrix} \quad E_1 = \begin{pmatrix} 0.1 & 0.1 \end{pmatrix}.
$$

With these data, solving LMIs (4.41) and (4.42) gives the set of feasible solutions

$$
\mathcal{A} = \begin{pmatrix} -0.6742 & 0.2465 \\ 0.1834 & 1.3844 \end{pmatrix} \quad \mathcal{B} = \begin{pmatrix} -0.4659 & 0.4917 \\ -0.6585 & -0.0411 \end{pmatrix}
$$

$$
\mathcal{C} = \begin{pmatrix} 14.1091 & 0 \\ 0 & 14.1091 \end{pmatrix} \quad R = \begin{pmatrix} -0.5055 & -0.0244 \\ -0.0244 & -0.2568 \end{pmatrix}
$$

$$
S = \begin{pmatrix} -0.5441 & 0.0453 \\ 0.0454 & -0.2957 \end{pmatrix}
$$

Using singular value decomposition of $I - RS$ yields M, N:

$$
M = \begin{pmatrix} 0.0376 & 0.7253 \\ 0.9247 & -0.0295 \end{pmatrix} \quad N = \begin{pmatrix} 0.0301 & 0.9995 \\ 0.9995 & -0.0301 \end{pmatrix},
$$

satisfying $MN^\top = I - RS$. With $R, S, M, N, \mathcal{A}, \mathcal{B}, \mathcal{C}$ given, solving (4.43) produces a set of stabilizing gains:

$$
K_A = \begin{pmatrix} 5.4032 & -14.7884 \\ -1.1938 & -11.8677 \end{pmatrix} \quad K_B = \begin{pmatrix} -0.6723 & -0.0263 \\ -0.4459 & 0.4927 \end{pmatrix}
$$

$$
K_C = \begin{pmatrix} 0.61889 & 19.4216 \\ 15.2329 & -0.7891 \end{pmatrix}.
$$

4.4   Notes

Robust stability and robust stabilizability problems are two major topics in control theory. They have in fact dominated the research effort of the control community for the last two decades. On these topics we can find a lot of references in the literature. Among the references dealing with robust stability for systems with delayed perturbations are studied using state transformation in [47, 99, 185, 190, 201, 212].

For the stabilization problem of a linear system with time delay, several kinds of stabilization have been considered. Among them we quote constant delay [77, 118, 149]; time-varying delay [77, 105]; constant delay and uncertainty [137, 175]; time-varying delay and parameter uncertainty [49, 108, 110]; and delay-independent stabilization [49, 108, 110, 206, 207, 215]; delay-dependent stabilization [118, 128, 155, 189, 190]. For the time-varying delay system, delay-dependent stability is dependent on the maximum value of the time derivative of the time-varying delay. Stability with discrete and distributed time delays is addressed by [118].

5
$\mathcal{H}_\infty$ Control and Filtering

Almost all practical systems are subject to external disturbances that can in some situations degrade system performance if their effects are not considered during the design phase. In the current literature there are many ways to eliminate the effects of the external disturbances. One of them is the $\mathcal{H}_\infty$ control technique. It consists of designing a suboptimal control that minimizes the effects of the external disturbance on the output. In other words, the problem can be explained as follows: Given a dynamical time delay system with exogenous input that belongs to $\mathcal{L}_2[0, \infty]$, design a controller that minimizes the $\mathcal{H}_\infty$-norm of the transfer function between the controlled output and the external disturbance, or at least guarantees that the $\mathcal{H}_\infty$-norm will not exceed a given level $\gamma > 0$.

In this chapter we will see how we can solve $\mathcal{H}_\infty$ control problems and provide some design algorithms for this problem. All the design algorithms we will develop later are based on LMI formalism, which makes them more useful.

The design of the state feedback controller requires access to the state of the system we are considering at each time t. Since this is not always possible because of some practical reasons and the cost of the sensors, it is preferable to resort to filtering to get a good estimate for the state vector at each t. Different techniques have been proposed in the literature. $\mathcal{H}_\infty$ theory is one of the approaches we can use to solve the filtering problem. In this case, the technique of the filtering problem can be stated as follows: Given a dynamical time delay system with exogenous input that belongs to $\mathcal{L}_2[0, \infty]$ and the measurement of the output, design a filter that gives an estimate of the unmeasured output such that the $\mathcal{H}_\infty$-norm of the transfer function between the estimation error and the exogenous input is minimized or at least kept below a certain given $\gamma > 0$.

$\mathcal{H}_\infty$-filtering is based mainly on $\mathcal{H}_\infty$ theory. Therefore, we have decided to put them together. In this chapter, we will develop some design filtering algorithms that can be used to solve the filtering we are considering. Once the two problems are solved, the control of any practical dynamical time delay system becomes solvable.

The goal of this chapter is to deal with the $\mathcal{H}_\infty$ control problem and the filtering problem for the class of continuous-time deterministic linear systems with time delay. The robust case of these problems will be treated in the next chapter. The chapter is organized as follows. In Section 5.1, the $\mathcal{H}_\infty$ control problem for the class of dynamical time delay systems is stated. A state feedback controller is considered to solve the problem. Complete access to the state vector is assumed. In Section 5.2, this assumption is relaxed, and an output feedback controller is used to stabilize the class of systems, while at the same time guaranteeing that the $\mathcal{H}_\infty$-norm of the transfer function between the controlled output and the external disturbance does not exceed a given level $\gamma > 0$. Section 5.3 presents a delay-dependent condition for the $\mathcal{H}_\infty$ control problem. A state feedback controller is designed in this case. Section 5.4 deals with the $\mathcal{H}_\infty$-filtering problem. Both delay-independent and delay-dependent conditions are developed to solve our filtering problem.

5.1 $\mathcal{H}_\infty$ Control Problem

Most often practical systems with time delay are subject to external disturbances that can degrade performance if they are not taken into account during the design phase. Many approaches have been proposed to deal with this problem. Among them we quote the linear quadratic Gaussian problem (LQG), which assumes that the external disturbance is Gaussian with a given mean and variance. The solution is then obtained by solving the appropriate Riccati equation (see [6] for more details). The other approach that we will adopt here consists of assuming that the disturbance belongs to $\mathcal{L}_2[0, \infty]$. This approach is referred to as $\mathcal{H}_\infty$ theory. Its purpose is to design a controller that minimizes the $\mathcal{H}_\infty$-norm of the transfer function between the controlled output $z(t)$ and the external disturbance $w(t)$, or at least guarantees that the $\mathcal{H}_\infty$-norm will not exceed a given level $\gamma > 0$. Mathematically, this control technique can be stated as follows: Given a positive $\gamma > 0$, find a controller such that

$$\sup_{w(t)\in\mathcal{L}_2[0,\infty]} \frac{\|z(t)\|_2^2}{\|w(t)\|_2^2} < \gamma^2.$$

Our goal in addressing the $\mathcal{H}_\infty$ control problem consists of

- developing sufficient conditions under which the unforced linear continuous-time system, that is, $u(t) \equiv 0$, will be stable and will guarantee the disturbance rejection level with level $\gamma > 0$;

- designing a state feedback controller that stabilizes the class of systems with time delay and guarantees the disturbance rejection of level $\gamma > 0$.

The sufficient conditions we will develop in the rest of this chapter in each case will be stated in LMI formalism which makes the conditions easier.

For this purpose, let us now consider a system described by the following dynamics:

$$\begin{cases} \dot{x}(t) = Ax(t) + A_d x(t - \tau) + B_1 w(t) + Bu(t) \\ x(t) = 0, \ t \leq 0 \\ z(t) = C_1 x(t) + D_{11} w(t) + D_{12} u(t) \\ y(t) = C_2 x(t) + D_{21} w(t) \end{cases} \qquad (5.1)$$

where $x(t) \in \mathbb{R}^n$ is the state vector, $w(t) \in \mathbb{R}^l$ is the square-integrable disturbance input, $u(t) \in \mathbb{R}^m$ is the control input, $z(t) \in \mathbb{R}^p$ is the controlled output, $y(t) \in \mathbb{R}^q$ is the measured output, $\tau > 0$ is the time delay and A, A_d, B_1, B, C_1, C_2, D_{11}, D_{12}, and D_{21} are constant matrices with appropriate dimensions. Also, we assume that (A, B, C_2) is stabilizable and detectable, and $w(\cdot) \in \mathcal{L}^2[0, \infty)$.

Definition 5.1 *A controller $\{u(t), t \geq 0\}$ is said to be a memoryless state feedback controller if it can be represented as $u(t) = Kx(t), t \geq 0$, where K is a constant matrix.*

Definition 5.2 *System (5.1) with $u(t) \equiv 0$ is said to be internally stable if it is stable when $w(t) \equiv 0$.*

The $\mathcal{H}_\infty$ control problem consists of choosing a control law $u(\cdot)$ guarantees that the closed-loop system is internally stable and at the same time satisfies

$$\|z(\cdot)\|_2 < \gamma \|w(\cdot)\|_2 \qquad (5.2)$$

where γ is a positive constant.

Now we come to develop some sufficient conditions for system (5.1) with $u(t) \equiv 0$ to have a given $\mathcal{H}_\infty$-norm γ, that is, one satisfying (5.2).

Let us now consider the unforced system, one in which $u(t) \equiv 0$ for all $t \geq 0$ and try to establish a test that allows us to check if our system is internally stable and assures the disturbance rejection of level γ. An answer to this question is given by the following theorem.

Theorem 5.1 *Let γ be a given positive constant. If there exist symmetric, positive-definite matrices $P > 0, Q > 0$ such that LMI*

$$
\begin{pmatrix}
A^\top P + PA + Q & PB_1 & C_1^\top & PA_d \\
B_1^\top P & -\gamma^2 I & D_{11}^\top & 0 \\
C_1 & D_{11} & -I & 0 \\
A_d^\top P & 0 & 0 & -Q
\end{pmatrix} < 0 \tag{5.3}
$$

holds, then system (5.1) with $u(t) \equiv 0$ is asymptotically stable and satisfies (5.2).

Proof: Let us define $\mathbf{x}_t \in \mathbb{C}[-\tau, 0]$ as $\mathbf{x}_t(s) = x(t+s), \quad s \in [-\tau, 0)$ and choose a Lyapunov functional candidate as

$$
V(\mathbf{x}_t) = x^\top(t)Px(t) + \int_{t-\tau}^t x^\top(s)Qx(s)ds, \tag{5.4}
$$

Then, in view of (5.1) we obtain

$$
\begin{aligned}
\dot{V}(\mathbf{x}_t, t) = {} & x^\top(t)[PA + A^\top P]x(t) + 2x^\top(t)PA_dx(t-\tau) \\
& + 2x^\top(t)PB_1w(t) + x^\top(t)Qx(t) - x^\top(t-\tau)Qx(t-\tau). \tag{5.5}
\end{aligned}
$$

From (5.3), we get

$$
\begin{pmatrix}
A^\top P + PA + Q & PA_d \\
A_d^\top P & -Q
\end{pmatrix} < 0,
$$

which proves that the system is internally stable using Theorem 3.1.

For any $T > 0$, let us define a performance function as follows:

$$
J_T = \int_0^T [z^\top(s)z(s) - \gamma^2 w^\top(s)w(s)]ds.
$$

Then we have

$$
J_T = \int_0^T \left[z^\top(s)z(s) - \gamma^2 w^\top(s)w(s) + \dot{V}(\mathbf{x}_s) \right] ds - \int_0^T \dot{V}(\mathbf{x}_s)ds,
$$

which, combined with (5.5), (5.1), and the fact that the initial condition is zero, yields

$$
\begin{aligned}
J_T &= \int_0^T \left[z^\top(s)z(s) - \gamma^2 w^\top(s)w(s) + \dot{V}(\mathbf{x}_s) \right] ds - V(\mathbf{x}_T) \\
&\leq \int_0^T \left[z^\top(s)z(s) - \gamma^2 w^\top(s)w(s) + \dot{V}(\mathbf{x}_s) \right] ds.
\end{aligned}
$$

Therefore, to prove (5.2) it suffices to establish that $J_T < 0$ for any $T > 0$. It follows from (5.5) that

$$
\int_0^T \left[z^\top(s)z(s) - \gamma^2 w^\top(s)w(s) + \dot{V}(\mathbf{x}_s) \right] ds \leq \int_0^T \left[\xi_t^\top \Theta_0 \xi_t \right] dt
$$

where

$$\xi_t^\top = (\ x^\top(t)\ \ w^\top(t)\ \ x^\top(t-\tau)\ )$$

$$\Theta_0 = \begin{pmatrix} PA + A^\top P + Q + C_1^\top C_1 & PB_1 + C_1^\top D_{11} & PA_d \\ B_1^\top P + D_{11}^\top C_1 & -\gamma^2 I + D_{11}^\top D_{11} & \mathbf{0} \\ A_d^\top P & \mathbf{0} & -Q \end{pmatrix}.$$

Using the Schur complement, it is easy to check that

$$\Theta_0 < 0$$

if and only if

$$\Theta_1 = \begin{pmatrix} \begin{bmatrix} PA + A^\top P + Q \\ +C_1^\top C_1 + PA_d Q^{-1} A_d^\top P \end{bmatrix} & PB_1 + C_1^\top D_{11} \\ B_1^\top P + D_{11}^\top C_1 & -\gamma^2 I + D_{11}^\top D_{11} \end{pmatrix} < 0.$$

Noting that Θ_1 can be rewritten as

$$\Theta_1 = \begin{pmatrix} \begin{bmatrix} PA + A^\top P + Q \\ +PA_d Q^{-1} A_d^\top P \end{bmatrix} & PB_1 \\ B_1^\top P & -\gamma^2 I \end{pmatrix}$$

$$+ \begin{pmatrix} C_1^\top \\ D_{11}^\top \end{pmatrix} (C_1\ D_{11}) < 0. \tag{5.6}$$

Using the Schur complement, we find that (5.3) is equivalent to (5.6). From the above derivation, we know that if there exist symmetric, positive-definite matrices $P > 0, Q > 0$ that satisfy (5.3), then they satisfy (5.6), which means that $J_T < 0$ for all $T > 0$. Therefore, we have

$$J_\infty = \int_0^\infty \left[z^\top(t)z(t) - \gamma^2 w^\top(t)w(t) \right] dt < 0,$$

yielding (5.2). This ends the proof.   □

Remark 5.1 *Theorem 5.1 can be used to check whether or not a given dynamical unforced system with time delay is internally stable and satisfies the disturbance rejection level $\gamma > 0$. Notice that the condition of this theorem is sufficient. Not satisfying this condition does not imply that the unforced system is unstable and does not satisfy the required disturbance rejection. Notice also that this condition does not depend on the system's time delays, a fact that makes this condition conservative.*

In practice it is desirable to compute the minimum disturbance rejection level γ that the unforced system under consideration can have. This minimum can be obtained by solving the linear optimization problem

Theorem 5.2 *Let the symmetric, the positive-definite matrices $P > 0, Q > 0$ and the positive scalar $\mu > 0$ be the solution of the following linear optimization problem*

$$\min_{P>0, Q>0, \mu>0}\ \mu$$

$$\text{s.t.} \quad \begin{pmatrix} A^\top P + PA + Q & PB_1 & C_1^\top & PA_d \\ B_1^\top P & -\mu I & D_{11}^\top & 0 \\ C_1 & D_{11} & -I & 0 \\ A_d^\top P & 0 & 0 & -Q \end{pmatrix} < 0,$$

then system (5.1) with $u(t) \equiv 0$ is asymptotically stable and

$$\|z(\cdot)\|_2 < \sqrt{\mu}\|w(\cdot)\|_2.$$

Remark 5.2 *Notice that this optimization problem is obtained from the results of Theorem 5.1 by replacing γ^2 by μ and then minimizing with respect to μ, P, and Q.*

Next, we address the synthesis of the state feedback controller, which can be used to stabilize an unstable system or improve its performance. The state feedback controller we will consider has the form (3.27). The goal of our design is to guarantee that the closed-loop system under this controller is asymptotically stable and satisfies (5.2).

Substituting (3.27) into (5.1) yields the dynamics of the closed-loop system

$$\begin{cases} \dot{x}(t) = \bar{A}x(t) + A_d x(t - \tau) + B_1 w(t) \\ x(t) = 0, \ t \le 0 \\ z(t) = \bar{C}_1 x(t) + D_{11} w(t) \end{cases} \tag{5.7}$$

where

$$\bar{A} = A + BK, \quad \bar{C}_1 = C_1 + D_{12}K.$$

According to Theorem 5.1, if there exist symmetric, positive-definite matrices $P > 0$ and $Q > 0$ such that

$$\begin{pmatrix} \bar{A}^\top P + P\bar{A} + Q & PB_1 & \bar{C}_1^\top & PA_d \\ B_1^\top P & -\gamma^2 I & D_{11}^\top & 0 \\ \bar{C}_1 & D_{11} & -I & 0 \\ A_d^\top P & 0 & 0 & -Q \end{pmatrix} < 0 \tag{5.8}$$

holds, then system (5.7) is asymptotically stable and satisfies (5.2). Using now the Schur complement, (5.8) is equivalent to

$$\begin{pmatrix} \bar{A}^\top P + P\bar{A} + PA_d Q^{-1} A_d^\top P & PB_1 & \bar{C}_1^\top & I \\ B_1^\top P & -\gamma^2 I & D_{11}^\top & 0 \\ \bar{C}_1 & D_{11} & -I & 0 \\ I & 0 & 0 & -Q^{-1} \end{pmatrix} < 0. \tag{5.9}$$

Let $X = P^{-1}$ and $U = Q^{-1}$. Pre- and post-multiplying (5.9) by $\text{diag}\{X, I, I, I\}$ and letting $KX = Y$ yields

$$\begin{pmatrix} \# & B_1 & XC_1^\top + Y^\top D_{12}^\top & X \\ B_1^\top & -\gamma^2 I & D_{11}^\top & 0 \\ C_1 X + D_{12}Y & D_{11} & -I & 0 \\ X & 0 & 0 & -U \end{pmatrix} < 0 \tag{5.10}$$

where $\# = XA^\top + Y^\top B^\top + AX + BY + A_d U A_d^\top$. Therefore, we obtain the following theorem.

Theorem 5.3 *Let γ be a given positive scalar. If there exist symmetric, positive-definite matrices $X > 0, U > 0$, and a matrix Y satisfying (5.10), then controller (3.27) with $K = YX^{-1}$ stabilizes system (5.1) and the closed-loop system under this controller satisfies (5.2).*

As we did for the unforced system, the optimal noise attenuation level can be obtained by solving the linear optimization problem

$$\min_{\mu>0,X>0,U>0,Y} \mu \tag{5.11}$$

s.t. (5.10) with γ^2 replaced by μ.

Corollary 5.1 *Let γ, $X > 0$, $U > 0$, and Y be a set of solutions of the optimization problem (5.11). Then, controller (3.27) with $K = YX^{-1}$ stabilizes system (5.1) and the closed-loop system under this controller satisfies (5.2) with the level $\nu = \sqrt{\gamma}$.*

To show the usefulness of the results of this section, let us consider the following numerical example.

Example 5.1 *Let us consider a dynamical time delay system as described by (5.1) and suppose that the system data are as follows:*

$$A = \begin{pmatrix} -2 & 1 \\ 1 & 1 \end{pmatrix} \quad A_d = \begin{pmatrix} 0.2 & 0.1 \\ 0.3 & 0.1 \end{pmatrix}$$

$$B_1 = \begin{pmatrix} 1 \\ 0 \end{pmatrix} \quad B = \begin{pmatrix} 1 \\ 1 \end{pmatrix}$$

$$C_1 = \begin{pmatrix} 1 & 0 \\ 0 & 1 \\ 0 & 0 \end{pmatrix} \quad C_2 = 1$$

$$D_{11} = \begin{pmatrix} 1 \\ 0 \\ 0 \end{pmatrix} \quad D_{12} = \begin{pmatrix} 0 \\ 0 \\ 1 \end{pmatrix}$$

$$D_{21} = 1.$$

With these data, solving the optimization problem (5.11) yields the following solution:

$$X = \begin{pmatrix} 0.4989 & -0.6368 \\ -0.6368 & -0.3052 \end{pmatrix} \quad Y = \begin{pmatrix} -0.8763 & -0.8763 \end{pmatrix}$$

$$U = \begin{pmatrix} 14.9726 & -11.3380 \\ -11.3380 & 12.3149 \end{pmatrix} \quad \gamma = 1.7526.$$

According to the corollary, controller (3.27) with

$$K = YX^{-1} = \begin{pmatrix} 0.5210 & 1.7842 \end{pmatrix}$$

stabilizes the system under study and the closed-loop system satisfies noise attenuation level $\gamma = 1.3274$.

All we have presented so far requires complete access to the state vector space. In the next section, we will drop this assumption and design an output feedback controller.

5.2   Output Feedback $\mathcal{H}_\infty$ Control Problem

In the previous section, we provided a method for designing a state feedback controller that stabilizes system (5.1) and guarantees the disturbance rejection level $\gamma > 0$ under the assumption that the system state is perfectly available for feedback. However, in most cases complete access to the system state is not possible. Therefore, an alternate method consists of using a dynamical output feedback controller. This section addresses the design of dynamical output feedback control of the form of (3.42).

Applying (3.42) to system (5.1), we obtain the following closed-loop system from w to z:

$$\begin{cases} \dot{\eta}(t) = \bar{A}\eta(t) + \bar{A}_d\eta(t-\tau) + \bar{B}_1 w(t) \\ z(t) = \bar{C}\eta(t) + \bar{D}w(t) \\ \eta(t) = 0, t \le 0 \end{cases} \tag{5.12}$$

where

$$\eta(t) = \begin{pmatrix} x(t) \\ \xi_t \end{pmatrix} \qquad\qquad \bar{A} = \begin{pmatrix} A + BK_DC_2 & BK_C \\ K_BC_2 & K_A \end{pmatrix}$$

$$\bar{A}_d = \begin{pmatrix} A_d & 0 \\ 0 & 0 \end{pmatrix} \qquad\qquad \bar{B}_1 = \begin{pmatrix} B_1 + BK_DD_{21} \\ K_BD_{21} \end{pmatrix}$$

$$\bar{C} = (C_1 + D_{12}K_DC_2 \quad D_{12}K_C) \quad \bar{D} = D_{11} + D_{12}K_DD_{21}.$$

Let us define a design parameter matrix K by

$$K \triangleq \begin{pmatrix} K_D & K_C \\ K_B & K_A \end{pmatrix}$$

and define the matrices A_0, A_1, I_0, B_0, B_{00}, C_0, C_{00}, D_1, and D_2 as follows:

$$A_0 = \begin{pmatrix} A & 0 \\ 0 & 0 \end{pmatrix} \qquad A_1 = \begin{pmatrix} A_d \\ 0 \end{pmatrix} \qquad I_0 = (\,I \quad 0\,)$$

$$B_0 = \begin{pmatrix} B_1 \\ 0 \end{pmatrix} \qquad B_{00} = \begin{pmatrix} B & 0 \\ 0 & I \end{pmatrix} \qquad C_0 = (\,C_1 \quad 0\,)$$

$$C_{00} = \begin{pmatrix} C_2 & 0 \\ 0 & I \end{pmatrix} \qquad D_1 = (\,D_{12} \quad 0\,) \qquad D_2 = \begin{pmatrix} D_{21} \\ 0 \end{pmatrix}.$$

With this notation, the parameter matrices of the closed-loop system can be rewritten as

$$\bar{A} = A_0 + B_{00}KC_{00} \quad \bar{A}_d = A_1 I_0 \qquad\qquad \bar{B}_1 = B_0 + B_{00}KD_2$$
$$\bar{C}_1 = C_0 + D_1 KC_{00} \quad \bar{D} = D_{11} + D_1 KD_2.$$

Using Theorem 5.1, we have that if there exist symmetric and positive-definite matrices $P > 0$ and $Q > 0$ such that the LMI

$$\begin{pmatrix} \bar{A}^\top P + P\bar{A} + I_0^\top QI_0 & P\bar{B}_1 & \bar{C}_1^\top & P\bar{A}_d \\ \bar{B}_1^\top P & -\gamma^2 I & \bar{D}^\top & 0 \\ \bar{C}_1 & \bar{D} & -I & 0 \\ \bar{A}_d^\top P & 0 & 0 & -Q \end{pmatrix} < 0, \qquad (5.13)$$

where $I_0 = (\ I \ \ \mathbf{0}\ )$, holds, then the closed-loop system (5.12) is asymptotically stable and satisfies (5.2). Obviously, (5.13) is nonlinear with respect to the design parameters K and P. Thus, (5.13) cannot be used to get the matrix K. One way to overcome this consists of using Lemma A.3 to eliminate parameter matrix K from (5.13), which leads to a feasible problem with respect to P, then solve the feasible problem to get P. With P given, solve (5.13) again to get the design parameter K.

Since the matrices $\bar{A}, \bar{A}_d, \bar{B}_1, \bar{C}$, and $\bar{D}$ in (5.13) are affine in K, condition (5.13) can be rewritten as

$$\Phi + \Sigma\Pi K\Theta^\top + \Theta K^\top \Pi^\top \Sigma^\top < 0, \qquad (5.14)$$

where

$$\Phi = \begin{pmatrix} A_0^\top P + PA_0 + I_0^\top QI_0 & PB_0 & C_0^\top & PA_1 \\ B_0^\top P & -\gamma^2 I & D_{11}^\top & 0 \\ C_0 & D_{11} & -I & 0 \\ A_1^\top P & 0 & 0 & -Q \end{pmatrix}$$

$$\Sigma = \begin{pmatrix} P & \mathbf{0} & \mathbf{0} & \mathbf{0} \\ \mathbf{0} & I & \mathbf{0} & \mathbf{0} \\ \mathbf{0} & \mathbf{0} & I & \mathbf{0} \\ \mathbf{0} & \mathbf{0} & \mathbf{0} & I \end{pmatrix},$$

$$\Pi = \begin{pmatrix} B_{00} \\ \mathbf{0} \\ D_1 \\ \mathbf{0} \end{pmatrix},$$

$$\Theta = \begin{pmatrix} C_{00}^\top \\ D_2^\top \\ \mathbf{0} \\ \mathbf{0} \end{pmatrix}.$$

According to Lemma A.3, we conclude that there exists a matrix K satisfying (5.14) if and only if

$$\Pi_\perp^\top \Sigma^{-1} \Phi \Sigma^{-1} \Pi_\perp < 0 \qquad (5.15)$$

$$\Theta_\perp^\top \Phi \Theta_\perp < 0 \tag{5.16}$$

where $\Pi_\perp$ and $\Theta_\perp$ are orthogonal complements of Π and Θ, respectively. Using Lemma A.3, we eliminate the design parameter matrix K from (5.14). To simplify (5.15) and (5.16), let us partition P and P^{-1} as

$$P = \begin{pmatrix} Y & N \\ N^\top & \# \end{pmatrix} \text{ and } P^{-1} = \begin{pmatrix} X & M \\ M^\top & \# \end{pmatrix}, \tag{5.17}$$

respectively, where $X, Y \in \mathbb{R}^{n \times n}$, $M, N \in \mathbb{R}^{n \times k}$, and $\#$ means the entry is irrelevant.

Let us choose

$$\begin{pmatrix} W_1 \\ W_2 \end{pmatrix} \text{ and } \begin{pmatrix} W_3 \\ W_4 \end{pmatrix}$$

as the orthogonal complements of

$$\begin{pmatrix} B \\ D_{12} \end{pmatrix} \text{ and } \begin{pmatrix} C_2^\top \\ D_{21}^\top \end{pmatrix},$$

respectively. Then we obtain

$$\Pi_\perp = \begin{pmatrix} W_1 & 0 & 0 \\ 0 & 0 & 0 \\ 0 & I & 0 \\ W_2 & 0 & 0 \\ 0 & 0 & I \end{pmatrix} \text{ and } \Theta_\perp = \begin{pmatrix} W_3 & 0 & 0 \\ 0 & 0 & 0 \\ W_4 & 0 & 0 \\ 0 & I & 0 \\ 0 & 0 & I \end{pmatrix}.$$

Consequently, (5.15) and (5.16) become

$$\mathcal{W}_0^\top \begin{pmatrix} XA^\top + AX + XQX & XC_1^\top & B_1^\top & A_d \\ C_1 X & -\gamma^2 I & D_{11}^\top & 0 \\ B_1 & D_{11} & -I & 0 \\ A_d^\top & 0 & 0 & -Q \end{pmatrix} \mathcal{W}_0 < 0 \tag{5.18}$$

$$\mathcal{W}_2^\top \begin{pmatrix} A^\top Y + YA + Q & YB_1 & C_1^\top & YA_d \\ B_1^\top Y & -\gamma^2 I & D_{11}^\top & 0 \\ C_1 & D_{11} & -I & 0 \\ A_d^\top Y & 0 & 0 & -Q \end{pmatrix} \mathcal{W}_2 < 0 \tag{5.19}$$

where

$$\mathcal{W}_0 = \begin{pmatrix} W_1 & 0 & 0 \\ W_2 & 0 & 0 \\ 0 & I & 0 \\ 0 & 0 & I \end{pmatrix} \text{ and } \mathcal{W}_2 = \begin{pmatrix} W_3 & 0 & 0 \\ W_4 & 0 & 0 \\ 0 & I & 0 \\ 0 & 0 & I \end{pmatrix}.$$

Since $Q > 0$, (5.18) is equivalent to

$$
\mathcal{W}_1^\top \begin{pmatrix} XA^\top + AX & XC_1^\top & B_1^\top & A_d & X \\ C_1 X & -\gamma^2 I & D_{11}^\top & 0 & 0 \\ B_1 & D_{11} & -I & 0 & 0 \\ A_d^\top & 0 & 0 & -Q & 0 \\ X & 0 & 0 & 0 & -Q^{-1} \end{pmatrix} \mathcal{W}_1 < 0 \qquad (5.20)
$$

where

$$
\mathcal{W}_1 = \begin{pmatrix} W_1 & 0 & 0 & 0 \\ W_2 & 0 & 0 & 0 \\ 0 & I & 0 & 0 \\ 0 & 0 & I & 0 \\ 0 & 0 & 0 & I \end{pmatrix}.
$$

The following theorem summarizes the results that give the existence condition of output feedback controller gains K_A, K_B, K_C, and K_D.

Theorem 5.4 *If there exist symmetric, positive-definite matrices $X > 0, Y > 0$ satisfying (5.19), (5.20) and*

$$
\begin{pmatrix} X & I \\ I & Y \end{pmatrix} \geq 0 \qquad (5.21)
$$

for some given symmetric, positive-definite matrix Q, then the γ-suboptimal $\mathcal{H}_\infty$ control problem is solvable; that is, there exist gains K_A, K_B, K_C, and K_D such that controller (3.42) stabilizes system (5.1) and the closed-loop system satisfies (5.2).

Proof: There exists matrix $P > 0$ satisfying (5.17) if and only if $X - Y^{-1} > 0$ holds, which is equivalent to (5.21). The rest of the proof follows from above discussion. □

Theorem 5.4 provides a sufficient condition for the existence of the stabilizing controller, rather than develops a design method. However, a stabilizing controller can be designed using the following algorithm.

Algorithm 5.1 *(Output Controller Design Algorithm)*

Step 1. Compute a feasible solution (X, Y) satisfying (5.19)-(5.21).

Step 2. Using the singular value decomposition, compute two full-column rank matrices $M, N \in \mathbb{R}^{n \times k}$ such that

$$
MN^\top = I - XY.
$$

Step 3. Get matrix P by solving the linear equation

$$
\Phi_2 = \begin{pmatrix} Y & I \\ N^\top & 0 \end{pmatrix} = P \begin{pmatrix} I & X \\ 0 & M^\top \end{pmatrix} = P\Phi_1. \qquad (5.22)
$$

Step 4. With P given, solve the feasible problem (5.13) to get K since it becomes an LMI.

Remark 5.3 *Linear equation (5.22) is always solvable when $Y > 0$ and M has full-column rank.*

In real life, one must reduce the disturbance effect. Therefore, to minimize the noise attenuation level γ we replace Step 1 in Algorithm 5.1 by the following step.

Step 1' Solve the linear optimization problem

$$\min_{\mu>0, X>0, Y>0} \mu$$

s.t. (5.20)-(5.21) with γ^2 replaced with μ

Since the order of the controller depends on the dimension of P, we can establish the following corollary.

Corollary 5.2 *Suppose that LMIs (5.20)-(5.21) have a set of feasible solutions $X > 0, Y > 0$. If $rank(I - XY) = k < n$, then there exists a controller (3.42) that stabilizes system (5.1), and the closed-loop system satisfies (5.2).*

To show the usefulness of this section, let us consider the following numerical example.

Example 5.2 *Let us consider a dynamical time delay system as described by (5.1) and suppose that the system data are as follows:*

$$A = \begin{pmatrix} -1.5 & 1 \\ 1 & -2.3 \end{pmatrix} \quad A_d = \begin{pmatrix} 0.2 & 0.1 \\ 0.3 & 0.1 \end{pmatrix} \quad B_1 = \begin{pmatrix} 1 \\ 0 \end{pmatrix}$$

$$B = \begin{pmatrix} 1 \\ 1 \end{pmatrix} \quad C1 = \begin{pmatrix} 0 & 1 \\ 1 & 1 \end{pmatrix} \quad C_2 = \begin{pmatrix} 0 & 3 \end{pmatrix}$$

$$D_{11} = \begin{pmatrix} 0 \\ 1 \end{pmatrix} \quad D_{12} = \begin{pmatrix} 0 \\ 1 \end{pmatrix} \quad D_{21} = 1.$$

Let $Q = I$. Then, using Algorithm 5.1 we get the feasible symmetric, positive-definite matrices of (5.20)-(5.21)

$$X = \begin{pmatrix} 1.5964 & -0.3873 \\ -0.3873 & 3.1783 \end{pmatrix} \quad and \quad Y = \begin{pmatrix} 2.8753 & -0.4128 \\ -0.4128 & 3.2310 \end{pmatrix}.$$

and the positive scalar $\mu = 5.0235$.

With the matrices X and Y and the use of the singular value decomposition, we obtain the invertible matrices M and N

$$N = \begin{pmatrix} 0.3387 & -0.9409 \\ -0.9409 & -0.3387 \end{pmatrix} \quad and \quad M = \begin{pmatrix} -3.0677 & 2.8812 \\ 9.6932 & 0.9118 \end{pmatrix},$$

which satisfy (5.21). Therefore, we get Φ_1 and Φ_2 as follows:

$$\Phi_1 = \begin{pmatrix} 1.0000 & 0 & 1.5964 & -0.3873 \\ 0 & 1.0000 & -0.3873 & 3.1783 \\ 0 & 0 & -3.0677 & 9.6932 \\ 0 & 0 & 2.8812 & 0.9118 \end{pmatrix}$$

$$\Phi_2 = \begin{pmatrix} 2.8753 & -0.4128 & 1.0000 & 0 \\ -0.4128 & 3.2310 & 0 & 1.0000 \\ 0.3387 & -0.9409 & 0 & 0 \\ -0.9409 & -0.3387 & 0 & 0 \end{pmatrix}.$$

With Φ_1 and Φ_2 given, solving (5.22) gives

$$P = \begin{pmatrix} 2.8753 & -0.4128 & 0.3387 & -0.9409 \\ -0.4128 & 3.2310 & -0.9409 & -0.3387 \\ 0.3387 & -0.9409 & 0.3196 & 0.0261 \\ -0.9409 & -0.3387 & 0.0261 & 0.5036 \end{pmatrix}.$$

With this P, solving (5.13) yields the solution

$$K = \begin{pmatrix} -0.7888 & 0.4475 & -0.5933 \\ 5.1577 & -8.0743 & -3.8820 \\ 2.7518 & -2.7856 & -5.8191 \end{pmatrix}$$

$$Q = \begin{pmatrix} 2.2091 & 0.4706 \\ 0.4706 & 0.7505 \end{pmatrix}.$$

Thus, we get a set of stabilizing gains as follows:

$$K_A = \begin{pmatrix} -8.0743 & -3.8820 \\ -2.7856 & -5.8191 \end{pmatrix} \quad K_B = \begin{pmatrix} 5.1577 \\ 2.7518 \end{pmatrix}$$
$$K_C = \begin{pmatrix} 0.4475 & -0.5933 \end{pmatrix} \quad K_D = (-0.7888).$$

The closed-loop system under controller (3.42) with this set of gains has noise attenuation level $\gamma = \sqrt{\mu} = 2.2413$.

5.3   Delay-Dependent $\mathcal{H}_\infty$ Control

The results of the previous section are independent of the time delay in the system, which means that these results hold for any time delay and thus are conservative. This section will address the delay-dependent $\mathcal{H}_\infty$ control problem. Our goal is to develop delay-dependent condition to overcome the conservatism of the previous section. The following theorem establishes a delay-dependent sufficient condition that can be used to check if system (5.1) is stable and satisfies a given noise attenuation level γ.

Theorem 5.5 *Let $\gamma > 0$ be a given positive constant. If there exist symmetric, positive-definite matrices $X > 0, Q_1 > 0, Q_2 > 0, Q_3 > 0$*

satisfying

$$\Theta_d = \begin{pmatrix} \# & B_1 & XC_1^\top & \mathcal{A} \\ B_1^\top & \begin{bmatrix} -\gamma^2 I+ \\ +\tau B_1^\top Q_3 B_1 \end{bmatrix} & D_{11}^\top & 0 \\ C_1 X & D_{11} & -I & 0 \\ \mathcal{A}^\top & 0 & 0 & -\mathcal{Q} \end{pmatrix} < 0 \qquad (5.23)$$

where

$$\# = [A + A_d]X + X[A + A_d]^\top + \tau A_d Q_1 A_d^\top + \tau A_d Q_2 A_d^\top$$
$$\mathcal{A} = \begin{pmatrix} XA^\top & XA_d^\top & A_d \end{pmatrix}$$
$$\mathcal{Q} = \begin{pmatrix} \frac{1}{\tau}Q_1 & 0 & 0 \\ 0 & \frac{1}{\tau}Q_2 & 0 \\ 0 & 0 & \frac{1}{\tau}Q_3 \end{pmatrix},$$

then system (5.1) with $u(t) \equiv 0$ is stable and satisfies (5.2).

Proof: Note that

$$x(t - \tau) = x(t) - \int_{t-\tau}^t \dot{x}(s)ds.$$

Substituting (5.1) with $u(t) \equiv 0$ into the above formula yields

$$x(t - \tau) = x(t) - \int_{t-\tau}^t [Ax(s) + A_d x(s - \tau) + B_1 w(s)]ds. \qquad (5.24)$$

Substituting now (5.24) into (5.1), we obtain

$$\begin{cases} \dot{x}(t) = [A + A_d]x(t) - A_d \int_{t-\tau}^t [Ax(s) + A_d x(s - \tau) \\ \qquad + B_1 w(s)]ds + B_1 w(t) \\ x(s) = 0, \ w(s) = 0, \ s \in [-2\tau, 0]. \end{cases} \qquad (5.25)$$

Let us define $\mathbf{x}_t \in \mathbb{C}[-2\tau, 0]$ with $\mathbf{x}_t(s) = x(t + s)$, $s \in [-2\tau, 0]$ and consider a candidate Lyapunov functional as follows:

$$V(\mathbf{x}_t) = V_0(\mathbf{x}_t) + V_1(\mathbf{x}_t), \qquad (5.26)$$

where

$$V_0(x(t)) = x^\top(t)Px(t)$$
$$V_1(\mathbf{x}_t) = \int_{-\tau}^0 \int_{t+\theta}^t x^\top(s)A^\top Q_1^{-1} Ax(s)dsd\theta$$
$$+ \int_{-\tau}^0 \int_{t-\tau+\theta}^t x^\top(s)A_d^\top Q_2^{-1} A_d x(s)dsd\theta$$
$$+ \int_{-\tau}^0 \int_{t+\theta}^t w^\top(s)B_1^\top Q_3 B_1 w(s)dsd\theta. \qquad (5.27)$$

Using (5.25), we obtain

$$\dot{V}_0(\mathbf{x}_t) = x^\top(t)\Big[P[A + A_d] + [A + A_d]^\top P\Big]x(t)$$

$$- 2x^\top(t)PA_d\Big[A \int_{t-\tau}^t x(s)ds + A_d \int_{t-\tau}^t x(s-\tau)ds$$

$$+ B_1 \int_{t-\tau}^t w(s)ds\Big] + 2x^\top(t)PB_1w(t). \tag{5.28}$$

Based on Lemma A.1, we obtain, respectively,

$$-2x^\top(t)PA_dA \int_{t-\tau}^t x(s)ds \leq \tau x^\top(t)PA_dQ_1A_d^\top Px(t)$$

$$+ \int_{t-\tau}^t x^\top(s)A^\top Q_1^{-1}Ax(s)ds \tag{5.29}$$

$$-2x^\top(t)PA_dA_d \int_{t-\tau}^t x(s-\tau)ds \leq \tau x^\top(t)PA_dQ_2A_d^\top Px(t)$$

$$+ \int_{t-\tau}^t x^\top(s-\tau)A_d^\top Q_2^{-1}A_dx(s-\tau)ds \tag{5.30}$$

$$-2x^\top(t)PA_dB_1 \int_{t-\tau}^t w(s)ds \leq \tau x^\top(t)PA_dQ_3^{-1}A_d^\top Px(t)$$

$$+ \int_{t-\tau}^t w^\top(s)B_1^\top Q_3B_1w(s)ds. \tag{5.31}$$

Moreover, simple computation gives

$$\dot{V}_1(\mathbf{x}_t) = \tau x^\top(t)A^\top Q_1^{-1}Ax(t) - \int_{t-\tau}^t x^\top(s)A^\top Q_1^{-1}Ax(s)ds$$

$$+ \tau x^\top(t)A_d^\top Q_2^{-1}A_dx(t) - \int_{t-\tau}^t x^\top(s-\tau)A_d^\top Q_2^{-1}A_dx(s-\tau)ds$$

$$+ \tau w^\top(t)B_1^\top Q_3B_1w(t) - \int_{t-\tau}^t w^\top(s)B_1^\top Q_3B_1w(s)ds. \tag{5.32}$$

Combining (5.29)-(5.32) yields

$$\dot{V}(\mathbf{x}_t) \leq x^\top(t)\Big[P[A + A_d] + [A + A_d]^\top P\Big]x(t)$$

$$+ \tau x^\top(t)[PA_dQ_1A_d^\top P + A^\top Q_1^{-1}A]x(t)$$

$$+ \tau x^\top(t)[PA_dQ_2A_d^\top P + A_d^\top Q_2^{-1}A_d]x(t)$$

$$+ \tau x^\top(t)PA_dQ_3^{-1}A_d^\top Px(t) + \tau w^\top(t)B_1^\top Q_3B_1w(t)$$

$$+ 2x^\top(t)PB_1w(t). \tag{5.33}$$

The proof of internal stability follows the same steps of the one of Theorem 5.1, so the details are omitted.

Let us now define the performance function by

$$J_T = \int_0^T [z^\top(t)z(t) - \gamma^2 w^\top(t)w(t)]dt.$$

Note that

$$J_T = \int_0^T [z^\top(t)z(t) - \gamma^2 w^\top(t)w(t) + \dot{V}(\mathbf{x}_t)]dt - \int_0^T \dot{V}(\mathbf{x}_t)dt$$

$$= \int_0^T [z^\top(t)z(t) - \gamma^2 w^\top(t)w(t) + \dot{V}(\mathbf{x}_t)]dt - V(\mathbf{x}_T)$$

$$\leq \int_0^T [z^\top(t)z(t) - \gamma^2 w^\top(t)w(t) + \dot{V}(\mathbf{x}_t)]dt. \qquad (5.34)$$

In view of (5.33), we obtain

$$z^\top(t)z(t) - \gamma^2 w^\top(t)w(t) + \dot{V}(\mathbf{x}_t) \leq \left(x^\top(t)\ w^\top(t)\right) \tilde{\Theta} \left(\begin{array}{c} x(t) \\ w(t) \end{array}\right), \qquad (5.35)$$

where

$$\tilde{\Theta} = \left(\begin{array}{cc} \left[\begin{array}{c} P[A+A_d] \\ +[A+A_d]^\top P \\ +\tau P A_d Q_1 A_d^\top P \\ +\tau A^\top Q_1^{-1} A \\ +\tau P A_d Q_2 A_d^\top P \\ +\tau A_d^\top Q_2^{-1} A_d \\ +\tau P A_d Q_3^{-1} A_d^\top P \\ +C_1^\top C_1 \end{array}\right] & PB_1 + C_1^\top D_{11} \\ B_1^\top P + D_{11}^\top C_1 & \left[\begin{array}{c} \tau B_1^\top Q_3 B_1 \\ +D_{11}^\top D_{11} - \gamma^2 I \end{array}\right] \end{array}\right). \qquad (5.36)$$

Let $X = P^{-1}$. Pre- and post-multiplying $\tilde{\Theta}$ by diag$\{X, I\}$ yields

$$\left(\begin{array}{cc} \left[\begin{array}{c} [A+A_d]X \\ +X[A+A_d]^\top \\ +\tau A_d Q_1 A_d^\top \\ +\tau X A^\top Q_1^{-1} A X \\ +\tau A_d Q_2 A_d^\top \\ +\tau X A_d^\top Q_2^{-1} A_d X \\ +\tau X A_d Q_3^{-1} A_d^\top X \\ +X C_1^\top C_1 X \end{array}\right] & B_1 + X C_1^\top D_{11} \\ B_1^\top + D_{11}^\top C_1 X & \left[\begin{array}{c} \tau B_1^\top Q_3 B_1 \\ +D_{11}^\top D_{11} - \gamma^2 I \end{array}\right] \end{array}\right) < 0$$

Using the Schur complement, the above inequality holds if and only if (5.23) is feasible. Above derivation yields that if the symmetric, positive-definite matrices $X > 0$, $Q_1 > 0$, $Q_2 > 0$ and $Q_3 > 0$ satisfy (5.23), then $P = X^{-1} > 0$, $Q_1 > 0$, $Q_2 > 0$, $Q_3 > 0$ satisfy $\tilde{\Theta} < 0$, which means that

$J_T < 0$ for any $T > 0$ and the results follow directly. This completes the proof of Theorem 5.5.   □

Based on Theorem 5.5, we can synthesize a memoryless controller (3.27) that stabilizes system (5.1) and guarantees that the closed-loop system satisfies (5.2). For this purpose, let us now show how we use the results of this theorem to design a memoryless controller. Substituting (3.27) into (5.1), we obtain the following dynamics for the closed-loop system:

$$\dot{x}(t) = \bar{A}x(t) + A_d x(t - \tau) + B_1 w(t) \tag{5.37}$$
$$z(t) = \bar{C}_1 x(t) + D_{11} w(t) \tag{5.38}$$

where $\bar{A} = A + BK, \bar{C}_1 = C_1 + D_{12}K$.

According to Theorem 5.5, we find that if there exist symmetric, positive-definite matrices $X > 0, Q_1 > 0, Q_2 > 0, Q_3 > 0$ satisfying

$$\left(\begin{array}{cccc} \left[\begin{array}{c} [\bar{A} + A_d]X \\ + X[\bar{A} + A_d]^\top \\ + \tau A_d Q_1 A_d^\top \\ + \tau A_d Q_2 A_d^\top \end{array} \right] & B_1 & X\bar{C}_1^\top & \mathcal{A} \\ B_1^\top & \left[\begin{array}{c} -\gamma^2 I \\ + \tau B_1^\top Q_3 B_1 \end{array} \right] & D_{11}^\top & 0 \\ \bar{C}_1 X & D_{11} & -I & 0 \\ \mathcal{A}^\top & 0 & 0 & -\mathcal{Q} \end{array} \right) < 0 \tag{5.39}$$

where

$$\mathcal{A} = \left(X\bar{A}^\top \ \ XA_d^\top \ \ A_d \right)$$

and $\mathcal{Q}$ is the same as defined in (5.23), then the closed-loop of system (5.1) is stable under controller (3.27) and satisfies (5.2). Therefore, we obtain the following theorem.

Theorem 5.6 *Let $\gamma > 0$ be a given scalar. If there exist symmetric, positive-definite matrices $X > 0, Q_1 > 0, Q_2 > 0, Q_3 > 0$, and a matrix Y satisfying*

$$\left(\begin{array}{cccc} \mathcal{H}_{11} & B_1 & \mathcal{H}_{13} & \mathcal{H}_{14} \\ B_1^\top & \left[\begin{array}{c} -\gamma^2 I \\ + \tau B_1^\top Q_3 B_1 \end{array} \right] & D_{11}^\top & 0 \\ \mathcal{H}_{13}^\top & D_{11} & -I & 0 \\ \mathcal{H}_{14}^\top & 0 & 0 & -\mathcal{Q} \end{array} \right) < 0 \tag{5.40}$$

where

$$\mathcal{H}_{11} = [A + A_d]X + X[A + A_d]^\top + BY + Y^\top B^\top + \tau A_d Q_1 A_d^\top$$
$$\qquad + \tau A_d Q_2 A_d^\top$$
$$\mathcal{H}_{13} = XC_1^\top + Y^\top D_{12}^\top$$
$$\mathcal{H}_{14} = \left(XA^\top + Y^\top B^\top, \ XA_d^\top, \ A_d \right)$$

$$Q = \text{diag}(\frac{1}{\tau}Q_1, \frac{1}{\tau}Q_2, \frac{1}{\tau}Q_3),$$

then the closed-loop system (5.1) under controller (3.27) with $K = YX^{-1}$ is stable and satisfies (5.2).

Proof: The proof follows from (5.39) by letting $Y = KX$.   □

Using the above theorem, a controller that stabilizes system (5.1) and optimizes the noise attenuation level $\gamma > 0$ can be designed using the results of the following theorem:

Theorem 5.7 *With notation as defined in Theorem 5.6, if symmetric, positive-definite matrices $X > 0$, $Q_1 > 0$, $Q_2 > 0$, $Q_3 > 0$, a matrix Y, and a positive scalar ρ_0 are the feasible solutions of the optimization problem*

$$\min_{X>0,Q_1>0,Q_2>0,Q_3>0,Y,\rho>0} \rho \tag{5.41}$$

$$s.t. \quad \begin{pmatrix} \mathcal{H}_{11} & B_1 & \mathcal{H}_{13} & \mathcal{H}_{14} \\ B_1^\top & \begin{bmatrix} -\rho I \\ + \tau B_1^\top Q_3 B_1 \end{bmatrix} & D_{11}^\top & 0 \\ \mathcal{H}_{13}^\top & D_{11} & -I & 0 \\ \mathcal{H}_{14}^\top & 0 & 0 & -Q \end{pmatrix} < 0,$$

then the closed-loop system (5.1) under controller (3.27) with $K = YX^{-1}$ is stable and satisfies (5.2) with noise attenuation level $\gamma = \sqrt{\rho_0}$.

To show the usefulness of the results of this section let us consider the following numerical example.

Example 5.3 *Let us consider a dynamical time delay system as described by (5.1) and suppose that the system has the following data:*

$$A = \begin{pmatrix} 5 & 1 \\ -1 & 3 \end{pmatrix} \quad A_d = \begin{pmatrix} 0.2 & 0.1 \\ 0.3 & 0.1 \end{pmatrix}$$

$$B_1 = \begin{pmatrix} 1 \\ 0 \end{pmatrix} \quad B = \begin{pmatrix} 1 \\ 1 \end{pmatrix}$$

$$C_1 = \begin{pmatrix} 1 & 0 \\ 0 & 1 \\ 1 & 1 \end{pmatrix} \quad D_{11} = \begin{pmatrix} 0 \\ 1 \\ 0 \end{pmatrix}$$

$$D_{12} = \begin{pmatrix} 0 \\ 0 \\ 1 \end{pmatrix} \quad \tau = 0.8.$$

With these data, solving (5.41) yields the following set of feasible solutions

$$X = \begin{pmatrix} 30.1338 & 55.5159 \\ 55.5159 & 113.2244 \end{pmatrix} \quad Y = \begin{pmatrix} -365.8745 & -478.9702 \end{pmatrix}$$

$$Q_1 = \begin{pmatrix} 476.1695 & 132.4043 \\ 132.4043 & 638.3926 \end{pmatrix} \quad Q_2 = \begin{pmatrix} 383.0046 & -65.5305 \\ -65.5305 & 293.6652 \end{pmatrix}$$

$$Q_3 = \begin{pmatrix} 363.7011 & -17.5074 \\ -17.5074 & 405.3836 \end{pmatrix}, \quad \rho_0 = 1371.7.$$

Based on Theorem 5.7, we conclude that the closed-loop system under controller (3.27) with $K = \begin{pmatrix} -44.9748 & 17.8217 \end{pmatrix}$ is stable and has noise attenuation level $\gamma = 37.0362$.

Remark 5.4 *In the previous section, we assumed complete access to the state vector or to the output vector and we designed the state feedback controller or the output feedback controller that stabilizes the class of systems we are dealing with. In the next section, we will restrict ourselves to unforced time delay systems and see how we can design a filter that estimates for instance the state vector that we can use later on to stabilize an unstable dynamical time delay system.*

5.4 $\mathcal{H}_\infty$-Filtering

The filtering problem is one of the control problems. It consists of determining an estimate $\hat{z}(t)$ of the controlled output $z(t)$ in some sense using the measurement of the measured output $y(t)$. $\mathcal{H}_\infty$ filtering is one of the approaches that we can use to solve our filtering problem. The problem can be stated as follows: Given a dynamical time delay system with exogenous input that belongs to $\mathcal{L}_2[0, \infty]$ and the measurement of the output, design a filter that gives an estimate of the unmeasured output such that the $\mathcal{H}_\infty$-norm of the transfer function between the estimation error, $e(t) = z(t) - \hat{z}(t)$ and the exogenous input, $w(t)$ is minimized or at least kept below a certain given $\gamma > 0$. Mathematically, it is stated as follows: given $\gamma > 0$, find a filter such that

$$\sup_{w \in \mathcal{L}_2[0,\infty]} \frac{\|z(t) - \hat{z}(t)\|_2^2}{\|w(t)_2^2} < \gamma^2.$$

Therefore, the $\mathcal{H}_\infty$-filtering problem can be regarded as a special $\mathcal{H}_\infty$ control problem.

Our goal in this section is to show how we can solve the filtering problem. Mainly, we establish delay-independent and delay-dependent conditions that we will use to design the filter that estimates the state vector for control feedback.

We treat the design of a linear time-invariant filter of order n for system (5.1) with $u(t) \equiv 0$. Consider the filter of the form

$$\begin{cases} \dot{\xi}_t = K_A \xi_t + K_B y(t) \\ \xi_0 = x_0, \ \xi_s = 0, s \in [-\tau, 0) \\ \hat{z}(t) = K_C \xi_t, \end{cases} \quad (5.42)$$

where ξ_t is the filter state, $y(t)$ is the output vector, $\hat{z}(t)$ is the estimation output vector, and K_A, K_B, and K_C are design matrices.

This filter should make the extended system $\{(x(t), \xi_t), t \geq 0\}$ asymptotically stable and the estimation error $e_t = z(t) - \hat{z}(t)$ should satisfy the condition

$$\|e\|_2 \leq \gamma \|w\|_2 \quad (5.43)$$

Let $\eta(t) = \begin{pmatrix} x(t) \\ \xi_t \end{pmatrix}$ denote the extended state. Then, this extended state satisfies the following dynamic

$$\begin{cases} \dot{\eta}(t) = \tilde{A}\eta(t) + \tilde{A}_d x(t - \tau) + \tilde{B}_1 w(t) \\ \eta(0) = 0, \end{cases} \quad (5.44)$$

where

$$\tilde{A} = \begin{pmatrix} A & 0 \\ K_B C_2 & K_A \end{pmatrix} \quad \tilde{A}_d = \begin{pmatrix} A_d \\ 0 \end{pmatrix} \quad \tilde{B}_1 = \begin{pmatrix} B_1 \\ K_B D_{21} \end{pmatrix}.$$

The filter error $e(t)$ can be represented as

$$e(t) = \tilde{L}\eta(t) + \tilde{T}w(t), \quad (5.45)$$

where

$$\tilde{L} = (C_1 \ - K_C) \quad \tilde{T} = D_{11}.$$

5.4.1 Delay-Independent $\mathcal{H}_\infty$ Filtering

Using Theorem 5.1, we conclude that if there exist symmetric, positive-definite matrices $0 < P \in \mathbb{R}^{2n \times 2n}$ and $Q > 0$ such that

$$\begin{pmatrix} \tilde{A}^\top P + P\tilde{A} + I_0^\top Q I_0 & P\tilde{B}_1 & \tilde{L}^\top & P\tilde{A}_d \\ \tilde{B}_1^\top P & -\gamma^2 I & \tilde{T}^\top & 0 \\ \tilde{L} & \tilde{T} & -I & 0 \\ \tilde{A}_d^\top P & 0 & 0 & -Q \end{pmatrix} < 0,$$

that is,

$$\begin{pmatrix} \tilde{A}^\top P + P\tilde{A} & P\tilde{B}_1 & \tilde{L}^\top & P\tilde{A}_d & I_0 \\ \tilde{B}_1^\top P & -\gamma^2 I & \tilde{T}^\top & 0 & 0 \\ \tilde{L} & \tilde{T} & -I & 0 & 0 \\ \tilde{A}_d^\top P & 0 & 0 & -Q & 0 \\ I_0^\top & 0 & 0 & 0 & -Q^{-1} \end{pmatrix} < 0 \quad (5.46)$$

with $I_0 = (I, 0)$, then the extended dynamic (5.44) is stable and satisfies (5.43).

Condition (5.46) provides a sufficient condition for the estimation $\hat{z}(t)$ to satisfy (5.43). However, (5.46) is obviously nonlinear in P and the design parameters K_A, K_B, K_C. To cast the filter design problem into the framework of LMI, let us suppose that P and P^{-1} can be partitioned as in (3.47) with M, N invertible, and let us define matrices Φ_1, Φ_2 as in (3.55). Then, direct matrix manipulation yields

$$P\Phi_1 = \Phi_2$$

and

$$\Phi_1^\top P\Phi_1 = \Phi_1^\top \Phi_2 = \begin{pmatrix} X & I \\ I & Y \end{pmatrix}.$$

If we define matrices $\mathcal{A}, \mathcal{B}$, and $\mathcal{C}$ by

$$\begin{cases} \mathcal{A} = YAX + NK_BC_2X + NK_AM^\top \\ \mathcal{B} = NK_B \\ \mathcal{C} = K_CM^\top, \end{cases} \tag{5.47}$$

then direct manipulation yields

$$\Phi_1^\top P\tilde{A}\Phi_1 = \Phi_2^\top \tilde{A}\Phi_1 = \begin{pmatrix} AX & A \\ \begin{bmatrix} YAX + NK_BC_2X \\ +NK_AM^\top \end{bmatrix} & YA + NK_BC_2 \end{pmatrix}$$

$$= \begin{pmatrix} AX & A \\ \mathcal{A} & YA + \mathcal{B}C_2 \end{pmatrix}.$$

From this it follows that

$$\Phi_1^\top[P\tilde{A} + \tilde{A}^\top P]\Phi_1 = \begin{pmatrix} AX + XA^\top & A + \mathcal{A}^\top \\ \mathcal{A} + A^\top & \begin{bmatrix} YA + A^\top Y^\top + \\ \mathcal{B}C_2 + C_2^\top \mathcal{B}^\top \end{bmatrix} \end{pmatrix} \overset{\triangle}{=} J_{11}$$

$$\Phi_1^\top P\tilde{B} = \Phi_2^\top \begin{pmatrix} B_1 \\ K_BD_{21} \end{pmatrix} = \begin{pmatrix} I & 0 \\ Y & N \end{pmatrix}\begin{pmatrix} B_1 \\ K_BD_{21} \end{pmatrix}$$

$$= \begin{pmatrix} B_1 \\ YB_1 + NK_BD_{21} \end{pmatrix} = \begin{pmatrix} B_1 \\ YB_1 + \mathcal{B}D_{21} \end{pmatrix} \overset{\triangle}{=} J_{12}$$

$$\Phi_1^\top \tilde{L}^\top = \begin{pmatrix} X & M \\ I & 0 \end{pmatrix}\begin{pmatrix} C_1^\top \\ -K_C^\top \end{pmatrix}$$

$$= \begin{pmatrix} XC_1^\top - MK_C^\top \\ C_1^\top \end{pmatrix} = \begin{pmatrix} XC_1^\top - \mathcal{C}^\top \\ C_1^\top \end{pmatrix} \overset{\triangle}{=} J_{13}$$

$$\Phi_1 P\bar{A}_d = \Phi_2^\top \bar{A}_d = \begin{pmatrix} I & 0 \\ Y & N \end{pmatrix}\begin{pmatrix} A_d \\ 0 \end{pmatrix}$$

$$= \begin{pmatrix} A_d \\ YA_d \end{pmatrix} \triangleq J_{14}$$

$$\Phi_1^\top I_0^\top = \begin{pmatrix} X \\ I \end{pmatrix} \triangleq J_{15}.$$

Due to the invertibility of M and N, matrices Φ_1 and Φ_2 are non-singular. Pre- and post-multiplying (5.46) by $\operatorname{diag}\{\Phi_1, I, I, I, I\}$ and $\operatorname{diag}\{\Phi_1^\top, I, I, I, I\}$ yields

$$\begin{pmatrix} J_{11} & J_{12} & J_{13} & J_{14} & J_{15} \\ J_{12}^\top & -\gamma^2 I & \tilde{T}^\top & 0 & 0 \\ J_{13}^\top & \tilde{T} & -I & 0 & 0 \\ J_{14}^\top & 0 & 0 & -Q & 0 \\ J_{15}^\top & 0 & 0 & 0 & -Q^{-1} \end{pmatrix} < 0. \tag{5.48}$$

Given positive-definite matrices X and Y and invertible matrices M and N, the gain matrices K_A, K_B, K_C can be uniquely determined by solving (5.47), that is,

$$\begin{cases} K_A = N^{-1} \left[\mathcal{A} - YAX - \mathcal{B}C_2 X \right] \left[M^\top \right]^{-1} \\ K_B = \mathcal{B}N^{-1} \\ K_C = \mathcal{C} \left[M^\top \right]^{-1}. \end{cases} \tag{5.49}$$

The main result of this section will be given in terms of the solution $X, Y, \mathcal{A}, \mathcal{B}, \mathcal{C}$ of (5.48) and

$$\begin{pmatrix} X & I \\ I & Y \end{pmatrix} > 0. \tag{5.50}$$

From above discussion, we can design a filter using the following procedure.

Algorithm 5.2 *(Filter Design Algorithm)*
Step 1. Solve the feasible problem (5.48) and (5.50) to get $X > 0, Y > 0$, and $\mathcal{A}, \mathcal{B}, \mathcal{C}$.
Step 2. Compute two invertible matrices M and N using the singular value decomposition of $(I - XY)$ and define the matrices Φ_1 and Φ_2 as in (3.55). Then, the matrix P is given by $P = \Phi_2 \Phi_1^{-1}$.
Step 3. Compute the gains $K_A, K_B,$ and K_C by solving (5.49).

Therefore, we get the following theorem.

Theorem 5.8   *(i) If there exist symmetric, positive-definite matrices $P > 0, K_A, K_B,$ and K_C satisfying (5.46), which implies that the extended system under filter (5.42) is asymptotically stable and the estimation error satisfies (5.43), then LMIs (5.48) and (5.50) have a set of feasible solutions $X > 0, Y > 0, \mathcal{A}, \mathcal{B},$ and $\mathcal{C}$.*

(ii) On the other hand, if LMIs (5.48) and (5.50) have a set of feasible solutions $X > 0, Y > 0, \mathcal{A}, \mathcal{B},$ and $\mathcal{C}$, then the extended system under

filter (5.42) with K_A, K_B, and K_C obtained using Algorithm 5.2 is asymptotically stable and the estimation error satisfies (5.43).

Proof: The proof of (i) follows directly from the above derivation. Now, we come to proof of (ii). Suppose that $X > 0, Y > 0, \mathcal{A}, \mathcal{B}$, and $\mathcal{C}$ are a set of solutions of (5.48) and (5.50). Note that for given $X > 0$ and $Y > 0$, there exists a matrix P satisfying (3.47) if and only if $X - Y^{-1} > 0$, which is ensured by (5.50). Let $P > 0, K_A, K_B$, and K_C be obtained from Algorithm 5.2. Then, it is easy to check that they satisfy (5.46). This completes the proof of (ii). $\qquad\square$

The noise attenuation level of the error system can be optimized by replacing the *Step 1* of Algorithm 5.2 with the following

Step 1': Solve the linear objective function minimization problem

$$\min_{\gamma>0,X>0,Y>0,\mathcal{A},\mathcal{B},\mathcal{C}} \gamma \qquad (5.51)$$

$$\text{s.t. (5.48) and (5.50)} \qquad (5.52)$$

to get $X > 0, Y > 0$ and $\mathcal{A}, \mathcal{B}, \mathcal{C}$.

Let us now consider a numerical example to show how we design a dynamical filter as it is proposed in this section.

Example 5.4 *Let us consider a dynamical time delay system as described by (5.1) with $u(t) \equiv 0$ and suppose that the system has the following data:*

$$A = \begin{pmatrix} -0.5 & 1 \\ -1 & -0.4 \end{pmatrix} \quad A_d = \begin{pmatrix} 0.2 & 0.1 \\ 0.3 & 0.1 \end{pmatrix} \quad B_1 = \begin{pmatrix} 1 \\ 0 \end{pmatrix}$$

$$B = \begin{pmatrix} 1 \\ 1 \end{pmatrix} \quad C_1 = \begin{pmatrix} 1 & 0 \\ 0 & 1 \\ 1 & 0 \end{pmatrix} \quad C_2 = \begin{pmatrix} 0 & 0.3 \end{pmatrix}$$

$$D_{11} = \begin{pmatrix} 0 \\ 1 \\ 0 \end{pmatrix} \quad D_{12} = \begin{pmatrix} 0 \\ 0 \\ 1 \end{pmatrix} \quad D_{21} = 1.$$

With this set of data, solving LMIs (5.52) gives the following solution:

$$X = \begin{pmatrix} 0.5551 & 0.0380 \\ 0.0380 & 0.4979 \end{pmatrix} \quad Y = \begin{pmatrix} 13.3534 & -0.3218 \\ -0.3218 & 23.7329 \end{pmatrix}$$

$$\mathcal{A} = \begin{pmatrix} -0.1718 & 0.0137 \\ 0.0137 & -2.0121 \end{pmatrix} \quad \mathcal{B} = \begin{pmatrix} -35.9102 \\ -467.1026 \end{pmatrix}$$

$$\mathcal{C} = \begin{pmatrix} 0.8878 & 0.5340 \\ -0.4546 & 0.2019 \\ 0.0600 & -0.4220 \end{pmatrix} \quad \gamma = 18.3111.$$

Using the matrices X, Y and the singular value decomposition of $(I - XY)$, we get the invertible matrices

$$M = \begin{pmatrix} -1.4098 & -6.2842 \\ -10.7780 & 0.8220 \end{pmatrix} \quad N = \begin{pmatrix} 0.1080 & 0.9941 \\ 0.9941 & -0.1080 \end{pmatrix}.$$

Solving equations (5.47), we obtain

$$K_A = \begin{pmatrix} -7.4692 & -0.3813 \\ -0.3981 & -0.6905 \end{pmatrix} \quad K_B = \begin{pmatrix} -468.2487 \\ 14.7516 \end{pmatrix}$$

$$K_C = \begin{pmatrix} -0.0941 & -0.0701 \\ -0.0262 & 0.0456 \\ 0.0653 & -0.0141 \end{pmatrix}.$$

Remark 5.5 *In this section, we have developed a delay-independent design algorithm $\mathcal{H}_\infty$ filter. As we said previously, the results of this algorithm are conservative. In the next subsection, we will overcome this conservatism by designing a delay-dependent $\mathcal{H}_\infty$ filter.*

5.4.2 Delay-Dependent $\mathcal{H}_\infty$ Filtering

This section deals with the delay-dependent $\mathcal{H}_\infty$-filtering problem, that is, design a delay-dependent filter (5.42) such that the system error (5.45) satisfies (5.43).

Two delay-dependent $\mathcal{H}_\infty$ filter design algorithms are proposed in this section. To get the first one, the following lemma is used.

Lemma 5.1 *Let K_A, K_B, K_C be given matrices and a given positive scalar. If there exist symmetric, positive-definite matrices $P > 0$, $Q_i > 0$, $i = 1, 2, 3$ satisfying*

$$\left(\begin{bmatrix} P[\tilde{A} + \tilde{A}_d I_0] \\ + [\tilde{A} + \tilde{A}_d I_0]^\top P \\ + \tau I_0^\top Q_1 I_0 + \tau I_0^\top Q_2 I_0 \end{bmatrix} \quad P\tilde{B}_1 \quad \tilde{L}^\top \quad \mathcal{T} \atop \begin{matrix} \tilde{B}_1^\top P & -\gamma^2 I + \tau Q_3 & D_{11}^\top & 0 \\ \tilde{L} & D_{11} & -I & 0 \\ \mathcal{T}^\top & 0 & 0 & -\frac{1}{\tau}\mathcal{Q} \end{matrix} \right) < 0 \quad (5.53)$$

where

$$\mathcal{T} = \begin{pmatrix} P\tilde{A}_d A & P\tilde{A}_d A_d & P\tilde{A}_d B_1 \end{pmatrix}$$

$$\mathcal{Q} = \begin{pmatrix} Q_1 & & \\ & Q_2 & \\ & & Q_3 \end{pmatrix}$$

then the error system $e(t)$ is asymptotically stable and satisfies (5.43).

Proof: Substituting

$$x(t - \tau) = x(t) - \int_{t-\tau}^t [Ax(s) + A_d\eta(s - \tau) + B_1 w(s)]ds$$

into (5.44) leads to the dynamic

$$\begin{cases} \dot{\eta}(t) = [\tilde{A} + \tilde{A}_d I_0]\eta(t) - \tilde{A}_d \int_{t-\tau}^t [Ax(s) + A_d x(s - \tau) \\ \qquad + B_1 w(s)]ds + \tilde{B}_1 w(t) \\ \eta(s) = 0, \ w(s) = 0, \ s \in [-2\tau, 0]. \end{cases} \tag{5.54}$$

Let us define $\hat{\eta}_t \in C[-2\tau, 0]$, with $\hat{\eta}_t(s) = \eta(t + s), s \in [-2\tau, 0]$, and let us consider a Lyapunov functional candidate as

$$V(\hat{\eta}_t) = V_0(\hat{\eta}(t)) + V_1(\hat{\eta}_t) \tag{5.55}$$

where

$$V_0(\hat{\eta}(t)) = \eta^\top(t)P\eta(t)$$

$$V_1(\hat{\eta}(t)) = \int_{-\tau}^0 \int_{t+\theta}^t x^\top(s)Q_1 x(s)dsd\theta$$

$$+ \int_{-\tau}^t \int_{t-\tau+\theta}^t x^\top(s)Q_2 x(s)dsd\theta$$

$$+ \int_{-\tau}^0 \int_{t+\theta}^t w^\top(s)Q_3 w(s)dsd\theta.$$

In view of (5.54), we obtain

$$\dot{V}_0(\eta(t)) = \eta^\top(t)\left[P[\tilde{A} + \tilde{A}_d I_0] + [\tilde{A} + \tilde{A}_d I_0]^\top P\right]\eta(t)$$

$$- 2\eta^\top(t)P\tilde{A}_d\left[A \int_{t-\tau}^t x(s)ds + A_d \int_{t-\tau}^t x(s - \tau)ds\right.$$

$$\left. + B_1 \int_{t-\tau}^t w(s)ds\right] + 2\eta^\top(t)P\tilde{B}_1 w(t). \tag{5.56}$$

Using Lemma A.1, we obtain, respectively,

$$-2\eta^\top(t)P\tilde{A}_d A \int_{t-\tau}^t x(s)ds \le \tau\eta^\top(t)[P\tilde{A}_d A Q_1^{-1} A^\top \tilde{A}_d^\top P]\eta(t)$$

$$+ \int_{t-\tau}^t x^\top(s)Q_1 x(s)ds \tag{5.57}$$

$$-2\eta^\top(t)P\tilde{A}_d A_d \int_{t-\tau}^t x(s - \tau)ds \le \tau\eta^\top(t)P\tilde{A}_d A_d Q_2^{-1} A_d^\top \tilde{A}_d^\top P\eta(t)$$

$$+ \int_{t-\tau}^t x^\top(s - \tau)Q_2 x(s - \tau)ds, \tag{5.58}$$

$$-2\eta^\top(t)P\tilde{A}_d B_1 \int_{t-\tau}^t w(s)ds \leq \tau\eta^\top(t)P\tilde{A}_d B_1 Q_3^{-1} B_1^\top \tilde{A}_d^\top P\eta(t)$$

$$+ \int_{t-\tau}^t w^\top(s)Q_3 w(s)ds. \qquad (5.59)$$

Combining (5.56)-(5.59) yields

$$\dot{V}_0(\hat{\eta}_t) \leq \eta^\top(t)\left[P[\tilde{A}+\tilde{A}_d I_0] + [\tilde{A}+\tilde{A}_d I_0]^\top P\right]\eta(t)$$
$$+ \tau\eta^\top(t)[P\tilde{A}_d A Q_1^{-1} A^\top \tilde{A}_d^\top P]\eta(t)$$
$$+ \int_{t-\tau}^t x^\top(s)Q_1 x(s)ds + \tau\eta^\top(t)P\tilde{A}_d A_d Q_2^{-1} A_d^\top \tilde{A}_d^\top P\eta(t)$$
$$+ \int_{t-\tau}^t x^\top(s-\tau)Q_2 x(s-\tau)ds$$
$$+ \tau\eta^\top(t)P\tilde{A}_d B_1 Q_3^{-1} B_1^\top \tilde{A}_d^\top P\eta(t)$$
$$+ \int_{t-\tau}^t w^\top(s)Q_3 W(s)ds + 2\eta^\top(t)P\tilde{B}_1 w(t). \qquad (5.60)$$

Simple computation gives

$$\dot{V}_1(\hat{\eta}_t) = -\int_{t-\tau}^t x^\top(s)Q_1 x(s)ds + \tau x^\top(t)Q_1 x(t)$$
$$- \int_{t-\tau}^t x^\top(s-\tau)Q_2 x(s-\tau)ds + \tau x^\top(t)Q_2 x(t)$$
$$- \int_{t-\tau}^t w^\top(s)Q_3 w(s)ds + \tau w^\top(t)Q_3 w(t). \qquad (5.61)$$

Combining now (5.60) and (5.61), we obtain finally the following:

$$\dot{V}(\hat{\eta}_t) \leq \eta^\top(t)\left[P[\tilde{A}+\tilde{A}_d I_0] + [\tilde{A}+\tilde{A}_d I_0]^\top P\right]\eta(t)$$
$$+ \tau\eta^\top(t)[P\tilde{A}_d A Q_1^{-1} A^\top \tilde{A}_d^\top P]\eta(t)$$
$$+ \tau\eta^\top(t)P\tilde{A}_d A_d Q_2^{-1} A_d^\top \tilde{A}_d^\top P\eta(t)$$
$$+ \tau\eta^\top(t)P\tilde{A}_d B_1 Q_3^{-1} B_1^\top \tilde{A}_d^\top P\eta(t) + 2\eta^\top(t)P\tilde{B}_1 w(t)$$
$$+ \tau x^\top(t)Q_1 x(t) + \tau x^\top(t)Q_2 x(t) + \tau w^\top(t)Q_3 w(t) \qquad (5.62)$$

The proof of stability follows the same steps of the one of Theorem 5.1. The details are omitted.

Now, let us define an $\mathcal{H}_\infty$ performance J_T as follows:

$$J_T = \int_0^T \left[e^\top(t)e(t) - \gamma^2 w^\top(t)w(t)\right] dt.$$

Then,

$$J_T = \int_0^T [e^\top(t)e(t) - \gamma^2 w^\top(t)w(t)]dt$$

$$= \int_0^T [e^\top(t)e(t) - \gamma^2 w^\top(t)w(t) + \dot{V}(\hat{\eta}(t)]dt - \int_0^T \dot{V}(\hat{\eta}(t)dt$$

$$\leq \int_0^T \begin{pmatrix} \eta^\top(t) & w^\top(t) \end{pmatrix} \hat{\Theta} \begin{pmatrix} \eta(t) \\ w(t) \end{pmatrix} dt - V(\hat{\eta}(T))$$

$$\leq \int_0^T \begin{pmatrix} \eta^\top(t) & w^\top(t) \end{pmatrix} \hat{\Theta} \begin{pmatrix} \eta(t) \\ w(t) \end{pmatrix} dt$$

where

$$\hat{\Theta} = \begin{pmatrix} \# & P\tilde{B}_1 + \tilde{L}^\top\tilde{T} \\ \tilde{B}_1^\top P + \tilde{T}\tilde{L} & -\gamma^2 I + \tau Q_3 + \tilde{T}^\top\tilde{T} \end{pmatrix}$$

with

$$\# = P[\tilde{A} + \tilde{A}_d I_0] + [\tilde{A} + \tilde{A}_d I_0]^\top P + \tau I_0^\top Q_1 I_0$$
$$+ \tau I_0^\top Q_2 I_0 + \tau P\tilde{A}_d A Q_1^{-1} A \tilde{A}_d P + \tau P\tilde{A}_d A_d Q_2^{-1} A_d^\top A_d^\top P$$
$$+ \tau P\tilde{A}_d B_1 Q_3^{-1} B_1^\top \tilde{A}_d^\top P + \tilde{L}^\top\tilde{L}.$$

Using now the Schur complement, it is easy to check that (5.53) implies $\hat{\Theta} < 0$ and consequently

$$J_T < 0 \quad \forall T > 0,$$

from which it follows that

$$J_\infty = \int_0^\infty e^\top(s)e(s) - \gamma^2 w^\top(s)w(s)ds < 0,$$

that is, $\|e\|_2 \leq \gamma\|w\|_2$. This completes the proof of Lemma 5.1.   □

Let us now see how we can design the filter gains. A way to do this is given below.

Using the Schur complement, (5.53) is equivalent to

$$\begin{pmatrix} \begin{bmatrix} P[\tilde{A} + \tilde{A}_d I_0] \\ + [\tilde{A} + \tilde{A}_d I_0]^\top P \end{bmatrix} & P\tilde{B}_1 & \tilde{L}^\top & \mathcal{T}_1 \\ \tilde{B}_1^\top P & -\gamma^2 I + \tau Q_3 & D_{11}^\top & 0 \\ \tilde{L} & D_{11} & -I & 0 \\ \mathcal{T}_1^\top & 0 & 0 & -\frac{1}{\tau}\mathcal{Q}_1 \end{pmatrix} < 0, \quad (5.63)$$

where

$$\mathcal{T}_1 = \begin{pmatrix} P\tilde{A}_d A & P\tilde{A}_d A_d & P\tilde{A}_d B_1 & I_0^\top & I_0^\top \end{pmatrix}$$
$$\mathcal{Q}_1 = \text{diag}\{Q_1, Q_2, Q_3, Q_1^{-1}, Q_2^{-1}\}.$$

Assume now that P and P^{-1} can be decomposed as in (3.47). Let the matrices Φ_1 and Φ_2 be defined by (3.55). Define the matrices $\mathcal{A}, \mathcal{B}, \mathcal{C}$ as follows:

$$\begin{cases} \mathcal{A} = Y[A + A_d]X + NK_BC_2X + NK_AM^\top \\ \mathcal{B} = NK_B \\ \mathcal{C} = K_CM^\top. \end{cases} \tag{5.64}$$

Direct manipulation gives

$$\Phi_1^\top P[\tilde{A} + \tilde{A}_1 I_0]\Phi_1 = \Phi_2^\top [\tilde{A} + \tilde{A}_1 I_0]\Phi_1$$
$$= \begin{pmatrix} \begin{bmatrix} Y[A + A_d]X + NK_BC_2X \\ + NK_AM^\top \end{bmatrix} & [A + A_d]X & A + A_d \\ & Y[A + A_d] + NK_BC_2 \end{pmatrix}$$
$$= \begin{pmatrix} [A + A_d]X & A + A_d \\ \mathcal{A} & Y[A + A_d] + \mathcal{B}C_2 \end{pmatrix},$$

from which it follows that

$$\Phi_1^\top P[\tilde{A} + \tilde{A}_1 I_0]\Phi_1 + \Phi_1^\top [\tilde{A} + \tilde{A}_1 I_0]^\top \Phi_1$$
$$= \begin{pmatrix} \begin{bmatrix} [A + A_d]X \\ + X[A + A_d]^\top \end{bmatrix} & A + A_d + \mathcal{A}^\top \\ A^\top + A_d^\top + \mathcal{A} & \begin{bmatrix} Y[A + A_d] \\ + [A + A_d]^\top P \\ + \mathcal{B}C_2 + C_2^\top \mathcal{B}^\top \end{bmatrix} \end{pmatrix} \triangleq T_{11}$$

$$\Phi_1^\top P\tilde{B}_1 = \Phi_2^\top \tilde{B}_1 = \begin{pmatrix} I & \mathbf{0} \\ Y & N \end{pmatrix} \begin{pmatrix} B_1 \\ K_BD_{21} \end{pmatrix}$$
$$= \begin{pmatrix} B_1 \\ YB_1 + NK_BD_{21} \end{pmatrix} = \begin{pmatrix} B_1 \\ YB_1 + \mathcal{B}D_{21} \end{pmatrix} \triangleq T_{12}$$

$$\Phi_1^\top \tilde{L}^\top = \begin{pmatrix} X & M \\ I & \mathbf{0} \end{pmatrix} \begin{pmatrix} C_1^\top \\ -K_C^\top \end{pmatrix}$$
$$= \begin{pmatrix} XC_1^\top - MK_C^\top \\ C_1^\top \end{pmatrix} = \begin{pmatrix} XC_1^\top - \mathcal{C}^\top \\ C_1^\top \end{pmatrix} \triangleq T_{13},$$

$$\Phi_1^\top \mathcal{T}_1 = \Phi_1^\top \begin{pmatrix} P\tilde{A}_dA & P\tilde{A}_dA_d & P\tilde{A}_dB_1 & I_0^\top & I_0^\top \end{pmatrix}$$
$$= \begin{pmatrix} A_dA & A_dA_d & A_dB_1 & X & X \\ YA_dA & YA_dA_d & YA_dB_1 & I & I \end{pmatrix} \triangleq T_{14}.$$

Therefore, pre- and post-multiplying (5.53) by $\mathrm{diag}\{\Phi_1^\top, I, I, I\}$ and $\mathrm{diag}\{\Phi_1, I, I, I\}$, respectively, yields

$$\begin{pmatrix} T_{11} & T_{12} & T_{13} & T_{14} \\ T_{12}^\top & -\gamma^2 I + \tau Q_3 & D_{11}^\top & \mathbf{0} \\ T_{13}^\top & D_{11} & -I & \mathbf{0} \\ T_{14}^\top & \mathbf{0} & \mathbf{0} & -\frac{1}{\tau}Q_1 \end{pmatrix} < 0. \tag{5.65}$$

Thus, if there exist symmetric, positive-definite matrices $X > 0$, $Y > 0$, and matrices $\mathcal{A}$, $\mathcal{B}$, $\mathcal{C}$ that satisfy (5.65), then, using singular value decomposition, we can get two matrices M, N satisfying

$$MN^\top = I - XY. \tag{5.66}$$

With $X > 0$, $Y > 0$, $\mathcal{A}$, $\mathcal{B}$, $\mathcal{C}$ given, solving (5.64) yields

$$\begin{cases} K_A = N^{-1} \left[\mathcal{A} - YAX - YA_dX - \mathcal{B}C_2X\right] \left[M^\top\right]^{-1} \\ K_B = N^{-1}\mathcal{B} \\ K_C = \mathcal{C}\left[M^{-1}\right]^\top. \end{cases} \tag{5.67}$$

Then direct computation gives

$$P\Phi_1 = \Phi_2 \tag{5.68}$$

and

$$\Phi_1^\top P \Phi_1 = \begin{pmatrix} X & I \\ I & Y \end{pmatrix} > 0. \tag{5.69}$$

From the above discussion, we obtain the following theorem that gives the first algorithm, which can be used to design a $\mathcal{H}_\infty$ filter.

Theorem 5.9 *Let γ be a given positive scalar. If there exist symmetric, positive-definite matrices $X > 0$, $Y > 0$, and matrices $\mathcal{A}$, $\mathcal{B}$, $\mathcal{C}$ satisfying (5.65) and (5.69), then there exist matrices K_A, K_B, K_C such that the error system of filter (5.42) is stable and satisfies (5.43).*

Proof: Let $X > 0, Y > 0, \mathcal{A}, \mathcal{B}, \mathcal{C}$ satisfy (5.65) and (5.69). For given matrices $X > 0, Y > 0$, there exists a symmetric, positive-definite matrix $P > 0$ satisfying (3.47) if and only if $X - Y^{-1} > 0$ holds, which is equivalent to (5.69). Let M, N be two full-column rank matrices obtained using the singular value decomposition of $(I - XY)$. Then, define Φ_1, Φ_2 as in (3.55) and solve (5.68) to get P. It is easy to check that K_A, K_B, K_C defined by (5.67) and P satisfy (5.53). Therefore, the proof of Theorem 5.53 follows from Lemma 5.1. $\qquad\square$

Let us now consider a numerical example to show the usefulness of the proposed first algorithm.

Example 5.5 *Let us consider a system described by (5.1) with $u(t) \equiv 0$ and suppose that the system has the following data:*

$$A = \begin{pmatrix} -1.5 & 1 \\ 1 & -2.3 \end{pmatrix} \quad A_d = \begin{pmatrix} 0.2 & 0.1 \\ 0.3 & 0.1 \end{pmatrix} \quad B_1 = \begin{pmatrix} 1 \\ 0 \end{pmatrix}$$

$$B = \begin{pmatrix} 1 \\ 1 \end{pmatrix} \quad C_1 = \begin{pmatrix} 0 & 1 \\ 1 & 1 \end{pmatrix} \quad C_2 = \begin{pmatrix} 0 & 3 \end{pmatrix}$$

$$D_{11} = \begin{pmatrix} 0 \\ 1 \end{pmatrix} \quad D_{12} = \begin{pmatrix} 0 \\ 0 \\ 1 \end{pmatrix} \quad D_{21} = 1.$$

With these data and letting $Q_1 = Q_2 = I$, solving (5.65) with γ^2 replaced with v and (5.69) at $\tau = 1.3$ yields a set of feasible solutions as follows:

$$X = \begin{pmatrix} 45.2895 & 18.0738 \\ 18.0738 & 33.3588 \end{pmatrix} \quad Y = \begin{pmatrix} 0.7619 & -0.5998 \\ -0.5998 & 4.6584 \end{pmatrix}$$

$$v = 149.4133 \qquad\qquad Q_3 = 68.9378$$

$$\mathcal{A} = \begin{pmatrix} -0.7132 & 0.7420 \\ -0.1743 & -8.5942 \end{pmatrix} \quad \mathcal{B} = \begin{pmatrix} -0.6716 \\ -15.6292 \end{pmatrix}$$

$$\mathcal{C} = \begin{pmatrix} 18.0620 & 33.4127 \\ 63.3885 & 51.4149 \end{pmatrix}.$$

Using singular value decomposition, we obtain the following matrices M, N, which satisfy (5.66):

$$N = \begin{pmatrix} 0.0170 & 0.9999 \\ 0.9999 & -0.0170 \end{pmatrix} \quad M = \begin{pmatrix} -57.4091 & -21.6914 \\ -143.4304 & 8.6822 \end{pmatrix}.$$

Substituting $X, Y, \mathcal{A}, \mathcal{B}, \mathcal{C}$ and M, N into (5.67), we obtain

$$K_A = \begin{pmatrix} -14.5053 & 3.2627 \\ 0.0833 & -4.9769 \end{pmatrix} \quad K_B = \begin{pmatrix} -15.6384 \\ -0.4052 \end{pmatrix}$$

$$K_C = \begin{pmatrix} -0.2442 & -0.1863 \\ -0.4614 & -1.7010 \end{pmatrix}.$$

According to Theorem 5.9, the error system of filter (5.42) satisfies (5.43) with $\gamma = \sqrt{v} = 12.2235$.

Let us now try to develop another design algorithm for delay-dependent $\mathcal{H}_\infty$-filter. For this purpose, using Theorem 5.5, we obtain the following proposition.

Proposition 5.1 *Let γ be a given positive scalar. If there exist symmetric, positive-definite matrices $P > 0$, $Q > 0$ satisfying*

$$\tilde{\Theta}_d = \begin{pmatrix} \# & P\tilde{B}_1 + \tilde{L}^\top D_{11} \\ \tilde{B}_1^\top P + D_{11}^\top \tilde{L} & \begin{bmatrix} \tau \tilde{B}_1^\top P \tilde{B}_1 \\ + D_{11}^\top D_{11} - \gamma^2 I \end{bmatrix} \end{pmatrix} < 0, \qquad (5.70)$$

where

$$\# = P[\tilde{A} + A_d I_0] + [\tilde{A} + A_d I_0]^\top P + \tau P A_d P^{-1} A_d^\top P$$
$$+ \tau \tilde{A}^\top P \tilde{A} + \tau P A_d Q A_d^\top P + \tau A_d^\top Q^{-1} A_d$$
$$+ \tau P A_d P^{-1} A_d^\top P + \tilde{L}^\top \tilde{L},$$

then the estimation error of system (5.44) is asymptotically stable and satisfies (5.43).

Proof: From the proof of Theorem 5.5, we conclude that if there exist symmetric, positive-definite matrices $P > 0, Q_i > 0, 1 \leq i \leq 3$, satisfying

$$\begin{pmatrix} \# & PB_1 + \tilde{L}^\top D_{11} \\ \tilde{B}_1^\top P + D_{11}^\top \tilde{L} & \begin{bmatrix} \tau \tilde{B}_1^\top Q_3 \tilde{B}_1 \\ + D_{11}^\top D_{11} - \gamma^2 I \end{bmatrix} \end{pmatrix} < 0,$$

where

$$\begin{aligned} \# &= P[A + A_d] + [A + A_d]^\top P + \tau P A_d Q_1 A_d^\top P \\ &\quad + \tau A^\top Q_1^{-1} A + \tau P A_d Q_2 A_d^\top P + \tau A_d^\top Q_2^{-1} A_d \\ &\quad + \tau P A_d Q_3^{-1} A_d^\top P + \tilde{L}^\top \tilde{L}, \end{aligned}$$

then the estimation error of system (5.44) is asymptotically stable and satisfies (5.43). Letting $Q_1 = P^{-1}, Q_3 = P$, and $Q = Q_2$ in above inequality yields (5.70). This completes the proof of Proposition 5.1.   □

Using now the Schur complement, we obtain the result of the following theorem that gives the second algorithm design.

Theorem 5.10 *Let γ be a given positive scalar. If there exist symmetric, positive-definite matrices $X > 0$, $Y > 0$, and matrices $\mathcal{K}_A$, $\mathcal{K}_B$, and K_C satisfying*

$$\begin{pmatrix} J_{11} & J_{12} & J_{13} & 0 & J_{15} \\ J_{12}^\top & -\gamma^2 I & D_{11}^\top & J_{24} & 0 \\ J_{13}^\top & D_{11} & -I & 0 & 0 \\ 0 & J_{24}^\top & 0 & -J_{44} & 0 \\ J_{15}^\top & 0 & 0 & 0 & -J_{55} \end{pmatrix} < 0, \tag{5.71}$$

then, the error system of (5.44) with $K_A = Y^{-1}\mathcal{K}_A, K_B = Y^{-1}\mathcal{K}_B$, and K_C is asymptotically stable and satisfies (5.43), where

$$J_{11} = \begin{pmatrix} \begin{bmatrix} XA + A^\top X + XA_d \\ + A_d X + \tau A_d U A_d^\top \end{bmatrix} & C_2^\top \mathcal{K}_B \\ \mathcal{K}_B^\top C_2 & \mathcal{K}_A + \mathcal{K}_A^\top \end{pmatrix}$$

$$J_{12} = \begin{pmatrix} XB_1 \\ \mathcal{K}_B D_{21} \end{pmatrix}$$

$$J_{13} = \begin{pmatrix} C_1^\top \\ -K_C^\top \end{pmatrix}$$

$$J_{15} = \begin{pmatrix} XA_d & [A + A_d]^\top X & C_2^\top \mathcal{K}_B^\top & XA_d & XA_d \\ 0 & 0 & \mathcal{K}_A^\top & 0 & 0 \end{pmatrix}$$

$$J_{24} = \begin{pmatrix} B_1^\top X & D_{21}^\top \mathcal{K}_B^\top \end{pmatrix}$$

$$J_{44} = \begin{pmatrix} X & 0 \\ 0 & Y \end{pmatrix}$$

$$J_{55} = \frac{1}{\tau} \text{diag}\{P, P, U, P\}.$$

Proof: Suppose that $P > 0, Q > 0, K_A, K_B$, and K_C satisfy (5.70). Using the Schur complement, (5.70) is equivalent to

$$
\left(
\begin{array}{ccccc}
\begin{bmatrix} P[\tilde{A} + \tilde{A}_d I_0] \\ + [\tilde{A} + \tilde{A}_d I_0]^\top P \\ + \tau \tilde{A}_d Q^{-1} \tilde{A}_d^\top \end{bmatrix} & P\tilde{B}_1 & \tilde{L}^\top & 0 & T_{15} \\
\tilde{B}_1^\top P & -\gamma^2 I & D_{11}^\top & \tilde{B}_1^\top P & 0 \\
\tilde{L} & D_{11} & -I & 0 & 0 \\
0 & P\tilde{B}_1 & 0 & -\frac{1}{\tau}P & 0 \\
T_{15}^\top & 0 & 0 & 0 & -T_{55}
\end{array}
\right) < 0,
$$

where

$$
T_{15} = \left(\begin{array}{cccc} P\tilde{A}_d & \tilde{A}^\top P & P\tilde{A}_d & P\tilde{A}_d \end{array} \right)
$$
$$
T_{55} = \frac{1}{\tau}\mathrm{diag}\{P, P, Q^{-1}, P\}.
$$

Suppose $P = \left(\begin{array}{cc} X & 0 \\ 0 & Y \end{array} \right)$ and let $YK_B = \mathcal{K}_B, YK_A = \mathcal{K}_A$, and $U = Q^{-1}$. Then, direct manipulation gives

$$
P[\tilde{A} + \tilde{A}_d I_0] + [\tilde{A} + \tilde{A}_d I_0]^\top P + \tau \tilde{A}_d Q^{-1} \tilde{A}_d^\top = J_{11}
$$
$$
P\tilde{B}_1 = J_{12}
$$
$$
\tilde{L}^\top = J_{13}
$$
$$
\left(P\tilde{A}_d \quad \tilde{A}^\top P \quad P\tilde{A}_d \quad P\tilde{A}_d \right) = J_{15}
$$
$$
\tilde{B}_1^\top P = J_{24}
$$

which means that $X, Y, \mathcal{K}_A, \mathcal{K}_B, K_C$ satisfy (5.65). From the above derivation, we obtain that if $X, Y, \mathcal{K}_A, \mathcal{K}_B, K_C$ satisfy (5.65), then $P = \left(\begin{array}{cc} X & 0 \\ 0 & Y \end{array} \right)$, $K_A = Y^{-1}\mathcal{K}_A$, $K_B = Y^{-1}\mathcal{K}_B$, and K_C satisfy (5.70). Therefore, the proof follows from Proposition 5.1. $\qquad\square$

5.5 Notes

The literature on $\mathcal{H}_\infty$ filter design for continuous-time systems with time delays mainly based on the Riccati algebraic equation can be divided into two classes: delay-dependent (see [76, 167]) and delay-independent (see [164, 177]). The output $\mathcal{H}_\infty$ control of a linear system with time-varying delay was studied by [109].

Recently, there has been a noticeable increase of interest in the problem of $\mathcal{H}_\infty$ optimal control of time delayed systems. The infinite horizon time-invariant control problem with delayed control input is treated by [7, 116, 197]. $\mathcal{H}_\infty$-filtering for a time-varying system with time delayed measurement is addressed in [166].

6

Robust $\mathcal{H}_\infty$ Control, Filtering, and Guaranteed Cost Control

In the previous chapter, we developed algorithms that can be used to design $\mathcal{H}_\infty$ controllers and $\mathcal{H}_\infty$ filters for dynamical linear systems with time delay. Since these algorithms are based on nominal systems, there is no guarantee that the robustness of system performance will be assured in the presence of uncertainties. To overcome this and avoid any trouble we may have, we should take into account system uncertainties during the analysis and the design phase. Therefore, the problems we studied in the previous chapter should be extended to cope with system uncertainties. Here we will consider norm-bounded uncertainties and deal with the robust $\mathcal{H}_\infty$ control problem and the robust $\mathcal{H}_\infty$-filtering problem.

The robust $\mathcal{H}_\infty$ control problem consists of designing a suboptimal controller that minimizes the effects of the external disturbance $w(t)$ on the output $z(t)$ for all admissible uncertainties. In other words, the problem can be stated as follows: Given a dynamical time delay system with exogenous input that belongs to $\mathcal{L}_2[0, \infty]$, design a controller that minimizes the $\mathcal{H}_\infty$-norm of the transfer function between the controlled output and the external disturbance, or at least guarantees that it will not exceed a given level $\gamma > 0$ for all admissible uncertainties. We are also interested in designing a controller that stabilizes the class of systems we are considering and at the same time guarantees that a chosen cost remains bounded by a certain given constant for all admissible uncertainties. This control problem is referred to as the guaranteed control problem.

The design of the state feedback controller requires access to the state of the system we consider at each time t. Since this is not always possible due to certain practical reasons and the cost of sensors, it is preferable to resort to filtering to get a good estimate for the state vector at each time t. Different techniques have been proposed in the literature; $\mathcal{H}_\infty$ theory is one of the approaches we can use to solve this filtering problem. In the case of the robust filtering problem, the technique can be stated as follows: Given an uncertain dynamical time delay system with exogenous input belonging to $\mathcal{L}_2[0, \infty]$, and given the measurement of the output, design a filter that gives an estimate of the unmeasured output such that the $\mathcal{H}_\infty$-norm of the transfer function between the estimation error and the exogenous input is minimized or kept below a certain given $\gamma > 0$ for all admissible uncertainties.

$\mathcal{H}_\infty$-filtering is mainly based on robust $\mathcal{H}_\infty$ control theory. Therefore, we have put them together. In this chapter, we will develop some design filtering algorithms that can be used to solve the filtering problem we are considering. Once the $\mathcal{H}_\infty$ control and robust $\mathcal{H}_\infty$ control are solved, the control of any practical dynamical time delay system becomes solvable.

The robust $\mathcal{H}_\infty$-filtering problem consists of designing a filter that asymptotically stabilizes the extended dynamics system and makes the error estimation satisfy a certain relationship. This problem has attracted a lot researchers from the control community and a lot of approaches have been developed. Along these contributions we cite the work of [135, 164, 165]

Our goal is to study the robust $\mathcal{H}_\infty$ control problem, the guaranteed cost control problem and the robust filtering problem. Algorithms based on the LMI framework will be developed to solve these two interesting control problems. In the rest of the chapter we will assume the required assumptions for each problem.

The chapter is organized as follows. In Section 6.1, the robust $\mathcal{H}_\infty$ control problem is stated and solved using the LMI framework. Delay-independent and delay-dependent conditions are developed. In Section 6.2, complete access to state vector is dropped and an output feedback controller that robustly stabilizes the uncertain dynamical time delay system and guarantees the disturbance rejection is designed. Section 6.3 deals with the robust $\mathcal{H}_\infty$ problem. In Section 6.4, we present the design of a guaranteed cost controller that robustly stabilizes the uncertain class of uncertain systems we are treating by assuring that a chosen cost remains bounded by a certain bound.

6.1   Robust $\mathcal{H}_\infty$ Control

This section deals with the robust $\mathcal{H}_\infty$ control problem for the class of dynamical linear systems with time delay and norm-bounded uncertainties. For this purpose, let us consider an uncertain linear dynamical system described by the following dynamics:

$$\begin{cases} \dot{x}(t) = A(t)x(t) + A_d(t)x(t-\tau) + B(t)u(t) + B_1 w(t) \\ x(s) = \phi(s), \ [-\tau, 0] \\ y(t) = C(t)x(t) + B_y(t)u(t) + C_d(t)x(t-\tau) + B_2 w(t) \\ z(t) = C_z x(t) + B_z u(t), \end{cases} \quad (6.1)$$

where $x(t) \in R^{n_1}$, $u(t) \in R^{n_2}$, $y(t) \in R^{n_3}$, $z(t) \in R^{n_4}$, $w(t) \in R^{n_5}$ denote, respectively, the state, the control input, the measured output, the system output, and the system disturbance $\phi(.)$ is the initial function and τ denotes the time delay in the system. The different matrices are given by

$$\begin{array}{ll} A(t) = A + D_a \Delta(t) E_a & B(t) = B + D_a \Delta(t) E_b \\ A_d(t) = A_d + D_d \Delta_d(t) E_d & C(t) = C + D_c \Delta(t) E_c \\ B_y(t) = B_y + D_{by} \Delta(t) E_{by} & C_d(t) = C_d + D_{cd} \Delta(t) E_{cd} \end{array}$$

with $A, B, A_d, B_1, B_2, B_y, C, C_z, C_d, D_{by}, E_{by}, D_{cd}, E_{cd}, D_a, E_a, E_b, D_d, E_d, E_w, D_c, E_c, C_z, B_z$ being known matrices with appropriate dimensions. $\Delta(t)$ and $\Delta_d(t)$ are time-varying unknown matrices satisfying

$$\begin{cases} \Delta^\top(t)\Delta(t) \leq I \\ \Delta_d^\top(t)\Delta_d(t) \leq I. \end{cases}$$

The system disturbance is assumed to belong to $\mathcal{L}_2[0, \infty)$.

This section addresses the $\mathcal{H}_\infty$ control problem of system (6.1). Before giving results on this subject, let us recall some definitions that we use extensively in the rest of this chapter. First of all, we define robust internal stability and then give the definition of the disturbance rejection with a given level $\gamma > 0$.

Definition 6.1 *System (6.1) with $u(t) \equiv 0$ is said to be robust internally stable if it is stable in the case of $w(t) \equiv 0$ for all admissible uncertainties.*

The γ-disturbance attenuation problem is to choose a control law $u(t)$ that stabilizes the nominal system, that is, that ensures that the closed-loop system is internally stable, and satisfies

$$\max_{0 \neq w \in \mathcal{L}_2[0,\infty)} (\|z\|_2^2 - \gamma^2 \|w\|_2^2) \leq 0, \quad \gamma > 0. \quad (6.2)$$

Definition 6.2 *A control law, $u(\cdot)$, internally stabilizes system (6.1) with γ-disturbance if the closed-loop system under this control law is internally stable and verifies noise attenuation level γ for all admissible uncertainties.*

Let us first see how can we check if a given unforced uncertain system with time delay and norm-bounded uncertainties (6.1) with $u(t) \equiv 0$ is stable and satisfies the disturbance rejection of level $\gamma > 0$. The following theorem gives a direct answer to this question.

Theorem 6.1 *Let γ be a given positive scalar and assume that $\phi(\cdot) = 0$. If there exist symmetric, positive-definite matrices $P > 0, Q > 0$ such that* *the following holds for all admissible uncertainties*

$$\Theta_0(t) = \begin{pmatrix} PA(t) + A^\top(t)P + Q + C_z^\top C_z & PA_d(t) & PB_1 \\ A_d^\top(t)P & -Q & 0 \\ B_1^\top P & 0 & -\gamma^2 I \end{pmatrix} < 0 \quad (6.3)$$

then system (6.1) with $u(t) \equiv 0$ is stable and satisfies noise attenuation level γ, that is, satisfies (5.2).

Proof: Obviously, (6.3) implies

$$\begin{pmatrix} PA(t) + A^\top(t)P + Q & PA_d(t) \\ A_d^\top(t)P & -Q \end{pmatrix} < 0,$$

which implies in turn that the unforced system with $w(t) \equiv 0$ is quadratically stable. Therefore, the robust stability follows from Theorem 4.1. Next, we come to establish (5.2).

Let us define $\mathbf{x}_t \in \mathbb{C}[-\tau, 0]$ by $\mathbf{x}_t(s) = x(t + s), s \in [-\tau, 0]$ and consider a Lyapunov functional candidate as follows:

$$V(\mathbf{x}_t) = x^\top(t)Px(t) + \int_{t-\tau}^t x^\top(s)Qx(s)ds. \quad (6.4)$$

Then, using (6.1), we have

$$\dot{V}(\mathbf{x}_t) = x^\top(t)[PA(t) + A^\top(t)P]x(t) + 2x^\top(t)PA_d(t)x(t - \tau) \\ + 2x^\top(t)PB_1w(t) + x^\top(t)Qx(t) - x^\top(t - \tau)Qx(t - \tau). \quad (6.5)$$

Let us now consider the performance function

$$J_T = \int_0^T \left[z^\top(t)z(t) - \gamma^2 w^\top(t)w(t) \right] dt.$$

Note that

$$z^\top(t)z(t) - \gamma^2 w^\top(t)w(t) + \dot{V}(\mathbf{x}_t) \\ = x^\top(t)[PA(t) + A^\top(t)P + C_z^\top C_z + Q]x(t) \\ + 2x^\top(t)PA_d(t)x(t - \tau) + 2x^\top(t)PB_1w(t) \\ - x^\top(t - \tau)Qx(t - \tau) - \gamma^2 w^\top(t)w(t) \\ = \xi^\top(t)\Theta_0(t)\xi(t)$$

with $\xi(t) = (x(t), x(t - \tau), w(t))^\top$.

Since

$$J_T = \int_0^T [z^\top(t)z(t) - \gamma^2 w^\top(t)w(t) + \dot{V}(\mathbf{x}_t)]dt - \int_0^T \dot{V}(\mathbf{x}_t)dt$$

$$= \int_0^T \xi_t^\top \Theta_0 \xi_t dt - V(\mathbf{x}_T)$$

$$\leq \int_0^T \xi_t^\top \Theta_0 \xi_t dt,$$

which is negative from (6.3), then, $J_T < 0$ holds for all $T > 0$, from which it follows that

$$J_\infty < 0,$$

This implies in turn that condition (5.2) holds and the proof of Theorem 6.1 is completed.   □

Theorem 6.1 develops a sufficient condition with which to test system (6.1) to see if it is stable, and verifies whether or not the noise attenuation level γ is satisfied. Unfortunately, since $\Theta_0(t)$ contains system uncertainties, it is not useful. The following theorem gives an LMI-based sufficient condition for $\Theta_0(t)$ to be negative-definite.

Theorem 6.2 *Let γ be a given positive scalar. If there exist symmetric, positive-definite matrices $P > 0, Q > 0$ and a scalar $\varepsilon > 0$ such that*

$$\left(\begin{array}{cccccc} \left[\begin{array}{c} PA + A^\top P \\ + Q + C_z^\top C_z \\ + \varepsilon E_a^\top E_a \end{array} \right] & PA_d & PB_1 & PD_a & PD_d \\ A_d^\top P & -Q + \varepsilon E_d^\top E_d & 0 & 0 & 0 \\ B_1^\top P & 0 & -\gamma^2 I & 0 & 0 \\ D_a^\top P & 0 & 0 & -\varepsilon I & 0 \\ D_d^\top P & 0 & 0 & 0 & -\varepsilon I \end{array} \right) < 0 \quad (6.6)$$

holds for all admissible uncertainties, then system (6.1) with $u(t) \equiv 0$ is internally stable and has noise attenuation level γ.

Proof: In view of Theorem 6.1, to prove Theorem 6.2 it suffices to establish that $\Theta_0(t)$ is negative-definite. Note that

$$\Theta_0(t) = \Theta_0 + \left(\begin{array}{cc} PD_a & PD_d \\ 0 & 0 \\ 0 & 0 \end{array} \right) \tilde{\Delta}(t) \left(\begin{array}{ccc} E_a & 0 & 0 \\ 0 & E_d & 0 \end{array} \right)$$

$$+ \left[\left(\begin{array}{cc} PD_a & PD_d \\ 0 & 0 \\ 0 & 0 \end{array} \right) \tilde{\Delta}^\top(t) \left(\begin{array}{ccc} E_a & 0 & 0 \\ 0 & E_d & 0 \end{array} \right) \right]^\top \quad (6.7)$$

where

$$\tilde{\Delta}(t) = \left(\begin{array}{cc} \Delta(t) & 0 \\ 0 & \Delta_d(t) \end{array} \right)$$

$$\Theta_0 = \left(\begin{array}{ccc} PA + A^\top + Q + C_z^\top C_z & PA_d & PB_1 \\ A_d^\top P & -Q & 0 \\ B_1^\top P & 0 & -\gamma^2 I \end{array} \right) < 0.$$

Therefore, using now Lemma A.5 we can conclude that $\Theta_0(t) < 0$ if and only if there exists an $\varepsilon > 0$ such that the following holds:

$$
\Theta_0 + \frac{1}{\varepsilon}
\begin{pmatrix}
PDD^\top P + PD_d D_d^\top P & 0 & 0 \\
0 & 0 & 0 \\
0 & 0 & 0
\end{pmatrix}
+ \varepsilon
\begin{pmatrix}
E_a^\top E_a & 0 & 0 \\
0 & E_d^\top E_d & 0 \\
0 & 0 & 0
\end{pmatrix}
$$

$$
=
\begin{pmatrix}
\begin{bmatrix} PA + A^\top + Q \\ + C_z^\top C_z + \varepsilon E_a^\top E_a \end{bmatrix} & PA_d & PB_1 \\
A_d^\top P & -Q + \varepsilon E_d^\top E_d & 0 \\
B_1^\top P & 0 & -\gamma^2 I
\end{pmatrix}
$$

$$
+ \frac{1}{\varepsilon}
\begin{pmatrix}
PD & PD_d \\
0 & 0 \\
0 & 0
\end{pmatrix}
\begin{pmatrix}
D_a^\top P & 0 & 0 \\
D_d^\top P & 0 & 0
\end{pmatrix}
< 0.
\tag{6.8}
$$

Using the Schur complement, (6.6) is equivalent to (6.8). This implies in turn that $\Theta_0(t) < 0$, which means that system (6.1) with $u(t) \equiv 0$ is stable and satisfies the given disturbance rejection for all admissible uncertainties. This completes the proof of Theorem 6.2. ☐

Theorem 6.2 gives us a condition that can be used to check whether or not the class of systems we are considering is stable and guarantees the disturbance rejection of level $\gamma > 0$ when the control is equal to zero. When the system is unstable and/or doesn't satisfy the disturbance rejection level $\gamma > 0$, we are interested in designing a controller that stabilizes the system and at the same time forces it to satisfy the given disturbance rejection level.

Let us now consider the problem of designing a memoryless controller (3.27) that stabilizes system (6.1) and ensures that the closed-loop system has a given noise attenuation level.

Before doing this, let us prove the following lemma, which will be useful in the design phase of our desired state feedback controller.

Lemma 6.1 *Let γ be a given positive scalar. If there exist symmetric, positive-definite matrices $X > 0$, $U > 0$ and a positive scalar ε such that*

$$
\begin{pmatrix}
\# & A_d U E_d^\top & X & X C_z^\top & X E_a^\top \\
E_d U A_d^\top & -\varepsilon I + E_d U E_d^\top & 0 & 0 & 0 \\
X & 0 & -U & 0 & 0 \\
C_z X & 0 & 0 & -I & 0 \\
E_a X & 0 & 0 & 0 & -\varepsilon I
\end{pmatrix}
< 0,
\tag{6.9}
$$

holds, where $\# = AX + XA^\top + \varepsilon D_a D_a^\top + \varepsilon D_d D_d^\top + \frac{1}{\gamma^2} B_1 B_1^\top + A_d U A_d^\top$, then system (6.1) with $u(t) \equiv 0$ is internally stable and has noise attenuation level γ.

Proof: By the same argument as in (6.8), we obtain that for any $\varepsilon > 0$,

$$\Theta_0 + \varepsilon \begin{pmatrix} PDD^\top P + PD_d D_d^\top P & 0 & 0 \\ 0 & 0 & 0 \\ 0 & 0 & 0 \end{pmatrix} + \frac{1}{\varepsilon} \begin{pmatrix} E_a^\top E_a & 0 & 0 \\ 0 & E_d^\top E_d & 0 \\ 0 & 0 & 0 \end{pmatrix}$$

$$< 0.$$

Using the Schur complement, the left-hand side of this inequality is negative-definite if and only if

$$PA + A^\top P + Q + C_z^\top C_z + \varepsilon PD_a D_a^\top P + \varepsilon PD_d D_d^\top P + \frac{1}{\varepsilon} E_a^\top E_a$$

$$+ \frac{1}{\gamma^2} PB_1 B_1^\top P + PA_d \left[Q - \frac{1}{\varepsilon} E_d^\top E_d \right]^{-1} A_d^\top P < 0.$$

Using the matrix inverse formula, we have

$$\left[Q - \frac{1}{\varepsilon} E_d^\top E_d \right]^{-1} = Q^{-1} + Q^{-1} E_d^\top \left[\varepsilon I - E_d Q^{-1} E_d^\top \right]^{-1} E_d Q^{-1}.$$

Therefore,

$$PA + A^\top P + Q + C_z^\top C_z + \varepsilon PD_a D_a^\top P + \varepsilon PD_d D_d^\top P$$

$$+ \frac{1}{\varepsilon} E_a^\top E_a + \frac{1}{\gamma^2} PB_1 B_1^\top P + PA_d \left[Q^{-1} \right.$$

$$\left. + Q^{-1} E_d^\top \left[\varepsilon I - E_d Q^{-1} E_d^\top \right]^{-1} E_d Q^{-1} \right] A_d^\top P < 0. \qquad (6.10)$$

Let $X = P^{-1}$ and $U = Q^{-1}$. Pre- and post-multiplying both sides of (6.10) by X yields

$$AX + XA^\top + XU^{-1}X + XC_z^\top C_z X + \varepsilon D_a D_a^\top$$

$$+ \varepsilon D_d D_d^\top + \frac{1}{\varepsilon} X E_a^\top E_a X + \frac{1}{\gamma^2} B_1 B_1^\top$$

$$+ A_d \left[U + U E_d^\top [\varepsilon I - E_d U E_d^\top]^{-1} E_d U \right] A_d^\top < 0.$$

Using the Schur complement we obtain LMI (6.9) from the above inequality.

From the above derivation, we obtain that if there exist symmetric, positive-definite matrices $X > 0$, $U > 0$ and a scalar $\varepsilon > 0$ satisfying (6.9), then $P = X^{-1} > 0$, $Q = U^{-1} > 0$, and ε ensure that $\Theta_0(t) < 0$; the proof of Lemma 6.1 follows from Theorem 6.1. $\qquad \square$

Now, we consider the design of a memoryless controller in the form of (3.27). Substituting (3.27) into (6.1) yields the dynamics of the closed-loop system as follows:

$$\begin{cases} \dot{x}(t) = \left[\bar{A} + D_a \Delta(t) \bar{E}_a \right] x(t) + A_d(t)x(t - \tau) + B_1 w(t) \\ z(t) = \bar{C}_z x(t), \end{cases} \qquad (6.11)$$

where

$$\bar{A} = A + BK, \quad \bar{C}_z = C_z + B_z K, \quad \bar{E}_a = E_a + E_b K.$$

Using Lemma 6.1, we find that if there exist symmetric, positive-definite matrices $X > 0$, $U > 0$ and a scalar $\varepsilon > 0$ such that the matrix inequality

$$\begin{pmatrix} \# & A_d U E_d^\top & X & X\bar{C}_z^\top & X\bar{E}_a^\top \\ E_d U A_d^\top & -\varepsilon I + E_d U E_d^\top & 0 & 0 & 0 \\ X & 0 & -U & 0 & 0 \\ \bar{C}_z X & 0 & 0 & -I & 0 \\ \bar{E}_a X & 0 & 0 & 0 & -\varepsilon I \end{pmatrix} < 0 \qquad (6.12)$$

holds, where $\# = \bar{A}X + X\bar{A}^\top + \varepsilon D_a D_a^\top + \varepsilon D_d D_d^\top + \frac{1}{\gamma^2} B_1 B_1^\top + A_d U A_d^\top$, then controller (3.27) internally stabilizes system (6.1) and ensures that the closed-loop system has noise attenuation level γ. Therefore, we get the following theorem.

Theorem 6.3 *Let γ be a given positive constant. If there exist symmetric, positive-definite matrices $X > 0$, $U > 0$, a matrix Y and a scalar $\varepsilon > 0$ such that*

$$\begin{pmatrix} \# & A_d U E_d^\top & \mathcal{H}_{13} \\ E_d U A_d^\top & -\varepsilon I + E_d U E_d^\top & 0 \\ \mathcal{H}_{13}^\top & 0 & -\mathcal{H}_{33} \end{pmatrix} < 0 \qquad (6.13)$$

holds, where

$$\# = AX + BY + XA^\top + Y^\top B^\top + \varepsilon D_a D_a^\top$$
$$+ \varepsilon D_d D_d^\top + \frac{1}{\gamma^2} B_1 B_1^\top + A_d U A_d^\top$$
$$\mathcal{H}_{13} = \left(X, \ XC_z^\top + Y^\top B_z^\top, \ XE_a^\top + Y^\top E_b^\top \right)$$
$$\mathcal{H}_{33} = \mathrm{diag}\{U, I, \varepsilon I\},$$

then controller (3.27) with $K = YX^{-1}$ internally stabilizes system (6.1) and ensures that the closed-loop system has noise attenuation level γ.

Proof: The proof follows directly by letting $Y = KX$ in (6.12). □

In Theorem 6.3, γ is assumed to be a given constant. In fact, the lower bound for γ that can be achieved by a state feedback memoryless controller (3.27) is obtained by solving the linear optimization problem

$$\min_{\varepsilon > 0, X > 0, Y, \mu > 0, U > 0} \mu \qquad (6.14)$$

$$\text{s.t.} \begin{pmatrix} \hat{\mathcal{H}}_{11} & A_d U E_d^\top & \hat{\mathcal{H}}_{13} \\ E_d U A_d^\top & -\varepsilon I + E_d U E_d^\top & 0 \\ \hat{\mathcal{H}}_{13}^\top & 0 & -\mathcal{H}_{33} \end{pmatrix} < 0, \qquad (6.15)$$

where

$$\hat{\mathcal{H}}_{11} = AX + BY + XA^\top + Y^\top B^\top$$
$$+ \varepsilon D_a D_a^\top + \varepsilon D_d D_d^\top + A_d U A_d^\top$$
$$\hat{\mathcal{H}}_{13} = \left(X, XC_z^\top + Y^\top B_z^\top, XE_a^\top + Y^\top E_b^\top, B_1\right)$$
$$\hat{\mathcal{H}}_{33} = \mathrm{diag}\{U, I, \varepsilon I, \mu I\}.$$

Remark 6.1 *To get (6.15) we have used the Schur complement and replaced γ^2 by μ.*

The following theorem gives the design algorithm for a memoryless controller that stabilizes the class of systems under study and assures the minimal disturbance rejection level of $\sqrt{\mu}$.

Theorem 6.4 *Let γ be a given positive scalar. If there exist symmetric, positive-definite matrices $X > 0$, $U > 0$, a matrix Y and two positive scalars $\mu > 0$, $\varepsilon > 0$ solution of (6.14-(6.15)), then controller (3.27) with gain $K = YX^{-1}$ stabilizes system (6.1) and ensures that the closed-loop system has noise attenuation level $\gamma = \sqrt{\mu}$.*

Proof: Letting $\gamma^2 = \mu$ and using the Schur complement, (6.15) follows from (6.13). Thus, the proof follows from Theorem 6.3. □

To illustrate the usefulness of the above theorem, let us consider a numerical example.

Example 6.1 *Let us consider a system described by (6.1) and suppose that the system data are as follows:*

$$A = \begin{pmatrix} -2 & 1 \\ 1 & 1 \end{pmatrix} \qquad D_a = \begin{pmatrix} 1 \\ 0 \end{pmatrix} \qquad E_a = \begin{pmatrix} 1 & 1 \end{pmatrix}$$

$$A_d = \begin{pmatrix} 0.20.1 \\ 0.3 \quad 0.1 \end{pmatrix} \qquad D_d = \begin{pmatrix} 0 \\ 1 \end{pmatrix} \qquad E_d = \begin{pmatrix} 1 & 0 \end{pmatrix}$$

$$B = \begin{pmatrix} 1 \\ 1 \end{pmatrix} \qquad E_b = \begin{pmatrix} 1 & -1 \end{pmatrix} \quad B_1 = \begin{pmatrix} 1 \\ 0 \end{pmatrix}$$

$$C_z = \begin{pmatrix} 1 & 0 \end{pmatrix}$$

With this set of data, solving (6.14) and (6.15) yields the solution

$$X = \begin{pmatrix} 0.5584 & -0.2904 \\ -0.2904 & 0.5135 \end{pmatrix} \quad Y = \begin{pmatrix} -0.0122 & -1.6782 \end{pmatrix}$$

$$U = \begin{pmatrix} 1.4400 & -0.0356 \\ -0.0356 & 1.9865 \end{pmatrix} \quad \mu = 1.9865 \ \varepsilon = 2.7065.$$

According to Theorem 6.4, we conclude that controller (3.27) with gain $K = YX^{-1} = \begin{pmatrix} -2.4394 & -4.6481 \end{pmatrix}$ internally stabilizes system (6.1) and ensures that the closed-loop system has noise attenuation level $\gamma = \sqrt{\mu} = 1.4094.$

All the results we previously developed in this section are delay-independent, which means that if the established condition holds, the system is then stable and satisfies the disturbance rejection with the desired level γ for any time delay in the system, and therefore the result is conservative. To overcome this, let us now try to see how we can develop delay-dependent conditions for robust $\mathcal{H}_\infty$ control. That is, how can we develop a sufficient delay-dependent condition such that (5.2) holds for all admissible uncertainties? For this purpose, let us assume that the following hold:

$$\Delta_d(t) = \Delta(t)$$
$$D_a = D_d.$$

Based on the assumption, the following theorem gives the first result concerning delay-dependence on robust $\mathcal{H}_\infty$ control for the class of systems we are considering in this chapter.

Theorem 6.5 *Let γ be a given positive constant. If there exist symmetric, positive-definite matrices $X > 0$, $U_1 > 0$, $U_2 > 0$, $U_3 > 0$ and scalars $\varepsilon_i > 0, 1 \le i \le 3$, such that*

$$\begin{pmatrix} \mathcal{G}_{11} & B_1 & \mathcal{G}_{13} & \mathcal{G}_{14} & \mathcal{G}_{15} & \mathcal{G}_{16} \\ B_1^\top & -\gamma^2 I + U_3 & 0 & 0 & 0 & 0 \\ \mathcal{G}_{13}^\top & 0 & -\mathcal{G}_{33} & \mathcal{G}_{34} & 0 & 0 \\ \mathcal{G}_{14}^\top & 0 & \mathcal{G}_{34}^\top & -\varepsilon_3 I & 0 & 0 \\ \mathcal{G}_{15}^\top & 0 & 0 & 0 & -\mathcal{G}_{55} & 0 \\ \mathcal{G}_{16}^\top & 0 & 0 & 0 & 0 & -\mathcal{G}_{66} \end{pmatrix} < 0 \qquad (6.16)$$

holds, where

$$\mathcal{G}_{11} = [A + A_d]X + X[A + A_d]^\top + \varepsilon_3 D_a D_a^\top$$
$$\mathcal{G}_{13} = \begin{pmatrix} A_d U_1 & A_d U_2 & A_d B_1 \end{pmatrix}$$
$$\mathcal{G}_{33} = \text{diag}\{\frac{1}{\tau}U_1, \frac{1}{\tau}U_2, U_3\}$$
$$\mathcal{G}_{14} = X[E_a^\top + E_d^\top]$$
$$\mathcal{G}_{34} = \begin{pmatrix} U_1 E_d^\top \\ U_2 E_d^\top \\ B_1^\top E_d^\top \end{pmatrix}$$
$$\mathcal{G}_{15} = \begin{pmatrix} X A^\top & X A_d^\top \end{pmatrix}$$
$$\mathcal{G}_{16} = \begin{pmatrix} X E_a^\top & X E_d^\top & X C_z^\top \end{pmatrix}$$
$$\mathcal{G}_{55} = \begin{pmatrix} U_1 - \varepsilon_1 D_a D_a^\top & 0 \\ 0 & U_2 - \varepsilon_2 D_d D_d^\top \end{pmatrix}$$
$$\mathcal{G}_{66} = \begin{pmatrix} \varepsilon_1 I & 0 & 0 \\ 0 & \varepsilon_2 I & 0 \\ 0 & 0 & I \end{pmatrix}$$

then system (6.1) with $u(t) \equiv 0$ is internally stable and satisfies (5.2).

Proof: We first establish (5.2). Substituting (6.1) with $u(t) \equiv 0$ into

$$x(t - \tau) = x(t) - \int_{t-\tau}^{t} \dot{x}(s)ds$$

gives

$$x(t - \tau) = x(t) - \int_{t-\tau}^{t} [A(s)x(s) + A_d(s)x(s - \tau) + B_1 w(s)]ds.$$

Substituting now (6.17) into (6.1), we obtain

$$\begin{cases} \dot{x}(t) = [A(t) + A_d(t)]x(t) - A_d(t) \int_{t-\tau}^{t} [A(s)x(s) + A_d(s)x(s - \tau) \\ \qquad + B_1 w(s)]ds + B_1 w(t) \\ x(s) = 0, \ w(s) = 0, \ s \in [-2\tau, 0]. \end{cases} \tag{6.17}$$

Let us define $\mathbf{x}_t \in \mathbb{C}[-2\tau, 0]$ with $\mathbf{x}_t(s) = x(t + s), s \in [-2\tau, 0]$ and consider a Lyapunov functional candidate as follows:

$$V(\mathbf{x}_t) = V_0(x(t)) + V_1(\mathbf{x}_t),$$

where

$$V_0(x(t)) = x^\top(t)Px(t)$$

$$V_1(\mathbf{x}_t) = \int_{-\tau}^{0} \int_{t+\theta}^{t} x^\top(s)A^\top(s)Q_1 A(s)x(s)dsd\theta$$

$$+ \int_{-\tau}^{0} \int_{t-\tau+\theta}^{t} x^\top(s)A_d^\top(s)Q_2 A_d(s)x(s)dsd\theta$$

$$+ \int_{-\tau}^{0} \int_{t+\theta}^{t} w^\top(s)Q_3 w(s)dsd\theta.$$

Using (6.17), we obtain

$$\dot{V}_0(\mathbf{x}_t) = x^\top(t) \left[P[A(t) + A_d(t)] + [A(t) + A_d(t)]^\top P \right] x(t)$$

$$- 2x^\top(t)PA_d(t) \left[\int_{t-\tau}^{t} A(s)x(s)ds \right.$$

$$+ \int_{t-\tau}^{t} A_d(s)x(s - \tau)ds + B_1 \int_{t-\tau}^{t} w(s)ds \right]$$

$$+ 2x^\top(t)PB_1 w(t).$$

Using Lemma A.5, we obtain, respectively

$$-2x^\top(t)PA_d(t) \int_{t-\tau}^{t} A(s)x(s)ds \leq \tau x^\top(t)PA_d(t)Q_1^{-1}A_d^\top(t)Px(t)$$

$$+ \int_{t-\tau}^{t} x^\top(s)A^\top(s)Q_1 A(s)x(s)ds \tag{6.18}$$

$$-2x^\top(t)PA_d(t)\int_{t-\tau}^t A_d(s)x(s-\tau)ds \le \tau x^\top(t)PA_d(t)Q_2^{-1}A_d^\top(t)Px(t)$$

$$+\int_{t-\tau}^t x^\top(s-\tau)A_d^\top(s)Q_2 A_d(s)x(s-\tau)ds \tag{6.19}$$

$$-2x^\top(t)PA_d(t)B_1\int_{t-\tau}^t w(s)ds \le \tau x^\top(t)PA_d(t)B_1 Q_3^{-1}B_1^\top A_d^\top(t)Px(t)$$

$$+\int_{t-\tau}^t w^\top(s)Q_3 w(s)ds. \tag{6.20}$$

Combining (6.18)-(6.20) yields

$$\begin{aligned}
\dot{V}_0(\mathbf{x}_t) = {}& x^\top(t)\left[P(A(t)+A_d(t))+(A(t)+A_d(t))^\top P\right]x(t)\\
& + \tau x^\top(t)PA_d(t)Q_1^{-1}A_d^\top(t)Px(t)\\
& + \int_{t-\tau}^t x^\top(s)A^\top(s)Q_1 A(s)x(s)ds\\
& + \tau x^\top(t)PA_d(t)Q_2^{-1}A_d^\top(t)Px(t)\\
& + \int_{t-\tau}^t x^\top(s-\tau)A_d^\top(s)Q_2 A_d(s)x(s-\tau)ds\\
& + \tau x^\top(t)PA_d(t)B_1 Q_3^{-1}B_1^\top A_d^\top(t)Px(t)\\
& + \int_{t-\tau}^t w^\top(s)Q_3 w(s)ds + 2x^\top(t)PB_1 w(t).
\end{aligned} \tag{6.21}$$

Moreover, simple computation gives

$$\begin{aligned}
\dot{V}_1(\mathbf{x}_t) = {}& \tau x^\top(t)A^\top(t)Q_1 A(t)x(t)\\
& - \int_{t-\tau}^t x^\top(s)A^\top(s)Q_1 A(s)x(s)ds\\
& + \tau x^\top(t)A_d^\top(t)Q_2 A_d(t)x(t)\\
& - \int_{t-\tau}^t x^\top(s-\tau)A_d^\top(s)Q_2 A_d(s)x(s-\tau)ds\\
& + \tau w^\top(t)Q_3 w(t) - \int_{t-\tau}^t w^\top(s)Q_3 w(s)ds.
\end{aligned} \tag{6.22}$$

Combining again (6.18)-(6.20) and (6.22) yields

$$\begin{aligned}
\dot{V}(\mathbf{x}_t) \le {}& x^\top(t)\left[P[A(t)+A_d(t)]+[A(t)+A_d(t)]^\top P\right]x(t)\\
& + \tau x^\top(t)[PA_d(t)Q_1^{-1}A_d^\top(t)P + A^\top(t)Q_1 A(t)]x(t)\\
& + \tau x^\top(t)[PA_d(t)Q_2^{-1}A_d^\top(t)P + A_d^\top(t)Q_2 A_d(t)]x(t)\\
& + \tau x^\top(t)PA_d(t)B_1 Q_3^{-1}B_1^\top A_d^\top(t)Px(t) + \tau w^\top(t)Q_3 w(t)\\
& + 2x^\top(t)PB_1 w(t).
\end{aligned} \tag{6.23}$$

Using now Lemma A.6, we obtain the following inequalities:

a) If we have a symmetric, positive-definite matrix $Q_1 > 0$ and a scalar $\varepsilon_1 > 0$ satisfying $Q_1^{-1} - \varepsilon_1 D_a D_a^\top > 0$, then the following holds:

$$A^\top(t) Q_1 A(t) \le A^\top [Q_1^{-1} - \varepsilon_1 D_a D_a^\top]^{-1} A + \frac{1}{\varepsilon_1} E_a^\top E_a. \qquad (6.24)$$

b) If we have a symmetric, positive-definite matrix $Q_2 > 0$ and a scalar $\varepsilon_2 > 0$ satisfying $Q_2^{-1} - \varepsilon_2 D_d D_d^\top > 0$, then the following holds:

$$A_d^\top(t) Q_2 A_d(t) \le A_d^\top [Q_2^{-1} - \varepsilon_2 D_d D_d^\top]^{-1} A_d + \frac{1}{\varepsilon_2} E_d^\top E_d. \qquad (6.25)$$

Combining now (6.23)-(6.25) yields

$$\dot{V}(\mathbf{x}_t) \le x^\top(t) \left[P[A(t) + A_d(t)] + [A(t) + A_d(t)]^\top P \right] x(t)$$
$$+ \tau x^\top(t) \left[P A_d(t) Q_1^{-1} A_d^\top(t) P + A^\top \left[Q_1^{-1} - \varepsilon_1 D_a D_a^\top \right]^{-1} A \right.$$
$$+ \left. \frac{1}{\varepsilon_1} E_a^\top E_a \right] x(t) + \tau x^\top(t) \left[P A_d(t) Q_2^{-1} A_d^\top(t) P \right.$$
$$+ \left. A_d^\top \left[Q_2^{-1} - \varepsilon_2 D_d D_d^\top \right]^{-1} A_d + \frac{1}{\varepsilon_2} E_d^\top E_d \right] x(t)$$
$$+ \tau x^\top(t) P A_d(t) B_1 Q_3^{-1} B_1^\top A_d^\top(t) P x(t) + \tau w^\top(t) Q_3 w(t)$$
$$+ 2 x^\top(t) P B_1 w(t).$$

Let $\mathcal{H}_0$ and $\mathcal{H}_1$ be defined by:

$$\mathcal{H}_0 = A^\top [Q_1^{-1} - \varepsilon_1 D_a D_a^\top]^{-1} A + \frac{1}{\varepsilon_1} E_a^\top E_a$$
$$+ A_d^\top [Q_2^{-1} - \varepsilon_2 D_d D_d^\top]^{-1} A_d + \frac{1}{\varepsilon_2} E_d^\top E_d$$

$$\mathcal{H}_1 = \mathcal{H}_0 + P[A(t) + A_d(t)] + [A(t) + A_d(t)]^\top P$$
$$+ \tau P A_d(t) Q_1^{-1} A_d^\top(t) P + \tau P A_d(t) Q_2^{-1} A_d^\top(t) P$$
$$+ \tau P A_d(t) B_1 Q_3^{-1} B_1^\top A_d^\top(t) P. \qquad (6.26)$$

Let us define the performance function J_T by

$$J_T = \int_0^T [z^\top(t) z(t) - \gamma^2 w^\top(t) w(t)] dt.$$

Note that

$$J_T = \int_0^T [z^\top(t) z(t) - \gamma^2 w^\top(t) w(t) + \dot{V}(\mathbf{x}_t)] dt - \int_0^T \dot{V}(\mathbf{x}_t) dt$$
$$= \int_0^T [z^\top(t) z(t) - \gamma^2 w^\top(t) w(t) + \dot{V}(\mathbf{x}_t)] dt - V(\mathbf{x}_T)$$
$$\le \int_0^T [z^\top(t) z(t) - \gamma^2 w^\top(t) w(t) + \dot{V}(\mathbf{x}_t)] dt.$$

In view of (6.23), we obtain

$$z^\top(t)z(t) - \gamma^2 w^\top(t)w(t) + \dot{V}(\mathbf{x}_t)$$
$$\leq \left(x^\top(t) \ w^\top(t)\right) \tilde{\Theta}(t) \begin{pmatrix} x(t) \\ w(t) \end{pmatrix}, \qquad (6.27)$$

where

$$\tilde{\Theta}(t) = \begin{pmatrix} \mathcal{H}_1 + C_z^\top C_z & PB_1 \\ B_1^\top P & -\gamma^2 I + \tau Q_3 \end{pmatrix}. \qquad (6.28)$$

Using the Schur complement, we obtain $\tilde{\Theta}(t) < 0$ if and only if

$$\begin{pmatrix} \# & PB_1 & PA_d(t) & PA_d(t) & PA_d(t)B_1 \\ B_1^\top P & -\gamma^2 I + \tau Q_3 & & & \\ A_d^\top(t)P & & -\frac{1}{\tau}Q_1 & & \\ A_d^\top(t)P & & & -\frac{1}{\tau}Q_2 & \\ B_1^\top A_d^\top(t)P & & & & -\frac{1}{\tau}Q_3 \end{pmatrix} < 0, \quad (6.29)$$

where $\# = \mathcal{H}_0 + C_z^\top C_z + P[A(t) + A_d(t)] + [A(t) + A_d(t)]^\top P$.
Note that the left-hand side of (6.29) can be written as

$$\begin{pmatrix} \begin{bmatrix} \mathcal{H}_0 + C_z^\top C_z \\ + P[A + A_d] \\ + [A + A_d]^\top P \end{bmatrix} & PB_1 & PA_dQ_1^{-1} & PA_dQ_2^{-1} & PA_dB_1 \\ B_1^\top P & -\gamma^2 I + \tau Q_3 & & & \\ Q_1^{-1}A_d^\top P & & -\frac{1}{\tau}Q_1^{-1} & & \\ Q_2^{-1}A_d^\top P & & & -\frac{1}{\tau}Q_2^{-1} & \\ B_1^\top A_d^\top P & & & & -\frac{1}{\tau}Q_3 \end{pmatrix}$$

$$+ \begin{pmatrix} PD_a \\ 0 \\ 0 \\ 0 \\ 0 \end{pmatrix} \Delta(t) \left(E_a + E_d \ \mathbf{0} \ E_dQ_1^{-1} \ E_dQ_2^{-1} \ E_dB_1\right)$$

$$+ \left[\begin{pmatrix} PD_a \\ 0 \\ 0 \\ 0 \\ 0 \end{pmatrix} \Delta(t) \left(E_a + E_d \ \mathbf{0} \ E_dQ_1^{-1} \ E_dQ_2^{-1} \ E_dB_1\right)\right]^\top.$$

Using Lemma A.5, we obtain that (6.29) holds if and only if there exists a scalar $\varepsilon_3 > 0$ such that

$$
\left(
\begin{array}{ccccc}
\begin{bmatrix} \mathcal{H}_0 + C_z^\top C_z \\ + P[A + A_d] \\ + [A + A_d]^\top P \end{bmatrix} & PB_1 & PA_dQ_1^{-1} & PA_dQ_2^{-1} & PA_dB_1 \\
B_1^\top P & -\gamma^2 I + \tau Q_3 & & & \\
Q_1^{-1}A_d^\top P & & -\frac{1}{\tau}Q_1^{-1} & & \\
Q_2^{-1}A_d^\top P & & & -\frac{1}{\tau}Q_2^{-1} & \\
B_1^\top A_d^\top P & & & & -\frac{1}{\tau}Q_3
\end{array}
\right)
$$

$$
+\, \varepsilon_3
\begin{pmatrix} PD_a \\ 0 \\ 0 \\ 0 \\ 0 \end{pmatrix}
\left(D_a^\top P\; 0\; 0\; 0\; 0 \right)
$$

$$
+\, \frac{1}{\varepsilon_3}
\begin{pmatrix} E_a^\top + E_d^\top \\ 0 \\ Q_1^{-1}E_d^\top \\ Q_2^{-1}E_d^\top \\ B_1^\top E_d^\top \end{pmatrix}
\Delta(t)
\left(E_a + E_d\; 0\; E_dQ_1^{-1}\; E_dQ_2^{-1}\; E_dB_1 \right) < 0,
$$

that is,

$$
\begin{pmatrix}
\mathcal{H}_2 & PB_1 & \mathcal{H}_{13} & \mathcal{H}_{14} \\
B_1^\top P & -\gamma^2 I + \tau Q_3 & 0 & 0 \\
\mathcal{H}_{13}^\top & 0 & -\mathcal{H}_{33} & \mathcal{H}_{34} \\
\mathcal{H}_{14}^\top & 0 & \mathcal{H}_{34} & -\varepsilon_3 I
\end{pmatrix} < 0, \tag{6.30}
$$

where

$$
\begin{aligned}
\mathcal{H}_2 &= P[A + A_d] + [A + A_d]^\top P \\
&\quad + A^\top [Q_1^{-1} - \varepsilon_1 D_a D_a^\top]^{-1} A + \frac{1}{\varepsilon_1} E_a^\top E_a \\
&\quad + A_d^\top [Q_2^{-1} - \varepsilon_2 D_d D_d^\top]^{-1} A_d + \frac{1}{\varepsilon_2} E_d^\top E_d + C_z^\top C_z \\
&\quad + \varepsilon_3 P D_a D_a^\top P \\
\mathcal{H}_{13} &= \left(PA_dQ_1^{-1} \quad PA_dQ_2^{-1} \quad PA_dB_1 \right) \\
\mathcal{H}_{33} &= \frac{1}{\tau} \begin{pmatrix} Q_1^{-1} & & \\ & Q_2^{-1} & \\ & & Q_3 \end{pmatrix} \\
\mathcal{H}_{14} &= E_a^\top + E_d^\top \\
\mathcal{H}_{34} &= \begin{pmatrix} Q_1^{-1} E_d^\top \\ Q_2^{-1} E_d^\top \\ B_1^\top E_d^\top \end{pmatrix}
\end{aligned}
$$

Let $X = P^{-1}$ and note that

$$
\begin{aligned}
X\mathcal{H}_2 X &= [A + A_d]X + X[A + A_d]^\top \\
&\quad + XA^\top[Q_1^{-1} - \varepsilon_1 D_a D_a^\top]^{-1}AX + \frac{1}{\varepsilon_1}XE_a^\top E_a X \\
&\quad + XA_d^\top[Q_2^{-1} - \varepsilon_2 D_d D_d^\top]^{-1}A_d X + \frac{1}{\varepsilon_2}XE_d^\top E_d X \\
&\quad + XC_z^\top C_z X + \varepsilon_3 D_a D_a^\top \\
&= \mathcal{G}_{11} + \mathcal{H}_{15}[\mathcal{H}_{55}]^{-1}\mathcal{H}_{15}^\top + \mathcal{H}_{16}[\mathcal{H}_{66}]^{-1}\mathcal{H}_{16}^\top,
\end{aligned}
$$

where

$$
\begin{aligned}
\mathcal{H}_{15} &= \left(XA^\top \ XA_d^\top \right) \\
\mathcal{H}_{16} &= \left(XE_a^\top \ XE_d^\top \ XC_z^\top \right) \\
\mathcal{H}_{55} &= \left(\begin{array}{cc} Q_1^{-1} - \varepsilon_1 D_a D_a^\top & 0 \\ 0 & Q_2^{-1} - \varepsilon_2 D_d D_d^\top \end{array} \right) \\
\mathcal{H}_{66} &= \left(\begin{array}{ccc} \varepsilon_1 I & 0 & 0 \\ 0 & \varepsilon_2 I & 0 \\ 0 & 0 & I \end{array} \right).
\end{aligned}
$$

Pre- and post-multiplying (6.30) by $\mathrm{diag}\{X, I, I, I\}$ and using the Schur complement yields that (6.30) is equivalent to

$$
\left(\begin{array}{cccccc}
\mathcal{G}_{11} & B_1 & X\mathcal{H}_{13} & X\mathcal{H}_{14} & \mathcal{H}_{15} & \mathcal{H}_{16} \\
B_1^\top & -\gamma^2 I + \tau Q_3 & 0 & 0 & 0 & 0 \\
\mathcal{H}_{13}^\top X & 0 & -\mathcal{H}_{33} & \mathcal{H}_{34} & 0 & 0 \\
\mathcal{H}_{14}^\top X & 0 & \mathcal{H}_{34}^\top & -\varepsilon_3 I & 0 & 0 \\
\mathcal{H}_{15}^\top & 0 & 0 & 0 & -\mathcal{H}_{55} & 0 \\
\mathcal{H}_{16}^\top & 0 & 0 & 0 & 0 & -\mathcal{H}_{66}
\end{array} \right) < 0. \quad (6.31)
$$

Let $U_1 = Q_1^{-1}, U_2 = Q_2^{-1}, U_3 = Q_3$ in (6.31) and note that

$$
X\mathcal{H}_{13} = \mathcal{G}_{13}, \ X\mathcal{H}_{14} = \mathcal{G}_{14}
$$
$$
\mathcal{H}_{13} = \mathcal{G}_{13}, \ \mathcal{H}_{13} = \mathcal{G}_{13}.
$$

Thus, (6.31) leads to (6.16).

Next, we proceed to prove that the system is internally stable. From the above derivation, if $X > 0, U_i > 0, 1 \le i \le 3$, and scalar $\varepsilon_i > 0, 1 \le i \le 3$ satisfy (6.16), we can conclude the internal stability. If we define $P = X^{-1}, Q_1 = U_1^{-1}, Q_2 = U_2^{-1}, Q_3 = U_3$, then $P, Q_i, 1 \le i \le 3$, satisfy

$$
\tilde{\Theta}(t) < 0
$$

for all admissible uncertainties, where $\tilde{\Theta}(t)$ is defined as in (6.28). Obviously, the above inequality implies

$$
\mathcal{H}_1 < 0
$$

with $\mathcal{H}_1$ being defined by (6.26).

Moreover, $\mathcal{H}_1 < 0$ implies $\dot{V}(\mathbf{x}_t) < 0$ if $w(t) \equiv 0$, which proves that the system under study is internally stable. This completes the proof of the theorem. $\qquad\square$

Let us now consider the delay-dependent robust $\mathcal{H}_\infty$ controller design problem and restrict ourselves to memoryless controller. Substituting (3.27) into (6.1) yields the dynamics of the closed-loop system as (6.11). Using Theorem 6.5, we obtain the following theorem.

Theorem 6.6 *Let symmetric, positive-definite matrices $X > 0$, $U_1 > 0$, $U_2 > 0$, $U_3 > 0$, a matrix Y and scalars $\varepsilon_i > 0, 1 \leq i \leq 3, \mu_0 > 0$ be a set of solutions of the optimization problem*

$$\min_{X>0,U_i>0,\varepsilon_i>0,1\leq i\leq 3,Y,\mu>0} \mu \tag{6.32}$$

s.t.

$$\begin{pmatrix} \mathcal{J}_{11} & B_1 & \mathcal{J}_{13} & \mathcal{J}_{14} & \mathcal{J}_{15} & \mathcal{J}_{16} \\ B_1^\top & -\gamma^2 I + \tau U_3 & 0 & 0 & 0 & 0 \\ \mathcal{J}_{13}^\top & 0 & -\mathcal{J}_{33} & \mathcal{J}_{34} & 0 & 0 \\ \mathcal{J}_{14}^\top & 0 & \mathcal{J}_{34}^\top & -\varepsilon_3 I & 0 & 0 \\ \mathcal{J}_{15}^\top & 0 & 0 & 0 & -\mathcal{J}_{55} & 0 \\ \mathcal{J}_{16}^\top & 0 & 0 & 0 & 0 & -\mathcal{J}_{66} \end{pmatrix} < 0 \tag{6.33}$$

where

$$\mathcal{J}_{11} = [A + A_d]X + BY + X[A + A_d]^\top + Y^\top B^\top + \varepsilon_3 D_a D_a^\top$$

$$\mathcal{J}_{13} = \begin{pmatrix} A_d U_1 & A_d U_2 & A_d B_1 \end{pmatrix}$$

$$\mathcal{J}_{33} = \text{diag}\{\frac{1}{\tau}U_1, \frac{1}{\tau}U_2, U_3\}$$

$$\mathcal{J}_{14} = X[E_a^\top + E_d^\top] + Y^\top E_b^\top$$

$$\mathcal{J}_{34} = \begin{pmatrix} U_1 E_d^\top \\ U_2 E_d^\top \\ B_1^\top E_d^\top \end{pmatrix}$$

$$\mathcal{J}_{15} = \begin{pmatrix} XA^\top + Y^\top B^\top & XA_d^\top \end{pmatrix}$$

$$\mathcal{G}_{16} = \begin{pmatrix} XE_a^\top + Y^\top E_b^\top & XE_d^\top & XC_z^\top + Y^\top B_z^\top \end{pmatrix}$$

$$\mathcal{J}_{55} = \begin{pmatrix} U_1 - \varepsilon_1 D_a D_a^\top & 0 \\ 0 & U_2 - \varepsilon_2 D_d D_d^\top \end{pmatrix}$$

$$\mathcal{G}_{66} = \begin{pmatrix} \varepsilon_1 I & 0 & 0 \\ 0 & \varepsilon_2 I & 0 \\ 0 & 0 & I \end{pmatrix}.$$

Then, controller (3.27) with $K = YX^{-1}$ internally stabilizes system (6.1) and the closed-loop system verifies noise attenuation level $\gamma = \sqrt{\mu_0}$.

Proof: Using Theorem 6.5, we obtain that for a given gain K, if there exist symmetric, positive-definite matrices $X > 0$, $U_i > 0, 1 \leq i \leq 3$, and scalars

$\varepsilon_i > 0, 1 \leq i \leq 3$ satisfying the LMI

$$\begin{pmatrix}
\bar{\mathcal{G}}_{11} & B_1 & \bar{\mathcal{G}}_{13} & \bar{\mathcal{G}}_{14} & \bar{\mathcal{G}}_{15} & \bar{\mathcal{G}}_{16} \\
B_1^\top & -\gamma^2 I + U_3 & 0 & 0 & 0 & 0 \\
\bar{\mathcal{G}}_{13}^\top & 0 & -\mathcal{G}_{33} & \mathcal{G}_{34} & 0 & 0 \\
\bar{\mathcal{G}}_{14}^\top & 0 & \mathcal{G}_{34}^\top & -\varepsilon_3 I & 0 & 0 \\
\bar{\mathcal{G}}_{15}^\top & 0 & 0 & 0 & -\mathcal{G}_{55} & 0 \\
\bar{\mathcal{G}}_{16}^\top & 0 & 0 & 0 & 0 & -\mathcal{G}_{66}
\end{pmatrix} < 0,$$

where

$$\bar{\mathcal{G}}_{11} = [\bar{A} + A_d]X + X[\bar{A} + A_d]^\top + \varepsilon_3 D_a D_a^\top$$
$$\bar{\mathcal{G}}_{14} = X[\bar{E}_a^\top + E_d^\top]$$
$$\bar{\mathcal{G}}_{15} = \begin{pmatrix} X\bar{A}^\top & XA_d^\top \end{pmatrix}$$
$$\bar{\mathcal{G}}_{16} = \begin{pmatrix} X\bar{E}_a^\top & XE_d^\top & X\bar{C}_z^\top \end{pmatrix},$$

then the closed-loop system under this control law is internally stable and satisfies (5.2). Letting $Y = KX$ and $\gamma^2 = \mu$ in the above inequality leads to (6.33). This completes the proof of Theorem 6.6. $\qquad\square$

To show the usefulness of the above controller design method, let us give a numerical example.

Example 6.2 *Consider a system described by (6.1) and suppose that the system data are as follows:*

$$A = \begin{pmatrix} 2 & 1 \\ 0 & 1.3 \end{pmatrix} \qquad A_d = \begin{pmatrix} 0.2 & 0.1 \\ 0.3 & -0.1 \end{pmatrix}$$

$$B = \begin{pmatrix} -1 & 0 \\ 0.1 & -1 \end{pmatrix} \qquad B_1 = \begin{pmatrix} 0 & 1 \\ 1 & 0 \end{pmatrix}$$

$$D_a = \begin{pmatrix} 1 & 0 \\ 1 & -1 \end{pmatrix} \qquad E_a = \begin{pmatrix} 0 & 1 \\ 1 & 0 \end{pmatrix}$$

$$E_b = \begin{pmatrix} 1 & -1 \\ 0 & 1 \end{pmatrix} \qquad E_d = \begin{pmatrix} 0.1 & 0 \\ 1 & 0 \end{pmatrix}$$

$$B_z = \begin{pmatrix} -1 & 0 \\ 0.1 & -1 \end{pmatrix} \qquad C_z = \begin{pmatrix} -1 & 0 \\ 0.1 & -1 \end{pmatrix}.$$

With these data, solving (6.33) at $\tau = 6.9980$ yields the following solution:

$$X = \begin{pmatrix} -0.5076 & -0.4164 \\ -0.4164 & -0.4207 \end{pmatrix} \qquad Y = \begin{pmatrix} 15.7021 & 4.2449 \\ 8.0412 & 22.0968 \end{pmatrix}$$

$$U_1 = \begin{pmatrix} 20.7334 & 11.5742 \\ 11.5742 & 54.1043 \end{pmatrix} \qquad U_2 = \begin{pmatrix} 7.1560 & -4.7415 \\ -4.7415 & 26.1926 \end{pmatrix}$$

$$U_3 = \begin{pmatrix} 21.0042 & -0.4652 \\ -0.4652 & 20.7636 \end{pmatrix} \qquad \varepsilon_1 = 1.1137, \ \varepsilon_2 = 4.6503$$

$$\varepsilon_3 = 55.1770 \quad \mu = 1.46.$$

According to Theorem 6.6, we conclude that controller (3.27) with

$$K = YX^{-1} = \begin{pmatrix} -120.5171 & 109.1985 \\ 144.9315 & -195.9778 \end{pmatrix}$$

internally stabilizes system (6.1) and the closed-loop system satisfies the noise attenuation level 1.21.

Remark 6.2 *All the results we presented so far require complete access of the state space vector. In case of violation of this assumption, the output feedback controller can be used. The next section deals with this problem and develops a design algorithm that can be used to compute the controller gains.*

6.2 Robust Output Feedback $\mathcal{H}_\infty$ Control

This section addresses the synthesis of a dynamical output feedback controller that stabilizes system (6.1). The controller is assumed to be of the following form:

$$\begin{cases} \dot{\xi}_t = K_A \xi_t + K_B y(t) \\ u(t) = K_C \xi_t \end{cases} \tag{6.34}$$

where $\xi_t \in \mathbb{R}^{n_1}$ is the controller state, $y(t)$ is the measured output, $u(t)$ is the control input, K_A, K_B, and K_C are constant matrices to be designed.

Applying this controller to system (6.1) yields the following extended dynamics:

$$\begin{cases} \dot{\eta}(t) = \bar{A}(t)\eta(t) + \bar{A}_d(t)I_0\eta(t-\tau) + \bar{B}_1 w(t) \\ z(t) = \bar{C}_z \eta(t) \end{cases} \tag{6.35}$$

where

$$\eta^\top(t) = \left(x^\top(t) \ \xi^\top(t) \right)$$

$$\bar{A}(t) = \begin{pmatrix} A(t) & B(t)K_C \\ K_B C(t) & K_A + K_B B_y K_C \end{pmatrix}$$

$$\bar{A}_d(t) = \begin{pmatrix} A_d(t) \\ K_B C_d(t) \end{pmatrix}$$

$$\bar{B}_1 = \begin{pmatrix} B_1 \\ K_B B_2 \end{pmatrix}$$

$$\bar{C}_z = (C_z \ B_z K_C)$$

$$I_0 = (I \ \mathbf{0}).$$

In this section, we assume the following:

$$D_a = D_1, \ D_d = D_2, \ D_c = D_{by} = D_3, \ D_{cd} = D_4,$$
$$E_a = E_c = E_1, \ E_{cd} = E_d = E_2, \ E_b = E_3.$$

Let us define the following matrices for simplicity

$$\bar{A} = \begin{pmatrix} A & BK_C \\ K_B C & K_A + K_B B_y K_C \end{pmatrix}$$

$$\bar{D}_a = \begin{pmatrix} D_1 \\ K_B D_3 \end{pmatrix}$$

$$\bar{E}_a = \begin{pmatrix} E_1 & E_3 K_C \end{pmatrix}$$

$$\bar{A}_d = \begin{pmatrix} A_d \\ K_B C_d \end{pmatrix}$$

$$\bar{D}_d = \begin{pmatrix} D_2 \\ K_B D_4 \end{pmatrix}$$

$$\bar{E}_d = E_2.$$

With the definition of these matrices, we obtain

$$\bar{A}(t) = \bar{A} + \bar{D}_a \Delta(t) \bar{E}_a, \quad \bar{A}_d(t) = \bar{A}_d + \bar{D}_d \Delta_d(t) \bar{E}_d.$$

Based on Theorem 6.2, we have the following proposition, which states how we can check whether or not the extended dynamics are stable and satisfy the desired disturbance level.

Proposition 6.1 *Let γ be a given positive constant. If there exist symmetric, positive-definite matrices $P > 0, Q > 0$, and a scalar $\varepsilon > 0$ such that*

$$\begin{pmatrix} \# & P\bar{A}_d & P\bar{B}_1 & P\bar{D}_a & P\bar{D}_d \\ \bar{A}_d^\top P & -Q + \varepsilon \bar{E}_d^\top \bar{E}_d & 0 & 0 & 0 \\ \bar{B}_1^\top P & 0 & -\gamma^2 I & 0 & 0 \\ \bar{D}_a^\top P & 0 & 0 & -\varepsilon I & 0 \\ \bar{D}_d^\top P & 0 & 0 & 0 & -\varepsilon I \end{pmatrix} < 0 \qquad (6.36)$$

holds, where

$$\# = P\bar{A} + \bar{A}^\top P + I_0^\top Q I_0 + \bar{C}_z^\top \bar{C}_z + \varepsilon \bar{E}_a^\top \bar{E}_a,$$

then system (6.35) is asymptotically stable and has noise attenuation level γ.

Let us now see how we can determine the minimum disturbance rejection that the class of systems we are considering can have when the output controller is used.

Using now the Schur complement, we obtain that (6.36) holds if and only if

$$
\begin{pmatrix}
P\bar{A} + \bar{A}P & P\bar{A}_d & P\bar{B}_1 & J_{14} & J_{15} \\
\bar{A}_d^\top P & -Q + \varepsilon E_d^\top E_d & 0 & 0 & 0 \\
\bar{B}_1^\top P & 0 & -\gamma^2 I & 0 & 0 \\
J_{14}^\top & 0 & 0 & -\varepsilon I & 0 \\
J_{15}^\top & 0 & 0 & 0 & -J_{55}
\end{pmatrix} < 0, \qquad (6.37)
$$

where

$$
J_{14} = \begin{pmatrix} P\bar{D}_a & P\bar{D}_d \end{pmatrix}
$$
$$
J_{15} = \begin{pmatrix} I_0^\top & \bar{C}_z^\top & \bar{E}_a^\top \end{pmatrix}
$$
$$
J_{55} = \begin{pmatrix} Q^{-1} & 0 & 0 \\ 0 & I & 0 \\ 0 & 0 & \frac{1}{\varepsilon}I \end{pmatrix}.
$$

Letting $\mu = \gamma^2$ in (6.37) leads to the following proposition.

Proposition 6.2 *Let K_A, K_B, K_C be a set of given gains. If there exist a symmetric, positive-definite matrix $P > 0$ and a positive scalar $\mu > 0$ such that, for some symmetric, positive-definite matrix $Q > 0$ and a positive scalar $\varepsilon > 0$,*

$$
\begin{pmatrix}
P\bar{A} + \bar{A}P & P\bar{A}_d & P\bar{B}_1 & J_{14} & J_{15} \\
\bar{A}_d^\top P & -Q + \varepsilon E_d^\top E_d & & & \\
\bar{B}_1^\top P & & -\mu I & & \\
J_{14}^\top & & & -\varepsilon I & \\
J_{15}^\top & & & & -J_{55}
\end{pmatrix} < 0 \qquad (6.38)
$$

holds, then controller (6.34) stabilizes system (6.1) and the closed-loop system has noise attenuation level $\gamma = \sqrt{\mu}$.

Proposition 6.2 develops a sufficient condition for a given controller of form (6.34) that stabilizes system (6.1). Next, we proceed with the synthesis problem of an output feedback controller that stabilizes the system under study and at the same time guarantees the desired disturbance rejection of level γ. For this purpose, let us suppose that the symmetric, positive-definite matrices P and P^{-1} are partitioned as in (3.47) and define the matrices Φ_1 and Φ_2 as in (3.55). Then, simple computation gives

$$
P\Phi_1 = \begin{pmatrix} Y & N \\ N^\top & Z_2 \end{pmatrix} \begin{pmatrix} X & I \\ M^\top & 0 \end{pmatrix}
$$
$$
= \begin{pmatrix} YX + NM^\top & Y \\ N^\top X + Z_2 M^\top & N^\top \end{pmatrix} = \Phi_2 \qquad (6.39)
$$

and

$$\Phi_1^\top P \Phi_1 = \begin{pmatrix} X & I \\ I & Y \end{pmatrix}.$$

Define a set of matrices $\mathcal{K}_A, \mathcal{K}_B$, and $\mathcal{K}_C$ as follows:

$$\begin{cases} \mathcal{K}_B = N K_B \\ \mathcal{K}_C = K_C M^\top \\ \mathcal{K}_A = Y A X + \mathcal{K}_B C X + Y B \mathcal{K}_C + N K_A M^\top + \mathcal{K}_B B_y \mathcal{K}_C. \end{cases} \quad (6.40)$$

Direct manipulation gives

$$\Phi_1^\top P \bar{A} \Phi_1 = \Phi_2^\top \bar{A} \Phi_1$$

$$= \begin{pmatrix} \begin{bmatrix} A X + B K_C M^\top \\ Y A X + N K_B C X \\ + Y B K_C M^\top + N K_A M^\top \\ + N K_B B_y K_C M^\top \end{bmatrix} & A \\ & Y A + N K_B C \end{pmatrix}$$

$$= \begin{pmatrix} A X + B \mathcal{K}_C & A \\ \mathcal{K}_A & Y A + \mathcal{K}_B C \end{pmatrix} \quad (6.41)$$

from which it follows that

$$\Phi_1^\top [P\bar{A} + \bar{A}^\top P]\Phi_1 =$$

$$\begin{pmatrix} \begin{bmatrix} A X + X A^\top + \\ B \mathcal{K}_C + \mathcal{K}_C^\top B^\top \end{bmatrix} & A + \mathcal{K}_A^\top \\ \mathcal{K}_A + A^\top & \begin{bmatrix} Y A + A^\top Y \\ + \mathcal{K}_B C + C^\top \mathcal{K}_B^\top \end{bmatrix} \end{pmatrix} \triangleq T_{11}$$

$$\Phi_1^\top P \bar{A}_d = \Phi_2^\top \begin{pmatrix} A_d \\ K_B C_d \end{pmatrix}$$

$$= \begin{pmatrix} A_d \\ Y A_d + N K_B C_d \end{pmatrix} = \begin{pmatrix} A_d \\ Y A_d + \mathcal{K}_B C_d \end{pmatrix} = T_{12}$$

$$\Phi_1^\top P \bar{B}_1 = \Phi_2^\top \begin{pmatrix} B_1 \\ K_B B_2 \end{pmatrix} = \begin{pmatrix} B_1 \\ Y B_1 + N K_B B_2 \end{pmatrix}$$

$$= \begin{pmatrix} B_1 \\ Y B_1 + \mathcal{K}_B B_2 \end{pmatrix} \triangleq T_{13}$$

and

$$\Phi_1^\top P \bar{D}_a = \Phi_2^\top \begin{pmatrix} D_1 \\ K_B D_3 \end{pmatrix}$$

$$= \begin{pmatrix} D_1 \\ Y D_1 + N K_B D_3 \end{pmatrix} = \begin{pmatrix} D_1 \\ Y D_1 + \mathcal{K}_B D_3 \end{pmatrix} \quad (6.42)$$

$$\Phi_1^\top P \bar{D}_d = \Phi_2^\top \begin{pmatrix} D_2 \\ K_B D_4 \end{pmatrix}$$

$$= \begin{pmatrix} D_2 \\ Y D_2 + N K_B D_4 \end{pmatrix} = \begin{pmatrix} D_2 \\ Y D_2 + \mathcal{K}_B D_4 \end{pmatrix}. \quad (6.43)$$

Combining (6.42) and (6.43) yields

$$\Phi_1^\top J_{14} = \begin{pmatrix} D_1 & D_2 \\ YD_1 + \mathcal{K}_B D_3 & YD_2 + \mathcal{K}_B D_4 \end{pmatrix} \triangleq T_{14}$$

$$\Phi_1^\top I_0^\top = \begin{pmatrix} X \\ I \end{pmatrix} \tag{6.44}$$

$$\Phi_1^\top \bar{C}_z^\top = \begin{pmatrix} X & M \\ I & \mathbf{0} \end{pmatrix} \begin{pmatrix} C_z^\top \\ K_C^\top B_z^\top \end{pmatrix}$$

$$= \begin{pmatrix} XC_z^\top + MK_C^\top B_z^\top \\ C_z^\top \end{pmatrix} \tag{6.45}$$

$$= \begin{pmatrix} XC_z^\top + \mathcal{K}_C^\top B_z^\top \\ C_z^\top \end{pmatrix} \tag{6.46}$$

$$\Phi_1^\top E_a^\top = \begin{pmatrix} X & M \\ I & \mathbf{0} \end{pmatrix} \begin{pmatrix} E_1^\top \\ K_C^\top E_3^\top \end{pmatrix}$$

$$= \begin{pmatrix} XE_1^\top + \mathcal{K}_C^\top E_3^\top \\ E_1^\top \end{pmatrix}. \tag{6.47}$$

Using now (6.44)-(6.47) yields

$$\Phi_1^\top J_{15} = \begin{pmatrix} X & XC_z^\top + \mathcal{K}_C^\top B_z^\top & XE_1^\top + \mathcal{K}_C^\top E_3^\top \\ I & C_z^\top & E_1^\top \end{pmatrix} \tag{6.48}$$

$$\triangleq T_{15}. \tag{6.49}$$

From the above discussion, we get the following theorem.

Theorem 6.7 (i) If there exist matrices K_A, K_B, K_C, a symmetric, positive-definite matrix $P > 0$, and a scalar $\mu > 0$ such that (6.38) holds for some given matrix $Q > 0$, and a scalar $\varepsilon > 0$, which implies that controller (6.34) stabilizes system (6.1) and the closed-loop system has noise attenuation level $\gamma = \sqrt{\mu}$, then LMIs

$$\begin{pmatrix} T_{11} & T_{12} & T_{13} & T_{14} & T_{15} \\ T_{12}^\top & -Q + \varepsilon E_2^\top E_2 & & & \\ T_{13}^\top & & -\mu I & & \\ T_{14}^\top & & & -\varepsilon I & \\ T_{15}^\top & & & & -J_{55} \end{pmatrix} < 0 \tag{6.50}$$

$$\begin{pmatrix} X & I \\ I & Y \end{pmatrix} > 0 \tag{6.51}$$

have a set of feasible solutions $X > 0, Y > 0, \mathcal{K}_A, \mathcal{K}_B, \mathcal{K}_C$.

(ii) On the other hand, if there exist matrices $X > 0$, $Y > 0$, $\mathcal{K}_A$, $\mathcal{K}_B$, $\mathcal{K}_C$ satisfying (6.50) and (6.51), then there exist matrices $P > 0$, K_A, K_B and K_C satisfying (6.38).

Proof: The proof of (i) follows directly from the previous discussion.

Now, we proceed to prove (ii). Let $X > 0, Y > 0, \mathcal{K}_A, \mathcal{K}_B, \mathcal{K}_C$ be a set of solutions of (6.50) and (6.51). For given $X > 0, Y > 0$, there exists matrix $P > 0$ satisfying (3.48) if and only if $X - Y^{-1} > 0$, which is equivalent to (6.51). Thus, there exists matrix $P > 0$ satisfying (3.48). Next, we construct such a P. Using singular value decomposition, we can obtain M, N satisfying $MN^\top = I - XY$. With X, Y, M, N, thus obtained, we can define Φ_1 and Φ_2 by (3.55). Using (6.39), we obtain

$$P = \Phi_2 \Phi_1^{-1}.$$

With $\mathcal{K}_A, \mathcal{K}_B, \mathcal{K}_C, X > 0, Y > 0$, M, and N given, solving (6.40) gives

$$
\begin{cases}
K_B = N^{-1}\mathcal{K}_B \\
K_C = \mathcal{K}_C[M^\top]^{-1} \\
K_A = N^{-1}\left[\mathcal{K}_A - YAX - \mathcal{K}_B CX - YB\mathcal{K}_C - \mathcal{K}_B B_y \mathcal{K}_C\right][M^\top]^{-1}.
\end{cases}
\tag{6.52}
$$

It is easy to check that K_A, K_C, K_C, and P defined above satisfy (6.38). This completes the proof of Theorem 6.7. □

To show the usefulness of the output feedback design method developed above, let us consider a numerical example.

Example 6.3 *Let us consider a system described by (6.1) and suppose that the system data are as follows:*

$$A = \begin{pmatrix} -2 & 1 \\ 0.3 & -1 \end{pmatrix} \quad A_d = \begin{pmatrix} 0.2 & 0.1 \\ 0.3 & -0.1 \end{pmatrix}$$

$$B = \begin{pmatrix} -1 & 0 \\ 0.1 & -1 \end{pmatrix} \quad B_1 = \begin{pmatrix} 0 & 1 \\ 1 & 0 \end{pmatrix}$$

$$B_2 = \begin{pmatrix} 1 & -1 \\ 1 & 0 \end{pmatrix} \quad C = \begin{pmatrix} -1 & 0 \\ 0 & -1 \end{pmatrix}$$

$$C_d = \begin{pmatrix} 0 & 1 \\ -1 & 0 \end{pmatrix} \quad C_z = \begin{pmatrix} 1 & 0 \\ -1 & 1 \end{pmatrix}$$

$$D_1 = \begin{pmatrix} 1 & 0 \\ 0 & 0 \end{pmatrix} \quad D_2 = \begin{pmatrix} 0 & 1 \\ 0 & 1 \end{pmatrix}$$

$$D_3 = \begin{pmatrix} 0 & 1 \\ 0 & 1 \end{pmatrix} \quad D_4 = \begin{pmatrix} 0 & 0 \\ 1 & 0 \end{pmatrix}$$

$$B_y = \begin{pmatrix} 0.1 & 0 \\ -1 & 0 \end{pmatrix} \quad E_1 = \begin{pmatrix} 0 & 1 \\ 1 & 0 \end{pmatrix}$$

$$E_3 = \begin{pmatrix} 0.1 & -1 \\ 0 & 0 \end{pmatrix} \quad E_2 = \begin{pmatrix} 0.1 & 0 \\ 0.1 & 0 \end{pmatrix}.$$

With these data and letting $Q = I, \varepsilon = 0.1$, solving LMIs (6.50) and (6.51) yields

$$X = \begin{pmatrix} 4.8147 & 0.6747 \\ 0.6747 & 0.4786 \end{pmatrix} \quad Y = \begin{pmatrix} 29.2682 & -15.5929 \\ -15.5929 & 36.1893 \end{pmatrix}$$

$\mu = 132.1377.$

Using singular value decomposition, we obtain the following matrices M, N satisfying $MN^\top = I - XY$

$$M = \begin{pmatrix} 0.9329 & -0.3601 \\ -0.3601 & -0.9329 \end{pmatrix} \quad N = \begin{pmatrix} -138.9587 & -0.6633 \\ -9.3733 & 9.8330 \end{pmatrix}.$$

With $X, Y, M, N, \mathcal{A}, \mathcal{B}, \mathcal{C}$ obtained, using (6.52) we obtain the following set of gains:

$$K_A = \begin{pmatrix} -34.8296 & 4.5874 \\ 6.7611 & -4.3101 \end{pmatrix} \quad K_B = \begin{pmatrix} -5.2998 & -1.5283 \\ -3.2006 & -7.5467 \end{pmatrix}$$

$$K_C = \begin{pmatrix} -1.1358 & 0.2288 \\ -0.2362 & 0.1136 \end{pmatrix}.$$

According to Theorem 6.7, we conclude that controller (6.34) stabilizes system (6.1) and that the closed-loop system satisfies noise attenuation level $\gamma = \sqrt{\mu} = 11.4951.$

6.3   Robust $\mathcal{H}_\infty$-Filtering

The design of a state feedback controller requires access to the state of the system being considered at each time t. Since this is not always possible for practical reasons and the cost of the sensors, it is preferable to resort to filtering to get a good estimate for the state vector at each time. Different techniques have been proposed in the literature and $\mathcal{H}_\infty$ theory is one of the approaches we can use to solve this filtering problem. In the case of the robust filtering problem, the technique can be stated as follows: Given a dynamical time delay system with exogenous input belonging to $\mathcal{L}_2 [0, \infty]$, and given the measurement of the output, design a filter that gives an estimate of the unmeasured output such that the $\mathcal{H}_\infty$-norm of the transfer function between the estimation error and the exogenous input is minimized or kept below a certain given $\gamma > 0$. Mathematically, this process is stated as follows: Given $\gamma > 0$, find a filter such that

$$\sup_{w \in \mathcal{L}_2 [0, \infty]} \frac{\|z(t) - \hat{z}(t)\|_2^2}{\|w(t)\|_2^2} < \gamma^2$$

for all admissible uncertainties.

Therefore, the $\mathcal{H}_\infty$-filtering problem can be regarded as a special $\mathcal{H}_\infty$ control problem.

This section deals with the $\mathcal{H}_\infty$-filtering problem of system (6.1) with $u(t) \equiv 0$, that is,

$$\begin{cases} \dot{x}(t) = A(t)x(t) + A_d(t)x(t-\tau) + B_1 w(t) \\ x(s) = \phi(s), \ [-\bar{\tau}, 0] \\ y(t) = C(t)x(t) + C_d(t)x(t-\tau) + B_2 w(t) \\ z(t) = C_z x(t) \end{cases} \tag{6.53}$$

where all the variables and the matrices will keep the same meanings as before.

For the rest of this section let us assume that the following hold:

$$A(t) = A + D_1\Delta(t)E_1 \quad A_d(t) = A_d + D_2\Delta_d(t)E_2$$
$$C(t) = C + D_3\Delta(t)E_1 \quad C_d(t) = C_d + D_4\Delta_d(t)E_2.$$

The goal of this section is to design a linear time-invariant filter of form of (5.42) such that the extended system $\{(x(t), \xi(t)), t \geq 0\}$ is internally stable and the estimation error

$$e(t) = z(t) - \hat{z}(t) \tag{6.54}$$

satisfies (5.43) for all admissible uncertainties.

Let $\eta^\top(t) = (x^\top(t) \ \xi^\top(t))$ be the state of the augmented system, then we have

$$\begin{cases} \dot{\eta}(t) = \hat{A}(t)\eta(t) + \hat{A}_d(t)x(t-\tau) + \hat{B}_1 w(t) \\ e(t) = \hat{C}\eta(t) \end{cases} \tag{6.55}$$

where

$$\hat{A}(t) = \begin{pmatrix} A(t) & \mathbf{0} \\ K_B C(t) & K_A \end{pmatrix}$$

$$\hat{A}_d(t) = \begin{pmatrix} A_d(t) \\ K_B C_d(t) \end{pmatrix}$$

$$\hat{B}_1 = \begin{pmatrix} B_1 \\ K_B B_2 \end{pmatrix}$$

$$\hat{C} = (C_z \ -K_C).$$

Let

$$\hat{A} = \begin{pmatrix} A & \mathbf{0} \\ K_B C & K_A \end{pmatrix} \quad \hat{A}_d = \begin{pmatrix} A_d \\ K_B C_d \end{pmatrix} \quad \hat{D}_a = \begin{pmatrix} D_1 \\ K_B D_3 \end{pmatrix}$$

$$\hat{E}_a = (E_1 \ \mathbf{0}) \quad \hat{D}_d = \begin{pmatrix} D_2 \\ K_B D_4 \end{pmatrix} \quad \hat{E}_d = E_2.$$

Then

$$\hat{A}(t) = \hat{A} + \hat{D}_a\Delta(t)\hat{E}_a \quad \hat{A}_d(t) = \hat{A}_d + \hat{D}_a\Delta_d(t)\hat{E}_d.$$

Assume $\phi(\cdot) = 0$. Using Theorem 6.2, we obtain the following proposition.

Proposition 6.3 *Let γ be a given positive scalar. If there exist symmetric, positive-definite matrices $P > 0$, $Q > 0$, and a scalar $\varepsilon > 0$ such that*

$$
\begin{pmatrix}
\# & P\hat{A}_d & P\hat{B}_1 & P\hat{D}_a & P\hat{D}_d \\
\hat{A}_d^\top P & -Q + \varepsilon \hat{E}_d^\top \hat{E}_d & 0 & 0 & 0 \\
\hat{B}_1^\top P & 0 & -\gamma^2 I & 0 & 0 \\
\hat{D}_a^\top P & 0 & 0 & -\varepsilon I & 0 \\
\hat{D}_d^\top P & 0 & 0 & 0 & -\varepsilon I
\end{pmatrix} < 0 \qquad (6.56)
$$

holds, where $\# = P\hat{A} + \hat{A}^\top P + I_0^\top Q I_0 + \hat{C}^\top \hat{C} + \varepsilon \hat{E}_a^\top \hat{E}_a$, $I_0 = (I, \mathbf{0})$, then the error system (6.54) is asymptotically stable and has noise attenuation level γ.

Proof: Consider the Lyapunov functional candidate

$$
V(\eta(s), t - \tau \le s \le t) = \eta^\top(t) P \eta(t) + \int_{t-\tau}^t \eta^\top(s) I_0^\top Q I_0 \eta(s) ds
$$

$$
= \eta^\top(t) P \eta(t) + \int_{t-\tau}^t x^\top(s) Q x(s) ds.
$$

The rest of the proof following the same line as in Theorem 6.1 and Theorem 6.2. □

For a filter (5.42) with K_A, K_B, K_C given, Proposition 6.3 provides a sufficient condition for the estimation error system to be stable and satisfies a given noise attenuation level γ. Since (6.56) is nonlinear with respect to the parameters K_A, K_B, K_C, and P, Proposition 6.3 cannot be used to design a filter. Next, we proceed to design a filter of the form of (5.42) by overcoming this nonlinearity. The following lemma gives a first attempt to solve the filtering problem.

Lemma 6.2 *Let γ be a given positive scalar. If there exist matrices K_A, K_B, K_C, and symmetric, positive-definite matrices $P > 0$, $Q > 0$, and a scalar $\varepsilon > 0$ satisfying (6.56), then the following LMIs have a set of feasible solutions $X > 0, Y > 0, \mathcal{A}, \mathcal{B}, \mathcal{C}$:*

$$
\begin{pmatrix} X & I \\ I & Y \end{pmatrix} > 0 \qquad (6.57)
$$

$$
\begin{pmatrix}
\mathcal{U}_{11} & \mathcal{U}_{12} & \mathcal{U}_{13} & \mathcal{U}_{14} \\
\mathcal{U}_{12}^\top & -Q + \varepsilon E_d^\top E_d & & \\
\mathcal{U}_{13}^\top & & -\mathcal{U}_{33} & \\
\mathcal{U}_{14}^\top & & & -\mathcal{U}_{44}
\end{pmatrix} < 0, \qquad (6.58)
$$

where

$$
\mathcal{U}_{11} = \begin{pmatrix}
AX + XA^\top & A + \mathcal{A}^\top \\
A + \mathcal{A}^\top & \begin{bmatrix} YA + \mathcal{B}C \\ +A^\top Y + \mathcal{C}^\top \mathcal{B}^\top \end{bmatrix}
\end{pmatrix}
$$

$$\mathcal{U}_{12} = \begin{pmatrix} A_d \\ YA_d + \mathcal{B}C_d \end{pmatrix}$$

$$\mathcal{U}_{13} = \begin{pmatrix} B_1 & D_1 & D_2 \\ YB_1 + \mathcal{B}B_2 & YD_1 + \mathcal{B}D_3 & YD_2 + \mathcal{B}D_4 \end{pmatrix}$$

$$\mathcal{U}_{33} = \text{diag}\{\gamma^2 I, \varepsilon I, \varepsilon I\}$$

$$\mathcal{U}_{14} = \Phi_1^\top \mathcal{F}_{14} = \begin{pmatrix} XC_z^\top - \mathcal{C}^\top & E_1^\top & X \\ C_z^\top & XE_1^\top & I \end{pmatrix}$$

$$\mathcal{U}_{44} = \text{diag}\{I, \frac{1}{\varepsilon}I, Q^{-1}\}.$$

Proof: Let $K_A, K_B, K_C, P > 0, Q > 0$, and the scalar $\varepsilon > 0$ satisfy (6.56). Using the Schur complement, we find that (6.56) holds if and only if the following holds:

$$\begin{pmatrix} P\hat{A} + \hat{A}P & P\hat{A}_d & \mathcal{F}_{13} & \mathcal{F}_{14} \\ \hat{A}_d^\top P & -Q + \varepsilon E_d^\top E_d & & \\ \mathcal{F}_{13}^\top & & -\mathcal{U}_{33} & \\ \mathcal{F}_{14}^\top & & & -\mathcal{U}_{44} \end{pmatrix} < 0, \qquad (6.59)$$

where

$$\mathcal{F}_{13} = \begin{pmatrix} P\hat{B}_1 & P\hat{D}_a & P\hat{D}_d \end{pmatrix}$$
$$\mathcal{F}_{14} = \begin{pmatrix} \hat{C}^\top & \hat{E}_a^\top & I_0^\top \end{pmatrix}.$$

Assume that matrices P and P^{-1} can be partitioned as in (3.47) with $X > 0, Y > 0$ being symmetric, positive-definite and M, N invertible and define matrices Φ_1, Φ_2 as in (3.55). Obviously, Φ_1, Φ_2 are invertible. Direct manipulation gives

$$P\Phi_1 = \Phi_2 \qquad (6.60)$$

and

$$\Phi_1^\top P\Phi_1 = \Phi_2^\top \Phi_1 = \begin{pmatrix} I & 0 \\ Y & N \end{pmatrix}\begin{pmatrix} X & I \\ M^\top & 0 \end{pmatrix}$$
$$= \begin{pmatrix} X & I \\ YX + NM^\top & Y \end{pmatrix} = \begin{pmatrix} X & I \\ I & Y \end{pmatrix}. \qquad (6.61)$$

Since Φ_1 is invertible, using the fact that P is symmetric, positive-definite matrix, (6.61) implies that X, Y satisfy (6.57). Pre- and post-multiplying (6.59) by $\Xi = \text{diag}\{\Phi_1^\top, I, I, I\}$ and $\Xi^\top$ yields

$$\begin{pmatrix} \Phi_1^\top[P\hat{A} + \hat{A}^\top P]\Phi_1 & \Phi_1^\top P\hat{A}_d & \Phi_1^\top \mathcal{F}_{13} & \Phi_1^\top \mathcal{F}_{14} \\ \hat{A}_d^\top P\Phi_1 & -Q + \varepsilon E_d^\top E_d & & \\ \mathcal{F}_{13}^\top \Phi_1 & & -\mathcal{U}_{33} & \\ \mathcal{F}_{14}^\top \Phi_1 & & & -\mathcal{U}_{44} \end{pmatrix} < 0. \quad (6.62)$$

Let us define the matrices $\mathcal{A}, \mathcal{B}, \mathcal{C}$ as follows:

$$\begin{cases} \mathcal{A} = YAX + NK_BCX + NK_AM^\top \\ \mathcal{B} = NB \\ \mathcal{C} = K_CM^\top. \end{cases} \tag{6.63}$$

Simple computation gives

$$\Phi_1^\top P\hat{A}\Phi_1 = \Phi_2^\top \hat{A}\Phi_1 = \begin{pmatrix} I & 0 \\ Y & N \end{pmatrix} \begin{pmatrix} A & 0 \\ K_BC & K_A \end{pmatrix} \begin{pmatrix} X & I \\ M^\top & 0 \end{pmatrix}$$

$$= \begin{pmatrix} AX & A \\ YAX + NK_BCX + NK_AM^\top & YA + NK_BC \end{pmatrix}$$

$$= \begin{pmatrix} AX & A \\ \mathcal{A} & YA + \mathcal{B}C \end{pmatrix},$$

from which it follows that

$$\Phi_1^\top[P\hat{A} + \hat{A}^\top P]\Phi_1 = \begin{pmatrix} AX + XA^\top & A + \mathcal{A}^\top \\ \mathcal{A} + A^\top & \begin{bmatrix} YA + \mathcal{B}C \\ +A^\top Y + C^\top \mathcal{B}^\top \end{bmatrix} \end{pmatrix} = \mathcal{U}_{11}$$

$$\Phi_1^\top P\hat{A}_d = \begin{pmatrix} I & 0 \\ Y & N \end{pmatrix} \begin{pmatrix} A_d \\ K_BC_d \end{pmatrix}$$

$$= \begin{pmatrix} A_d \\ YA_d + NK_BC_d \end{pmatrix}$$

$$= \begin{pmatrix} A_d \\ YA_d + \mathcal{B}C_d \end{pmatrix} = \mathcal{U}_{12}$$

$$\Phi_1^\top P\hat{B}_1 = \Phi_2^\top \hat{B}_1 = \begin{pmatrix} I & 0 \\ Y & N \end{pmatrix} \begin{pmatrix} B_1 \\ K_BB_2 \end{pmatrix}$$

$$= \begin{pmatrix} B_1 \\ YB_1 + NK_BB_2 \end{pmatrix}$$

$$= \begin{pmatrix} B_1 \\ YB_1 + \mathcal{B}B_2 \end{pmatrix} \tag{6.64}$$

$$\Phi_1^\top P\hat{D}_a = \Phi_2^\top \hat{D}_a = \begin{pmatrix} I & 0 \\ Y & N \end{pmatrix} \begin{pmatrix} D_1 \\ K_BD_3 \end{pmatrix}$$

$$= \begin{pmatrix} D_1 \\ YD_1 + NK_BD_3 \end{pmatrix} = \begin{pmatrix} D_1 \\ YD_1 + \mathcal{B}D_3 \end{pmatrix} \tag{6.65}$$

$$\Phi_1^\top P\hat{D}_d = \Phi_2^\top \hat{D}_d = \begin{pmatrix} I & 0 \\ Y & N \end{pmatrix} \begin{pmatrix} D_2 \\ K_BD_4 \end{pmatrix}$$

$$= \begin{pmatrix} D_2 \\ YD_2 + NK_BD_4 \end{pmatrix} = \begin{pmatrix} D_2 \\ YD_2 + \mathcal{B}D_4 \end{pmatrix}. \tag{6.66}$$

Combining now (6.64) - (6.66) yields

$$\Phi_1^\top \mathcal{F}_{13} = \begin{pmatrix} B_1 & D_1 & D_2 \\ YB_1 + \mathcal{B}B_2 & YD_1 + \mathcal{B}D_3 & YD_2 + \mathcal{B}D_4 \end{pmatrix} = \mathcal{U}_{13}. \tag{6.67}$$

Moreover,

$$\Phi_1^\top \hat{C}^\top = \begin{pmatrix} X & M \\ I & 0 \end{pmatrix} \begin{pmatrix} C_z^\top \\ -K_C^\top \end{pmatrix} = \begin{pmatrix} XC_z^\top - MK_C^\top \\ C_z^\top \end{pmatrix}$$
$$= \begin{pmatrix} XC_z^\top - \mathcal{C}^\top \\ C_z^\top \end{pmatrix} \tag{6.68}$$

$$\Phi_1^\top \hat{E}_a^\top = \begin{pmatrix} X & M \\ I & 0 \end{pmatrix} \begin{pmatrix} E_1^\top \\ 0 \end{pmatrix} = \begin{pmatrix} XE_1^\top \\ E_1^\top \end{pmatrix} \tag{6.69}$$

$$\Phi_1^\top I_0^\top = \begin{pmatrix} X & M \\ I & 0 \end{pmatrix} \begin{pmatrix} I \\ 0 \end{pmatrix} = \begin{pmatrix} X \\ I \end{pmatrix}. \tag{6.70}$$

Using (6.68)-(6.70) yields

$$\Phi_1^\top \mathcal{F}_{14} = \begin{pmatrix} XC_z^\top - \mathcal{C}^\top & XE_1^\top & X \\ C_z^\top & E_1^\top & I \end{pmatrix}. \tag{6.71}$$

Combining the above derivation and (6.62) completes the proof of Lemma 6.2. □

From above lemma, we obtain the following theorem.

Theorem 6.8 *(i) Let γ be a given positive scalar. If there exist matrices K_A, K_B, K_C, $P > 0$, $Q > 0$ and scalar $\varepsilon > 0$ satisfying (6.56), which means that the estimation error of filter (5.42) is internally stable and satisfies (5.43), then LMIs (6.57) and (6.58) have a set of feasible solutions $X > 0, Y > 0$, $\mathcal{A}, \mathcal{B}, \mathcal{C}$.*
(ii) On the other hand, if LMIs (6.57) and (6.58) have a set of feasible solutions $X > 0, Y > 0$, $\mathcal{A}, \mathcal{B}, \mathcal{C}$ for a given $Q > 0$ and $\varepsilon > 0$, then there exist matrices K_A, K_B, K_C such that the estimation error system of filter (5.42) is stable and satisfies (5.43).

Proof: The proof of (i) is given in Lemma 6.2. We come to the proof of (ii). In view of Proposition 6.3, to prove (ii) it suffices to find K_A, K_B, K_C, and $P > 0$ satisfying (6.38). Let $X > 0, Y > 0$, $\mathcal{A}, \mathcal{B}, \mathcal{C}$ be a set of feasible solutions of LMIs (6.57) and (6.58). For given matrices $X > 0, Y > 0$, there exists matrix P satisfying

$$P = \begin{pmatrix} Y & \# \\ \# & \# \end{pmatrix} \quad P^{-1} = \begin{pmatrix} X & \# \\ \# & \# \end{pmatrix} \tag{6.72}$$

where $\#$ denotes appropriate matrices if and only if $X - Y^{-1} > 0$, which is ensured by (6.57). Using singular value decomposition, we can obtain matrices M, N with full rank satisfying

$$MN^\top = I - XY.$$

With X, Y, M, N, $\mathcal{A}, \mathcal{B}, \mathcal{C}$ given, solving matrix equation (6.63) gives K_A, K_B, K_C as follows:

$$\begin{cases} K_A = N^{-1} \left[\mathcal{A} - YAX - \mathcal{B}CX \right] \left[M^\top \right]^{-1} \\ K_B = N^{-1} \mathcal{B} \\ K_C = \mathcal{C} \left[M^\top \right]^{-1}. \end{cases} \tag{6.73}$$

Define matrices Φ_1 and Φ_2 as in (3.55), and let P be the solution of

$$P\Phi_1 = \Phi_2.$$

It is easy to check that K_A, K_B, K_C and P defined above satisfy (6.38). This completes the proof of Theorem 6.8. $\qquad\qquad\square$

6.4 Guaranteed Cost Control

This section deals with the guaranteed cost control problem for uncertain linear systems with constant time delay. It consists of designing a controller that robustly stabilizes system (2.1) and at the same guarantees a given bound for the cost function. Mathematically, the technique can be formulated as follows: Given a dynamical system with the appropriate assumptions, find a controller $u(.)$ that robustly stabilizes the system and at the same time guarantees that the cost is bounded for all admissible uncertainties $\Delta_1(t)$ and $\Delta_2(t)$. That is,

$$J = \int_0^\infty \left[x^\top(t) R_1 x(t) + u^\top(t) R_2 u(t) \right] dt$$

with R_1 and R_2 being symmetric, positive-definite matrices. For this purpose let us consider a system described by the following dynamics:

$$\begin{cases} \dot{x}(t) = A(t)x(t) + B(t)u(t) + A_d(t)x(t - \tau) \\ x(s) = \phi(s), \ s \in [-\tau, 0] \\ y(t) = C(t)x(t) + B_2(t)u(t) + C_d(t)x(t - \tau) \end{cases} \tag{6.74}$$

where $x(t) \in \mathbb{R}^{n_1}, y(t) \in \mathbb{R}^{n_2}, u(t) \in \mathbb{R}^{n_3}$ denote the state, the output and the control input of the system, respectively, and the matrices $A(t), A_d(t), B(t), B_2(t), C(t), C_d(t)$ are defined as follows:

$$\begin{array}{ll} A(t) = A + D_1 \Delta(t) E_1 & B(t) = B + D_1 \Delta(t) E_2 \\ A_d(t) = A_d + D_2 \Delta_d(t) E_3 & C(t) = C + D_3 \Delta(t) E_1 \\ B_2(t) = B_2 + D_3 \Delta(t) E_2 & C_d(t) = C_d + D_3 \Delta_d(t) E_3 \end{array}$$

where $A, B, A_d, C, C_d, B_2, D_1, D_2, D_3, E_1, E_2, E_3$ are matrices with appropriate dimensions.

For system (6.74), let us consider the following cost function

$$J = \int_0^\infty \left[x^\top(t) R_1 x(t) + u^\top(t) R_2 u(t) \right] dt, \tag{6.75}$$

where $R_1 \geq 0, R_2 \geq 0$ are symmetric, positive-definite matrices. The goal of this section is to design a state feedback controller and an output feedback controller that stabilize system (6.74) and assure that the cost function defined by (6.75) satisfies a given upper bound for all admissible uncertainties.

6.4.1   Guaranteed Cost Bound

Before tackling the design problem of these two controllers, let us first consider the case of the unforced systems, that is, $u(t) \equiv 0$ and try to develop a test that can be used to check if the unforced system is stable and the corresponding cost is bounded. We are also interested in the optimal bound for the cost function. The following theorem gives such a test and an upper bound for the cost function defined by (6.75).

Theorem 6.9 *If there exist symmetric, positive-definite matrices $P > 0$, $Q > 0$ such that*

$$\Theta_g(t) = \begin{pmatrix} PA(t) + A^\top(t)P + R_1 + Q & PA_d(t) \\ A_d^\top(t)P & -Q \end{pmatrix} < 0 \qquad (6.76)$$

holds for all admissible uncertainties, then system (6.74) with $u(t) \equiv 0$ is stable; moreover, the cost defined by (6.75) satisfies

$$J \leq \mathrm{tr}[Px_0 x_0^\top] + \mathrm{tr}\left[Q \int_{-\tau}^0 \phi(s)\phi^\top(s)ds\right] \qquad (6.77)$$

for all admissible uncertainties, where $x_0 = \phi(0)$.

Proof: For any $t \geq 0$, define $\mathbf{x}_t \in \mathbb{C}[-\tau, 0]$ by $\mathbf{x}_t = x(t+s), \quad s \in [-\tau, 0]$. Let us consider a Lyapunov functional candidate as follows:

$$V(\mathbf{x}_t) = x^\top(t)Px(t) + \int_{t-\tau}^t x^\top(s)Qx(s)ds,$$

where P and Q are symmetric, positive-definite matrices.
Then, we have

$$\dot{V}(\mathbf{x}_t) = x^\top(t)[PA(t) + A^\top(t)P]x(t) + 2x^\top(t)PA_d(t)x(t-\tau) \\ + x^\top(t)Qx(t) - x^\top(t-\tau)Qx(t-\tau),$$

yielding

$$\dot{V}(\mathbf{x}_t) + x^\top(t)R_1 x(t) = \left(x^\top(t) \; x^\top(t-\tau)\right) \Theta_g(t) \begin{pmatrix} x(t) \\ x(t-\tau) \end{pmatrix}$$

$$< 0. \qquad (6.78)$$

Therefore, (6.76) implies that system (6.74) with $u(t) \equiv 0$ is asymptotically stable. Using now (6.78) and the fact that the system is stable, we obtain

$$\int_0^\infty x^\top(t)R_1x(t)dt \le -\int_0^\infty \dot{V}(\mathbf{x}_t)dt$$

$$= V(\mathbf{x}_0) = x^\top(0)Px(0) + \int_{-\tau}^0 \phi^\top(s)Q\phi(s)ds,$$

yielding (6.77). This ends the proof of Theorem (6.9)   □

Theorem 6.9 establishes a sufficient condition for system (6.74) with $u(t) \equiv 0$ to be stable, while the corresponding cost function defined by (6.75) verifies upper bound (6.77). However, (6.76) contains system uncertainties and thus it is not useful. The following theorem provides an equivalent sufficient condition for (6.76) to hold for all admissible uncertainties in the system.

Theorem 6.10 *If there exist symmetric, positive-definite matrices $P > 0$, $Q > 0$, and a scalar $\varepsilon > 0$ such that*

$$\left(\begin{array}{cccc} \begin{bmatrix} PA + A^\top P \\ + R_1 + Q \\ + \varepsilon E_1^\top E_1 \end{bmatrix} & PA_d & PD_1 & PD_2 \\ A_d^\top P & -Q + \varepsilon E_3^\top E_3 & & \\ D_1^\top P & & -\varepsilon I & \\ D_2^\top P & & & -\varepsilon I \end{array} \right) < 0 \qquad (6.79)$$

holds, then system (6.74) with $u(t) \equiv 0$ is asymptotically stable; moreover, the cost defined by (6.75) satisfies (6.77) for all admissible uncertainties.

Proof: Using Theorem 6.9, to prove Theorem 6.10 it suffices to show that the symmetric, positive-definite matrices $P > 0$, $Q > 0$, and the positive scalar $\varepsilon > 0$ satisfying (6.79) must satisfy also (6.76).

Noting that

$$\Theta_g(t) = \left(\begin{array}{cc} PA + A^\top P + R_1 + Q & PA_d \\ A_d^\top P & -Q \end{array} \right)$$

$$+ \left(\begin{array}{cc} PD_1 & PD_2 \\ \mathbf{0} & \mathbf{0} \end{array} \right) \left(\begin{array}{cc} \Delta(t) & \mathbf{0} \\ \mathbf{0} & \Delta_d(t) \end{array} \right) \left(\begin{array}{cc} E_1 & \mathbf{0} \\ \mathbf{0} & E_3 \end{array} \right)$$

$$+ \left(\begin{array}{cc} E_1^\top & \mathbf{0} \\ \mathbf{0} & E_3^\top \end{array} \right) \left(\begin{array}{cc} \Delta^\top(t) & \mathbf{0} \\ \mathbf{0} & \Delta_d^\top(t) \end{array} \right) \left(\begin{array}{cc} D_1^\top P & \mathbf{0} \\ D_2^\top P & \mathbf{0} \end{array} \right)$$

$$(6.80)$$

and using Lemma A.5, we obtain that $\Theta_g(t) < 0$ if and only if

$$\left(\begin{array}{cc} PA + A^\top P + R_1 + Q & PA_d \\ A_d^\top P & -Q \end{array} \right)$$

$$+ \frac{1}{\varepsilon} \left(\begin{array}{cc} PD_1 & PD_2 \\ \mathbf{0} & \mathbf{0} \end{array} \right) \left(\begin{array}{cc} D_1^\top P & \mathbf{0} \\ D_2^\top P & \mathbf{0} \end{array} \right) + \varepsilon \left(\begin{array}{cc} E_1^\top E_1 & \mathbf{0} \\ \mathbf{0} & E_3^\top E_3 \end{array} \right) < 0$$

holds for some $\varepsilon > 0$, that is,

$$
\begin{pmatrix} PA + A^\top P + R_1 + Q + \varepsilon E_1^\top E_1 & PA_d \\ A_d^\top P & -Q + \varepsilon E_3^\top E_3 \end{pmatrix}
$$
$$
+ \frac{1}{\varepsilon} \begin{pmatrix} PD_1 & PD_2 \\ 0 & 0 \end{pmatrix} \begin{pmatrix} D_1^\top P & 0 \\ D_2^\top P & 0 \end{pmatrix} < 0.
$$

Using the Schur complement, the above matrix inequality holds if and only if (6.79) holds. Therefore, (6.79) ensures that the symmetric, positive-definite matrices P, $Q > 0$, and the positive scalar $\varepsilon > 0$ satisfy (6.76). This completes the proof of Theorem 6.10.   □

Based on Theorem 6.10, the optimal guaranteed cost bound can be obtained by solving the linear optimization problem

$$
\min_{P>0,Q>0,\varepsilon>0} \quad tr[Px_0 x_0^\top] + tr\left[Q \int_{-\tau}^0 \phi(s)\phi^\top(s)ds\right] \qquad (6.81)
$$
$$
\text{s.t. } (6.79)
$$

Remark 6.3 *The upper bound of the cost index obtained in this section depends on the initial condition x_0. To remove this dependence, we can assume that x_0 is a random vector satisfying $\mathbb{E}[x_0 x_0^\top] = $ a constant matrix.*

6.4.2   State Feedback Control

Now we are ready to design a state feedback controller that stabilizes the class of systems we are considering in this section and also assures a given upper bound for the chosen cost function. For this purpose let us consider the memoryless state feedback controller (3.27) in order to stabilize system (6.74) and guarantee a given bound for the cost function defined by (6.75) for all admissible uncertainties.

The following lemma states some preliminary results that will be used later on to prove the following theorems.

Lemma 6.3 *If there exist symmetric, positive-definite matrices $P > 0$ and $Q > 0$ and a positive scalar $\varepsilon > 0$ such that*

$$
\begin{pmatrix} \begin{bmatrix} PA + A^\top P + R_1 + Q \\ +\varepsilon P[D_1 D_1^\top + D_2 D_2^\top]P \end{bmatrix} & PA_d & E_1^\top & 0 \\ A_d^\top P & -Q & 0 & E_3^\top \\ E_1 & 0 & -\varepsilon I & 0 \\ 0 & E_3 & 0 & -\varepsilon I \end{pmatrix} < 0 \qquad (6.82)
$$

holds, then system (6.74) with $u(t) \equiv 0$ is asymptotically stable and moreover, the cost defined by (6.75) satisfies (6.77) for any admissible uncertainties.

Proof: Based on Theorem 6.9, to prove Lemma 6.3 it suffices to prove that if the matrices $P > 0, Q > 0$, and the scalar $\varepsilon > 0$ satisfy (6.82), then they satisfy also (6.76). Replacing ε by $1/\varepsilon$ in (6.80), we obtain

$$\begin{pmatrix} PA + A^\top P + R_1 + Q & PA_d \\ A_d^\top P & -Q \end{pmatrix}$$

$$+ \varepsilon \begin{pmatrix} PD_1 & PD_2 \\ 0 & 0 \end{pmatrix} \begin{pmatrix} D_1^\top P & 0 \\ D_2^\top P & 0 \end{pmatrix} + \frac{1}{\varepsilon} \begin{pmatrix} E_1^\top E_1 & 0 \\ 0 & E_3^\top E_3 \end{pmatrix}$$

$$= \begin{pmatrix} \begin{bmatrix} PA + A^\top P + R_1 + Q \\ +\varepsilon P[D_1 D_1^\top + D_2 D_2^\top]P \end{bmatrix} & PA_d \\ A_d^\top P & -Q \end{pmatrix}$$

$$+ \frac{1}{\varepsilon} \begin{pmatrix} E_1^\top E_1 & 0 \\ 0 & E_3^\top E_3 \end{pmatrix}.$$

Using the Schur complement, the above matrix is negative-definite if and only if (6.82) holds. Thus (6.82) implies (6.76). This completes the proof of Lemma 6.3. $\qquad\square$

Let us now consider the design of the gain matrix K of the state feedback controller. Substituting (3.27) into (6.74) yields the dynamics of the closed-loop system described by

$$\dot{x}(t) = \bar{A}(t)x(t) + A_d(t)x(t - \tau), \qquad (6.83)$$

where $\bar{A}(t) = \bar{A} + D_1\Delta(t)\bar{E}_a$ with $\bar{A} = A + BK$ and $\bar{E}_a = E_1 + E_2 K$.

The cost function is

$$J = \int_0^\infty [x^\top(t)R_1 x(t) + u^\top(t)R_2 u(t)]dt = \int_0^\infty [x^\top(t)\hat{R}x(t)]dt$$

where $\hat{R} = R_1 + K^\top R_2 K$.

Using Lemma 6.3, we get the following theorem.

Theorem 6.11 *If there exist symmetric, positive-definite matrix $X > 0$, matrix Y, and scalar $\varepsilon > 0$ such that*

$$\begin{pmatrix} \# & A_d & * & 0 & X & XS_1^\top & Y^\top S_2^\top \\ A_d^\top & -Q & 0 & E_3^\top & & & \\ E_1 X + E_2 Y & 0 & -\varepsilon I & & & & \\ 0 & E_3 & & -\varepsilon I & & & \\ X & & & & -Q^{-1} & & \\ S_1 X & & & & & -I & \\ S_2 Y & & & & & & -I \end{pmatrix} < 0, \quad (6.84)$$

holds for some symmetric, positive-definite $Q > 0$, where

$$\# = AX + XA^\top + BY + Y^\top B^\top + \varepsilon D_1 D_1^\top + \varepsilon D_2 D_2^\top$$

$$S_1 = R_1^{1/2}$$

$$S_2 = R_2^{1/2},$$

then controller (3.27) with $K = YX^{-1}$ stabilizes system (6.74) and the cost function of the closed-loop system satisfies

$$J \le \operatorname{tr}\left[X^{-1}x_0 x_0^\top\right] + \operatorname{tr}\left[Q \int_{-\tau}^0 \phi(s)\phi^\top(s)ds\right]$$

for all admissible uncertainties.

Proof: Using Lemma 6.3, we know that if there exist symmetric, positive-definite matrices $P > 0$, $Q > 0$, and a positive scalar $\varepsilon > 0$ satisfying

$$\left(\begin{bmatrix} P\bar{A} + \bar{A}^\top P + \hat{R} + Q \\ + \varepsilon P[D_1 D_1^\top + D_2 D_2^\top]P \end{bmatrix} \quad PA_d \quad \bar{E}_1^\top \quad 0 \atop A_d^\top P \quad -Q \quad 0 \quad E_3^\top \atop \bar{E}_1 \quad 0 \quad -\varepsilon I \quad 0 \atop 0 \quad E_3 \quad 0 \quad -\varepsilon I\right) < 0, \qquad (6.85)$$

then the closed-loop system is asymptotically stable and, moreover, the cost defined by (6.75) satisfies (6.77). Let $X = P^{-1}$. Pre- and post-multiplying both sides of (6.85) by $\operatorname{diag}\{X, I, I, I\}$ yields

$$\left(\begin{bmatrix} \bar{A}X + X\bar{A}^\top \\ + X\hat{R}X + XQX \\ + \varepsilon[D_1 D_1^\top + D_2 D_2^\top] \end{bmatrix} \quad A_d \quad X\bar{E}_1^\top \quad 0 \atop A_d^\top \quad -Q \quad 0 \quad E_3^\top \atop \bar{E}_1 X \quad 0 \quad -\varepsilon I \quad 0 \atop 0 \quad E_3 \quad 0 \quad -\varepsilon I\right) < 0. \qquad (6.86)$$

Letting $Y = KX$ in (6.86) and using the Schur complement, we obtain (6.84). This completes the proof. $\square$

To show the validity of the above controller design method, let us provide the following numerical example.

Example 6.4 *Consider a dynamical system with dynamics described by (6.74) and suppose that the system has the following data:*

$$A = \begin{pmatrix} 2 & 1 \\ 0 & 1.3 \end{pmatrix} \qquad A_d = \begin{pmatrix} 0.2 & 0.1 \\ 0.3 & -0.1 \end{pmatrix}$$

$$B = \begin{pmatrix} -1 & 0 \\ 0.1 & -1 \end{pmatrix} \qquad D_1 = \begin{pmatrix} 1 & 0 \\ 1 & -1 \end{pmatrix}$$

$$D_2 = \begin{pmatrix} 0 & -1 \\ 0 & 1 \end{pmatrix} \qquad E_1 = \begin{pmatrix} 0 & 1 \\ 1 & 0 \end{pmatrix}$$

$$E_3 = \begin{pmatrix} 1 & -1 \\ 0 & 1 \end{pmatrix} \qquad E_2 = \begin{pmatrix} 0.1 & 0 \\ 1 & 0 \end{pmatrix}$$

$$R_1 = R_2 = I.$$

With this set of data, solving (6.84) gives

$$X = \begin{pmatrix} -1.6969 & -0.0484 \\ -0.0484 & -0.6454 \end{pmatrix} \quad Y = \begin{pmatrix} 1.6527 & -0.0382 \\ -0.0107 & 1.6900 \end{pmatrix}.$$

$$\varepsilon = 0.8929.$$

Therefore, according to Theorem 6.11 controller (3.27) with

$$K = YX^{-1} = \begin{pmatrix} -0.9777 & 0.1325 \\ 0.0813 & -2.6247 \end{pmatrix}$$

stabilizes system (6.74) and guarantees cost bound (6.77).

6.4.3   Output Feedback Control

Under the assumption that the system state is completely accessible to feedback, we developed a state feedback controller that stabilizes system (6.1) and guarantees an upper bound for the cost function defined by (6.75). When the system state is not available for feedback, we can use output feedback control to stabilize our class of systems and guarantee that the cost is bounded.

This subsection deals with the problem of designing an output feedback controller in the form of (6.34) that stabilizes system (6.74).

Substituting (6.34) into (6.74) yields the dynamics of the closed-loop system as follows:

$$\dot{\eta}(t) = \bar{A}(t)\eta(t) + \bar{A}_d(t)I_0\eta(t-\tau)$$
$$= \bar{A}(t)\eta(t) + \bar{A}_d(t)x(t-\tau), \qquad (6.87)$$

where

$$\eta^\top(t) = \begin{pmatrix} x^\top(t) & \xi^\top(t) \end{pmatrix}$$
$$\bar{A}(t) = \begin{pmatrix} A(t) & B(t)K_C \\ K_B C(t) & K_A + K_B B_2(t)K_C \end{pmatrix}$$
$$\bar{A}_d(t) = \begin{pmatrix} A_d(t) \\ K_B C_d(t) \end{pmatrix}.$$

The cost function defined by (6.75) becomes

$$J = \int_0^\infty \left[x^\top(t)R_1 x(t) + \xi^\top(t)R_2\xi(t) \right] dt$$
$$= \int_0^\infty \eta^\top(t)\hat{R}\eta(t) dt \qquad (6.88)$$

where

$$\hat{R} = \begin{pmatrix} R_1 & \mathbf{0} \\ \mathbf{0} & K_C^\top R_2 K_C \end{pmatrix}.$$

For notation simplicity, let us define the following matrices:

$$\bar{A} = \left(\begin{array}{cc} A & BK_C \\ K_B C & K_A + K_B B_2 K_C \end{array} \right)$$

$$\bar{D}_a = \left(\begin{array}{c} D_1 \\ K_B D_3 \end{array} \right)$$

$$\bar{E}_a = \left(\begin{array}{cc} E_1 & E_2 K_C \end{array} \right)$$

$$\bar{A}_d = \left(\begin{array}{c} A_d \\ K_B C_d \end{array} \right)$$

$$\bar{D}_d = \left(\begin{array}{c} D_2 \\ K_B D_3 \end{array} \right)$$

$$\bar{E}_d = E_3.$$

With this notation, it is easy to check that

$$\bar{A}(t) = \bar{A} + \bar{D}_a \Delta(t) \bar{E}_a$$
$$\bar{A}_d(t) = \bar{A}_d + \bar{D}_d \Delta_d(t) \bar{E}_d.$$

Using Lemma 6.3, we obtain that if there exist symmetric, positive-definite matrices $P > 0$, $Q > 0$, and a positive scalar $\varepsilon > 0$ such that the following holds

$$\left(\begin{array}{cccc} \left[\begin{array}{c} P\bar{A} + \bar{A}^\top P + \hat{R} + I_0^\top Q I_0 \\ + \varepsilon P[\bar{D}_a \bar{D}_a^\top + \bar{D}_d \bar{D}_d^\top]P \end{array} \right] & P\bar{A}_d & \bar{E}_a^\top & 0 \\ \bar{A}_d^\top P & -Q & 0 & \bar{E}_d^\top \\ \bar{E}_a & 0 & -\varepsilon I & 0 \\ 0 & \bar{E}_d & 0 & -\varepsilon I \end{array} \right) < 0, \qquad (6.89)$$

then system (6.87) is stable; moreover, the cost defined by (6.88) satisfies (6.77) for all admissible uncertainties. Using the Schur complement, we get the following lemma.

Lemma 6.4 *If there exists a symmetric, positive-definite matrix $P > 0$ such that*

$$\left(\begin{array}{ccccc} P\bar{A} + \bar{A}^\top P & P\bar{A}_d & \bar{E}_a^\top & 0 & \mathcal{D}_{15} \\ \bar{A}_d^\top P & -Q & 0 & \bar{E}_d^\top & 0 \\ \bar{E}_a & 0 & -\varepsilon I & 0 & 0 \\ 0 & \bar{E}_d & 0 & -\varepsilon I & 0 \\ \mathcal{D}_{15}^\top & 0 & 0 & 0 & -\mathcal{D}_{55} \end{array} \right) < 0, \qquad (6.90)$$

holds for some symmetric, positive-definite matrix $Q > 0$, and a positive scalar $\varepsilon > 0$, where

$$\mathcal{D}_{15} = \left(\begin{array}{ccc} I_0^\top & P\bar{D}_a & P\bar{D}_d & \hat{S}^\top \end{array} \right)$$

$$\mathcal{D}_{55} = \text{diag}\{Q^{-1}, \frac{1}{\varepsilon}I, \frac{1}{\varepsilon}I, I\}$$

with

$$\hat{S} = \left[\hat{R}\right]^{1/2} = \begin{pmatrix} R_1^{1/2} & \mathbf{0} \\ \mathbf{0} & R_2^{1/2}K_C \end{pmatrix},$$

then system (6.87) is stable and, moreover, the cost defined by (6.88) satisfies

$$J \leq \eta^\top(0)P\eta(0) + \int_{-\tau}^0 \phi(s)Q\phi^\top(s)ds \qquad (6.91)$$

for all admissible uncertainties.

Lemma 6.5 If there exists matrix $Q > 0$ and scalar $\varepsilon > 0$, such that (6.90) has a feasible solution $P > 0$, then the following LMIs have a set of feasible solutions $X > 0, Y > 0, \mathcal{A}, \mathcal{B}, \mathcal{C}$:

$$\begin{pmatrix} X & I \\ I & Y \end{pmatrix} > 0 \qquad (6.92)$$

$$\begin{pmatrix} \mathcal{G}_{11} & \mathcal{G}_{12} & \mathcal{G}_{13} & \mathbf{0} & \mathcal{G}_{15} \\ \mathcal{G}_{12}^\top & -Q & \mathbf{0} & \bar{E}_d^\top & \mathbf{0} \\ \mathcal{G}_{13}^\top & \mathbf{0} & -\varepsilon I & \mathbf{0} & \mathbf{0} \\ \mathbf{0} & \bar{E}_d & \mathbf{0} & -\varepsilon I & \mathbf{0} \\ \mathcal{G}_{15}^\top & \mathbf{0} & \mathbf{0} & \mathbf{0} & -\mathcal{D}_{55} \end{pmatrix} < 0, \qquad (6.93)$$

where

$$\mathcal{G}_{11} = \begin{pmatrix} \begin{bmatrix} AX + B\mathcal{C}^\top \\ +\mathcal{C}^\top B^\top + XA^\top \end{bmatrix} & A + \mathcal{A} \\ A + A^\top & \begin{bmatrix} YA + A^\top Y \\ +\mathcal{B}C + C^\top \mathcal{B}^\top \end{bmatrix} \end{pmatrix}$$

$$\mathcal{G}_{12} = \begin{pmatrix} A_d \\ YA_d + \mathcal{B}C_d \end{pmatrix}$$

$$\mathcal{G}_{13} = \begin{pmatrix} XE_1^\top + \mathcal{C}^\top E_2^\top \\ E_1^\top \end{pmatrix}$$

$$\mathcal{G}_{15} = \begin{pmatrix} X & D_1 & D_2 & X[R_1^{1/2}]^\top & \mathcal{C}^\top[R_2^{1/2}]^\top \\ I & YD_1 + \mathcal{B}D_3 & \mathbf{0} & YD_2 + NK_BD_3 & \left[R_1^{1/2}\right]^\top \end{pmatrix}$$

Proof: Let K_A, K_B, K_C be a given set of gains and suppose that the matrices $P > 0$, $Q > 0$, and the scalar $\varepsilon > 0$ satisfy (6.90). Let us suppose that P and P^{-1} can be partitioned as in (3.47) with M, N invertible and X and Y symmetric, positive-definite matrices. Define matrices Φ_1, Φ_2 as in (3.55). Then, Φ_1, Φ_2 are invertible. Note that $\Phi_1^\top P\Phi_1 = \begin{pmatrix} X & I \\ I & Y \end{pmatrix}$, which yields (6.92). Let $\Xi = \text{diag}\{\Phi_1, I, I, I, I\}$. Pre- and post-multiplying

(6.4) yields

$$\begin{pmatrix} \Phi_1^\top[P\bar{A}+\bar{A}^\top P]\Phi_1 & \Phi_1^\top P\bar{A}_d & \Phi_1^\top \bar{E}_a^\top & 0 & \Phi_1^\top \mathcal{D}_{15} \\ \bar{A}_d^\top P\Phi_1 & -Q & 0 & \bar{E}_d^\top & 0 \\ \bar{E}_a\Phi_1 & 0 & -\varepsilon I & 0 & 0 \\ 0 & \bar{E}_d & 0 & -\varepsilon I & 0 \\ \mathcal{D}_{15}^\top\Phi_1 & 0 & 0 & 0 & -\mathcal{D}_{55} \end{pmatrix} < 0. \quad (6.94)$$

Define matrices

$$\begin{cases} \mathcal{A} = YAX + NK_BCX + YBK_CM^\top + N[K_A + K_BB_2K_C]M^\top \\ \mathcal{B} = NK_B \\ \mathcal{C} = K_CM^\top. \end{cases} \quad (6.95)$$

Next, we come to prove that X, Y and $\mathcal{A}, \mathcal{B}, \mathcal{C}$ defined above are a set of feasible solutions to LMIs (6.92)-(6.93).

Note that

$$P\Phi_1 = \Phi_2.$$

Direct computation gives

$$\Phi_1^\top P\bar{A}\Phi_1 = \begin{pmatrix} \begin{bmatrix} AX + BK_CM^\top \\ YAX + NK_BCX \\ + YBK_CM^\top \\ + N[K_A + K_BB_2K_C]M^\top \end{bmatrix} & A \\ & YA + NK_BC \end{pmatrix}$$

$$= \begin{pmatrix} AX + \mathcal{B}\mathcal{C} & A \\ \mathcal{A} & YA + \mathcal{B}\mathcal{C} \end{pmatrix},$$

from which it follows that

$$\Phi_1^\top[P\bar{A}+\bar{A}^\top P]\Phi_1$$

$$= \begin{pmatrix} \begin{bmatrix} AX + \mathcal{B}\mathcal{C} \\ +\mathcal{C}^\top\mathcal{B}^\top + XA^\top \end{bmatrix} & A + \mathcal{A}^\top \\ A + \mathcal{A}^\top & \begin{bmatrix} YA + \mathcal{B}\mathcal{C} \\ +\mathcal{C}^\top\mathcal{B}^\top \end{bmatrix} \end{pmatrix} = \mathcal{G}_{11} \quad (6.96)$$

$$\Phi_1^\top P\bar{A}_d = \Phi_2^\top \bar{A}_d = \begin{pmatrix} A_d \\ YA_d + NK_BC_d \end{pmatrix}$$

$$= \begin{pmatrix} A_d \\ YA_d + \mathcal{B}C_d \end{pmatrix} = \mathcal{G}_{12}, \quad (6.97)$$

$$\Phi_1^\top \bar{E}_a = \begin{pmatrix} X & M \\ I & 0 \end{pmatrix}\begin{pmatrix} E_1^\top \\ K_C^\top E_2^\top \end{pmatrix}$$

$$= \begin{pmatrix} XE_1^\top + MK_C^\top E_2^\top \\ E_1^\top \end{pmatrix}$$

$$= \begin{pmatrix} XE_1^\top + \mathcal{C}^\top E_2^\top \\ E_1^\top \end{pmatrix} = \mathcal{G}_{13} \quad (6.98)$$

and

$$\Phi_1^\top I_0 = \begin{pmatrix} X & M \\ I & 0 \end{pmatrix} \begin{pmatrix} I \\ 0 \end{pmatrix} = \begin{pmatrix} X \\ I \end{pmatrix} \tag{6.99}$$

$$\Phi_1^\top P\bar{D}_a = \Phi_2^\top \bar{D}_a = \begin{pmatrix} I & 0 \\ Y & N \end{pmatrix} \begin{pmatrix} D_1 \\ K_B D_3 \end{pmatrix}$$

$$= \begin{pmatrix} D_1 \\ YD_1 + NK_B D_3 \end{pmatrix} = \begin{pmatrix} D_1 \\ YD_1 + \mathcal{B}D_3 \end{pmatrix} \tag{6.100}$$

$$\Phi_1^\top P\bar{D}_d = \Phi_2^\top \bar{D}_d = \begin{pmatrix} I & 0 \\ Y & N \end{pmatrix} \begin{pmatrix} D_2 \\ K_B D_3 \end{pmatrix}$$

$$= \begin{pmatrix} D_2 \\ YD_2 + NK_B D_3 \end{pmatrix} = \begin{pmatrix} D_2 \\ YD_2 + \mathcal{B}D_3 \end{pmatrix} \tag{6.101}$$

$$\Phi_1^\top \hat{S}^\top = \begin{pmatrix} X & M \\ I & 0 \end{pmatrix} \begin{pmatrix} [R_1^{1/2}]^\top & 0 \\ 0 & K_C^\top [R_2^{1/2}]^\top \end{pmatrix}$$

$$= \begin{pmatrix} X[R_1^{1/2}]^\top & MK_C^\top [R_2^{1/2}]^\top \\ \left[R_1^{1/2}\right]^\top & 0 \end{pmatrix}$$

$$= \begin{pmatrix} X[R_1^{1/2}]^\top & C^\top [R_2^{1/2}]^\top \\ \left[R_1^{1/2}\right]^\top & 0 \end{pmatrix}. \tag{6.102}$$

Combining (6.99)-(6.102) yields

$$\Phi_1^\top \mathcal{D}_{15} = \mathcal{G}_{15}. \tag{6.103}$$

Therefore, (6.93) follows from (6.94). □

The following theorem provides a method for designing a stabilizing output feedback controller.

Theorem 6.12 *If LMIs (6.92) and (6.93) have a set of feasible solutions $X > 0, Y > 0, \mathcal{A}, \mathcal{B}, \mathcal{C}$ for a given matrix $Q > 0$ and a positive scalar ε, then there exist matrices $P > 0$, K_A, K_B, K_C satisfying (6.90), which means that the closed-loop system under controller (6.34) is asymptotically stable and the cost function defined by (6.88) satisfies*

$$J \leq \eta^\top(0)P\eta(0) + \int_{-\tau}^0 \phi^\top(s)Q\phi(s)ds \tag{6.104}$$

for all admissible uncertainties.

Proof: Let $X > 0, Y > 0, \mathcal{A}, \mathcal{B}, \mathcal{C}$ satisfy LMIs (6.92)-(6.93). Using singular value decomposition, we can get invertible matrices M, N satisfying

$$MN^\top = I - XY.$$

With X, Y, M, N given, construct Φ_1, Φ_2 as in (3.55). Then, we can define $P = \Phi_2\Phi_1^{-1}$ and K_A, K_B, K_C as follows:

$$\begin{cases} K_A = N^{-1}[\mathcal{A} - YAX - \mathcal{B}CX - YB\mathcal{C} - \mathcal{B}B_2\mathcal{C}][M^\top]^{-1} \\ K_B = N^{-1}\mathcal{B} \\ K_C = \mathcal{C}[M^\top]^{-1}. \end{cases} \tag{6.105}$$

It is easy to check that K_A, K_B, K_C, and P defined above satisfying (6.90).
□

Let us now show the usefulness of the results of this theorem through the following numerical example.

Example 6.5 *Let us consider a dynamical time delay system as described by (6.74) and suppose that the system data are as follows:*

$$A = \begin{pmatrix} 2 & 0 \\ 0 & 0.3 \end{pmatrix} \qquad A_d = \begin{pmatrix} 0.2 & 0.1 \\ 0.3 & -0.1 \end{pmatrix}$$

$$B = \begin{pmatrix} -1 & 0 \\ 0.1 & -1 \end{pmatrix} \qquad B_1 = \begin{pmatrix} 0 & 1 \\ 1 & 0 \end{pmatrix}$$

$$B_2 = \begin{pmatrix} 0 & 1 \\ 1 & 0 \end{pmatrix} \qquad D_1 = \begin{pmatrix} 1 & 0 \\ 1 & -1 \end{pmatrix}$$

$$D_2 = \begin{pmatrix} 0 & -1 \\ 0 & 1 \end{pmatrix} \qquad D_3 = \begin{pmatrix} 1 & -1 \\ 0 & 0 \end{pmatrix}$$

$$E_1 = \begin{pmatrix} 0 & 1 \\ 1 & 0 \end{pmatrix} \qquad E_2 = \begin{pmatrix} 1 & -1 \\ 0 & 1 \end{pmatrix}$$

$$C = \begin{pmatrix} 1 & 0 \\ 0 & 1 \end{pmatrix} \qquad C_d = \begin{pmatrix} 0 & -1 \\ 0 & 1 \end{pmatrix}$$

$$C_z = \begin{pmatrix} 0 & -1 \\ 0 & 1 \end{pmatrix} \qquad R_1 = R_2 = I.$$

Let $Q = I$ and $\varepsilon = 1.2$. Solving LMIs (6.92)-(6.93) yields a set of solutions as follows:

$$X = \begin{pmatrix} 0.0871 & -0.0474 \\ -0.0474 & 0.2169 \end{pmatrix} \qquad Y = \begin{pmatrix} -0.0710 & 0.1818 \\ 0.1818 & 0.3861 \end{pmatrix}$$

$$\mathcal{A} = \begin{pmatrix} -2.4461 & -0.2276 \\ -0.2276 & -0.8193 \end{pmatrix} \qquad \mathcal{B} = \begin{pmatrix} -0.4945 & -0.5611 \\ -0.5611 & -1.4863 \end{pmatrix}$$

$$\mathcal{C} = \begin{pmatrix} 1.0806 & 0.2227 \\ 0.2227 & 0.8691 \end{pmatrix}.$$

Using singular value decomposition, we obtain the following matrices M, N:

$$M = \begin{pmatrix} 0.9942 & -0.2032 \\ -0.2250 & -0.8981 \end{pmatrix} \qquad N = \begin{pmatrix} 0.9802 & -0.1979 \\ -0.1979 & -0.9802 \end{pmatrix},$$

which satisfies $MN^{\top} = I - XY$. With $\mathcal{A}, \mathcal{B}, \mathcal{C}$ and M, N obtained, using (6.105) we get the gains

$$K_A = \begin{pmatrix} -2.0752 & 0.0257 \\ -0.9930 & 1.1133 \end{pmatrix}$$

$$K_B = \begin{pmatrix} -0.3737 & -0.2559 \\ 0.6479 & 1.5679 \end{pmatrix}$$

$$K_C = \begin{pmatrix} 0.9857 & -0.4949 \\ 0.0249 & -0.9740 \end{pmatrix}$$

$$\Phi_1 = \begin{pmatrix} 0.0871 & -0.0474 & 1.0 & 0 \\ -0.0474 & 0.2169 & 0 & 1.0 \\ 0.9942 & -0.2250 & 0 & 0 \\ -0.2032 & -0.8981 & 0 & 0 \end{pmatrix}$$

$$\Phi_2 = \begin{pmatrix} 1.0 & 0 & -0.0710 & 0.1818 \\ 0 & 1.0 & 0.1818 & 0.3861 \\ 0 & 0 & 0.9802 & -0.1979 \\ 0 & 0 & -0.1979 & -0.9802 \end{pmatrix}.$$

Solving $P\Phi_2 = \Phi_1$ gives

$$P = \begin{pmatrix} -0.0710 & 0.1818 & 0.9802 & -0.1979 \\ 0.1818 & 0.3861 & -0.1979 & -0.9802 \\ 0.9802 & -0.1979 & -0.1100 & -0.0719 \\ -0.1979 & -0.9802 & -0.0719 & -0.2083 \end{pmatrix}.$$

Therefore, in view of Theorem 6.12, the designed controller stabilizes system (6.74) and guarantees the cost bound (6.104).

6.5 Notes

Delay-dependent robust $\mathcal{H}_\infty$ control for linear systems, with norm-bounded uncertainties and time-varying delays is addressed in [43]. Output feedback $\mathcal{H}_\infty$ control for linear systems without with time-varying delay is addressed in [109]. Delay-dependent robust filtering for linear systems with convex polytopic uncertainties is covered in [164], while [72] studies the suboptimal state feedback guaranteed cost control for uncertain time delay systems, and output feedback guaranteed cost control is studied in [73].

Part II

Stochastic Control

7
Stochastic Time Delay Systems

During the past decades, we have seen more and more examples showing the importance of dynamical systems subject to abrupt variations in their structures. This is partly due to the fact that very often dynamical systems are inherently vulnerable to component failures or repairs, sudden environmental disturbances, changing subsystem interconnections, abrupt variations in the operating point of a nonlinear plant, and so on. The class of Markov jump linear systems (MJLS) is an example of this class of dynamical systems, and it represents an important class of stochastic dynamical systems which, in several situations, is most appropriate for modeling the above phenomenon.

Roughly speaking, an MJLS is a hybrid system with a state vector that has two components, $x(t)$ and r_t. The first one is in general referred to as the state and the second one is referred to as the mode. In its operation, the jump linear system jumps abruptly from one mode to another in a random way that makes this class of systems a stochastic one. The switching between the modes is governed by a continuous-time Markov process with discrete and finite state space. When the system mode is fixed, the system evolves like a deterministic linear system.

Krasovskii and Lidskii [120, 121] were the first to study this class of systems. They state the formalism of this class of systems. Sworder and his coauthors [11, 191, 192, 193, 195, 194] also contributed to this class of systems mainly with regard to optimal control and stochastic estimation. Wonham [211] was also one of the pioneer researchers who contributed to the development of the Markov jump linear systems. During the last two decades we have seen an increasing interest in this class of systems, mainly for the advantage in modeling dynamical practical systems. Mariton [142] summarizes his contribution with that of his coworkers on the subject. Loparo, Chizeck, and their coauthors have contributed to different problems, among them the stability and the stabilizability problems (see [78, 79, 92, 111].

Boukas and his coauthors [4, 5, 8, 9, 10, 14, 15, 17, 18, 19, 20, 23, 24, 26, 27, 28, 29, 30, 31, 32, 33, 34, 35, 36, 37, 52, 55, 67, 179, 180, 181, 182, 183] extensively contributed to problems like stability, stabilizability, $\mathcal{H}_\infty$ control, estimation, guaranteed cost control, and to the robustness of these techniques. Costa and his coauthors [50, 51, 53, 54, 56, 57, 58, 59, 62] also contributed to different problems for this class of systems. Other contributions can be found in [1, 2, 63, 65, 66, 68, 69, 74, 75, 80, 81, 82, 150, 151, 152, 168, 169].

More recently the class of Markov jump linear systems has attracted researchers from the control community. Benjelloun et al. [10] were the first to deal with such a class of systems. Other researchers like Cao and Lam [40] have made contributions to the class of dynamical linear systems with Markovian jump and time delay.

7.1   Class of Dynamical Systems with Markov Jump and Time Delay

A mathematical representation of the class of systems with Markov jump and time delay is described by the following dynamics:

$$\begin{cases} \dot{x}(t) = A(r_t,t)x(t) + A_d(r_t,t)x(t-\tau) + B_1(r_t,t)w(t) + B(r_t,t)u(t) \\ z(t) = C_1(r_t,t)x(t) + D_{11}(r_t,t)w(t) + D_{12}(r_t,t)u(t) \\ y(t) = C_2(r_t,t)x(t) + D_{21}(r_t,t)w(t) \\ x(t) = \phi(t), \ t \in [-\tau, 0] \end{cases} \tag{7.1}$$

where $x(t) \in \mathbb{R}^n$ is the state vector at time t, $u(t) \in \mathbb{R}^m$ is the control input vector at time t, $w(t) \in \mathbb{R}^l$ is the square-integrable disturbance input vector at time t, $z(t) \in \mathbb{R}^p$ is the controlled output vector at time t, $y(t) \in \mathbb{R}^q$ is the measured output vector at time t, τ is the time delay, $\phi(.)$ is the initial function, and $A(r_t,t)$, $A_d(r_t,t)$, $B_1(r_t,t)$, $B(r_t,t)$, $C_1(r_t,t)$, $C_2(r_t,t)$, $D_{11}(r_t,t)$, $D_{12}(r_t,t)$, and $D_{21}(r_t,t)$ are as follows:

$$A(r_t,t) = A(r_t) + D_A(r_t)\Delta_1(r_t,t)E_A(r_t)$$
$$A_d(r_t,t) = A_d(r_t) + D_{A_d}(r_t)\Delta_2(r_t,t)E_{A_d}(r_t)$$
$$B_1(r_t,t) = B_1(r_t) + D_{B_1}(r_t)\Delta_1(r_t,t)E_{B_1}(r_t)$$
$$B(r_t,t) = B(r_t) + D_B(r_t)\Delta_1(r_t,t)E_B(r_t)$$
$$C_1(r_t,t) = C_1(r_t) + D_{C_1}(r_t)\Delta_1(r_t,t)E_{C_1}(r_t)$$
$$C_2(r_t,t) = C_2(r_t) + D_{C_2}(r_t)\Delta_1(t)(r_t,t)E_{C_2}(r_t)$$

$$D_{11}(r_t, t) = D_{11}(r_t) + D_{D_{11}}(r_t)\Delta_1(r_t, t)E_{D_{11}}(r_t)$$
$$D_{12}(r_t, t) = D_{12}(r_t) + D_{D_{12}}(r_t)\Delta_1(r_t, t)E_{D_{12}}(r_t)$$
$$D_{21}(r_t, t) = D_{21}(r_t) + D_{D_{21}}(r_t)\Delta_1(r_t, t)E_{D_{21}}(r_t)$$

with $A(r_t)$, $A_d(r_t)$, $B_1(r_t)$, $B(r_t)$, $C_1(r_t)$, $C_2(r_t)$, $D_{11}(r_t)$, $D_{12}(r_t)$, and $D_{21}(r_t)$ constant matrices with appropriate dimensions $\Delta_1(r_t, t)$ and $\Delta_2(r_t, t)$ are unknown time-varying matrices of appropriate dimensions representing the parameter uncertainties in the system and satisfying

$$\Delta_1^\top(r_t, t)\Delta_1(r_t, t) \leq I$$
$$\Delta_2^\top(r_t, t)\Delta_2(r_t, t) \leq I,$$

which are called admissible uncertainties.

The switching between the different modes of the stochastic systems is described by the probability transitions

$$P[r_{t+\Delta t} = j | r_t = i] = \begin{cases} \lambda_{ij}\Delta t + o(\Delta t) & \text{if } i \neq j \\ 1 + \lambda_{ii}\Delta t + o(\Delta t) & \text{otherwise} \end{cases} \tag{7.2}$$

with $\lambda_{ij} \geq 0$ for all $i \neq j$, $\lambda_{ii} = -\sum_{j \neq i} \lambda_{ij}$ and $\lim_{\Delta t \to 0} \frac{o(\Delta t)}{\Delta t} = 0$.

Remark 7.1 *Notice that when $\mathcal{S}$ is reduced to one element, the class of systems under consideration in this part becomes the one we studied in Part I.*

Remark 7.2 *Notice that the jump rates λ_{ij} for all i and j in $\mathcal{S}$ that we are considering in this book are constant. A direct extension of this class of systems consists of taking the jump rates dependent of the system state $x(t)$ and the control $u(t)$. In this case, the control problems become difficult. For more details on this, we refer the reader to Boukas [14, 17, 25].*

Our goal in this part is to study the following problems and their robustness:

- stochastic stability;

- stochastic stabilizability using memoryless and memory controllers and output feedback control;

- the $\mathcal{H}_\infty$ control problem; and

- the estimation problem.

We will also deal with the guaranteed cost control problem.

7.2   Definitions

Let us now define the concepts of stochastic stability, stochastic stabilizability, $\mathcal{H}_\infty$ control, the filtering problem, and so on. For this purpose, let us denote by $x(t, \phi(.), r_0)$ the solution of the nominal system (7.1) when the initial conditions are $\phi(s), s \in [-\tau, 0]$ and $r(0) = r_0$.

For system (7.1), when $\Delta_1(r_t, t) \equiv 0$ and $\Delta_2(r_t, t) \equiv 0$, we have the following definitions.

Definition 7.1 *System (7.1) with $u(t) \equiv 0$ and $w(t) \equiv 0$ is said to be*

(i) *stochastically stable (SS) if there exists a constant $T(r_0, \phi(\cdot))$ such that*

$$\mathbb{E}\left[\int_0^\infty \|x(t)\|^2 dt | \phi(\cdot), r_0\right] \leq T(r_0, \phi(\cdot));$$

(ii) *mean square stable (MSS) if*

$$\lim_{t \to \infty} \mathbb{E}\|x(t)\|^2 = 0$$

holds for any initial condition $(r_0, \phi(\cdot))$; and

(iii) *mean exponentially stable (MES) if there exist constants $\alpha > 0, \beta > 0$ such that the following holds for any initial conditions $(r_0, \phi(.))$:*

$$\mathbb{E}\left[\|x(t)\|^2 | \phi(.), r_0\right] \leq \alpha \|\phi(.)\| e^{-\beta t}.$$

Obviously, MES implies MSS and SS.

When the class of systems is subject to admissible uncertainties, the concept of stochastic stability becomes robust stochastic stability and is defined for system (7.1), by the following definition.

Definition 7.2 *System (7.1) with $u(t) \equiv 0$ and $w(t) \equiv 0$ is said to be*

(i) *robustly stochastically stable (RSS) if there exists a constant $T(r_0, \phi(\cdot))$ such that*

$$\mathbb{E}\left[\int_0^\infty \|x(t)\|^2 dt | \phi(\cdot), r_0\right] \leq T(r_0, \phi(\cdot))$$

holds for all admissible uncertainties; and

(ii) *robust mean exponentially stable (RMES) if there exist constants $\alpha > 0, \beta > 0$ such that the following holds:*

$$\mathbb{E}\left[\|x(t)\| | \phi(\cdot), r_0\right] \leq \alpha \|\phi(\cdot)\| e^{-\beta t}.$$

Obviously, RMES implies RSS.

Our aim is to develop delay-independent and delay-dependent conditions that we can use to to check if our class of systems is SS, and RSS. It is preferable that these conditions are in LMI formalism to allow the use of existing powerful tools like Matlab LMI toolbox and Scilab.

It can happen that some existing systems are by construction stochastically unstable, thereby telling us that we should stabilize them and make them useful for industrial applications. Let us now define the concept of stochastic stabilizability we will use later on.

Definition 7.3 *System (7.1) is said to be stabilizable in the SS sense if there exists a state feedback controller*

$$u(t) = K(r_t)x(t) \tag{7.3}$$

such that the closed-loop system is SS, where $K(i), i \in \mathcal{S}$ are constant gain matrices.

In the presence of admissible uncertainties we have the following definition.

Definition 7.4 *System (7.1) is said to be robustly stabilizable in the RSS sense if there exists a stable feedback controller (7.3) such that the closed-loop system is RSS, where $K(i), i \in \mathcal{S}$ are constant gain matrices.*

In a similar way, we will develop delay-independent and delay-dependent conditions in the LMI formalism.

In some circumstances dynamical systems are subject to some external disturbances, whose effect on the system performances we would like to minimize. Some alternates have been proposed in the literature to tackle such problem. Among these approaches we quote the $\mathcal{H}_\infty$ control problem (technique), which, under some appropriate assumptions, consists of rejecting the effect of external disturbances at a certain desired level. In general, we are interested in answering the following questions:

- Is the system under consideration stable and does it guarantee the disturbance rejection at the desired level $\gamma > 0$?

- How can we stabilize an unstable system and at the same time guarantee the disturbance rejection with the given level $\gamma > 0$?

- How can we design a controller that will ensure the class of systems considered in this book has the desired performance?

The $\mathcal{H}_\infty$ control problem can be stated in a similar way as in Part I, except the norm is replaced by

$$\|z(t)\|_2 = \left\{ \mathbb{E}\left[\int_0^\infty z_t^\top z_t dt \right] \right\}^{1/2}.$$

The following definitions will be used in the rest of the book.

Definition 7.5 *Let $\gamma > 0$ be a positive constant. System (7.1) is said to be stochastically stable with $\gamma-$disturbance attenuation if there exists a* *constant $M(r_0, \phi(\cdot))$ with $M(r_0, 0) = 0$, such that*

$$\|z\|_2 \triangleq \mathbb{E}\left[\int_0^\infty z_t^\top z_t dt\right]^{1/2} \leq \left[\gamma^2\|w\|_2^2 + M(r_0, \phi(\cdot))\right]^{1/2}. \qquad (7.4)$$

Definition 7.6 *System (7.1) is said to be stochastically stabilizable with γ-disturbance if there exist matrices $K(i), i \in \mathcal{S}$, such that the closed-loop system under control (7.3) satisfies (7.4).*

Remark 7.3 *In the presence of admissible uncertainties, robust stability and robust stabilizability with γ-disturbance are defined in the same manner as previously.*

The goal of Part II, besides the development of control algorithms, is to design a linear filter of order n of the form

$$\begin{cases} \dot{\hat{x}}(t) = K_A(r_t)\hat{x}(t) + K_B(r_t)y_t \\ \hat{x}(0) = x_0 \triangleq \phi(0), \ \hat{x}(s) = 0, \ s \in [-\tau, 0] \\ \hat{z}(t) = K_C(r_t)\hat{x}(t) \end{cases} \qquad (7.5)$$

which can ensure that the extended system $\{(x^\top(t), \hat{x}^\top(t))^\top, t \geq 0\}$ is SS and the estimation error $e_t = \hat{z}_t - z_t$ satisfies the condition

$$\|e\|_2 \leq \gamma\|w\|_2, \qquad (7.6)$$

where γ is a given positive scalar.

If (7.6) is satisfied, the error system of the filter (7.5) is said to verify the noise attenuation level γ. The designed robust filter is also considered.

All these problems are addressed in Part II, along with other problems like guaranteed cost control problem and some control problems for the class of nonlinear dynamical stochastic systems with time delay. We will try to use LMI formalism and also provide many numerical examples to illustrate the usefulness of the proposed results.

7.3   Objective of Part II

The objective of Part II is to address these problems for the class of stochastic dynamical systems with time delay. This part is organized into six chapters.

Chapter 8 covers stochastic stability and stochastic stabilizability of the class of linear systems with Markov jumps and time delay. Delay-independent and delay-dependent conditions are established to check the stochastic stability and the stochastic stabilizability of the class of systems under study. The output feedback stabilization of this class of systems is also treated. Finally, the stability of the class of systems with Markov jumps and mode-dependent time delay is considered.

Chapter 9 considers the class of uncertain linear systems with Markov jumps and time delay. Norm-bounded type uncertainty is covered in this chapter. The same problems covered in Chapter 8 are treated here to establish sufficient conditions that guarantee the robustness of these techniques.

Chapter 10 deals with the $\mathcal{H}_\infty$ control problem, the filtering problem, and the output feedback guaranteed cost control problem. We will also deal with the robustness of these techniques.

Chapter 11 considers robust $\mathcal{H}_\infty$ control and guaranteed cost control problems for the class of uncertain systems with Markov jumps and time delay. Norm-bounded uncertainties are considered.

Chapter 12 treats the nonlinear dynamical systems with time delay and Gaussian noise. The stability problem and the stabilization problems are considered. Linear and nonlinear state feedback controllers are designed to stochastically stabilize the class of systems we are considering in this chapter.

7.4 Notes

For more details on the basic theory of MJLS, for example, controllability, observability, stability, stabilizability, and optimal control, the reader is referred to [79, 111, 142, 211] and the references therein.

8

Stability and Stabilizability of Markov Jump Systems

This chapter deals with stability and stabilizability problems of the class of linear systems with Markov jumps and time delay. For this purpose, consider a hybrid system with N modes, that is, $\mathcal{S} = \{1, 2, \cdots, N\}$. Mode switching is assumed to be governed by a continuous-time Markov process $\{r_t, t \geq 0\}$ taking values in the state space $\mathcal{S}$ and having the following infinitesimal generator

$$\Lambda = (\lambda_{ij}), \ i, j \in \mathcal{S},$$

where $\lambda_{ij} \geq 0, \forall j \neq i, \lambda_{ii} = -\sum_{j \neq i} \lambda_{ij}$. Then, the mode transition probabilities are described as

$$P[r_{t+\Delta} = j | r_t = i] = \begin{cases} \lambda_{ij}\Delta + o(\Delta), & j \neq i \\ 1 + \lambda_{ii}\Delta + o(\Delta), & j = i \end{cases}$$

where $\lim_{\Delta \to 0} o(\Delta)/\Delta = 0$.

Let $x(t) \in \mathbb{R}^n$ be the continuous-time state of the system, which satisfies the dynamics

$$\begin{cases} \dot{x}(t) = A(r_t)x(t) + A_1(r_t)x(t - \tau) + B(r_t)u(t) \\ y(t) = C(r_t)x(t) \end{cases} \tag{8.1}$$

where $A(r_t), A_1(r_t) \in \mathbb{R}^{n \times n}, B(r_t) \in \mathbb{R}^{n \times k}, C(r_t) \in \mathbb{R}^{m \times n}$ are given matrices with appropriate dimensions, $\tau > 0$ represents the delay in the system, and $u(t) \in \mathbb{R}^k, y(t) \in \mathbb{R}^m$ denote the control input and output of the system, respectively. The initial condition of the system is specified as $(r_0, \phi(\cdot))$ with $x(s) = \phi(s) \in L_2[-\tau, 0]$, with $x_0 = x(0) = \phi(0)$.

For system (8.1), we have the following definitions:

Definition 8.1 *System (8.1) with $u(t) \equiv 0$ is said to be*

(i) stochastically stable (SS) if there exists a constant $T(r_0, \phi(\cdot))$ such that

$$\mathbb{E}\left[\int_0^\infty \|x(t)\|^2 |dt(\phi(\cdot), r_0)\right] \leq T(r_0, \phi(\cdot));$$

(ii) mean square stable (MSS) if

$$\lim_{t \to \infty} \mathbb{E}\left[\|x(t)\|^2 |(r_0, \phi(\cdot)\right] = 0$$

holds for any initial condition $(r_0, \phi(\cdot))$; and

(iii) mean exponentially stable (MES) if there exist constants $\alpha(r_0, \phi(\cdot)) > 0, \beta > 0$ such that

$$\mathbb{E}\left[\|x(t)\|^2 |(r_0, \phi(\cdot))\right] \leq \alpha(r_0, \phi(\cdot))e^{-\beta t}.$$

Obviously, MES implies MSS and SS.

Definition 8.2 *System (8.1) is said to be stabilizable in the SS(MES, MSQS) sense if there exists state feedback controller, that is,*

$$u(t) = K(r_t)x(t) \tag{8.2}$$

such that the closed-loop system is SS(MES, MSQS), where $K(i), i \in \mathcal{S}$ are constant gain matrices.

Remark 8.1 *The control law (8.2) assumes complete access to the state $x(t)$ and to the mode r_t at time t.*

In this chapter, we study the stability and stabilizability of the class of continuous-time linear systems with Markov jumps and time delays. We will develop delay-independent and delay-dependent sufficient conditions that can be used to check whether a system of the class we are considering in this chapter is stable or stabilizable. The stability of this class of systems under dynamical output feedback control is considered, and delay-independent and delay-dependent conditions are also studied. For the stabilizability problem, different types of control law are considered.

This chapter is organized as follows. In Section 8.1, the delay-independent conditions for stability and stabilizability are developed. Section 8.2 overcomes the conservatism of delay-independent sufficient conditions by establishing delay-dependent conditions. In section 8.3, output feedback control is studied and both delay-independence and delay-dependence are developed. Section 8.4 considers the stability and stabilizability of the class of Markov linear systems with mode-dependent time delay.

8.1   Delay-Independent Stability and Stabilizability

The goal of this section consists of developing the delay-independent conditions for stability and stabilizability of system (8.1).

8.1.1   Stochastic Stability

Let us first establish sufficient conditions for system (8.1) with $u(t) \equiv 0$ to be SS.

Theorem 8.1 *If there exist symmetric, positive-definite matrices* $P = (P(1), \cdots, P(N)) > 0$ *and* $Q > 0$ *satisfying the LMIs*

$$A^\top(i)P(i) + P(i)A(i) + Q + \sum_{j=1}^{N} \lambda_{ij} P(j) + P(i) \overset{\triangle}{=} \Xi_0(i) < 0 \quad (8.3)$$

$$A_1^\top(i)P(i)A_1(i) \le Q, \quad (8.4)$$

then system (8.1) with $u(t) \equiv 0$ *is SS.*

Proof: Let $\mathbb{C}[-\tau, 0]$ be the space of continuous functions on the interval $[-\tau, 0]$ and for any $\mathbf{x} \in \mathbb{C}[-\tau, 0]$, define $\|\mathbf{x}\| = \sup_{-\tau \le s \le 0} \|x(s)\|$. Obviously, the evolution of $x(t)$ depends on $x(s), t - \tau \le s \le t$, which means that $\{(x(t), r_t), t \ge 0\}$ is not a Markov process. To cast our model into the framework of Markov systems, let us define a process $\mathbf{x}(t)$ taking values in $\mathbb{C}[-\tau, 0]$ by

$$\mathbf{x}_s(t) = x(s + t), \ t - \tau \le s \le t. \quad (8.5)$$

Then, $\{(\mathbf{x}(t), r_t), t \ge 0\}$ is a strong Markov process.

Let $P = (P(1), \cdots, P(N)) > 0$ and $Q > 0$ satisfy (8.3) and (8.4). Consider the Lyapunov functional candidate with the following form:

$$V(\mathbf{x}(t), r_t) = x^\top(t)P(r_t)x(t) + \int_{-\tau}^{0} x^\top(t + \theta)Qx(t + \theta)d\theta. \quad (8.6)$$

Let $\mathcal{A}$ be the infinitesimal generator of $\{(\mathbf{x}(t), r_t), t \ge 0\}$ (see Appendix D). Then, the expression of this infinitesimal generator is given by

$$\mathcal{A}V(\mathbf{x}(t), r_t) = \dot{x}^\top(t)P(r_t)x(t) + x^\top(t)\left[\sum_{j=1}^{N} \lambda_{r_t j} P(j)\right]x(t)$$

$$+ x^\top(t)P(r_t)\dot{x}(t) + x^\top(t)Qx(t) - x^\top(t - \tau)Qx(t - \tau)$$

$$= x^\top(t)\left[A^\top(r_t)P(r_t) + Q + P(r_t)A(r_t)\right.$$

$$\left.+ \sum_{j=1}^{N} \lambda_{r_t j} P(j)\right]x(t) + 2x^\top(t)P(t)A_1(r_t)x(t - \tau)$$

$$- x^\top(t - \tau)Qx(t - \tau). \quad (8.7)$$

Noting that

$$2x^\top(t)P(r_t)\,A_1(r_t)x(t-\tau)$$
$$= -[x(t) - A_1(r_t)x(t-\tau)]^\top P(r_t)[x(t) - A_1(r_t)x(t-\tau)]$$
$$\quad + x^\top(t)P(r_t)x(t) + x^\top(t-\tau)A_1^\top(r_t)P(r_t)A_1(r_t)x(t-\tau),$$

we obtain

$$\mathcal{A}V(\mathbf{x}(t), r_t)$$

$$\leq x^\top(t)\left[A^\top(r_t)P(r_t) + P(r_t)A(r_t) + Q + P(r_t) + \sum_{j=1}^{N}\lambda_{r_t j}P(j)\right]x(t)$$

$$\quad - x^\top(t-\tau)Qx(t-\tau) - [x(t) - A_1(r_t)x(t-\tau)]^\top$$
$$\quad \times P(r_t)[x(t) - A_1(r_t)x(t-\tau)] + x^\top(t-\tau)A_1^\top(r_t)P(r_t)A_1(r_t)x(t-\tau)$$

$$\leq x^\top(t)\left[A^\top(r_t)P(r_t) + P(r_t)A(r_t) + Q\right.$$

$$\quad \left. + P(r_t) + \sum_{j=1}^{N}\lambda_{r_t j}P(j)\right]x(t)$$

$$\quad - x^\top(t-\tau)[Q - A_1^\top(r_t)P(r_t)A_1(r_t)]x(t-\tau)$$
$$\leq x^\top(t)\Xi_0(r_t)x(t). \tag{8.8}$$

The last inequality follows from (8.4). Therefore, we obtain

$$\mathcal{A}V(\mathbf{x}(t), r_t) \leq -\min_{i \in \mathcal{S}}\{\lambda_{\min}(-\Xi_0(i))\}x^\top(t)x(t),$$

which combined with Dynkin's formula (see Appendix D) yields

$$\mathbb{E}[V(\mathbf{x}(t), r_t)] - \mathbb{E}[V(\mathbf{x}(0), r_0)] = \mathbb{E}\left[\int_0^t \mathcal{A}V(\mathbf{x}(s), r_s)ds|(r_0, \phi(\cdot))\right]$$

$$\leq -\min_{i \in \mathcal{S}}\{\lambda_{\min}(-\Xi_0(i))\}\mathbb{E}\left[\int_0^t x^\top(s)x(s)ds|(r_0, \phi(\cdot))\right],$$

implying, in turn,

$$\min_{i \in \mathcal{S}}\{\lambda_{\min}(-\Xi_0(i))\}\mathbb{E}\left[\int_0^t x^\top(s)x(s)ds(r_0, \phi(\cdot))\right]$$
$$\leq \mathbb{E}[V(\mathbf{x}(0), r_0)] - \mathbb{E}[V(\mathbf{x}(t), r_t)]$$
$$\leq \mathbb{E}[V(\mathbf{x}(0), r_0)].$$

This yields that

$$\mathbb{E}\left[\int_0^t x^\top(s)x(s)ds|(r_0, \phi(\cdot))\right] \leq \frac{\mathbb{E}[V(\mathbf{x}(0), r_0)]}{\min_{i \in \mathcal{S}}\{\lambda_{\min}(-\Xi_0(i))\}}$$

holds for any $t > 0$. This proves Theorem 8.1. $\square$

We are now in a position to synthesize a memoryless state feedback controller (8.2) that stabilizes system (8.1) in the SS sense.

Theorem 8.2 *If there exist symmetric, positive-definite matrices* $X = (X_1, \cdots, X_N) > 0, U > 0$ *satisfying*

$$
\begin{pmatrix}
\# & X_i & S_i(X) \\
X_i & -U & \mathbf{0} \\
S_i^\top(X) & \mathbf{0} & -\mathcal{X}_i
\end{pmatrix} < 0 \tag{8.9}
$$

$$
\begin{pmatrix}
U & UA_1^\top(i) \\
A_1(i)U & X_i
\end{pmatrix} > 0, \tag{8.10}
$$

where $\# = A(i)X_i + B(i)Y_i + X_iA^\top(i) + Y_i^\top B^\top(i) + \lambda_{ii}X_i + X_i$, *then controller (8.2) with* $K(i) = Y_iX_i^{-1}$ *stabilizes system (8.1) in the SS sense,*

$$
S_i(X) = \left(\sqrt{\lambda_{i1}}X_i \quad \cdots \quad \sqrt{\lambda_{ii-1}}X_i \; \sqrt{\lambda_{ii+1}}X_i \quad \cdots \quad \sqrt{\lambda_{iN}}X_i \right) \tag{8.11}
$$

$$
\mathcal{X}_i = \mathrm{diag}\{X_1, \cdots, X_{i-1}, X_{i+1}, \cdots, X_N\}, \tag{8.12}
$$

Proof: Substituting (8.2) into (8.1) yields the dynamics of the closed-loop system described as follows:

$$
\dot{x}(t) = \bar{A}(r_t)x(t) + A_1(r_t)x(t - \tau).
$$

Using Theorem 8.1, to prove that controller (8.2) stabilizes system (8.1) in the stochastic sense it suffices to show that there exist symmetric, positive-definite matrices $P = (P(1), \cdots, P(N)) > 0$ and $Q > 0$ satisfying

$$
\bar{A}^\top(i)P(i) + P(i)\bar{A}(i) + Q + P(i) + \sum_{j=1}^{N} \lambda_{ij}P(j) < 0 \tag{8.13}
$$

$$
A_1^\top(i)P(i)A_1(i) \le Q. \tag{8.14}
$$

Let $X_i = P^{-1}(i)$ and $U = Q^{-1}$. Pre- and post-multiplying (8.13) by X_i yields

$$
X_i\bar{A}^\top(i) + \bar{A}(i)X_i + X_iU^{-1}X_i + \lambda_{ii}X_i + X_i \left[\sum_{j \ne i} \lambda_{ij}X_j^{-1} \right] X_i + X_i < 0.
$$

Noticing that $X_i[\sum_{j \ne i} \lambda_{ij}X_j^{-1}]X_i = S_i(X)\mathcal{X}_iS_i^\top(X)$, letting $Y_i = K(i)X_i$ and using the Schur complement, the above inequality is equivalent to (8.9). Likewise, (8.14) is equivalent to

$$
\begin{pmatrix}
U^{-1} & A_1^\top(i) \\
A_1(i) & X_i
\end{pmatrix} > 0.
$$

Pre- and post-multiplying both sides of the above inequality by $\mathrm{diag}\{U, I\}$ yields (8.10). From the above derivation, we conclude that if $X = (X_1, \cdots, X_N) > 0$, $Y = (Y_1, \cdots, Y_N)$, and $U > 0$ satisfy (8.9) and (8.10), then $P(i) = X_i^{-1}, K(i) = Y_iX_i^{-1}$, and $Q = U^{-1}$ satisfying (8.13)-(8.14).

This completes the proof of Theorem 8.2.   □

To illustrate the validness of the above algorithm, let us give a numerical example as follows.

Example 8.1 *Consider a system described by (8.1) and suppose that the system has the following data:* $\mathcal{S} = \{1, 2\}$, $\Lambda = \begin{pmatrix} -2 & 2 \\ 3 & -3 \end{pmatrix}$, *and*

$$A(1) = \begin{pmatrix} 2 & 0.1 \\ 0 & 1 \end{pmatrix} \qquad A_1(1) = \begin{pmatrix} 0.8 & 0 \\ 0.3 & 0.2 \end{pmatrix}$$

$$A(2) = \begin{pmatrix} 0.5 & 0 \\ 0.1 & 2.1 \end{pmatrix} \qquad A_1(2) = \begin{pmatrix} 0.4 & 0 \\ 0.3 & 0.2 \end{pmatrix}$$

$$B(1) = \begin{pmatrix} 1 \\ 1 \end{pmatrix} \qquad B(2) = \begin{pmatrix} 0 \\ 1 \end{pmatrix}.$$

With these data, solving LMIs (8.9) and (8.10) yields the following set of feasible solutions:

$$X_1 = \begin{pmatrix} 1.2598 & 1.3253 \\ 1.3253 & 1.4572 \end{pmatrix} \qquad X_2 = \begin{pmatrix} 0.2017 & 0.2195 \\ 0.2195 & 0.8462 \end{pmatrix}$$

$$U = \begin{pmatrix} 0.7429 & 1.5356 \\ 1.5356 & 4.1386 \end{pmatrix}.$$

Therefore, according to Theorem 8.2, controller (8.2) with

$$K(1) = Y_1 X_1^{-1} = \begin{pmatrix} -26.5188 & 12.9090 \end{pmatrix}$$
$$K(2) = Y_2 X_2^{1} = \begin{pmatrix} 19.5490 & -20.0016 \end{pmatrix}$$

stabilizes system (8.1) in the SS sense.

8.1.2   Mean Exponential Stability

Next, we proceed with the establishment of the sufficient conditions for the MES of system (8.1) with $u(t) \equiv 0$. To this end, let us assume the following.

Assumption 8.1 *There exists a symmetric matrix M such that the state trajectory of system (8.1) satisfies the following for all $\theta \in [-\tau, 0]$:*

$$\|x(t + \theta)\|^2 \le (x^{\top}(t) \ x^{\top}(t - \tau)) M \begin{pmatrix} x(t) \\ x(t - \tau) \end{pmatrix}. \tag{8.15}$$

Remark 8.2 *Assumption 8.1 means that the system state between t and $t - \tau$ is bounded by a function of $x(t)$ and $x(t-\tau)$. Setting $M = \begin{pmatrix} h^2 & 0 \\ 0 & 0 \end{pmatrix}$ with $h > 0$ leads to the Assumption 4 in [137], that is,*

$$\|x(t + \theta)\| \le h\|x(t)\|, \tag{8.16}$$

which was pointed out not to be restrictive since h can be chosen arbitrarily. Assumption 8.1 is more realistic than (8.16). In fact, in the case of a stable system, (8.16) restricts the convergence rate of the system.

Definition 8.3 *System (8.1) with $u(t) \equiv 0$ is called mean square quadratically stable (MSQS) if there exist symmetric, positive-definite matrices $P = (P(1), \cdots, P(N)) > 0$ and $Q > 0$ such that the following holds for every $i \in \mathcal{S}$:*

$$\Theta(i) \triangleq \begin{pmatrix} J(i) & P(i)A_1(i) \\ 3A_1^\top(i)P(i) & -Q \end{pmatrix} < 0, \ \forall i \in \mathcal{S}, \tag{8.17}$$

where $J(i) = A^\top(i)P(i) + P(i)A(i) + \sum_{j=1}^N \lambda_{ij}P(j) + Q$.

The following theorem provides the relationship between MSQS and MES.

Theorem 8.3 *Under Assumption 8.1, if system (8.1) with $u(t) \equiv 0$ is MSQS, then it is MES.*

Proof: Let us consider the Lyapunov functional candidate defined by (8.6). Employing (8.7), we obtain

$$\mathcal{A}V(\mathbf{x}(t), r_t) = \hat{x}^\top(t)\Theta(r_t)\hat{x}(t), \tag{8.18}$$

where $\hat{x}^\top(t) = (x^\top(t), x^\top(t-\tau))$. Therefore,

$$\frac{\mathcal{A}V(\mathbf{x}(t), r_t)}{V(\mathbf{x}(t), r_t)} = \frac{\hat{x}^\top(t)\Theta(r_t)\hat{x}(t)}{x^\top(t)P(r_t)x(t) + \int_{-\tau}^0 x(t+\theta)Qx(t+\theta)d\theta} < 0. \tag{8.19}$$

Noting that $\|\hat{x}(t)\| \geq \|x(t)\|$ and using Assumption 8.1, we have

$$x^\top(t)P(r_t)x(t) + \int_{-\tau}^0 x^\top(t+\theta)Qx(t+\theta)d\theta$$

$$\leq \lambda_{\max}(P(r_t))\|x_t\|^2 + \int_{-\tau}^0 \lambda_{\max}(Q)\|x(t+\theta)\|^2 d\theta$$

$$\leq [\lambda_{\max}(P(r_t)) + \tau\lambda_{\max}(Q)\lambda_{\max}(M)]\|\hat{x}(t)\|^2 \tag{8.20}$$

where $\lambda_{\max}(X)$ denotes the maximum eigenvalue of X. Combining now (8.19) and (8.20) gives

$$\frac{\mathcal{A}V(\mathbf{x}(t), r_t)}{V(\mathbf{x}(t), r_t)} \leq -\varrho_0 \triangleq -\min_{r_t \in \mathcal{S}} \left[\frac{\lambda_{\min}(-\Theta(r_t))}{\lambda_{\max}(P(r_t)) + \tau\lambda_{\max}(Q)\lambda_{\max}(M)} \right],$$

yielding, in turn,

$$\mathbb{E}[\mathcal{A}V(\mathbf{x}(t), r_t)] \leq -\varrho_0\mathbb{E}[V(\mathbf{x}(t), r_t)]. \tag{8.21}$$

Using now (8.21) and Dynkin's formula, we get

$$\mathbb{E}\left[V(\mathbf{x}(t), r_t)|(r_0, \phi(\cdot))\right] = \mathbb{E}\left[V(\mathbf{x}(0), r_0)\right] + \mathbb{E}\left[\int_0^t \mathcal{A}V(\mathbf{x}(s), r_s)ds\right]$$

$$\leq \mathbb{E}\left[V(\mathbf{x}(0), r_0)\right] - \varrho_0 \int_0^t \mathbb{E}[V(\mathbf{x}(s), r_s)]ds$$

which combined with the Gronwall–Bellman lemma yields

$$\mathbb{E}[V(\mathbf{x}(t), r_t)] \leq e^{-\varrho_0 t}\mathbb{E}[V(\mathbf{x}(0), r_0)]. \tag{8.22}$$

Noting that

$$\min_{r_t \in \mathcal{S}}[\lambda_{\min}(P(r_t)]\mathbb{E}[\|x(t)\|^2] \leq \mathbb{E}[V(\mathbf{x}(t), r_t)]$$

and using (8.22) we get finally the inequality

$$\mathbb{E}[\|x(t)\|^2] \leq \frac{1}{\min_{i \in \mathcal{S}}[\lambda_{\min}(P(i))]}e^{-\varrho_0 t}\mathbb{E}[V(\mathbf{x}(0), r_0)],$$

which implies that the system is MES. This ends the proof of Theorem 8.3. □

Example 8.2 *Consider a dynamical system with Markov jump parameters and time delay with dynamics described by (8.1). Suppose that the system data are as follows:*

$$\Lambda = \left(\begin{array}{cc} -3 & 3 \\ 1 & -1 \end{array}\right)$$

$$A(1) = \left(\begin{array}{cc} -1.8 & 0.4 \\ 0 & -1.2 \end{array}\right) \quad A(2) = \left(\begin{array}{cc} 2 & 0 \\ -1 & -3 \end{array}\right)$$

$$A_1(1) = \left(\begin{array}{cc} 0.4 & 0 \\ -0.1 & 0 \end{array}\right) \quad A_1(2) = \left(\begin{array}{cc} -0.2 & 0 \\ -0 & -1.5 \end{array}\right).$$

With these data, solving the feasible problem (8.17) gives

$$P(1) = \left(\begin{array}{cc} -0.33748 & -0.047199 \\ -0.047199 & 1.0416 \end{array}\right) \quad P(2) = \left(\begin{array}{cc} -1.460 & -0.1478 \\ -0.1478 & 0.62855 \end{array}\right)$$

$$Q = \left(\begin{array}{cc} 2.06887 & 0.1385 \\ 0.1385 & 1.7838 \end{array}\right).$$

This means that the system under consideration is MSQS.

Theorem 8.3 shows that MSQS provides an LMI-based sufficient condition for system (8.1) to be MES, which can be easily verified using the LMI toolbox. With the stability condition developed in Theorem 8.3, the stabilizability problem under controller (8.2) can be solved using this. Substituting (8.2) into (8.1) gives the following dynamic of the closed-loop system:

$$\dot{x}(t) = [A(r_t) + B(r_t)K(r_t)]x(t) + A_1(r_t)x(t - \tau). \tag{8.23}$$

Utilizing Theorem 8.3, a controller that stabilizes system (8.1) in the MSQS sense can be provided. The following theorem states such result.

Theorem 8.4 *If there exist symmetric, positive-definite matrices $X = (X_1, \cdots, X_N) > 0, U > 0$, and matrices $Y = (Y_1, \cdots, Y_N)$, such that the LMIs*

$$\begin{pmatrix} J_1(i) & X_i & S_i(X) \\ X_i & -U & \mathbf{0} \\ S_i^\top(X) & \mathbf{0} & -\mathcal{X}_i \end{pmatrix} < 0 \qquad (8.24)$$

are feasible for every $i \in \mathcal{S}$, where $S_i(X)$ and $\mathcal{X}_i$ are defined as in (8.11) and (8.12), respectively, and

$$J_1(i) = A(i)X_i + X_i A^\top(i) + B(i)Y_i + Y_i^\top B^\top(i)$$
$$+ A_1(i)U A_1^\top(i) + \lambda_{ii} X_i,$$

then controller (8.2) with $K(i) = Y_i X_i^{-1}, i \in \mathcal{S}$ stabilizes system (8.1) in the MSQS sense.

Proof: Consider the closed-loop system (8.23). Using Theorem 8.3, the closed-loop system (8.23) is MSQS if and only if there exist symmetric, positive-definite matrices $P = (P(1), \cdots, P(N)) > 0, Q > 0$ such that

$$\begin{pmatrix} \bar{J}(i) & P(i)A_1(i) \\ A_1^\top(i)P(i) & -Q \end{pmatrix} < 0, \ \forall i \in \mathcal{S} \qquad (8.25)$$

holds for every $i \in \mathcal{S}$, where $\bar{J}(i) = \bar{A}^\top(i)P(i) + P(i)\bar{A}(i) + \sum_{j=1}^{N} \lambda_{ij} P(j) + Q$ with $\bar{A}(i) = A(i) + B(i)K(i)$. In view of Lemma A.2 (the Schur complement), (8.25) is equivalent to

$$\bar{J}(i) + P(i)A_1(i)Q^{-1}A_1^\top(i)P(i) < 0, \ \forall i \in \mathcal{S}. \qquad (8.26)$$

Pre- and post-multiplying (8.26) by $P^{-1}(i)$ and letting $X_i = P^{-1}(i), Y_i = K(i)X_i, U = Q^{-1}$ yields

$$A(i)X_i + X_i A^\top(i) + B(i)Y_i + Y_i^\top B^\top(i) + A_1(i)Q^{-1}A_1^\top(i)$$

$$+ X_i Q X_i + X_i \left[\sum_{j=1}^{N} \lambda_{ij} X_j^{-1} \right] X_i < 0, \ \forall i \in \mathcal{S}, \qquad (8.27)$$

which, combined with the Schur complement, yields (8.24). Thus, if $X_i > 0, U > 0$, and Y_i are a set of feasible solutions to (8.24), then $P(i) = X_i^{-1}, i = 1, \cdots, N, Q = U^{-1}$, and $K(i) = Y_i X_i^{-1}, i = 1, \cdots, N$, satisfy (8.25). This ends the proof of Theorem 8.4. $\qquad \square$

Using Lemma A.3, some sufficient conditions for stabilizability of system (8.1) can now be developed.

Theorem 8.5 *System (8.1) is stabilizable in the MSQS sense if and only if the LMIs*

$$
\mathcal{B}_i^\top
\begin{pmatrix}
\tilde{J}(i) & X_i & S_i(X) \\
X_i & -U & \mathbf{0} \\
S_i^\top(X) & \mathbf{0} & -\mathcal{X}_i
\end{pmatrix}
\mathcal{B}_i < 0, \ \forall i \in \mathcal{S}
\tag{8.28}
$$

are feasible with respect to $X_i > 0, U > 0$, and Y_i, where

$$
\tilde{J}(i) = A(i)X_i + X_i A^\top(i) + A_1(i)U A_1^\top(i) + \lambda_{ii} X_i
$$
$$
\mathcal{B}_i = \mathrm{diag}\{B_\perp(i), I, I\}
$$

or

$$
\begin{pmatrix}
\tilde{J}(i) + \sigma B(i)B^\top(i) & X_i & S_i(X) \\
X_i & -U & \mathbf{0} \\
S_i^\top(X) & \mathbf{0} & -\mathcal{X}_i
\end{pmatrix}
< 0, \ \forall i \in \mathcal{S}
\tag{8.29}
$$

are feasible for some given $\sigma > 0$.

Proof: Note that

$$
\begin{pmatrix}
J_1(i) & X_i & S_i(X) \\
X_i & -U & \mathbf{0} \\
S_i^\top(X) & \mathbf{0} & -\mathcal{X}_i
\end{pmatrix}
=
\begin{pmatrix}
\tilde{J}(i) & X_i & S_i(X) \\
X_i & -U & \mathbf{0} \\
S_i^\top(X) & \mathbf{0} & -\mathcal{X}_i
\end{pmatrix}
$$
$$
+
\begin{pmatrix}
B(i) \\
\mathbf{0} \\
\mathbf{0}
\end{pmatrix}
Y_i \, (I \ 0 \ 0) +
\left[
\begin{pmatrix}
B(i) \\
\mathbf{0} \\
\mathbf{0}
\end{pmatrix}
Y_i \, (I \ 0 \ 0)
\right]^\top
\tag{8.30}
$$

and

$$
\begin{pmatrix}
B(i) \\
\mathbf{0} \\
\mathbf{0}
\end{pmatrix}_\perp
=
\begin{pmatrix}
B_\perp(i) & \mathbf{0} \\
\mathbf{0} & I \\
\mathbf{0} & I
\end{pmatrix},
\quad
\begin{pmatrix}
I \\
0 \\
0
\end{pmatrix}_\perp
=
\begin{pmatrix}
0 & 0 \\
I & 0 \\
0 & I
\end{pmatrix}.
$$

Thus, using Lemma A.3, we have that X_i, Y_i, U satisfy (8.24) if and only if (8.28) holds. This proves the first part of Theorem 8.5. In view of (ii) of Lemma A.3, (8.28) is equivalent to (8.29). This ends the proof of Theorem 8.5. ☐

Using Lemma A.3, the matrices $Y_i, i \in \mathcal{S}$, which contain $N \times n \times k$ unknown parameters, are eliminated from (8.24). In this way the computation burden is reduced significantly; however, no stabilizing controller is provided.

Example 8.3 *Consider a jump linear system with dynamics described by (8.1). The Markov process governing the mode switching has the following state space $\mathcal{S} = \{1, 2\}$ and the generator Λ given by*

$$
\Lambda =
\begin{pmatrix}
-2 & 2 \\
3 & -3
\end{pmatrix}.
$$

The system parameters are

$$A(1) = \begin{pmatrix} -2 & 0.1 \\ 0 & -1 \end{pmatrix} \quad A(2) = \begin{pmatrix} -2.5 & 0 \\ 0.1 & -2.1 \end{pmatrix}$$

$$A_1(1) = \begin{pmatrix} 0.8 & 0 \\ 0.3 & 0.2 \end{pmatrix} \quad A_1(2) = \begin{pmatrix} 0.4 & 0 \\ 0.3 & 0.2 \end{pmatrix}$$

$$B(1) = \begin{pmatrix} 1 \\ 1 \end{pmatrix} \quad B(2) = \begin{pmatrix} 0 \\ 1 \end{pmatrix}.$$

Solving (8.24) gives the following results:

$$X_1 = \begin{pmatrix} 2.4236 & 0.1214 \\ 0.1214 & 2.1106 \end{pmatrix} \quad X_2 = \begin{pmatrix} 2.0833 & -0.2747 \\ -0.2747 & 2.3273 \end{pmatrix}$$

$$Y_1 = \begin{pmatrix} -1.8732 & -1.5833 \end{pmatrix} \quad Y_2 = \begin{pmatrix} 0.0780 & -6.0852 \end{pmatrix}$$

$$U = \begin{pmatrix} 1.023 & 0.1365 \\ 0.1365 & 3.0504 \end{pmatrix}.$$

This means that this system is stabilizable in the SS sense and the stabilizing gain matrices are given by

$$K(1) = (-0.7375 \quad -0.70777) \ \text{and} \ K(2) = (-0.3121 \quad -2.65157).$$

Remark 8.3 *In this section we developed delay-independent conditions that can be used to check the stochastic stability and/or the stochastic stabilizability of the class of systems under study. These conditions are restrictive since they don't depend on the time delay. In the next section we will focus on the development of conditions that depend on the time delay and try to determine what interval will allow the delay to keep the system stable or stabilizable in the stochastic sense.*

8.2 Delay-Dependent Stability and Stabilizability

Theorem 8.3 develops a sufficient condition for the stability of system (8.1), and Theorem 8.4 provides a design method of the controller that stabilizes system (8.1). These results are all delay-independent, which means that these conditions hold for any $\tau \in [0, \infty)$. Thus, the results of the previous section are conservative. To reduce such conservatism, this section develops delay-dependent stability and stabilizability sufficient conditions for system (8.1).

8.2.1 Delay-Dependent Stochastic Stability

Lemma 8.1 *If there exist symmetric, positive-definite matrices $P = (P(1), \cdots, P(N)) > 0, Q > 0, Q_1 > 0$ satisfying for every $i \in \mathcal{S}$, the inequalities*

$$\Theta_0(i) \stackrel{\Delta}{=} [A(i) + A_1(i)]^\top P(i) + P(i)[A(i)$$

$$+ A_1(i)] + \sum_{j=1}^{N} \lambda_{ij} P(j) + \tau[A^\top(i)QA(i) + Q_1]$$

$$+ 2\tau P(i)A_1(i)Q^{-1}A_1^\top(i)P(i) < 0 \qquad (8.31)$$

$$A_1^\top(i)QA_1(i) \le Q_1, \qquad (8.32)$$

then system (8.1) with $u(t) \equiv 0$ is SS.

Proof: Note that

$$x(t - \tau) = x(t) - \int_{t-\tau}^{t} \dot{x}(s)ds. \qquad (8.33)$$

Substituting (8.1) with $u(t) \equiv 0$ into (8.33) yields

$$x(t - \tau) = x(t) - \int_{-\tau}^{0} [A(r_{t+s})x(t+s) + A_1(r_{t+s})x(t+s-\tau)]ds. \quad (8.34)$$

In turn, substituting (8.34) into (8.1) gives the following dynamics:

$$\begin{cases} \dot{x}(t) = [A(r_t) + A_1(r_t)]x(t) - \int_{t-\tau}^{t} [A_1(r_t)A(r_\theta)x(\theta) \\ \qquad + A_1(r_t)A_1(r_\theta)x(\theta - \tau)] \, d\theta \\ x(t) = \phi(t), \ r_t = r_0, \ t \in [-2\tau, 0]. \end{cases} \qquad (8.35)$$

Since the stability of system (8.1) results from system (8.35), we can proceed with the development of the delay-dependent stability conditions for system (8.35). Let us define a process $\{\mathbf{x}(t), t \ge 0\}$ taking values in $\mathbb{C}[-2\tau, 0]$ as follows:

$$\mathbf{x}_s(t) = x(s+t), \ s \in [t - 2\tau, t].$$

Then, $\{(\mathbf{x}(t), r_t), t \ge 0\}$ is a Markov process with initial state $(\phi(\cdot), r_0)$. Consider the following Lyapunov functional candidate:

$$V(\mathbf{x}(t), r_t) = x^\top(t)P(r_t)x(t) + V_1(\mathbf{x}(t), \mathbf{r}(t)), \qquad (8.36)$$

where $\mathbf{r}(t) = (r_s, t - 2\tau \le s \le t)$ and

$$V_1(\mathbf{x}(t), \mathbf{r}(t)) = \int_{-\tau}^{0} \int_{t+\theta}^{t} x^\top(s)Q_0(r_s)x(s)dsd\theta$$

$$+ \int_{-\tau}^{0} \int_{t-\tau+\theta}^{t} x^\top(s)Q_1x(s)dsd\theta, \qquad (8.37)$$

where $Q_0(r_s) = A^\top(r_s)QA(r_s)$.

Let $\mathcal{A}$ be the weak generator of $\{(\mathbf{x}(t), \mathbf{r}(t)), t \ge 0\}$. Then

$$\mathcal{A}V(\mathbf{x}(t), r_t) = \mathcal{A}V_1(\mathbf{x}(t), \mathbf{r}_t) + \mathcal{A}(x^\top(t)P(r_t)x(t))$$

$$= \tau x^\top(t)[Q_0(r_t) + Q_1]x(t)$$

$$- \int_{-\tau}^{0} x^\top(t+\theta)Q_0(r_{t+\theta})x(t+\theta)d\theta$$

$$-\int_{-\tau}^{0} x^{\top}(t-\tau+\theta)Q_1 x(t-\tau+\theta)d\theta$$

$$+ x^{\top}(t)\Big[[A(r_t) + A_1(r_t)]^{\top}P(r_t)$$

$$+P(r_t)[A(r_t) + A_1(r_t)] + \sum_{j=1}^{N}\lambda_{r_t j}P(j)\Big]x(t) + h(t) \quad (8.38)$$

where

$$h(t) = -2\int_{-\tau}^{0} x^{\top}(t)P(r_t)\Big[A_1(r_t)A(r_{t+\theta})x(t+\theta)$$

$$+ A_1(r_t)A_1(r_{t+\theta})x(t+\theta-\tau)\Big]d\theta.$$

Using Lemma A.1 and (8.32), we have

$$-2\int_{-\tau}^{0} x^{\top}(t)P(r_t)A_1(r_t)A_1(r_{t+\theta})x(t+\theta-\tau)d\theta$$

$$\leq \tau x^{\top}(t)P(r_t)A_1(r_t)Q^{-1}A_1^{\top}(r_t)P(r_t)x(t)$$

$$+ \int_{-\tau}^{0} x^{\top}(t+\theta-\tau)A_1^{\top}(r_{t+\theta})QA_1(r_{t+\theta})x(t+\theta-\tau)d\theta$$

$$\leq \tau x^{\top}(t)P(r_t)A_1(r_t)Q^{-1}A_1^{\top}(r_t)P(r_t)x(t)$$

$$+ \int_{-\tau}^{0} x^{\top}(t+\theta-\tau)Q_1 x(t+\theta-\tau)d\theta. \quad (8.39)$$

The last inequality results from (8.32).

Moreover,

$$-\int_{-\tau}^{0} x^{\top}(t+\theta)Q_0(r_{t+\theta})x(t+\theta)d\theta$$

$$-2\int_{-\tau}^{0} x^{\top}(t)P(r_t)A_1(r_t)A(r_{t+\theta})x(t+\theta)d\theta$$

$$= -\int_{-\tau}^{0} \big[Q^{-1}A_1^{\top}(r_t)P(r_t)x(t) + A(r_{t+\theta})x(t+\theta)\big]^{\top}$$

$$\times Q\big[Q^{-1}A_1^{\top}(r_t)P(r_t)x(t) + A(r_{t+\theta})x(t+\theta)\big]d\theta$$

$$+ \tau x^{\top}(t)P(r_t)A_1(r_t)Q^{-1}A_1^{\top}(r_t)P(r_t)x(t). \quad (8.40)$$

Combining (8.38)-(8.40) yields

$$\mathcal{A}V(\mathbf{x}(t), r_t) \leq x^{\top}(t)\Theta_0(r_t)x(t) \leq -\min_{i\in\mathcal{S}}\{\lambda_{\min}(\Theta_0(i))\}x^{\top}(t)x(t). \quad (8.41)$$

The rest of the proof follows the same line as in the proof of Theorem 8.1. $\square$

Lemma 8.1 gives a sufficient condition for the stability of system (8.1). However, it is not easy to verify the feasibility of $\Theta_0(i) < 0$. Using the Schur complement, we get the following equivalent condition that can be easily checked using LMI toolbox.

Theorem 8.6 *If there exist symmetric, positive-definite matrices* $P = (P(1), \cdots, P(N)), Q,$ *and* Q_1 *such that*

$$\begin{pmatrix} J(i) & \tau P(i)A_1(i) \\ \tau A_1^\top(i)P(i) & -\frac{\tau}{2}Q \end{pmatrix} < 0 \qquad (8.42)$$

$$A_1^\top(i)QA_1(i) \leq Q_1, \ \forall i \in \mathcal{S} \qquad (8.43)$$

hold, where

$$J(i) = [A(i) + A_1(i)]^\top P(i) + P(i)[A(i) + A_1(i)]$$
$$+ \sum_{j=1}^{N} \lambda_{ij}P(j) + \tau[A^\top(i)QA(i) + Q_1],$$

then system (8.1) with $u(t) \equiv 0$ *is SS.*

Proof: Using the Schur complement, (8.42) is equivalent to (8.31) and thus the proof follows from Lemma 8.1.   □

The above theorem shows that (8.42) and (8.43) ensure the stochastic stability of system (8.1). The following theorem indicates that they, in fact, ensure the MES of (8.1) with $u(t) \equiv 0$ under some conditions. Before giving the theorem, let us introduce the following assumption.

Assumption 8.2 *There exists a scalar* $\varrho > 0$ *such that the state trajectory of system (8.1) satisfies*

$$\|x(t + \theta)\| \leq \varrho\|x(t)\|, \ -2\tau \leq \theta \leq 0. \qquad (8.44)$$

Theorem 8.7 *Under Assumption 8.2, if there exist symmetric, positive-definite matrices* $P = (P(1), \cdots, P(N)) > 0, Q > 0, Q_1 > 0$ *satisfying (8.31) and (8.32), then system (8.1) with* $u(t) \equiv 0$ *is MES.*

Proof: Let us consider a Lyapunov functional defined as in (8.36). Using Assumption 8.2, we obtain

$$V(\mathbf{x}(t), r_t) \leq x^\top(t)P(r_t)x(t) + \xi\|x(t)\|^2 \qquad (8.45)$$

where

$$\xi = \varrho\tau^2 \left[\max_{r_t \in \mathcal{S}} \left(\frac{1}{2}\lambda_{\max}(Q_0(r_t)) + \frac{1}{2}\lambda_{\max}(Q_1) \right) \right].$$

Using now (8.41) and (8.45) yields

$$\frac{\mathcal{A}V(\mathbf{x}(t), r_t)}{V(\mathbf{x}(t), r_t)} \leq \frac{x^\top(t)\Theta_0(r_t)x(t)}{x^\top(t)P(r_t)x(t) + \xi\|x(t)\|^2} \leq -\gamma$$

where

$$\gamma \overset{\Delta}{=} \min_{r_t \in \mathcal{S}} \left[\frac{\lambda_{\min}(-\Theta_0(r_t))}{\lambda_{\max}(P(r_t)) + \xi} \right] > 0,$$

which implies the following:

$$\mathbb{E}\left[\int_0^t \mathcal{A}V(\mathbf{x}(s), \mathbf{r}_s)ds \right] \leq -\gamma \mathbb{E}\left[\int_0^t V(\mathbf{x}(s), \mathbf{r}_s)ds \right]. \tag{8.46}$$

Using Dynkin's formula and the Gronwall–Bellman lemma, it follows from (8.46) that

$$\mathbb{E}[V(\mathbf{x}(t), r_t)] \leq e^{-\gamma t} \mathbb{E}\left[x_0^\top P(r_0)x_0 + \int_{-\tau}^0 \int_\theta^0 \phi(s)^\top Q_0(r_0)\phi(s)dsd\theta \right.$$

$$\left. + \int_{-\tau}^0 \int_{-\tau+\theta}^0 \phi(s)^\top Q_1\phi(s)dsd\theta \right]. \tag{8.47}$$

Moreover, it is easy to check that

$$\mathbb{E}[\|x(t)\|^2] \leq \frac{1}{\min_{i \in \mathcal{S}} \lambda_{\min}(P(i))} \mathbb{E}\left[x^\top(t)P(r_t)x(t) \right]$$

$$\leq \frac{1}{\min_{i \in \mathcal{S}} \lambda_{\min}(P(i))} \mathbb{E}\left[V(\mathbf{x}(t), \mathbf{r}_t) \right]. \tag{8.48}$$

Combining (8.47) and (8.48) proves that system (8.1) is MES. This completes the proof of Theorem 8.7.   □

Example 8.4 *Consider a jump linear system with time delay and two modes:* $\mathcal{S} = \{1, 2\}$. *The system parameters are as follows:*

$$\Lambda = \begin{pmatrix} -2 & 2 \\ 3 & -3 \end{pmatrix}$$

$$A(1) = \begin{pmatrix} -2.737 & -.6077 \\ -.737 & -1.7 \end{pmatrix} \quad A(2) = \begin{pmatrix} -2.5 & 0 \\ -.212 & -4.75 \end{pmatrix}$$

$$A_1(1) = \begin{pmatrix} 0.8 & 0 \\ 0.3 & 0.2 \end{pmatrix} \quad A_1(2) = \begin{pmatrix} 0.4 & 0 \\ 0.3 & 0.2 \end{pmatrix}.$$

Employing a one-dimensional search method, we get an upper bound for τ *(namely,* $\tau = 0.56$*) for which (8.42) and (8.43) remain feasible. With this value for the delay, solving the feasible problems (8.42) and (8.43) gives:*

$$P(1) = \begin{pmatrix} 1.4257 & -0.206 \\ -0.206 & 1.003 \end{pmatrix} \quad P(2) = \begin{pmatrix} 0.845 & -0.06319 \\ -0.06319 & 0.476 \end{pmatrix}$$

$$Q = \begin{pmatrix} 0.4279 & -0.48 \\ -0.48 & 0.152 \end{pmatrix} \quad Q_1 = \begin{pmatrix} 3.795 & -0.269 \\ -0.269 & 4.204 \end{pmatrix}.$$

Therefore, this system is SS for any $\tau \in [0, 0.56)$.

8.2.2   Delay-Dependent Stabilizability

Utilizing the stability test of Theorem 8.6, a controller of the form (8.2) that stabilizes system (8.1) can be designed and some sufficient conditions for delay-dependent stabilizability can be provided. The following theorem establishes such results.

Theorem 8.8 *If there exist symmetric, positive-definite matrices* $X = (X_1, \cdots, X_N) > 0, U > 0, U_1 > 0,$ *and matrices* $Y = (Y_1, \cdots, Y_N),$ *such that*

$$
\begin{pmatrix}
J_0(i) + 2\tau A_1(i)UA_1^\top(i) & * & * & * \\
\tau[A(i)X_i + B(i)Y_i] & -\tau U & 0 & 0 \\
\tau X_i & 0 & -\tau U_1 & 0 \\
S_i^\top(X) & 0 & 0 & -\mathcal{X}_i
\end{pmatrix} < 0 \quad (8.49)
$$

$$
\begin{pmatrix}
-U_1 & U_1 A_1^\top(i) \\
A_1(i)U_1 & -U
\end{pmatrix} < 0, \ \forall i \in \mathcal{S} \quad (8.50)
$$

hold for every $i \in \mathcal{S},$ *where* $J_0(i) = [A(i) + A_1(i)]X_i + X_i[A^\top(i) + A_1^\top(i)] + B(i)Y_i + Y_i^\top B^\top(i) + \lambda_{ii}X_i,$ *then system (8.1) is stochastically stable under controller (8.2) and* $K(i) = Y_i X_i^{-1}, i \in \mathcal{S}$ *are a set of stabilizing gains.*

Proof: Suppose that $K(i), i \in \mathcal{S}$ are given gains. Substituting controller (8.2) into (8.1) yields the following dynamics for the closed-loop system:

$$
\dot{x}(t) = [A(r_t) + B(r_t)K(r_t)]x(t) + A_1(r_t)x(t - \tau). \quad (8.51)
$$

Thus, utilizing Lemma 8.1, we conclude that the closed-loop system is SS if there exist symmetric, positive-definite matrices $P = (P(1), \cdots, P(N)) > 0, Q > 0, Q_1 > 0$ satisfying

$$
\bar{\Theta}_0(i) < 0, \ \forall i \in \mathcal{S} \quad (8.52)
$$

$$
A_1^\top(i)QA_1(i) < Q_1, \ \forall i \in \mathcal{S} \quad (8.53)
$$

with $\bar{\Theta}_0(i)$ obtained from $\Theta_0(i)$ by replacing $A(i)$ with $A(i) + B(i)K(i)$. Pre- and post-multiplying $\bar{\Theta}_0(i)$ by $P^{-1}(i)$ yields

$$
P^{-1}(i)[\bar{A}(i) + A_1(i)]^\top + [\bar{A}(i) + A_1(i)]P^{-1}(i)
$$

$$
+ \lambda_{ii}P^{-1}(i) + P^{-1}(i)\left[\sum_{j \neq i}\lambda_{ij}P(j)\right]P^{-1}(i)
$$

$$
+ \tau P^{-1}(i)[\bar{A}^\top(i)Q\bar{A}(i) + Q_1]P^{-1}(i)
$$

$$
+ 2\tau A_1(i)Q^{-1}A_1^\top(i) < 0. \quad (8.54)
$$

Multiplying now both sides of (8.53) by Q_1^{-1} produces

$$
Q_1^{-1}A_1^\top(i)QA_1(i)Q_1^{-1} - Q_1^{-1} < 0, \ \forall i \in \mathcal{S}. \quad (8.55)
$$

Letting $X_i = P^{-1}(i), Y_i = K(i)X_i, i \in \mathcal{S}$ and $U = Q^{-1}, U_1 = Q_1^{-1}$ and utilizing the Schur complement, we get (8.49) and (8.50) from (8.54) and (8.55), respectively. Thus, if $X_i > 0, Y_i, i \in \mathcal{S}, U > 0$, and $U_1 > 0$ are a set of feasible solutions of (8.49) and (8.50), then matrices $P(i) = X_i^{-1}, K(i) = Y_i X_i^{-1}, Q = U^{-1}, Q_1 = U_1^{-1}$ satisfy (8.52) and (8.53), which completes the proof of Theorem 8.8. $\qquad\square$

Using Lemma A.3, we can eliminate $Y_i, i \in \mathcal{S}$ from (8.49) and get the following sufficient condition for stabilizability of system (8.1).

Theorem 8.9 *There exists a controller in the form of controller (8.2) stochastically stabilizing system (8.1), if there exist symmetric, positive-definite matrices* $X = (X_1, \cdots, X_N) > 0$ *and* $U > 0, U_1 > 0$ *satisfying*

$$\mathcal{B}_i^\top \Phi(i) \mathcal{B}_i < 0 \tag{8.56}$$

$$\begin{pmatrix} -U_1 & U_1 A_1^\top(i) \\ A_1(i)U_1 & -U \end{pmatrix} < 0, \ \forall i \in \mathcal{S}, \tag{8.57}$$

where

$$\Phi(i) = \begin{pmatrix} J_1(i) & \tau X_i A^\top(i) & \tau X_i & S_i(X) \\ \tau A(i)X_i & -\tau U & 0 & 0 \\ \tau X_i & 0 & -\tau U_1 & 0 \\ S_i^\top(X) & 0 & 0 & -\mathcal{X}_i \end{pmatrix},$$

$$J_1(i) = [A(i) + A_1(i)]X_i + X_i[A^\top(i) + A_1(i)]$$
$$+ 2\tau A_1(i)U A_1^\top(i) + \lambda_{ii} X_i$$

$$\mathcal{B}_i = \mathrm{diag}\{B_\perp(i), \frac{1}{\tau}B_\perp(i), I, I\}$$

or

$$\begin{pmatrix} J_1(i) & \tau X_i A^\top(i) & \tau X_i & S_i(X) \\ \tau A(i)X_i & -\tau U & 0 & 0 \\ \tau X_i & 0 & -\tau U_1 & 0 \\ S_i^\top(X) & 0 & 0 & -\mathcal{X}_i \end{pmatrix}$$
$$+ \sigma \begin{pmatrix} B(i)B^\top(i) & \tau B(i)B^\top(i) & 0 & 0 \\ \tau B(i)B^\top(i) & \tau^2 B(i)B^\top(i) & 0 & 0 \\ 0 & 0 & 0 & 0 \\ 0 & 0 & 0 & 0 \end{pmatrix} < 0 \tag{8.58}$$

and (8.50) hold for some $\sigma > 0$.

Proof: Noting that

$$\begin{pmatrix} J_0(i) + 2\tau A_1(i)U A_1^\top(i) & * & * & * \\ \tau[A(i)X_i + B(i)Y_i] & -\tau U & 0 & 0 \\ \tau X_i & 0 & -\tau U_1 & 0 \\ S_i^\top(X) & 0 & 0 & -\mathcal{X}_i \end{pmatrix}$$

$$= \Phi(i) + \begin{pmatrix} B(i) \\ \tau B(i) \\ 0 \\ 0 \end{pmatrix} Y_i \begin{pmatrix} I & 0 & 0 & 0 \end{pmatrix}$$

$$+ \begin{pmatrix} I \\ 0 \\ 0 \\ 0 \end{pmatrix} Y_i^\top \begin{pmatrix} B^\top(i) & \tau B^\top(i) & 0 & 0 \end{pmatrix}$$

and

$$\begin{pmatrix} B(i) \\ \tau B(i) \\ 0 \\ 0 \end{pmatrix}_\perp = \mathcal{B}_i.$$

Using (i) of Lemma A.3 we show that (8.56) is equivalent to (8.49). Moreover, the equivalence between (8.58) and (8.56) follows from (ii) of Lemma A.3. This completes the proof of Theorem 8.9. ☐

8.2.3 Maximization of the Time Delay

Stability and stabilizability of system (8.1) are based on the assumption that time delay is a given constant. If we assume that the time delay is constant but unknown, then an interesting problem is to design a controller that stabilizes the system and maximizes the upper bound of the time delay. This optimization problem is stated as follows:

$$\max_{\substack{\tau > 0, X = (X_1, \cdots, X_N) > 0, \\ Y = (Y_1, \cdots, Y_N), \\ U > 0, U_1 > 0}} \tau$$

s.t. (8.49), (8.50).

Note that (8.49) is equivalent to

$$\begin{pmatrix} -\mathcal{X}_i & S_i^\top(X) & 0 & 0 \\ S_i(X) & J_0(i) + 2\tau A_1^\top(i)U A_1(i) & * & * \\ 0 & \tau(A(i)X_i + B(i)Y_i) & -\tau U & 0 \\ 0 & \tau X_i & 0 & -\tau U_1 \end{pmatrix} < 0. \qquad (8.59)$$

Let $\gamma = 1/\tau$. Then (8.59) is equivalent to

$$\begin{pmatrix} 0 & 0 & 0 & 0 \\ 0 & 2A_1^\top(i)U A_1(i) & * & * \\ 0 & A(i)X_i + B(i)Y_i & -U & 0 \\ 0 & X_i & 0 & -U_1 \end{pmatrix} < \begin{pmatrix} \Gamma & 0 & 0 \\ 0 & 0 & 0 \\ 0 & 0 & 0 \end{pmatrix}, \qquad (8.60)$$

$$\Gamma \leq \gamma \begin{pmatrix} X_i & -S_i^\top(X) \\ -S_i(X) & -J_0(X) \end{pmatrix}. \tag{8.61}$$

Therefore, we get the following theorem.

Theorem 8.10 *Let* $X = (X_1, \cdots, X_N) > 0, Y = (Y_1, \cdots, Y_N), U > 0, U_1 > 0$ *and* $\gamma_0 > 0$ *be a set of solutions of the following GEVP*

$$\min_{\gamma, X>0, Y, U>0, U_1>0, \Gamma>0} \gamma. \tag{8.62}$$

$$\text{s.t. } (8.60), (8.61), (8.50).$$

Then, the controller (8.2) with $K(i) = Y_i X_i^{-1}, i \in \mathcal{S}$ *stabilizes system (8.1) for any* $\tau \in [0, 1/\gamma_0]$.

To illustrate the results of Theorem 8.10, let us consider the following numerical example.

Example 8.5 *Consider a two-mode system described by (8.1) with parameters as follows:*

$$\Lambda = \begin{pmatrix} -5 & 5 \\ 3 & -3 \end{pmatrix}$$

$$A(1) = \begin{pmatrix} -1 & -1 \\ 0 & 1 \end{pmatrix} \quad A(2) = \begin{pmatrix} 2 & 0 \\ 1 & -1 \end{pmatrix}$$

$$A_1(1) = \begin{pmatrix} 0.2 & 0 \\ 0 & 0.1 \end{pmatrix} \quad A_1(2) = \begin{pmatrix} -0.4 & 0 \\ 0.3 & 0 \end{pmatrix}$$

$$B(1) = \begin{pmatrix} 1 \\ 0 \end{pmatrix} \quad B(2) = \begin{pmatrix} 0 \\ -1 \end{pmatrix}.$$

Solving the optimization problem (8.62) yields

$$K(1) = (0.9996 \ 1.7486) \ and \ K(2) = (-1.6816 \ -0.9924)$$

and the upper bound for the time delay is $\tau = 6.5599$.

8.2.4 Stabilization of MJLS with Delayed Control

The above results are based on the assumption that the control input doesn't include delay effect. This subsection considers the stabilization problem of MJLS with delayed state and delayed control. That is, we consider a system with the following dynamics:

$$\dot{x}(t) = A(r_t)x(t) + B(r_t)u(t) + A_1(r_t)x(t - \tau)$$
$$+ B_1(r_t)u(t - \tau). \tag{8.63}$$

If controller (8.2) is used, then the closed-loop system is described by

$$\dot{x}(t) = [A(r_t) + B(r_t)K(r_t)]x(t)$$
$$+ [A_1(r_t) + B_1(r_t)K(t - \tau)]x(t - \tau). \tag{8.64}$$

The delay-dependent sufficient condition for stochastic stability of system (8.64) can be established using Lemma 8.1, where $\Theta_0(i)$ is replaced by $\hat{\Theta}_1(r_t)$. The matrix $\hat{\Theta}_1(r_t)$ is obtained from $\Theta_0(r_t)$ by replacing $A(r_t)$ and $A_1(r_t)$ by $\hat{A}(r_t) = A(r_t)+B(r_t)K(r_t)$ and $\hat{A}_1(r_t) = A_1(r_t)+B_1(r_t)K(r_{t-\tau})$, which means that in this case one has to handle $K(r_t)$ and $K(r_{t-\tau})$ at the same time. Since $\{r_t, t \geq 0\}$ is a stochastic process, this seems impossible. Here, we will restrict our design to a controller with nonswitching gain,

$$u(t) = Kx(t). \tag{8.65}$$

Remark 8.4 *To stabilize jump linear systems, in most cases a controller of form (8.2) is used. However, in some cases (for example, when the plant state $x(t)$ is perfectly available but the mode observation is not present) in order to design a state feedback controller the mode has to be estimated, which will increase the mode uncertainties. To overcome this shortcoming, a state feedback controller (8.65) is often used (see [142]).*

Substituting (8.65) into (8.63) and using Lemma 8.1 and the Schur complement, we have the following proposition.

Proposition 8.1 *Let $\gamma = 1/\tau$ and suppose that there exist symmetric, positive-definite matrices $P = (P(1), \cdots, P(N)) > 0, Q > 0, Q_1 > 0$, and scalars v, δ satisfying*

$$\begin{pmatrix} J(i) & * & * & * \\ \left[A_1^\top(i) + B_1(i)K\right]^\top P(i) & -\frac{\gamma}{2}Q & 0 & 0 \\ Q\left[A(i) + B(i)K\right] & 0 & -\gamma Q & 0 \\ Q_1 & 0 & 0 & -\gamma Q_1 \end{pmatrix} < \delta I \tag{8.66}$$

$$\begin{pmatrix} Q_1 & \left[A_1(i) + B_1(i)K\right]^\top Q \\ Q\left[A_1(i) + B_1(i)K\right] & Q \end{pmatrix} > 0 \tag{8.67}$$

where $J(i) = [\bar{A}(i) + \bar{A}_1(i)]^\top P(i) + P(i)[\bar{A}(i) + \bar{A}_1(i)] + \sum_{j=1}^N \lambda_{ij}P(j)$. Let K, γ_0, δ_0 be the solution of the following optimization problem

$$\mathcal{P}_0 : \quad \min_{\substack{K, (P(1), \cdots, P(N)) > 0, \\ Q > 0, Q_1 > 0}} \quad \gamma + \delta$$

$$\text{s.t. (8.66), (8.67).}$$

If $\delta_0 < 0$, then the controller (8.65) stabilizes system (8.1) in the SS sense for all $\tau \in [0, 1/\gamma_0]$.

Proof: From Lemma 8.1, it follows that the closed-loop system is SS if

$$\bar{\Theta}_0(i) < 0 \tag{8.68}$$

and

$$\bar{A}_1(i)Q\bar{A}_1(i) < Q_1 \tag{8.69}$$

hold for some $P(i) > 0, i \in \mathcal{S}$ and $Q > 0, Q_1 > 0$, where $\bar{\Theta}_0(i)$ is obtained from $\Theta_0(i)$ by replacing $A(i)$ and $A_1(i)$ by $\bar{A}(i)$ and $\bar{A}_1(i)$, respectively. Since $\delta_0 < 0$, (8.68) and (8.69) can be obtained from (8.66) and (8.67), respectively. This completes the proof. $\square$

Obviously (8.66) is nonlinear in K and $P(i), i \in \mathcal{S}$. Thus, $\mathcal{P}_0$ cannot be directly solved by LMI toolbox. However, the following iterative algorithm can provide a suboptimal solution.

Algorithm 8.1 *(Control Design Algorithm)*

Step 1: Choose an initial $P = (P(1), \cdots, P(N)) > 0$ and an error bound $\varepsilon > 0$.

Step 2: For a set of given matrices $P = (P(1), \cdots, P(N)) > 0$, solve

$$\mathcal{P}_1 : \min_K \quad \gamma + \delta$$
$$s.t. \ (8.66), \ (8.67)$$

and denote the optimal solution by γ_1 and δ_1.

Step 3: With K obtained in Step 2, solve

$$\mathcal{P}_2 : \min_{(P(1),\cdots,P(N))>0,Q>0,Q_1>0} \quad \gamma + \delta$$
$$s.t. \ (8.66), \ (8.67)$$

and denote the optimal solution by γ_2 and δ_2.

Step 4 If $|\gamma_2 - \gamma_1| < \varepsilon$ and $\delta_2 < 0$, stop. Otherwise, go to Step 2 with $P = (P(1), \cdots, P(N))$ obtained in Step 3.

Let us now give a numerical example to show the validity of this algorithm.

Example 8.6 *Consider a two-mode system with delayed input as described by (8.63). The system parameters are chosen as follows:*

$$\Lambda = \begin{pmatrix} -3 & 3 \\ 2 & -2 \end{pmatrix}$$

$$A(1) = \begin{pmatrix} -2 & 0 \\ 0 & -0.9 \end{pmatrix} \quad A(2) = \begin{pmatrix} -2.3 & 0 \\ 0 & -1 \end{pmatrix},$$

$$B(1) = \begin{pmatrix} 2 & 0 \\ 0 & 1 \end{pmatrix} \quad B(2) = \begin{pmatrix} 2 & 0 \\ 0 & 1 \end{pmatrix}$$

$$A_1(1) = \begin{pmatrix} -1 & 0 \\ -1 & -1 \end{pmatrix} \quad A_1(2) = \begin{pmatrix} -1.5 & 0 \\ -1 & 0.3 \end{pmatrix}$$

$$B_1(1) = \begin{pmatrix} 1 & 0 \\ 0 & -0.5 \end{pmatrix} \quad B_1(2) = \begin{pmatrix} 0.8 & 0 \\ 0 & 1 \end{pmatrix}.$$

Applying Algorithm 8.1 yields

$$P(1) = \begin{pmatrix} 0.7692 & 0.0563 \\ 1.0864 & 1.0864 \end{pmatrix} \quad P(2) = \begin{pmatrix} 0.8761 & -0.2631 \\ -0.2631 & 0.8443 \end{pmatrix}$$

$$K = \begin{pmatrix} 0.0404 & 0.5112 \\ 0.5799 & 0.291 \end{pmatrix}$$

and the upper bound of the time delay is $\tau = 0.6207$.

Sometimes, access to the state vector $x(t)$ is not possible and therefore, state feedback cannot be applied. An alternative method consists of using an output feedback controller. The next section will deal with such a controller, and time delay dependent sufficient stabilizability will be developed.

8.3   Output Feedback Stabilization

In this section, we study the stability of system (8.1) under a dynamical output feedback controller

$$\begin{cases} \dot{\xi}(t) = K_A(r_t)\xi(t) + K_B(r_t)y(t) \\ u(t) = K_C(r_t)\xi(t), \ t \geq 0, \end{cases} \tag{8.70}$$

where $\xi(t) \in \mathbb{R}^n$ is the controller state, $y(t)$ is the output of system (8.1), $K_A(r_t), K_B(r_t), K_C(r_t)$ are design parameter matrices with appropriate dimensions to be determined.

Applying controller (8.70) to system (8.1) yields the following dynamics for the closed-loop system:

$$\begin{cases} \dot{\eta}(t) = \bar{A}(r_t)\eta(t) + \bar{A}_1(r_t)x(t - \tau) \\ \eta(t) = \left(\phi^\top(t), \ 0\right), \ t \in [-\tau, 0] \end{cases} \tag{8.71}$$

where

$$\eta(t) = \begin{pmatrix} x(t) \\ \xi(t) \end{pmatrix} \quad \bar{A}_1(i) = \begin{pmatrix} A_1(i) \\ 0 \end{pmatrix}$$

$$\bar{A}(i) = \begin{pmatrix} A(i) & B(i)K_C(i) \\ K_B(i)C(i) & K_A(i) \end{pmatrix}.$$

Based on the delay-independent and the delay-dependent stability conditions developed in Section 8.1 and Section 8.2, this section addresses the problem of designing a dynamical output feedback controller of the form (8.70) that stabilizes system (8.1) in the SS and MES sense.

8.3.1   Delay-Independent Output Feedback Stabilization

If we assume that $(x(t), \xi(t))$ satisfy Assumption 8.1, by the same derivation as in Theorem 8.3, the closed-loop system (8.71) is MES if there exist symmetric, positive matrices $0 < P(i) \in R^{2n \times 2n}, i \in \mathcal{S}, 0 < Q \in R^{n \times n}$ satisfying

$$
\begin{pmatrix} J_0(i) & P(i)\bar{A}_1(i) \\ \bar{A}_1^\top(i)P(i) & -Q \end{pmatrix} < 0, \tag{8.72}
$$

where $J_0(i) = \bar{A}^\top(i)P(i) + P(i)\bar{A}(i) + \sum_{j=1}^N \lambda_{ij}P(j) + I_0^\top Q I_0$ with $I_0 = (\, I \quad 0 \,)$. In view of the Schur complement, we have the following proposition:

Proposition 8.2 *System (8.1) is stabilizable in the MSQS sense under control law (8.70) if there exist $K_A(i), K_B(i), K_C(i)$ and $P(i) > 0, i \in \mathcal{S}$ such that*

$$
\Theta(i) = \begin{pmatrix} J(i) & P(i)\bar{A}_1(i) & I_0^\top \\ \bar{A}_1(i)P(i) & -Q & 0 \\ I_0 & 0 & -Q^{-1} \end{pmatrix} < 0, \ i \in \mathcal{S}, \tag{8.73}
$$

holds for some $Q > 0$, where $J(i) = \bar{A}^\top(i)P(i) + P(i)\bar{A}(i) + \sum_{j=1}^N \lambda_{ij}P(j)$.

Proposition 8.2 provides a sufficient condition for system (8.71) to be MSQS. (8.73) is nonlinear with respect to $K_A(i), K_B(i), K_C(i)$ and $P(i)$, thus the conditions established in Proposition 8.2 cannot be solved directly by the LMI toolbox. However, an LMI-based equivalent condition for (8.73) to be feasible can be established. To this end, let us introduce a lemma and suppose that $P(i), i \in \mathcal{S}$ has the following partition:

$$
P(i) = \begin{pmatrix} P_{1i} & P_{2i} \\ P_{2i}^\top & P_{3i} \end{pmatrix}. \tag{8.74}
$$

Without loss of generality, we can assume that P_{2i} is nonsingular, which is formulated as a lemma as follows.

Lemma 8.2 *If (8.73) has a set of solutions $P(i), i \in \mathcal{S}$, then it has another set of solutions $\tilde{P}(i), i \in \mathcal{S}$ with $\tilde{P}_{2i}$ invertible.*

Proof: Let $P(i), i \in \mathcal{S}$ satisfy (8.73). Suppose that $P(i)$ is partitioned as in (8.74) with P_{2i} singular. In such case, perturb P_{2i} to make it full rank and denote the resulting $P(i)$ by $\tilde{P}(i)$. Due to its strict nature, (8.73) remains valid for $\tilde{P}(i), i \in \mathcal{S}$ when the perturbation is small enough. Thus, we can assume without loss of generality that $P(i), i \in \mathcal{S}$ is of form (8.74) with P_{2i} invertible. $\square$

Lemma 8.3 *If there exist matrices $K_A(i), K_B(i), K_C(i)$, and symmetric, positive-definite matrices $P = (P(1), \cdots, P(N)) > 0$ satisfying (8.73), then the LMIs*

$$
\begin{pmatrix} \Gamma_{11}(i) & \Gamma_{12}(i) & \Gamma_{13}(i) & S_i(Y) \\ \Gamma_{12}^\top(i) & -Q & 0 & 0 \\ \Gamma_{13}^\top(i) & 0 & -Q^{-1} & 0 \\ S_i^\top(Y) & 0 & 0 & -\mathcal{Y}_i \end{pmatrix} < 0 \tag{8.75}
$$

$$\begin{pmatrix} Y_i & I \\ I & X_i \end{pmatrix} > 0 \qquad (8.76)$$

have a set of feasible solutions $X_i > 0, Y_i > 0, H_i, G_i, M_i,$ *where*

$$\Gamma_{11}(i) = \begin{pmatrix} W_{11}(i) & M_i^\top + A(i) \\ M_i + A^\top(i) & W_{22}(i) \end{pmatrix}$$

with

$$W_{11}(i) = A(i)Y_i + Y_iA^\top(i) + B(i)G_i + G_i^\top B^\top(i) + \lambda_{ii}Y_i$$

$$W_{22}(i) = A^\top(i)X_i + X_iA(i) + H_iC(i) + C^\top(i)H_i^\top + \sum_{j=1}^N \lambda_{ij}X_j$$

$$\Gamma_{12}(i) = \begin{pmatrix} A_1(i) \\ X_iA_1(i) \end{pmatrix}$$

$$\Gamma_{13}(i) = \begin{pmatrix} Y_i \\ I \end{pmatrix}$$

$$S_i(Y) = \begin{pmatrix} \sqrt{\lambda_{i1}}Y_i & \cdots & \sqrt{\lambda_{ii-1}}Y_i & \sqrt{\lambda_{ii+1}}Y_i & \cdots & \sqrt{\lambda_{iN}}Y_i \\ 0 & \cdots & 0 & 0 & \cdots & 0 \end{pmatrix}$$

$$\mathcal{Y}_i = \text{diag}\{Y_1, \cdots, Y_{i-1}, Y_{i+1}, \cdots, Y_N\}.$$

Proof: Let $P(i)$, $K_A(i)$, $K_B(i)$ and $K_C(i)$, $i \in \mathcal{S}$ be a set of feasible solutions of (8.73). Partition $P(i)$ as in (8.74) with P_{2i} invertible. Define matrices

$$\begin{cases} X_i = P_{1i} \\ Y_i = [P_{1i} - P_{2i}P_{3i}^{-1}P_{2i}^\top]^{-1} \\ G_i = -K_C(i)P_{3i}^{-1}P_{2i}^\top Y_i \\ H_i = P_{2i}K_B(i) \\ M_i = X_iA(i)Y_i + X_iB(i)G_i + H_iC(i)Y_i \\ \qquad - P_{2i}K_A(i)P_{3i}^{-1}P_{2i}^\top Y_i + \sum_{j=1}^N \lambda_{ij}[P_{1j} - P_{2j}P_{3i}^{-1}P_{2i}^\top]Y_i \end{cases} \qquad (8.77)$$

and

$$S_i = \begin{pmatrix} Y_i & I \\ -P_{3i}^{-1}P_{2i}^\top Y_i & \mathbf{0} \end{pmatrix}. \qquad (8.78)$$

Obviously, S_i is invertible. Therefore, we have

$$0 < S_i^\top P(i)S_i = \begin{pmatrix} Y_i & I \\ I & P_{1i} \end{pmatrix} = \begin{pmatrix} Y_i & I \\ I & X_i \end{pmatrix}, \qquad (8.79)$$

yielding (8.76).

Pre- and post-multiplying $\Theta(i)$ by $\Xi_i^\top \overset{\Delta}{=} \text{diag}\{S_i^\top, I, I\}$ and Ξ_i yields

$$\Xi_i^\top \Theta(i)\Xi_i < 0, \ \forall i \in S. \qquad (8.80)$$

Direct manipulation gives

$$S_i^\top P(i)\bar{A}(i)S_i = \begin{pmatrix} A(i)Y_i + B(i)G_i & A(i) \\ \begin{bmatrix} P_{1i}A(i)Y_i + H_iC(i)Y_i \\ +X_iB(i)G_i \\ -P_{2i}K_A(i)P_{3i}^{-1}P_{2i}^\top Y_i \end{bmatrix} & \begin{bmatrix} P_{1i}A(i) \\ +H_iC(i) \end{bmatrix} \end{pmatrix} \quad (8.81)$$

$$S_i^\top P(j)S_i$$
$$= \begin{pmatrix} \begin{bmatrix} Y_i[P_{1j} - P_{2i}P_{3j}^{-1}P_{2j}^\top \\ -P_{2j}P_{3i}^{-1}P_{2i}^\top \\ +P_{2i}P_{3i}^{-1}P_{3j}P_{3i}^{-1}P_{2i}^\top]Y_i \\ P_{1j}Y_i - P_{2j}P_{3i}^{-1}P_{2i}^\top Y_i \end{bmatrix} & Y_iP_{1j} - Y_iP_{2i}P_{3i}^{-1}P_{2j}^\top \\ & P_{1j} \end{pmatrix}$$
$$= \begin{pmatrix} \begin{aligned} Y_i\,&[Y_j^{-1} + [P_{2i}P_{3i}^{-1}P_{3j} - P_{2j}] \\ &\times P_{3j}^{-1}[P_{2i}P_{3i}^{-1}P_{3j} - P_{2j}]^\top]\,Y_i \\ &P_{1j}Y_i - P_{2j}P_{3i}^{-1}P_{2i}^\top Y_i \end{aligned} & Y_iP_{1j} - Y_iP_{2i}P_{3i}^{-1}P_{2j}^\top \\ & P_{1j} \end{pmatrix} (8.82)$$

Combining (8.81) and (8.82) yields

$$S_i^\top J(i)S_i = \begin{pmatrix} \begin{bmatrix} A(i)Y_i + Y_iA^\top(i) \\ + B(i)G_i + \lambda_{ii}Y_i \\ + G_i^\top B^\top(i) \end{bmatrix} & M_i^\top + A(i) \\ M_i + A^\top(i) & \begin{bmatrix} A^\top(i)P_{1i} + P_{1i}A(i) \\ +H_iC(i) + C^\top(i)H_i^\top \\ +\sum_{j=1}^N \lambda_{ij}P_{1j} \end{bmatrix} \end{pmatrix}$$
$$+ \sum_{j=1}^N \lambda_{ij} \begin{pmatrix} \begin{aligned} Y_i[Y_j^{-1} &+ [P_{2i}P_{3i}^{-1}P_{3j} - P_{2j}] \\ &\times P_{3j}^{-1}[P_{2i}P_{3i}^{-1}P_{3j} - P_{2j}]^\top] \end{aligned} & 0 \\ 0 & 0 \end{pmatrix}$$
$$= \Gamma_{11}(i) + \sum_{j=1,j\neq i}^N \lambda_{ij} \begin{pmatrix} Y_i\left[Y_j^{-1}\right]Y_i & 0 \\ 0 & 0 \end{pmatrix}$$
$$+ \sum_{j=1,j\neq i}^N \lambda_{ij} \begin{pmatrix} \begin{aligned} Y_i[P_{2i}P_{3i}^{-1}P_{3j} &- P_{2j}]P_{3j}^{-1} \\ &\times[P_{2i}P_{3i}^{-1}P_{3j} - P_{2j}]^\top Y_i \end{aligned} & 0 \\ 0 & 0 \end{pmatrix}. \quad (8.83)$$

Moreover, simple manipulation yields

$$S_i^\top P(i) = \begin{pmatrix} I & 0 \\ P_{1i} & P_{2i} \end{pmatrix}$$

$$S_i^\top P(i)\bar{A}_1(i) = \begin{pmatrix} I & 0 \\ P_{1i} & P_{2i} \end{pmatrix}\begin{pmatrix} A_1(i) \\ 0 \end{pmatrix}$$
$$= \begin{pmatrix} A_1(i) \\ X_iA_1(i) \end{pmatrix} = \Gamma_{12}(i) \quad (8.84)$$

$$S_i^\top I_0 = \begin{pmatrix} Y_i & -Y_i P_{2i} P_{3i}^{-1} \\ I & \mathbf{0} \end{pmatrix} \begin{pmatrix} I \\ \mathbf{0} \end{pmatrix}$$

$$= \begin{pmatrix} Y_i \\ I \end{pmatrix} = \Gamma_{13}(i). \tag{8.85}$$

Noting now the fact that

$$\sum_{j=1, j\neq i}^{N} \lambda_{ij} [P_{2i} P_{3i}^{-1} P_{3j} - P_{2j}] P_{3j}^{-1} [P_{2i} P_{3i}^{-1} P_{3j} - P_{2j}]^\top \geq 0,$$

it follows from (8.83)-(8.85) that X_i, Y_i, M_i, H_i, G_i are a set of feasible solutions to (8.75). This completes the proof of Lemma 8.3. □

With Lemma 8.3, we get the following necessary and sufficient condition for (8.73) to be feasible.

Theorem 8.11 *There exist matrices $K_A(i), K_B(i), K_C(i)$ and symmetric, positive-definite matrices $P = (P(1), \cdots, P(N))$ satisfying (8.73) if and only if the LMIs*

$$\begin{pmatrix} W_{11}(i) + A_1(i)Q^{-1}A_1^\top(i) & Y_i & S_i(Y) \\ Y_i & -Q^{-1} & \mathbf{0} \\ S_i^\top(Y) & \mathbf{0} & -\mathcal{Y}_i \end{pmatrix} < 0 \tag{8.86}$$

$$\begin{pmatrix} W_{22}(i) + Q & X_i A_1(i) \\ A_1^\top(i) X_i & -Q \end{pmatrix} < 0 \tag{8.87}$$

$$\begin{pmatrix} Y_i & I \\ I & X_i \end{pmatrix} > 0 \tag{8.88}$$

hold for some given $Q > 0$.

Proof: *Necessity.* Suppose that there exist matrices $K_A(i), K_B(i), K_C(i)$ and symmetric, positive-definite matrices $P = (P(1), \cdots, P(N)) > 0$ satisfying (8.73), then Lemma 8.3 implies that LMIs (8.75)-(8.76) have a set of feasible solutions X_i, Y_i, H_i, G_i, M_i. From the Schur complement, it follows that (8.75) is equivalent to

$$\Gamma_{11}(i) + \begin{pmatrix} \sum_{j\neq i} Y_i Y_j^{-1} Y_i & \mathbf{0} \\ \mathbf{0} & \mathbf{0} \end{pmatrix} + \Gamma_{13}(i)Q\Gamma_{13}^\top(i) + \Gamma_{12}(i)Q^{-1}\Gamma_{12}^\top(i)$$

$$= \begin{pmatrix} W_{11}(i) + \sum_{j\neq i} \lambda_{ij} Y_i Y_j^{-1} Y_i & A(i) + M_i^\top \\ A^\top(i) + M_i & W_{22}(i) \end{pmatrix} + \begin{pmatrix} Y_i \\ I \end{pmatrix} Q (Y_i\ I)$$

$$+ \begin{pmatrix} A_1(i) \\ X_i A_1(i) \end{pmatrix} Q^{-1} (A_1^\top(i)\ A_1^\top(i) X_i) < 0, \ \forall i \in \mathcal{S},$$

that is,

$$\left(\begin{bmatrix} W_{11}(i) + \sum_{j \neq i} \lambda_{ij} Y_i Y_j^{-1} Y_i \\ + Y_i Q Y_i + A_1(i) Q^{-1} A_1^\top(i) \\ A^\top(i) + M_i + Q Y_i \\ + X_i A_1(i) Q^{-1} A_1^\top(i) \end{bmatrix} \quad \begin{bmatrix} A(i) + M_i^\top + Y_i Q \\ + A_1(i) Q^{-1} A_1^\top(i) X_i \\ W_{22}(i) + Q \\ + X_i A_1(i) Q^{-1} A_1^\top(i) X_i \end{bmatrix} \right) < 0$$

which in turn implies

$$W_{11}(i) + \sum_{j=1, j \neq i}^N \lambda_{ij} Y_i Y_j^{-1} Y_i + Y_i Q Y_i + A_1(i) Q^{-1} A_1^\top(i) < 0, \quad (8.89)$$

$$W_{22}(i) + Q + X_i A_1(i) Q^{-1} A_1^\top(i) X_i < 0. \quad (8.90)$$

Using now the Schur complement, we conclude that (8.89) and (8.90) are equivalent to (8.86) and (8.87), respectively. This proves the necessary condition.

Sufficiency. Suppose that $X_i > 0, Y_i > 0, H_i$, and G_i are a set of feasible solutions to (8.75)-(8.76). Let $P(i)$ be defined by

$$P(i) = \begin{pmatrix} X_i & Y_i^{-1} - X_i \\ Y_i^{-1} - X_i & X_i - Y_i^{-1} \end{pmatrix}. \quad (8.91)$$

Then it follows from (8.76) that

$$P(i) > 0, \ i \in \mathcal{S}.$$

Let us consider a set of gains $K_A(i), K_B(i), K_C(i)$ defined by

$$\begin{cases} K_A(i) = [X_i - Y_i^{-1}]^{-1} \left[A^\top(i) + Q Y_i + X_i A_1^\top(i) Q^{-1} A_1(i) \\ \qquad + X_i A(i) Y_i + X_i B(i) G_i + H_i C(i) Y_i \\ \qquad + \sum_{j=1}^N \lambda_{ij} Y_j^{-1} Y_i \right] Y_i^{-1} \\ K_B(i) = [Y_i^{-1} - X_i]^{-1} H_i \\ K_C(i) = G_i Y_i^{-1}. \end{cases} \quad (8.92)$$

Set

$$S_i = \begin{pmatrix} Y_i & Y_i \\ I & 0 \end{pmatrix}$$

$$\Xi_i = \mathrm{diag}\{S_i, I, I\}.$$

Substituting $K_A(i), K_B(i), K_C(i), P(i), i \in \mathcal{S}$ into $\Theta(i)$ and using a similar argument as in the proof of necessity, it is easy to check that

$$\Xi_i^\top \Theta(i) \Xi_i < 0 \quad (8.93)$$

is equivalent to

$$\left(\begin{bmatrix} W_{11}(i) + \sum_{j \neq i} \lambda_{ij} Y_i Y_j^{-1} Y_i \\ + Y_i Q Y_i + A_1(i) Q^{-1} A_1^\top(i) \end{bmatrix} \quad \mathbf{0} \\ \mathbf{0} \quad \begin{bmatrix} W_{22}(i) + Q \\ + X_i A_1(i) Q^{-1} A_1^\top(i) X_i \end{bmatrix} \right) < 0$$

which results from (8.86) and (8.87). This terminates the proof of Theorem 8.11.   □

For stability of jump linear systems in the MSQS sense, the presence of the jump term $\sum_{j=1}^{N} \lambda_{ij} P(j)$ in $\Theta(i)$ makes it difficult to design a dynamical compensator for system (8.1) using LMI. Theorem 8.11 gives a solution by restricting $-P_{2i} = P_{3i} = X_i - Y_i^{-1}$, which increases the conservatism of the design method. If we choose $P(i)$ to be diagonal block matrices, we can get an easy design technique. The theorem gives such results.

Theorem 8.12 Let $X_i > 0, Y_i > 0, Q > 0, U_i, V_i, W_i, i \in \mathcal{S}$ satisfy the following LMI for every $i \in \mathcal{S}$

$$
\begin{pmatrix}
\# & W_i + C^\top(i)V_i^\top & X_i A_1(i) \\
W_i^\top + V_i C(i) & U_i + U_i^\top + \sum_{j=1}^{N} \lambda_{ij} Y_j & 0 \\
A_1^\top(i) X_i & 0 & -Q
\end{pmatrix} < 0 \qquad (8.94)
$$

where $\# = X_i A(i) + A^\top(i) X_i + \sum_{j=1}^{N} \lambda_{ij} X_j + Q$, and the matrix equation

$$
X_i B(i) Z_i = W_i \qquad (8.95)
$$

has a solution Z_i for every $i \in \mathcal{S}$. Then system (8.1) is stable under control (8.70) in the MSQS sense and

$$
\begin{cases}
K_A(i) = Y_i^{-1} U_i \\
K_B(i) = Y_i^{-1} V_i \\
K_C(i) = Z_i \ (a \ solution \ to \ (8.95))
\end{cases} \qquad (8.96)
$$

are a set of stabilizing compensator parameters.

Proof: Suppose $X_i > 0, Y_i > 0, Q > 0, U_i, V_i, W_i, i \in \mathcal{S}$ satisfy LMIs (8.94). Let $P(i) = \begin{pmatrix} X_i & 0 \\ 0 & Y_i \end{pmatrix}$. Then, it is easy to check that the LMIs (8.94) guarantees that $P(i)$ and $K_A(i), K_B(i), K_C(i)$ defined by (8.96) satisfy (8.72), which completes the proof of Theorem 8.12.   □

Example 8.7 *Consider a time delay system with two modes as described by (8.1). The system parameters are as follows:*

$$\Lambda = \begin{pmatrix} -3 & 3 \\ 5 & -5 \end{pmatrix}$$

$$A(1) = \begin{pmatrix} -2 & 0 \\ 0 & -0.9 \end{pmatrix} \quad A(2) = \begin{pmatrix} -2.1 & -0.1 \\ -0.1 & -1 \end{pmatrix}$$

$$B(1) = \begin{pmatrix} 0 & 1 \\ 1 & 1 \end{pmatrix} \quad B(2) = \begin{pmatrix} 1 & -1 \\ 0 & -1 \end{pmatrix}$$

$$C(1) = \begin{pmatrix} 1 & 1 \\ 0 & 1 \end{pmatrix} \quad C(2) = \begin{pmatrix} 1 & -1 \\ 1 & 0 \end{pmatrix}$$

$$A_1(1) = \begin{pmatrix} -1 & 0 \\ -1 & -1 \end{pmatrix} \quad A_1(2) = \begin{pmatrix} -1 & 0 \\ -1 & 0 \end{pmatrix}.$$

Using Theorem 8.12 yields a stabilizing output feedback control with gain matrices:

$$K_A(1) = \begin{pmatrix} -12.14 & 12.4649 \\ -17.9302 & -4.4035 \end{pmatrix} \quad K_A(2) = \begin{pmatrix} -9.3794 & 36.9783 \\ -30.5791 & 2.0048 \end{pmatrix}$$

$$K_B(1) = \begin{pmatrix} 8.7683 & 0.064 \\ -4.1228 & 3.7966 \end{pmatrix} \quad K_B(2) = \begin{pmatrix} 12.0526 & -5.8708 \\ 6.3876 & -.5553 \end{pmatrix}$$

$$K_C(1) = \begin{pmatrix} -20.2355 & -26.3268 \\ -2.6979 & 22.2539 \end{pmatrix} \quad K_C(2) = \begin{pmatrix} 22.7720 & 27.8474 \\ 24.8376 & 3.4871 \end{pmatrix}.$$

8.3.2 Delay-Dependent Output Feedback

This subsection utilizes the sufficient delay-dependent stability conditions established in Theorem 8.6 to develop a sufficient delay-dependent condition for the dynamical output feedback controller (8.70) to stabilize the system (8.1) in the MSQS sense. As in the previous subsection, we have the following proposition.

Proposition 8.3 *For a set of given matrices $K_A(i), K_B(i), K_C(i)$, if there exist symmetric, positive-definite matrices $0 < P = (P(1), \cdots, P(N)) \in \mathbb{R}^{2n \times 2n}, 0 < U \in \mathbb{R}^{n \times n}$ such that the inequalities*

$$\begin{pmatrix} J(i) & I_0^\top A^\top(i) & P(i)\bar{A}_1(i) & I_0^\top \\ A(i)I_0 & -\frac{1}{\tau}Q^{-1} & 0 & 0 \\ \bar{A}_1^\top(i)P(i) & 0 & -\frac{1}{2\tau}Q & 0 \\ I_0 & 0 & 0 & -\frac{1}{\tau}U \end{pmatrix} \triangleq \Theta_d(i) < 0 \quad (8.97)$$

$$\begin{pmatrix} U & UA_1^\top(i) \\ A_1(i)U & Q^{-1} \end{pmatrix} > 0, \ i \in \mathcal{S} \quad (8.98)$$

hold for some given $Q > 0$, where $J(i) = [\bar{A}(i) + I_0^\top A_1(i)I_0]^\top P(i) + P(i)[\bar{A}(i) + I_0^\top A_1(i)I_0] + \sum_{j=1}^N \lambda_{ij}P(j)$, then the closed-loop system (8.71) is MSQS.

Proof: Substituting (8.34) into (8.71) gives the following dynamics:

$$\begin{cases} \dot{\eta}(t) = [\bar{A}(r_t) + I_0^\top A_1(r_t)I_0]\eta(t) - \int_{t-\tau}^t [\bar{A}_1(r_t)A(r_\theta)x(\theta) \\ \qquad + \bar{A}_1(r_t)A_1(r_\theta)x(\theta-\tau)]\,d\theta \\ \eta(t) = [\phi^\top(t), \mathbf{0}]^\top, \ r_t = r_0, \ t \in [-2\tau, 0]. \end{cases} \tag{8.99}$$

By the same argument as in Section 4.1, the stability of system (8.71) follows from the stability of system (8.99). Let us define the process $\{\eta(t), t \geq 0\} \in \mathbb{C}[-2\tau, 0]$ as follows:

$$\bar{\eta}_s(t) = \eta(s+t), s \in [t-2\tau, t]\}.$$

Then $\{(\bar{\eta}(t), r_t), t \geq 0\}$ is a strong Markov process with initial state $(\phi(\cdot), r_0)$.

Let $P = (P(1), \cdots, P(N)), Q, U$ be a set of feasible solutions to (8.97) and (8.98) and let $Q_1 = U^{-1}$. Consider the following Lyapunov functional candidate:

$$V(\bar{\eta}(t), r_t) = \eta^\top(t)P(r_t)\eta(t) + V_1(\bar{\eta}(t), r_t)$$

where $r_t = (r_s, t-2\tau \leq s \leq t)$ and

$$V_1(\bar{\eta}(t), \mathbf{r}_t) = \int_{-\tau}^0 \int_{t+\theta}^t x^\top(s)Q_0(r_s)x(s)\,ds\,d\theta$$

$$+ \int_{-\tau}^0 \int_{t-\tau+\theta}^t x^\top(s)Q_1x(s)\,ds\,d\theta,$$

where $Q_0(r_s) = A^\top(r_s)QA(r_s)$.

Let $\mathcal{A}$ be the weak generator of $\{(\bar{\eta}(t), \mathbf{r}_t), t \geq 0\}$. Then

$$\begin{aligned} \mathcal{A}V(\bar{\eta}(t), \mathbf{r}_t) &= \mathcal{A}V_1(\bar{\eta}(t), \mathbf{r}_t) + \mathcal{A}(\eta^\top(t)P(r_t)\eta(t)) \\ &= \tau x^\top(t)[Q_0(r_t) + Q_1]x(t) \\ &\quad - \int_{-\tau}^0 x^\top(t+\theta)Q_0(r_{t+\theta})x(t+\theta)\,d\theta \\ &\quad - \int_{-\tau}^0 x^\top(t-\tau+\theta)Q_1x(t-\tau+\theta)\,d\theta \\ &\quad + \eta^\top(t)\Big[[\bar{A}(r_t) + I_0^\top A_1(r_t)I_0]^\top P(r_t) \\ &\quad + P(r_t)[\bar{A}(r_t) + I_0^\top A_1(r_t)I_0] \\ &\quad + \sum_{j=1}^N \lambda_{r_t j}P(j)\Big]\eta(t) + h_1(t), \end{aligned} \tag{8.100}$$

where

$$h_1(t) = -2\int_{-\tau}^0 \eta^\top(t)P(r_t)\Big[\bar{A}_1(r_t)A(r_{t+\theta})x(t+\theta)$$

$$+ \bar{A}_1(r_t)A_1(r_{t+\theta})x(t+\theta-\tau)\Big]d\theta.$$

Note that

$$-\int_{-\tau}^{0} x^\top(t+\theta)Q_0(r_{t+\theta})x(t+\theta)d\theta$$

$$-2\int_{-\tau}^{0} \eta^\top(t)P(r_t)\bar{A}_1(r_t)A(r_{t+\theta})x(t+\theta)d\theta$$

$$= -\int_{-\tau}^{0} \left[Q^{-1}\bar{A}_1^\top(r_t)P(r_t)\eta(t) + A(r_{t+\theta})x(t+\theta)\right]^\top$$

$$\times Q\left[Q^{-1}\bar{A}_1^\top(r_t)P(r_t)\eta(t) + A(r_{t+\theta})x(t+\theta)\right]d\theta$$

$$+ \tau\eta^\top(t)P(r_t)\bar{A}_1(r_t)Q^{-1}\bar{A}_1^\top(r_t)P(r_t)\eta(t). \qquad (8.101)$$

Moreover, using the Schur complement and (8.98), we have

$$A_1^\top(i)QA_1(i) < Q_1,$$

which combined with Lemma A.1 yields

$$-2\int_{-\tau}^{0} \eta^\top(t)P(r_t)\bar{A}_1(r_t)A_1(r_{t+\theta})x(t+\theta-\tau)d\theta$$

$$\leq \tau\eta^\top(t)P(r_t)\bar{A}_1(r_t)Q^{-1}\bar{A}_1^\top(r_t)P(r_t)\eta(t)$$

$$+ \int_{-\tau}^{0} x^\top(t+\theta-\tau)A_1^\top(r_{t+\theta})QA_1(r_{t+\theta})x(t+\theta-\tau)d\theta$$

$$\leq \tau x^\top(t)P(r_t)\bar{A}_1(r_t)Q^{-1}\bar{A}_1^\top(r_t)P(r_t)x(t)$$

$$+ \int_{-\tau}^{0} x^\top(t+\theta-\tau)Q_1x(t+\theta-\tau)d\theta. \qquad (8.102)$$

Combining now (8.100)-(8.102) produces

$$\mathcal{A}V(\bar{\eta}(t),r_t) \leq \eta^\top(t)\tilde{\Theta}(r_t)\eta(t) \qquad (8.103)$$

where

$$\tilde{\Theta}(r_t) = J(r_t) + \tau I_0^\top[A^\top(r_t)QA(r_t) + Q_1]I_0$$

$$+ 2\tau P(r_t)\bar{A}_1(r_t)Q^{-1}\bar{A}_1^\top(r_t)P(r_t)$$

which is negative-definite by using (8.97). The rest of the proof is the same as the one of Lemma 8.1, and it is omitted here. This completes the proof of Proposition 8.3. $\qquad\square$

Proposition 8.3 provides a sufficient condition for a given controller to stabilize system (8.1). However, when $K_A(i), K_B(i), K_C(i)$ are unknown parameters, (8.97) is nonlinear in $K_A(i), K_B(i), K_C(i)$ and $P = (P(1), \cdots, P(N))$ and the problem is not tractable. To cast the controller design problem into the LMI framework, the following theorem gives an equivalent condition for (8.97).

Theorem 8.13 *There exist matrices* $K_A(i), K_B(i), K_C(i)$, $Q > 0, U > 0$ *and symmetric, positive-definite matrices* $P = (P(1), \cdots, P(N)) > 0$ *such that (8.97) holds if and only if there exist symmetric, positive-definite matrices* $X = (X_1, \cdots, X_N) > 0, Y = (Y_1, \cdots, Y_N) > 0$, *and matrices* $H = (H_1, \cdots, H_N), G = (G_1, \cdots, G_N)$ *such that*

$$
\begin{pmatrix}
V_{11}(i) + 2\tau A_1(i)Q^{-1}A_1^\top(i) & \tau Y_i A^\top(i) & \tau Y_i & S_i(Y) \\
\tau A(i)Y_i & -\tau Q^{-1} & 0 & 0 \\
\tau Y_i & 0 & -\tau U & 0 \\
S_i^\top(Y) & 0 & 0 & -\mathcal{Y}_i
\end{pmatrix} < 0 \quad (8.104)
$$

$$
\begin{pmatrix}
V_{22}(i) + \tau A^\top(i)QA(i) & \tau X_i A_1(i) & \tau I \\
\tau A_1^\top(i)X_i & -\tau Q & 0 \\
\tau I & 0 & -\tau U
\end{pmatrix} < 0 \quad (8.105)
$$

$$
\begin{pmatrix}
Y_i & I \\
I & X_i
\end{pmatrix} > 0 \quad (8.106)
$$

hold, where

$$
S_i(Y) = \begin{pmatrix}
\sqrt{\lambda_{i1}}Y_i & \cdots & \sqrt{\lambda_{ii-1}}Y_i & \sqrt{\lambda_{ii+1}}Y_i & \cdots & \sqrt{\lambda_{iN}}Y_i \\
0 & \cdots & 0 & 0 & \cdots & 0
\end{pmatrix}
$$

$$
\mathcal{Y}_i = \text{diag}\{Y_1, \cdots, Y_{i-1}, Y_{i+1}, \cdots, Y_N\}
$$

$$
V_{11}(i) = [A(i) + A_1(i)]Y_i + Y_i[A(i) + A_1(i)]^\top + B(i)G_i \\
+ G_i^\top B^\top(i) + \lambda_{ii}Y_i
$$

$$
V_{22}(i) = [A(i) + A_1(i)]^\top X_i + X_i[A(i) + A_1(i)]
$$

$$
+ H_i C(i) + C^\top(i)H_i^\top + \sum_{j=1}^{N} \lambda_{ij}X_j.
$$

Proof: *Necessity.* Suppose $P(i), K_A(i), K_B(i), K_C(i), i \in \mathcal{S}$, are a set of feasible solutions of (8.97). Partition $P(i)$ as in (8.74). In view of Lemma 8.2, we can assume that P_{2i} is invertible without loss of generality. Let us define the following matrices:

$$
\begin{cases}
X_i = P_{1i} \\
Y_i = [P_{1i} - P_{2i}P_{3i}^{-1}P_{2i}^\top]^{-1} \\
G_i = -K_C(i)P_{3i}^{-1}P_{2i}^\top Y_i \\
H_i = P_{2i}K_B(i) \\
M_i = X_i[A(i) + A_1(i)]Y_i + X_i B(i)G_i + H_i C(i)Y_i \\
\quad - P_{2i}K_A(i)P_{3i}^{-1}P_{2i}^\top Y_i + \sum_{j=1}^{N} \lambda_{ij}[P_{1j} - P_{2j}P_{3j}^{-1}P_{2j}^\top]Y_i.
\end{cases} \quad (8.107)
$$

Now, we proceed to show that X_i, Y_i, G_i, H_i satisfy (8.104)-(8.106). To this end, let us define S_i as in (8.78). Since S_i is invertible, (8.106) follows from (8.79).

To prove (8.104)-(8.105), let us show that the LMIs

$$
\begin{pmatrix}
\Phi_{11}(i) & \Phi_{12}(i) & \Phi_{13}(i) & \Phi_{14}(i) & S_i(Y) \\
\Phi_{12}^\top(i) & -\frac{1}{\tau}Q^{-1} & 0 & 0 & 0 \\
\Phi_{13}^\top(i) & 0 & -\frac{1}{2\tau}Q & 0 & 0 \\
\Phi_{14}^\top(i) & 0 & 0 & -\frac{1}{\tau}U & 0 \\
S_i^\top(Y) & 0 & 0 & 0 & -\mathcal{Y}_i
\end{pmatrix} < 0 \qquad (8.108)
$$

are feasible for X_i, Y_i, M_i, G_i, H_i, where

$$
\Phi_{11}(i) = \begin{pmatrix}
V_{11}(i) & M_i^\top + A(i) + A_1(i) \\
M_i + A^\top(i) + A_1^\top(i) & V_{22}(i)
\end{pmatrix}
$$

$$
\Phi_{12}(i) = \begin{pmatrix} Y_i A^\top(i) \\ A^\top(i) \end{pmatrix} \quad
\Phi_{13}(i) = \begin{pmatrix} A_1(i) \\ X_i A_1(i) \end{pmatrix} \quad
\Phi_{14}(i) = \begin{pmatrix} Y_i \\ I \end{pmatrix}.
$$

In fact, pre- and post-multiplying $\Theta_d(i)$ by $\Xi_i^\top \triangleq \mathrm{diag}\{S_i^\top, I, I, I\}$ and Ξ_i yields

$$
\Xi_i^\top \Theta(i) \Xi_i < 0, \ \forall i \in S. \qquad (8.109)
$$

Moreover, direct manipulation gives

$$
S_i^\top P(i) = \begin{pmatrix} I & 0 \\ P_{1i} & P_{2i} \end{pmatrix}
$$

$$
S_i^\top P(i)[\bar{A}(i) + I_0^\top A_1(i) I_0] S_i =
\begin{pmatrix}
[A(i) + A_1(i)]Y_i + B(i)G_i & A(i) + A_1(i) \\
T_1 & \begin{bmatrix} P_{1i}[A(i) + A_1(i)] \\ + H_i C(i) \end{bmatrix}
\end{pmatrix}, \qquad (8.110)
$$

where $T_1 = P_{1i}[A(i) + A_1(i)]Y_i + H_i C(i)Y_i + X_i B(i)G_i - P_{2i}K_A(i)P_{3i}^{-1}P_{2i}^\top Y_i$.

$$
S_i^\top P(j) S_i = \begin{pmatrix}
\begin{bmatrix} Y_i[P_{1j} - P_{2i}P_{3j}^{-1}P_{2j}^\top] \\ - P_{2j}P_{3i}^{-1}P_{2i}^\top \\ + P_{2i}P_{3i}^{-1}P_{3j}P_{3i}^{-1}P_{2i}^\top]Y_i \end{bmatrix} & Y_i P_{1j} - Y_i P_{2i}P_{3i}^{-1}P_{2j}^\top \\
P_{1j}Y_i - P_{2j}P_{3i}^{-1}P_{2i}^\top Y_i & P_{1j}
\end{pmatrix}
$$

$$
= \begin{pmatrix}
T_2 & Y_i P_{1j} - Y_i P_{2i}P_{3i}^{-1}P_{2j}^\top \\
P_{1j}Y_i - P_{2j}P_{3i}^{-1}P_{2i}^\top Y_i & P_{1j}
\end{pmatrix}, \qquad (8.111)
$$

where $T_2 = Y_i \left[Y_j^{-1} + [P_{2i}P_{3i}^{-1}P_{3j} - P_{2j}]P_{3j}^{-1}[P_{2i}P_{3i}^{-1}P_{3j} - P_{2j}]^\top \right] Y_i$.

Combining now (8.110) and (8.111) yields

$$
S_i^\top J(i) S_i = \left(
\begin{bmatrix}
\begin{bmatrix}
[A(i) + A_1(i)] Y_i \\
+ Y_i [A(i) + A_1(i)]^\top \\
+ B(i) G_i + G_i^\top B^\top(i)
\end{bmatrix}
& M_i^\top + A(i) + A_1(i) \\[2ex]
M_i + [A(i) + A_1(i)]^\top
&
\begin{bmatrix}
[A(i) + A_1(i)]^\top P_{1i} \\
+ P_{1i}[A(i) + A_1(i)] \\
+ H_i C(i) + C^\top(i) H_i^\top \\
+ \sum_{j=1}^N \lambda_{ij} P_{1j}
\end{bmatrix}
\end{bmatrix}
\right)
$$

$$
+ \sum_{j=1}^N \lambda_{ij}
\left(
\begin{matrix}
Y_i \left[Y_j^{-1} + [P_{2i} P_{3i}^{-1} P_{3j} - P_{2j}] P_{3j}^{-1} \right. \\
\left. \times [P_{2i} P_{3i}^{-1} P_{3j} - P_{2j}]^\top \right] Y_i & 0 \\[2ex]
0 & 0
\end{matrix}
\right)
$$

$$
= \Phi_{11}(i) + \sum_{j=1,\,j\neq i}^N \lambda_{ij}
\left(
\begin{matrix}
Y_i \left[Y_j^{-1} \right] Y_i & 0 \\
0 & 0
\end{matrix}
\right)
$$

$$
+ \sum_{j=1,\,j\neq i}^N \lambda_{ij}
\left(
\begin{matrix}
Y_i [P_{2i} P_{3i}^{-1} P_{3j} - P_{2j}] P_{3j}^{-1} \\
\times [P_{2i} P_{3i}^{-1} P_{3j} - P_{2j}]^\top Y_i & 0 \\
0 & 0
\end{matrix}
\right)
\tag{8.112}
$$

Furthermore, simple manipulation yields

$$
S_i^\top I_0^\top A^\top(i) =
\left(
\begin{matrix}
Y_i & -Y_i P_{2i} P_{3i}^{-1} \\
I & 0
\end{matrix}
\right)
\left(
\begin{matrix}
A^\top(i) \\
0
\end{matrix}
\right)
$$

$$
= \left(
\begin{matrix}
Y_i A^\top(i) \\
A^\top(i)
\end{matrix}
\right) = \Phi_{12}(i)
\tag{8.113}
$$

$$
S_i^\top P(i) \bar{A}_1(i) =
\left(
\begin{matrix}
I & 0 \\
P_{1i} & P_{2i}
\end{matrix}
\right)
\left(
\begin{matrix}
A_1(i) \\
0
\end{matrix}
\right)
$$

$$
= \left(
\begin{matrix}
A_1(i) \\
X_i A_1(i)
\end{matrix}
\right) = \Phi_{13}(i),
\tag{8.114}
$$

$$
S_i^\top I_0^\top =
\left(
\begin{matrix}
Y_i & -Y_i P_{2i} P_{3i}^{-1} \\
I & 0
\end{matrix}
\right)
\left(
\begin{matrix}
I \\
0
\end{matrix}
\right)
$$

$$
= \left(
\begin{matrix}
Y_i \\
I
\end{matrix}
\right) = \Phi_{14}(i).
\tag{8.115}
$$

Noting that

$$
\sum_{j=1,\,j\neq i}^N \lambda_{ij} [P_{2i} P_{3i}^{-1} P_{3j} - P_{2j}] P_{3j}^{-1} [P_{2i} P_{3i}^{-1} P_{3j} - P_{2j}]^\top \geq 0,
$$

it follows from (8.109), (8.112)-(8.115) that $X_i, Y_i, M_{ij}, H_i, G_i$ are a set of feasible solutions to (8.108).

In view of the Schur complement, (8.108) is equivalent to

$$\Phi_{11}(i) + \begin{pmatrix} \sum_{j \neq i} \lambda_{ij} Y_i Y_j^{-1} Y_i & 0 \\ 0 & 0 \end{pmatrix} + \tau \Phi_{12}(i) Q \Phi_{12}^\top(i)$$

$$+ 2\tau \Phi_{13}(i) Q^{-1} \Phi_{13}^\top(i) + \tau \Phi_{14}(i) U^{-1} \Phi_{14}^\top(i) \overset{\Delta}{=} Z_i < 0. \qquad (8.116)$$

Designate $Z_i = \begin{pmatrix} Z_{1i} & Z_{2i} \\ Z_{2i}^\top & Z_{3i} \end{pmatrix}$. Then, simple computation gives

$$Z_{1i} = V_{11}(i) + \sum_{j \neq i} \lambda_{ij} Y_i Y_j^{-1} Y_i + \tau Y_i A^\top(i) Q A(i) Y_i$$

$$+ 2\tau A_1(i) Q^{-1} A_1^\top(i) + \tau Y_i U^{-1} Y_i$$

$$Z_{2i} = M_i^\top + A(i) + A_1(i) + \tau Y_i A^\top(i) Q A(i)$$

$$+ 2\tau A_1(i) Q^{-1} A_1^\top(i) X_i + \tau Y_i U^{-1}$$

$$Z_{3i} = V_{22}(i) + \tau A^\top(i) Q A(i) + \tau U^{-1}$$

$$+ 2\tau X_i A_1(i) Q^{-1} A_1^\top(i) X_i.$$

Obviously, (8.116) implies that

$$Z_{1i} < 0, \ Z_{3i} < 0,$$

which are equivalent to (8.104) and (8.105), respectively. This proves the necessity.

Sufficiency. Let us assume that $X_i > 0, Y_i > 0$, H_i and G_i are a set of feasible solutions to (8.104)-(8.106). Let $P(i)$ be defined as in (8.74) with $P_{1i} = X_i, P_{3i} = X_i - Y_i^{-1}, P_{2i} = -P_{3i}$. Then it follows from (8.106) that

$$P(i) > 0, \ i \in \mathcal{S}.$$

Let us construct a set of gain matrices $K_A(i), K_B(i), K_C(i)$ as follows:

$$\begin{cases} K_A(i) = [X_i - Y_i^{-1}]^{-1} \Big[[A(i) + A_1(i)]^\top Y_i^{-1} + \tau A^\top(i) Q A(i) \\ \qquad + 2\tau X_i A_1(i) Q^{-1} A_1^\top(i) Y_i^{-1} + \tau U^{-1} + X_i[A(i) + A_1(i)] \\ \qquad + X_i B(i) G_i Y_i^{-1} + H_i C(i) + \sum_{j=1}^N \lambda_{ij} Y_j^{-1} \Big], \\ K_B(i) = [Y_i^{-1} - X_i]^{-1} H_i \\ K_C(i) = G_i Y_i^{-1}. \end{cases} \qquad (8.117)$$

Set

$$S_i = \begin{pmatrix} Y_i & Y_i \\ I & 0 \end{pmatrix}, \ \Xi_i = \text{diag}\{S_i, I, I, I\}.$$

Substituting $K_A(i), K_B(i), K_C(i), P(i), i \in \mathcal{S}$ into $\Theta_d(i)$ and using similar argument as in the proof of necessity, it is easy to check that

$$\Xi_i^\top \Theta_d(i) \Xi_i < 0$$

which is equivalent to

$$\begin{pmatrix} J_1 & 0 \\ 0 & J_2 \end{pmatrix} < 0 \tag{8.118}$$

where

$$J_1 = V_{11}(i) + \sum_{j \neq i} \lambda_{ij} Y_i Y_j^{-1} Y_i + \tau Y_i A^\top(i) Q A(i) Y_i$$
$$+ \tau Y_i U^{-1} Y_i + 2\tau A_1(i) Q^{-1} A_1^\top(i)$$
$$J_2 = V_{22}(i) + \tau A^\top(i) Q A(i) + 2\tau X_i A_1(i) Q^{-1} A_1^\top(i) X_i + \tau U^{-1}.$$

Using the Schur complement, (8.118) is equivalent to (8.104) and (8.105). This concludes the proof of Theorem 8.13. □

From Proposition 8.3 and Theorem 8.13, we have the following theorem.

Theorem 8.14 *If there exist symmetric, positive-definite matrices $Q > 0, U > 0$ such that LMIs (8.104), (8.105), (8.106), and (8.98) have a set of feasible solutions X_i, Y_i, H_i, G_i, then the output feedback controller with parameters $K_A(i), K_B(i), K_C(i)$ defined by (8.117) stabilizes system (8.1) in the SS sense.*

The problem of designing an output feedback controller with gain matrices of the form of (8.117) that stabilizes system (8.1) in the SS sense and the closed-loop system with the maximum delay can be cast into a linear objective optimization problem.

Let $v = 1/\tau$. Then it is easy to check that LMIs (8.104), (8.105), (8.106), and (8.98) are equivalent, respectively, to

$$\begin{pmatrix} V_{11}(i) & Y_i A^\top(i) & Y_i & A_1(i) & S_i(Y) \\ A(i)Y_i & -vQ^{-1} & 0 & 0 & 0 \\ Y_i & 0 & -vU & 0 & 0 \\ A_1^\top(i) & 0 & 0 & -\frac{v}{2}Q & 0 \\ S_i^\top(Y) & 0 & 0 & 0 & -\mathcal{Y}_i \end{pmatrix} < 0 \tag{8.119}$$

$$\begin{pmatrix} V_{22}(i) & X_i A_1(i) & I & A^\top(i) \\ A_1^\top(i)X_i & -vQ & 0 & 0 \\ I & 0 & -vU & 0 \\ A(i) & 0 & 0 & -vQ^{-1} \end{pmatrix} < 0 \tag{8.120}$$

$$\begin{pmatrix} Y_i & I \\ I & X_i \end{pmatrix} > 0 \tag{8.121}$$

$$\begin{pmatrix} U & U A_1^\top(i) \\ A_1(i)U & Q^{-1} \end{pmatrix} > 0. \tag{8.122}$$

Therefore, we get the following theorem.

Theorem 8.15 *Let* v_0, X_i, Y_i, G_i, H_i *be the solution of the following optimization problem:*

$$\min_{v>0, X_i>0, Y_i>0, G_i, H_i} v \tag{8.123}$$

$$s.t. \ (8.119)\text{-}(8.122).$$

Then the closed-loop system under output feedback controller with gains defined by (8.117) with $\tau = 1/v_0$ *is SS for any* $\tau \in [0, 1/v_0]$.

Proof: Suppose v_0, X_i, Y_i, G_i, H_i are the solutions to (8.123) and define gains by (8.117) with $\tau = 1/v_0$. Then, from the proof of the sufficiency of Theorem 8.13 we conclude that $K_A(i), K_B(i), K_C(i)$, and $P(i)$ with $P_{1i} = X_i, P_{3i} = X_i - Y_i^{-1}, P_{2i} = -P_{3i}$ satisfy (8.97) in case of $\tau = 1/v_0$. To prove this theorem, it suffices to show that (8.97) remains valid for all $\tau \in [0, 1/v_0]$. In fact, noting that $\Theta_d(i)$ is monotone with respect to τ in the sense of negative-definiteness, it is easy to check that (8.97) holds for any $\tau \in [0, 1/v_0]$. This ends the proof of Theorem 8.15. $\qquad\square$

In Theorem 8.15, we can choose $Q = \varepsilon_1 I, U = \varepsilon_2 I$ and get the optimal $\varepsilon_1, \varepsilon_2$ by two-dimensional search.

In case of v being a known scalar, matrix equalities (8.119)-(8.122) represent a feasible problem with respect to X_i, Y_i, G_i, H_i, U. The upper bound of the time delay can be found by the following algorithm.

Algorithm 8.2 *(Control Design Algorithm)*

Step 1. Set initial v and step length h.

Step 2. Solve feasible problem (8.119)-(8.122) to get X_i, Y_i, H_i, G_i, U.

Step 3. If Step 2 gives a feasible solution, set $v = v - h$ and go to Step 2, otherwise stop.

Example 8.8 *Consider a two-mode system as described by (8.1) with parameters as follows:*

$$\Lambda = \begin{pmatrix} -3 & 3 \\ 5 & -5 \end{pmatrix}$$

$$A(1) = \begin{pmatrix} -1 & 0 \\ 0 & -2 \end{pmatrix} \quad A(2) = \begin{pmatrix} 1 & 0 \\ -1 & -2.3 \end{pmatrix}$$

$$B(1) = \begin{pmatrix} 1 \\ 1 \end{pmatrix} \quad B(2) = \begin{pmatrix} 1 \\ -1 \end{pmatrix}$$

$$C(1) = \begin{pmatrix} 1 & 0 \end{pmatrix} \quad C(2) = \begin{pmatrix} 0 & -1 \end{pmatrix}$$

$$A_1(1) = \begin{pmatrix} 0.4 & 0 \\ -.1 & 0 \end{pmatrix} \quad A_1(2) = \begin{pmatrix} -0.2 & 0 \\ -0 & -0.51 \end{pmatrix}.$$

Solving the linear objective minimization problem (8.123) produces the following gain matrices:

$$K_A(1) = 10^3 \begin{pmatrix} -4.069 & -0.4959 \\ 0.343 & -0.495 \end{pmatrix}$$

$$K_A(2) = 10^3 \begin{pmatrix} -2.758 & 25.32 \\ 27.589 & -2.96 \end{pmatrix}$$

$$K_B(1) = \begin{pmatrix} 353.07 \\ -883.68 \end{pmatrix}, K_B(2) = \begin{pmatrix} -5.05414 \\ -4.2779 \end{pmatrix}$$

$$K_C(1) = (-539.98 \;\; -493.76), \;\; K_C(2) = (-2758.82 \;\; 2532.16)$$

and an upper bound of time delay, $\tau = 2.011$.

Using Algorithm 8.2 yields the following gains:

$$K_A(1) = \begin{pmatrix} -602.46 & -722.86 \\ -527.19 & -722.88 \end{pmatrix} \qquad K_A(2) = \begin{pmatrix} -209.68 & 146.25 \\ 213.92 & -9.19 \end{pmatrix}$$

$$K_B(1) = \begin{pmatrix} 602.308 \\ -151.58 \end{pmatrix} \qquad K_B(2) = \begin{pmatrix} -45.62 \\ -728.91 \end{pmatrix}$$

$$K_C(1) = (\;-5423.11 \;\; -7227.205\;) \quad K_C(2) = (\;-211.30 \;\; 191.51\;)$$

$$U = \begin{pmatrix} 5.896 & 2.605 \\ 2.6054 & 3.512 \end{pmatrix}$$

and an upper bound for time delay, $\tau = 3.0303$.

Example 8.8 shows that Algorithm 8.2 yields less conservative result than Theorem 8.15. This is due to the fact that in Algorithm 8.2, Q is full symmetric matrix instead of a diagonal block one in Theorem 8.15.

As we have seen in the introduction some practical systems can have mode-dependent time delays. The goal of the next section is to address the stability and the stabilization problems.

8.4 Stability and Stabilizability of JLS with Mode-Dependent Time Delays

Previous sections dealt with jump linear systems (JLS) with constant delay. The goal of this section is to consider the stability of JLS with mode-dependent time delays. This class of systems is described by the following differential-difference equations:

$$\begin{cases} \dot{x}(t) = A(r_t)x(t) + A_1(r_t)x(t - \tau_{r_t}) + B(r_t)u(t) \\ x(s) = \phi(s), \; s \in [-\bar{\tau}, 0] \end{cases} \qquad (8.124)$$

where $x(t), u(t), A(r_t), A_1(r_t), B(r_t), C(r_t)$ are as defined in (8.1), τ_{r_t} denotes the time delay when the system is in mode r_t and $\bar{\tau} = \max\{\tau_i, i \in \mathcal{S}\}$.

Theorem 8.16 *If there exist symmetric, positive-definite matrices* $P = (P(1), \cdots, P(N)) > 0, Q > 0$ *satisfying*

$$A^\top(i)P(i) + P(i)A(i) + \sum_{j=1}^N \lambda_{ij}P(j) + \varrho Q + P \stackrel{\Delta}{=} \Theta_{md}(i) < 0 \quad (8.125)$$

$$A_1^\top(i)P(i)A_1(i) \leq Q, \quad (8.126)$$

where $\varrho = 1 + \bar{\lambda}(\tau - \underline{\tau}), \underline{\tau} = \min\{\tau_i, i \in \mathcal{S}\}, \bar{\tau} = \max\{\tau_i, i \in \mathcal{S}\}, \bar{\lambda} = \max\{|\lambda_{ii}|, i \in \mathcal{S}\}$, *then system (8.124) with* $u(t) \equiv 0$ *is SS.*

Proof: Let $\mathbb{C}[-\tau_i, 0]$ be the space of continuous functions on interval $[-\tau_i, 0]$ and put

$$\mathbb{C}_0 = \bigcup_{i \in \mathcal{S}} \mathbb{C}[-\tau_i, 0] \times \{i\}.$$

Since $\{(x(t), r_t), t \geq 0\}$ is non-Markov, to cast our model into the framework of Markov system let us define a process $\{(\mathbf{x}(t), r_t), t \geq 0\}$ taking values in $\mathbb{C}_0$ as follows:

$$\mathbf{x}_s(t) = x(s+t), t - \tau_{r_t} \leq s \leq t.$$

Then, $\{(\mathbf{x}(t), r_t), t \geq 0\}$ is a strong Markov process with state space $\mathbb{C}_0$ (see Appendix D).

Let us consider a Lyapunov functional candidate as follows:

$$V(\mathbf{x}(t), r_t) = x^\top(t)P(r_t)x(t) + V_1(\mathbf{x}(t), r_t) + V_2(\mathbf{x}(t), r_t) \quad (8.127)$$

where

$$V_1(\mathbf{x}(t), r_t) = \int_{t-\tau_{r_t}}^t x^\top(s)Qx(s)ds$$

$$V_2(\mathbf{x}(t), r_t) = \int_{-\bar{\tau}}^{-\underline{\tau}} \int_{t+s}^t x^\top(\theta)\bar{\lambda}Qx(\theta)d\theta ds.$$

Let $\mathcal{A}$ be the weak infinitesimal generator of $\{(\mathbf{x}(t), r_t), t \geq 0\}$ (see Appendix D). Then

$$\mathcal{A}V(\mathbf{x}(t), r_t) = x^\top(t)\left[A^\top(r_t)P(r_t) + P(r_t)A(r_t) + \sum_{j=1}^N \lambda_{r_tj}P(j)\right]x(t)$$

$$+ 2x^\top(t)P(r_t)A_1(r_t)x(t - \tau_{r_t})$$

$$+ \mathcal{A}V_1(\mathbf{x}(t), r_t) + \mathcal{A}V_2(\mathbf{x}(t), r_t). \quad (8.128)$$

Suppose $r_t = i \in \mathcal{S}, \mathbf{x} \in \mathbb{C}[-\tau_i, 0]$; then

$$\mathbb{E}[V_1(\mathbf{x}(t + \Delta), r_{t+\Delta})\Big|(\mathbf{x}(t), r_t) = (\mathbf{x}, i)]$$

$$= \sum_{j \neq i} \mathbb{E}\left[I_{\{r_{t+\Delta}=j\}} \int_t^{t+\Delta} x^\top(s)Qx(s)ds \Big| (\mathbf{x}(t), r_t) = (\mathbf{x}, i) \right]$$

$$+ \sum_{j \neq i} \mathbb{E}\left[I_{\{r_{t+\Delta}=j\}} \int_{t+\Delta-\tau_j}^t x^\top(s)Qx(s)ds \Big| (\mathbf{x}(t), r_t) = (\mathbf{x}, i) \right]$$

$$+ \mathbb{E}\left[I_{\{r_{t+\Delta}=i\}} \int_{t+\Delta-\tau_i}^{t+\Delta} x^\top(s)Qx(s)ds \Big| (\mathbf{x}(t), r_t) = (\mathbf{x}, i) \right]$$

$$\triangleq I + II + III \tag{8.129}$$

where $I_{\{.\}}$ is an indicator function.

Note that

$$0 \leq I = O(\Delta^2),$$

$$II = \sum_{j \neq i} P[r_{t+\Delta} = j | r_t = i] \int_{t+\Delta-\tau_j}^t x^\top(s)Qx(s)ds$$

$$= \sum_{j \neq i} (\lambda_{ij}\Delta + o(\Delta)) \int_{t+\Delta-\tau_j}^t x^\top(s)Qx(s)ds$$

$$III = P[r_{t+\Delta} = i | r_t = i] \mathbb{E}\left[\int_{t+\Delta-\tau_i}^{t+\Delta} x^\top(s)Qx(s)ds \right.$$

$$\left. \Big| (\mathbf{x}(t), r_t) = (\mathbf{x}, i), r_{t+\Delta} = i \right]$$

$$= (1 + \lambda_{ii}\Delta + o(\Delta)) \int_{t+\Delta-\tau_i}^{t+\Delta} x^\top(s)Qx(s)ds.$$

Thus, we have

$$\frac{1}{\Delta}\left[\mathbb{E}[V_1(\mathbf{x}(t+\Delta), r_{t+\Delta}|(\mathbf{x}(t), r_t) = (\mathbf{x}, i)] - V(\mathbf{x}(t), r_t) \right]$$

$$= \frac{1}{\Delta}\left[\int_{t+\Delta-\tau_{r_t}}^{t+\Delta} x^\top(s)Qx(s)ds - \int_{t-\tau_i}^t x^\top(s)Qx(s)ds \right]$$

$$+ \lambda_{ii} \int_{t+\Delta-\tau_i}^{t+\Delta} x^\top(s)Qx(s)ds + \sum_{j \neq i} \lambda_{ij} \int_{t+\Delta-\tau_j}^{t+\Delta} x^\top(s)Qx(s)ds + o(\Delta),$$

$$\tag{8.130}$$

from which it follows that

$$\mathcal{A}V_1(\mathbf{x}(t), r_t) = x^\top(t)Qx(t) - x^\top(t-\tau_{r_t})Qx(t-\tau_{r_t})$$

$$+ \sum_{j \in \mathcal{S}} \lambda_{r_t j} \int_{t-\tau_j}^t x^\top(s)Qx(s)ds. \tag{8.131}$$

Direct computation gives

$$\mathcal{A}V_2(\mathbf{x}(t), r_t) = \bar{\lambda}(\bar{\tau} - \underline{\tau})x^\top(t)Qx(t) - \bar{\lambda}\int_{t-\bar{\tau}}^{t-\underline{\tau}} x^\top(s)Qx(s)ds. \quad (8.132)$$

Note that

$$\sum_{j=1}^{N}\lambda_{ij}\int_{t-\tau_j}^{t} x^\top(s)Qx(s)ds$$

$$= \sum_{j\neq i}\lambda_{ij}\int_{t-\tau_j}^{t} x^\top(s)Qx(s)ds + \lambda_{ii}\int_{t-\tau_i}^{t} x^\top(s)Qx(s)ds$$

$$\leq \sum_{j\neq i}\lambda_{ij}\int_{t-\bar{\tau}}^{t} x^\top(s)Qx(s)ds + \lambda_{ii}\int_{t-\underline{\tau}}^{t} x^\top(s)Qx(s)ds$$

$$\leq |\lambda_{ii}|\int_{t-\bar{\tau}}^{t-\underline{\tau}} x^\top(s)Qx(s)ds$$

$$\leq \bar{\lambda}\int_{t-\bar{\tau}}^{t-\underline{\tau}} x^\top(s)Qx(s)ds. \quad (8.133)$$

Combining (8.128)-(8.133) yields

$$\mathcal{A}V(\mathbf{x}(t), r_t) = x^\top(t)\left[A^\top(i)P(i) + P(i)A(i) + \sum_{j=1}^{N}\lambda_{ij}P(j)\right]x(t)$$
$$+ 2x^\top(t)P(r_t)A_1(r_t)x(t - \tau_{r_t}) + x^\top(t)Qx(t)$$
$$- x^\top(t - \tau_{r_t})Qx(t - \tau_{r_t})$$
$$+ \bar{\lambda}(\bar{\tau} - \underline{\tau})x^\top(t)Qx(t). \quad (8.134)$$

Note that

$$2x^\top(t)P(r_t)A_1(r_t)x(t - \tau) = x^\top(t)P(r_t)x(t) - (x(t)$$
$$- A_1(r_t)x(t - \tau_{r_t}))^\top P(r_t)(x(t) - A_1(r_t)x(t - \tau_{r_t}))$$
$$+ x^\top(t - \tau_{r_t})A_1^\top(r_t)P(r_t)A_1(r_t)x(t - \tau_{r_t}).$$

Combining this with (8.134) and (8.126) yields

$$\mathcal{A}V(\mathbf{x}(t), r_t) \leq x^\top(t)\Theta_{\mathrm{md}}(r_t)x(t) \leq -\min_{i\in\mathcal{S}}\{\lambda_{\min}(-\Theta_{\mathrm{md}}(i))\}x^\top(t)x(t).$$

The rest of the proof is the same as in Theorem 8.1. This completes the proof of Theorem 8.16. □

Next, we proceed to establish a sufficient condition for system (8.124) to be MES. To this end, let us introduce an assumption.

Assumption 8.3 *There exists a symmetric matrix M such that the state trajectory of system (8.1) satisfies the following for $-\bar{\tau} \leq \theta \leq 0$:*

$$\|x(t+\theta)\|^2 \leq \left(x^\top(t)\ x^\top(t-\tau)\right) M \begin{pmatrix} x(t) \\ x(t-\tau) \end{pmatrix}. \tag{8.135}$$

Theorem 8.17 *Suppose that Assumption 8.3 holds for system (8.124). Then, system (8.124) with $u(t) \equiv 0$ is MES if there exist symmetric, positive-definite matrices $P = (P(1), \cdots, P(N)) > 0, Q > 0$ such that the LMIs*

$$\begin{pmatrix} \# & P(i)A_1(i) \\ A_1^\top(i)P(i) & -Q \end{pmatrix} < 0, \quad \forall i \in \mathcal{S} \tag{8.136}$$

hold, where

$$\# = A^\top(i)P(i) + P(i)A(i) + \sum_{j=1}^{N} \lambda_{ij}P(j) + [1 + (\bar{\tau} - \underline{\tau})\bar{\lambda}]Q.$$

Proof: Note that (8.136) ensures (8.134) is negative. The rest of the proof is the same as that of Theorem 8.3, and it is omitted here. □

Remark 8.5 *When $\tau_i \equiv \tau, i \in \mathcal{S}$, then $\underline{\tau} = \bar{\tau}$, and Theorem 8.17 reduces to Theorem 8.3.*

Example 8.9 *Consider a system similar to the one described by (8.1) with the following parameters: $\mathcal{S} = \{1, 2\}, \tau_1 = 1, \tau_2 = 2$,*

$$\Lambda = \begin{pmatrix} -3 & 3 \\ 5 & -5 \end{pmatrix}$$

$$A(1) = \begin{pmatrix} -5 & 0 \\ 0.4 & -3 \end{pmatrix} \qquad A(2) = \begin{pmatrix} -1 & 0.1 \\ 0 & -2 \end{pmatrix}$$

$$A_1(1) = \begin{pmatrix} -0.6 & -0.1 \\ 0 & -0.2 \end{pmatrix} \qquad A_1(2) = \begin{pmatrix} -0.5 & 0 \\ 0.1 & -0.2 \end{pmatrix}.$$

With this set of data, solving (8.136) yields

$$P(1) = \begin{pmatrix} 19.9739 & 1.3329 \\ 1.3329 & 33.0846 \end{pmatrix} \quad P(2) = \begin{pmatrix} 35.6562 & 1.1229 \\ 1.1229 & 38.5782 \end{pmatrix}$$

$$Q = \begin{pmatrix} 20.3074 & -0.2784 \\ -0.2784 & 25.3219 \end{pmatrix}.$$

Thus, according to Theorem 8.17, the system under study is stochastically stable.

Next, we come to design a memoryless state feedback controller

$$u(t) = K(r_t)x(t) \tag{8.137}$$

that stabilizes system (8.124) in the stochastic sense.

Theorem 8.18 *If there exist symmetric, positive-definite matrices* $X = (X_1, \cdots, X_N) > 0$, *matrices* $Y = (Y_1, \cdots, Y_N)$, *and* $U > 0$ *satisfying*

$$\begin{pmatrix} \# & X_i & S_i(X) \\ X_i & -\frac{1}{1+(\bar{\tau}-\underline{\tau})\lambda}U & 0 \\ S_i^\top(X) & 0 & -\mathcal{X}_i \end{pmatrix} < 0, \quad i \in \mathcal{S} \qquad (8.138)$$

where

$$\# = A(i)X_i + X_iA^\top(i) + \lambda_{ii}X_i + B(i)Y_i + Y_i^\top B^\top(i) + A_1(i)UA_1^\top(i),$$

$$S_i(X) = \left(\sqrt{\lambda_{i1}}X_i \; \cdots \; \sqrt{\lambda_{ii-1}}X_i \; \sqrt{\lambda_{ii+1}}X_i \; \cdots \; \sqrt{\lambda_{iN}}X_i \right)$$

$$\mathcal{X}_i = \mathrm{diag}\{X_1, \cdots, X_{i-1}, X_{i+1}, \cdots, X_N\},$$

then controller (8.137) with $K(i) = Y_iX_i^{-1}, i \in \mathcal{S}$ *stabilizes system (8.124).*

Proof: For a set of given gains $K(i), i \in \mathcal{S}$, applying (8.137) to system (8.124) yields the closed-loop system:

$$\dot{x}(t) = [A(r_t) + B(r_t)K(r_t)]x(t) + A_1(r_t)x(t - \tau_{r_t}). \qquad (8.139)$$

In view of Theorem 8.17, the closed-loop system (8.139) is stochastically stable if there exist symmetric, positive-definite matrices $P = (P(1), \cdots, P(N)) > 0, Q > 0$ such that

$$\begin{pmatrix} \#_1 & P(i)A_1(i) \\ A_1^\top(i)P(i) & -Q \end{pmatrix} < 0, \quad \forall i \in \mathcal{S}, \qquad (8.140)$$

where

$$\#_1 = [A(i) + B(i)K(i)]^\top P(i) + P(i)[A(i) + B(i)K(i)] + \sum_{j=1}^N \lambda_{ij}P(j) + \varrho Q.$$

Using the Schur complement, (8.140) holds if and only if

$$\#_1 + P(i)A_1(i)Q^{-1}A_1^\top(i)P(i) < 0 \qquad (8.141)$$

is satisfied. Pre- and post-multiplying the left-hand side of (8.141) by $P^{-1}(i)$ and letting $X(i) = P(i)$ and $K(i)X_i = Y_i$ and $Q^{-1} = U$ yields

$$A(i)X_i + X_iA^\top(i) + B(i)Y_i + Y_i^\top B^\top(i) + \lambda_{ii}X_i$$

$$+ \varrho Q + X_i\left[\sum_{j\neq i}\lambda_{ij}X_j^{-1}\right]X_i + A_1(i)UA_1^\top(i) < 0, \qquad (8.142)$$

which implies (8.138). Therefore, from the above derivation we conclude that if there exist symmetric, positive-definite matrices $X = (X_1, \cdots, X_N) > 0$, $U > 0$, and matrices $Y = (Y_1, \cdots, Y_N)$ satisfying (8.138), then $P(i) = X_i^{-1}, K(i) = Y_iX_i^{-1}, i \in \mathcal{S}$, and $Q = U^{-1}$ satisfy (8.140). This completes the proof of Theorem 8.18. $\square$

Example 8.10 *Consider a system as described by (8.1) with the following parameters:* $S = \{1, 2\}, \tau_1 = 1, \tau_2 = 2,$

$$\Lambda = \begin{pmatrix} -3 & 3 \\ 5 & -5 \end{pmatrix},$$

$$A(1) = \begin{pmatrix} 5 & 0 \\ 0.4 & -3 \end{pmatrix} \qquad A(2) = \begin{pmatrix} -1 & 0.1 \\ 0 & 2 \end{pmatrix}$$

$$A_1(1) = \begin{pmatrix} -0.6 & -0.1 \\ 0 & -0.2 \end{pmatrix} \qquad A_1(2) = \begin{pmatrix} -0.5 & 0 \\ 0.1 & -0.2 \end{pmatrix}.$$

Solving (8.138) gives

$$X_1 = \begin{pmatrix} 108.3442 & -69.7202 \\ -69.7202 & 2581.1263 \end{pmatrix} \qquad X_2 = \begin{pmatrix} 5.9591 & -47.6995 \\ -47.6995 & 2117.0642 \end{pmatrix}$$

$$Y_1 = \begin{pmatrix} -6529.1036 & 1359.6862 \end{pmatrix} \qquad Y_2 = \begin{pmatrix} 110.8039 & -11889.713 \end{pmatrix}$$

$$U = \begin{pmatrix} 35.6053 & -212.8383 \\ -212.8383 & 1428.9981 \end{pmatrix}.$$

Therefore, from Theorem 8.18 it follows that the closed-loop system is stochastically stable under controller (8.137) with

$$K(1) = (-60.9836 \ -1.1205), \quad K(2) = (-32.16 \ -6.3407).$$

8.5   Notes

For MJLS, the theory of stability, optimal control, and H_∞-control as well as applications, can be found in current literature, e.g., [1, 21, 31, 63, 79, 86, 112, 143, 151, 168].

There are many publications studying the stabilization problem of linear systems with time delay. Several kinds of stabilizations are considered: constant delay [77, 118, 149]; time-varying delay [77, 105]; constant delay and uncertainty [137, 175]; time-varying delay and parameter uncertainty [49, 108, 110]; delay-independent stabilization [49, 108, 110, 206, 207, 215]; delay-dependent stabilization [118, 128, 155, 189, 190]. For the time-varying delay system, delay-dependent stability is dependent on the maximum value of the time derivative of the time-varying delay. Stability with discrete and distributed time delays is addressed by [118].

For jump linear systems with time delay, [8] addressed mean square stochastic stability, and [26] developed the delay-dependent stability and H_∞ control using the LMI approach. Delay-dependent output feedback stabilization was adapted from [19].

The results of Section 8.1, 8.2 can be extended to time delay JLS with polytopic uncertain parameters. However, the results on output feedback control become invalid for systems with polytopic uncertainties (see [23]).

9

Robust Stability and Stabilizability of Jump Linear Uncertain Systems with Time Delay

In chapter 8, we dealt with the class of time delay systems with Markov jumps, and we developed results for stability and the stabilizability problems. The dynamic was assumed to be free of uncertainties. But since in practice uncertainties cannot be neglected, this chapter deals with the robust stability and stabilizability of the class of linear uncertain system with Markov jumps and time delay. For this purpose, consider a hybrid system with N modes, that is, $\mathcal{S} = \{1, 2, \cdots, N\}$. Mode switching is governed by a continuous-time Markov process $\{r_t, t \geq 0\}$ as defined previously.

Let $x(t) \in \mathbb{R}^n$ be the continuous state of the system satisfying the dynamics

$$\begin{cases} \dot{x}(t) = A(r_t, t)x(t) + B(r_t, t)u(t) + A_d(r_t, t)x(t - \tau) \\ x(s) = \phi(s), \ s \in [-\tau, 0] \end{cases} \tag{9.1}$$

where τ is the delay in the system, $u(t) \in \mathbb{R}^m$ is the control input, and

$$A(r_t, t) = A(r_t) + D_a(r_t)\Delta(r_t, t)E_a(r_t)$$
$$B(r_t, t) = B(r_t) + D_a(r_t)\Delta(r_t, t)E_b(r_t)$$
$$A_d(r_t, t) = A_d(r_t) + D_d(r_t)\Delta_d(r_t, t)E_d(r_t)$$

with $A(r_t) \in \mathbb{R}^{n \times n}$, $A_d(r_t) \in \mathbb{R}^{n \times n}$, $B(r_t) \in \mathbb{R}^{n \times m}$, $D_a(r_t)$, $E_a(r_t)$, $E_b(r_t)$, $D_d(r_t)$, and $E_d(r_t)$ being known matrices with appropriate dimensions. $\Delta(r_t, t)$ and $\Delta_d(r_t, t)$ are unknown time-varying matrices of appropriate dimensions that represent the parameter uncertainties in the system. In what follows, the matrix I will always be used to denote the identity matrix, whose dimension can be determined from the context and thus will not be specified.

The uncertainties $\Delta(r_t, t)$ and $\Delta_d(r_t, t)$ will be admissible if they satisfy the following conditions:

$$\begin{cases} \Delta^\top(r_t, t)\Delta(r_t, t) \leq I \\ \Delta_d^\top(r_t, t)\Delta_d(r_t, t) \leq I. \end{cases}$$

The initial state is assumed to be $(r_0, x(s)) = (r_0, \phi(s))$ for $-\tau \leq s \leq 0$, with $\phi(\cdot) \in \mathcal{L}_2[-\tau, 0]$.

For system (9.1), we have the following definitions.

Definition 9.1 *System (9.1) with $u(t) \equiv 0$ is said to be*

(i) *robustly stochastically stable (SS) if there exists a constant $T(r_0, \phi(\cdot))$ such that*

$$\mathbb{E}\left[\int_0^\infty \|x(t)\|^2 dt \,|\, (r_0, \phi(\cdot))\right] \leq T(r_0, \phi(\cdot))$$

holds for all admissible uncertainties;

(ii) *mean exponentially stable (MES) if there exist constants $\alpha(r_0, \phi(\cdot)) > 0$, and $\beta > 0$ such that*

$$\mathbb{E}\left[\|x(t)\| \,|\, (r_0, \phi(\cdot))\right] \leq \alpha(r_0, \alpha(\cdot))e^{-\beta t}.$$

holds.

Obviously, MES implies SS.

Definition 9.2 *System (9.1) is said to be robustly stabilizable in the SS sense if there exists a stable feedback controller (8.2) such that the closed-loop system is SS.*

In this chapter, we study the robust stability and robust stabilizability of the class of continuous-time uncertain linear systems with Markov jumps and time delays. We will develop delay-independent and delay-dependent sufficient conditions that guarantee the stochastic stability of the class of systems under study when $u(t) \equiv 0$ or under different types of control laws. Robust output feedback is also investigated. Finally, the robust stability and robust stabilizability of systems with Markov jumps and mode-dependent time delays are also studied.

The rest of this chapter is organized as follows. In section 9.1, we develop delay-independent conditions for stability and stabilizability. Section 9.2 establishes delay-dependent conditions for robust stability and stabilizability. Section 9.3 treats output feedback control. Section 9.4 considers a more general class of systems by considering discrete and distributed time delays. Stability and the stabilizability problems are both solved in this case. Finally, Section 9.5 considers mode-dependent time delay systems with Markov jumps, and the problems of their robust stability and stabilizability are solved.

9.1   Delay-Independent Robust Stability and Stabilizability

Let us first develop conditions that we can use to check if system (9.1) with $u(t) \equiv 0$ is SS. For the stabilizability problem, we will assume that the state vector $x(t)$ and the mode r_t are available for feedback at each time t.

Let us start by developing delay-independent sufficient conditions for robust stability and stabilizability of the class of systems we are considering. The following theorem states the first result in this direction.

Theorem 9.1 *If there exist symmetric, positive-definite matrices $P = (P(1), \cdots, P(N)) > 0$, and $Q > 0$ such that the following holds for any $r_t \in S$ and for all admissible uncertainties:*

$$A^\top(r_t, t)P(r_t) + P(r_t)A(r_t, t) + Q$$

$$+ \sum_{j=1}^{N} \lambda_{r_t j} P(j) + P(r_t) \stackrel{\triangle}{=} \Xi_0(r_t, t) < 0 \qquad (9.2)$$

$$A_d^\top(r_t, t)P(r_t)A_d(r_t, t) < Q, \qquad (9.3)$$

then system (9.1) with $u(t) \equiv 0$ is robust SS.

Proof: The proof can be adapted from Theorem 8.1. $\square$

The following theorem provides an LMI-based sufficient condition for the system under study to be robustly SS.

Theorem 9.2 *If there exist symmetric, positive-definite matrices $P = (P(1), \cdots, P(N)) > 0$, $Q > 0$, and positive scalars $\varepsilon_i, \gamma_i, i \in S$, satisfying for every $i \in S$ the LMIs*

$$\begin{pmatrix} J(i) + \varepsilon_i E_a^\top(i)E_a(i) & P(i)D_a(i) \\ D_a^\top(i)P(i) & -\varepsilon_i I \end{pmatrix} < 0 \qquad (9.4)$$

$$\begin{pmatrix} -Q + \gamma_i E_d^\top(i)E_d(i) & A_d^\top(i)P(i) & 0 \\ P(i)A_d(i) & -P(i) & P(i)D_d(i) \\ 0 & D_d^\top(i)P(i) & -\gamma_i I \end{pmatrix} < 0, \qquad (9.5)$$

then system (9.1) with $u(t) \equiv 0$ is SS, where $J(i) = A^\top(i)P(i) + P(i)A(i) + Q + \sum_{j=1}^{N} \lambda_{ij} P(j) + P(i)$.

Proof: To prove this theorem, it suffices to prove that (9.2) is equivalent to (9.4), and (9.3) is equivalent to (9.5). In fact, noticing that $\Xi_0(r_t, t)$ can be rewritten as

$$\Xi_0(r_t, t) = J(r_t) + P(r_t)D_a(r_t)\Delta(r_t, t)E_a(r_t)$$
$$+ E_a^\top(r_t)\Delta^\top(r_t, t)D_a^\top(r_t)P(r_t),$$

we find that (9.2) holds for all admissible uncertainties if and only if there exist $\varepsilon_i > 0$ satisfying

$$J(i) + \varepsilon_i E_a^\top(i)E_a(i) + \frac{1}{\varepsilon_i}P(i)D_a(i)D_a^\top(i)P(i) < 0.$$

Using the Schur complement yields that the above inequality is equivalent to (9.4). Furthermore, using the Schur complement we find that (9.3) is equivalent to

$$\begin{pmatrix} -Q & A_d^\top(r_t,t)P(r_t) \\ P(r_t)A_d(r_t,t) & -P(r_t) \end{pmatrix} < 0. \tag{9.6}$$

Note that the left-hand side of the above inequality can be rewritten as

$$\begin{pmatrix} -Q & A_d^\top(r_t)P(r_t) \\ P(r_t)A_d(r_t) & -P(r_t) \end{pmatrix} + \begin{pmatrix} 0 \\ P(r_t)D_a(r_t) \end{pmatrix} \Delta(r_t,t)\,(E_d(r_t)\;0)$$
$$+ \begin{pmatrix} 0 \\ E_d^\top(r_t) \end{pmatrix} \Delta^\top(r_t,t)\,(0\;D_a^\top(i)P(i)).$$

Likewise, using Lemma A.5 we obtain that (9.6) holds if and only if there exist scalars $\gamma_i > 0$ such that

$$\begin{pmatrix} -Q & A_d^\top(i)P(i) \\ P(i)A_d(i) & -P(i) \end{pmatrix} + \gamma_i \begin{pmatrix} E_d^\top(i)E_d(i) & 0 \\ 0 & 0 \end{pmatrix}$$
$$+ \frac{1}{\gamma_i}\begin{pmatrix} 0 \\ P(i)D_a(i) \end{pmatrix}(0\;D_a^\top(i)P(i)) < 0.$$

Using the Schur complement, the above inequality is equivalent to (9.5). This concludes the proof of Theorem 9.2.   □

Theorem 9.2 can be used to design a state feedback controller in the form of (8.2) that stabilizes system (9.1) in the SS sense. Substituting (8.2) into (9.1) yields the dynamic of the closed-loop system:

$$\dot{x}(t) = \bar{A}(r_t,t)x(t) + A_d(r_t,t)x(t-\tau),$$

where $\bar{A}(r_t,t) = \bar{A}(r_t) + D_a(r_t)\Delta(r_t,t)\bar{E}_a(r_t)$ with $\bar{A}(r_t) = A(r_t) + B(r_t)K(r_t)$, $\bar{E}_a(r_t) = E_a(r_t) + E_b(r_t)K(r_t)$. For a given controller (8.2), using Theorem 9.2, we find that the closed-loop system is robust SS if there exist symmetric, positive-definite matrices $P = (P(1), \cdots, P(N)) > 0$, $Q > 0$, and scalars $\varepsilon_i > 0, \gamma_i > 0, i \in \mathcal{S}$ such that the following inequalities hold for every $i \in \mathcal{S}$:

$$\begin{pmatrix} \bar{J}(i) + \varepsilon_i \bar{E}_a^\top(i)\bar{E}_a(i) & P(i)D_a(i) \\ D_a^\top(i)P(i) & -\varepsilon_i I \end{pmatrix} < 0 \tag{9.7}$$

$$\begin{pmatrix} -Q + \gamma_i E_d^\top(i)E_d(i) & A_d^\top(i)P(i) & 0 \\ P(i)A_d(i) & -P(i) & P(i)D_a(i) \\ 0 & D_a^\top(i)P(i) & -\gamma_i I \end{pmatrix} < 0, \tag{9.8}$$

where $\bar{J}(i)$ is obtained from $J(i)$ by replacing $A(i)$ and $E_a(i)$ by $\bar{A}(i)$ and $\bar{E}_a(i)$, respectively. Using the Schur complement twice yields that (9.7) is equivalent to

$$\left(\begin{array}{cc} \bar{J}(i) + \frac{1}{\varepsilon_i} P(i) D_a(i) D_a^\top(i) P(i) & \bar{E}_a^\top(i) \\ \bar{E}_a(i) & -\frac{1}{\varepsilon_i} I \end{array} \right) < 0.$$

Let $X_i = P^{-1}(i)$ and $U = Q^{-1}$. Pre- and post-multiplying both sides of the above inequality by $\mathrm{diag}\{X_i, I\}$ yields

$$\left(\begin{array}{cc} X_i \bar{J}(i) X_i + \frac{1}{\varepsilon_i} D_a(i) D_a^\top(i) & X_i \bar{E}_a^\top(i) \\ \bar{E}_a(i) X_i & -\frac{1}{\varepsilon_i} I \end{array} \right) < 0. \tag{9.9}$$

Note that

$$X_i \bar{J}(i) X_i = \bar{A}(i) X_i + X_i \bar{A}^\top(i) + \lambda_{ii} X_i + X_i$$

$$+ X_i U^{-1} X_i + X_i \left[\sum_{j \neq i} \lambda_{ij} X_j^{-1} \right] X_i.$$

Letting $\rho_{1i} = 1/\varepsilon_i, Y_i = K(i) X_i$, and using the Schur complement, we obtain that (9.9) is equivalent to

$$\left(\begin{array}{cccc} J_1(i) & X_i E_a^\top(i) + Y_i^\top E_b^\top(i) & X_i & S_i(X) \\ \star & -\rho_{1i} I & 0 & 0 \\ \star & 0 & -U & 0 \\ \star & 0 & 0 & -\mathcal{X}_i \end{array} \right) < 0, \tag{9.10}$$

where $J_1(i) = A(i) X_i + B(i) Y_i + X_i A^\top(i) + Y_i^\top B^\top(i) + \lambda_{ii} X_i + X_i + \rho_{1i} D_a(i) D_a^\top(i)$, and $S_i(X)$ and $\mathcal{X}_i$ are defined by (8.11) and (8.12).

Likewise, using Schur complement twice gives that (9.8) is equivalent to

$$\left(\begin{array}{ccc} -Q & A_d^\top(i) P(i) & E_d^\top(i) \\ P(i) A_d(i) & -P(i) + \frac{1}{\gamma_i} P(i) D_a(i) D_a^\top(i) P(i) & 0 \\ E_d(i) & 0 & -\frac{1}{\gamma_i} I \end{array} \right) < 0.$$

Pre- and post-multiplying both sides of the above inequality by $\mathrm{diag}\{U, X_i, I\}$ and letting $\rho_{2i} = 1/\gamma_i$ leads to

$$\left(\begin{array}{ccc} -U & U A_d^\top(i) & U E_d^\top(i) \\ A_d(i) U & -X_i + \rho_{2i} D_a(i) D_a^\top(i) & 0 \\ E_d(i) U & 0 & -\rho_{2i} I \end{array} \right) < 0. \tag{9.11}$$

Therefore, from the above derivation we get a controller design algorithm, which is given by the following theorem.

Theorem 9.3 *If there exist symmetric, positive-definite matrices $X_i > 0$, $U > 0$, and scalars $\rho_{1i} > 0, \rho_{2i} > 0$ satisfying LMIs (9.10) and (9.11) for every $i \in \mathcal{S}$, then controller (8.2) with $K(i) = Y_i X_i^{-1}, i \in \mathcal{S}$, robustly stabilizes system (9.1).*

This theorem provides an algorithm to design a memoryless state feedback controller of form (8.2) that stabilizes system (9.1) in the robust SS sense. To show its validity, let us give a numerical example.

Example 9.1 *Consider a system described by (9.1) and suppose that the system parameters are as follows:* $\mathcal{S} = \{1, 2\}$, $\Lambda = \begin{pmatrix} -8 & 8 \\ 5 & -5 \end{pmatrix}$,

$$A(1) = \begin{pmatrix} 1 & 0.1 \\ 0 & 1 \end{pmatrix} \qquad A(2) = \begin{pmatrix} 1 & 0 \\ 0.2 & -2 \end{pmatrix}$$

$$A_d(1) = \begin{pmatrix} -0.3 & 0 \\ 0 & -0.2 \end{pmatrix} \qquad A_d(2) = \begin{pmatrix} 0 & 0.5 \\ -0.1 & -0.5 \end{pmatrix}$$

$$B(1) = \begin{pmatrix} 1 \\ 0.1 \end{pmatrix} \qquad B(2) = \begin{pmatrix} -1 \\ 0.1 \end{pmatrix}.$$

$$D_a(1) = \begin{pmatrix} 0.1 \\ 0.3 \end{pmatrix} \qquad D_a(2) = \begin{pmatrix} 0 \\ 0.3 \end{pmatrix}$$

$$D_d(1) = \begin{pmatrix} 0.1 \\ 0.3 \end{pmatrix} \qquad D_d(2) = \begin{pmatrix} 0 \\ 0.3 \end{pmatrix}$$

$$E_a(1) = \begin{pmatrix} 0.1 & 0.2 \end{pmatrix} \quad E_b(1) = [0.3]$$

$$E_a(2) = (0.1 \ 0.2) \qquad E_b(2) = [0.2]$$

$$E_d(1) = (0 \ 0.1) \qquad E_d(2) = (0 \ -0.1).$$

With these data, solving LMIs (9.10) and (9.11) yields the following set of feasible solutions:

$$X_1 = \begin{pmatrix} 201.5799 & -58.3033 \\ -58.3033 & 77.2494 \end{pmatrix} \qquad X_2 = \begin{pmatrix} 274.4536 & -112.5623 \\ -112.5623 & 128.3290 \end{pmatrix}$$

$$U = \begin{pmatrix} 674.7582 & -329.6896 \\ -329.6896 & 484.2614 \end{pmatrix}$$

$$\varepsilon_{11} = 75.3861, \ \varepsilon_{12} = 219.9227, \qquad \varepsilon_{21} = 224.8657, \ \varepsilon_{22} = 201.0124,$$

$$Y_1 = \begin{pmatrix} -347.6774 & -77.5440 \end{pmatrix} \qquad Y_2 = \begin{pmatrix} 1050.8 & -4232 \end{pmatrix}$$

Therefore, according to Theorem 9.3 we conclude that controller (8.2) with

$$K(1) = (-2.5778 \ -2.9494) \ \text{and} \ K(2) = (3.8670 \ 0.0938)$$

stabilizes the system under study.

Next, we proceed with the establishment of conditions that guarantee robust MES for system (9.1).

Let us consider delay-independent stability. For this purpose, let us give the following definition.

Definition 9.3 *System (9.1) with $u(t) \equiv 0$ is called mean square quadratically stable (MSQS) if there exist symmetric, positive-definite matrices*

$P = (P(1), \cdots, P(N)) > 0, Q > 0$ *such that*

$$\Theta(r_t, t) \triangleq \begin{pmatrix} J(r_t, t) & P(r_t)A_d(r_t, t) \\ A_d^\top(r_t, t)P(r_t) & -Q \end{pmatrix} < 0 \qquad (9.12)$$

holds for each $r_t \in \mathcal{S}$ *for all admissible uncertainties, where* $J(r_t, t) = A^\top(r_t, t)P(r_t) + P(r_t)A(r_t, t) + \sum_{j=1}^{N} \lambda_{r_t j} P(j) + Q$.

The following theorem provides the relationship between MSQS and MES.

Theorem 9.4 *Suppose Assumption 8.2 holds for (9.1). If system (9.1) with* $u(t) \equiv 0$ *is MSQS, then it is robustly MES.*

Proof: The proof can be adapted from Theorem 8.3. $\qquad\square$

Theorem 9.4 provides a sufficient condition for system (9.1) with $u(t) \equiv 0$ to be robustly SS. However, $\Theta(r_t, t)$ contains the system uncertainties and therefore it is difficult to check if a system is MSQS using Definition 9.12. The following theorem gives an equivalent condition for MSQS, which can be verified using LMI toolbox.

Theorem 9.5 *System (9.1) with* $u(t) \equiv 0$ *is MSQS if there exist symmetric, positive-definite matrices* $P = (P(1), \cdots, P(N))$ *and* Q, *and a scalar* $\varepsilon > 0$ *such that the following LMI holds for every* $i \in \mathcal{S}$:

$$\begin{pmatrix} \# & P(i)A_d(i) & P(i)D_a(i) & P(i)D_d(i) \\ * & -Q + \varepsilon E_d^\top(i)E_d(i) & 0 & 0 \\ * & 0 & -\varepsilon I & 0 \\ * & 0 & 0 & -\varepsilon I \end{pmatrix} < 0, \qquad (9.13)$$

where $\# = A^\top(i)P(i) + P(i)A(i) + \sum_{j=1}^{N} \lambda_{ij} P(j) + Q + \varepsilon E_a^\top(i)E_a(i)$.

Proof: By definition, to prove Theorem 9.5 it suffices to establish the equivalence between (9.12) and (9.13). Note that

$$\Theta(r_t, t) = \begin{pmatrix} J_0(r_t) & P(r_t)A_d(r_t) \\ A_d^\top(r_t)P(r_t) & -Q \end{pmatrix}$$
$$+ \begin{pmatrix} P(r_t)D_a(r_t) & PD_d(r_t) \\ 0 & 0 \end{pmatrix} \begin{pmatrix} \Delta(r_t, t) & 0 \\ 0 & \Delta_d(r_t, t) \end{pmatrix}$$
$$\times \begin{pmatrix} E_a(r_t) & 0 \\ 0 & E_d(r_t) \end{pmatrix} + \begin{pmatrix} E_a^\top(r_t) & 0 \\ 0 & E_d^\top(r_t) \end{pmatrix}$$
$$\times \begin{pmatrix} \Delta^\top(r_t, t) & 0 \\ 0 & \Delta_d^\top(r_t, t) \end{pmatrix} \begin{pmatrix} D_a^\top(r_t)P(r_t) & 0 \\ D_d^\top(r_t)P(r_t) & 0 \end{pmatrix} \qquad (9.14)$$

where $J_0(r_t) = A^\top(r_t)P(r_t) + P(r_t)A(r_t) + \sum_{j=1}^{N} \lambda_{r_t j} P(j) + Q$.

In view of Lemma A.5, we conclude that (9.12) holds for all admissible uncertainties if there exists a scalar $\varepsilon > 0$ such that

$$
\begin{pmatrix} J_0(r_t) & P(r_t)A_d(r_t) \\ A_d^\top(r_t)P(r_t) & -Q \end{pmatrix}
$$
$$
+ \frac{1}{\varepsilon} \begin{pmatrix} P(r_t)D_a(r_t) & P(r_t)D_d(r_t) \\ 0 & 0 \end{pmatrix} \begin{pmatrix} D_a^\top(r_t)P(r_t) & 0 \\ D_d^\top(r_t)P(r_t) & 0 \end{pmatrix}
$$
$$
+ \varepsilon \begin{pmatrix} E_a^\top(r_t)E_a(r_t) & 0 \\ 0 & E_d^\top(r_t)E_d(r_t) \end{pmatrix} < 0,
$$

that is,

$$
\begin{pmatrix} J_0(r_t) + \varepsilon E_a^\top(r_t)E_a(r_t) & P(r_t)A_d(r_t) \\ A_d^\top(r_t)P(r_t) & -Q + \varepsilon E_d^\top(r_t)E_d(r_t) \end{pmatrix}
$$
$$
+ \begin{pmatrix} P(r_t)D_a(r_t) & P(r_t)D_d(r_t) \\ 0 & 0 \end{pmatrix} \begin{pmatrix} \frac{1}{\varepsilon}I & 0 \\ 0 & \frac{1}{\varepsilon}I \end{pmatrix} \begin{pmatrix} D_a^\top(r_t)P(r_t) & 0 \\ D_d^\top(r_t)P(r_t) & 0 \end{pmatrix} < 0,
$$

which is equivalent to (9.13) by the Schur complement. This completes the proof of Theorem 9.5. □

To show the usefulness of these results, let us consider the following numerical example.

Example 9.2 *Let us consider a system with two modes and for which the dynamics of the continuous state is described by (9.1) with $u(t) \equiv 0$. Assume the parameters are* $\Lambda = \begin{pmatrix} -3 & 3 \\ 5 & -5 \end{pmatrix}$ *and*

$$A(1) = \begin{pmatrix} -5 & 0 \\ 0 & -2 \end{pmatrix} \qquad A(2) = \begin{pmatrix} -2 & 0.8 \\ 0.2 & -2 \end{pmatrix}$$

$$A_d(1) = \begin{pmatrix} -0.3 & 0 \\ 0 & -0.2 \end{pmatrix} \qquad A_d(2) = \begin{pmatrix} 0 & 0.5 \\ -0.1 & -0.5 \end{pmatrix}$$

$$B(1) = \begin{pmatrix} 1 \\ 0.1 \end{pmatrix} \qquad B(2) = \begin{pmatrix} -1 \\ 0.1 \end{pmatrix}$$

$$D(1) = \begin{pmatrix} 1 \\ 0.3 \end{pmatrix} \qquad D(2) = \begin{pmatrix} 0 \\ 0.3 \end{pmatrix}$$

$$D_d(1) = \begin{pmatrix} 0.2 \\ 1 \end{pmatrix} \qquad D_d(2) = \begin{pmatrix} 0 \\ 1 \end{pmatrix}$$

$$E_a(1) = \begin{pmatrix} 0.1 & 0.2 \end{pmatrix} \qquad E_a(2) = \begin{pmatrix} 1 & 0.2 \end{pmatrix}$$

$$E_b(1) = [0.3] \qquad E_b(2) = [0.2]$$
$$E_d(1) = (0\ 1) \qquad E_d(2) = (0\ -1).$$

With this set of data, solving the LMI (9.13) yields

$$P(1) = \begin{pmatrix} 6.776 & 0.4518 \\ 0.4518 & 16.8532 \end{pmatrix} \qquad P(2) = \begin{pmatrix} 12.16423 & 2.1159 \\ 2.1159 & 17.1058 \end{pmatrix}$$

$$Q = \begin{pmatrix} 25.9694 & -0.6139 \\ -0.6139 & 40.2653 \end{pmatrix} \qquad \varepsilon = 22.1407.$$

This satisfies the requirements of Theorem 9.5, and therefore, the system under study is robustly MSQS.

Let us now turn to the design of a memoryless controller.

9.1.1 Stabilization under Memoryless Controller

We next come to consider the problem of designing a memoryless controller (8.2) that stabilizes system (9.1) in the MSQS sense.

Applying (8.2) to (9.1) yields the closed-loop system

$$\dot{x}(t) = [\bar{A}(r_t) + D_a(r_t)\Delta(r_t, t)\bar{E}_a(r_t)]x(t)$$
$$+ [A_d(r_t) + D_d(r_t)\Delta_d(r_t, t)E_d(r_t)]x(t - \tau) \qquad (9.15)$$

where

$$\bar{A}(r_t) = A(r_t) + B(r_t)K(r_t)$$
$$\bar{E}_a(r_t) = E_a(r_t) + E_b(r_t)K(r_t).$$

For a given controller (8.2), we have the following lemma.

Lemma 9.1 *For a given set of gains $K = (K(1), \cdots, K(N))$, the closed-loop system (9.15) is MSQS if there exist symmetric, positive-definite matrices $P = (P(1), \cdots, P(N)) > 0, Q > 0$, and a scalar $\eta > 0$ such that the following holds for each $i \in S$:*

$$\bar{J}_0(i) + \eta P(i)\Big[D_a(i)D_a^\top(i) + D_d(i)D_d^\top(i)\Big]P(i)$$

$$+ \frac{1}{\eta}\bar{E}_a^\top(i)\bar{E}_a(i) + P(i)A_d(i)Q^{-1}A_d^\top(i)P(i) + P(i)A_d(i)Q^{-1}E_d^\top(i)$$

$$\times \Big[\eta I - E_d(i)Q^{-1}E_d^\top(i)\Big]^{-1} E_d(i)Q^{-1}A_d^\top(i)P(i) < 0, \qquad (9.16)$$

where $\bar{J}_0(i) = \bar{A}^\top(i)P(i) + P(i)\bar{A}(i) + \sum_{j=1}^{N} \lambda_{ij}P(j) + Q.$

Proof: By the same argument as in the proof of Theorem 9.5, it is easy to check that the closed-loop system (9.15) is MSQS if there exist symmetric, positive-definite matrices $P = (P(1), \cdots, P(N)) > 0, Q > 0$ and a scalar

$\eta > 0$ satisfying

$$
\begin{pmatrix} \bar{J}_0(i) & P(i)A_d(i) \\ A_d^\top(i)P(i) & -Q \end{pmatrix} + \eta \begin{pmatrix} P(i)D_a(i) & P(i)D_d(i) \\ 0 & 0 \end{pmatrix}
$$

$$
\times \begin{pmatrix} D_a^\top(i)P(i) & 0 \\ D_d^\top(i)P(i) & 0 \end{pmatrix} + \frac{1}{\eta} \begin{pmatrix} \bar{E}_a^\top(i)\bar{E}_a(i) & 0 \\ 0 & E_d^\top(i)E_d(i) \end{pmatrix} < 0
$$

that is,

$$
\begin{pmatrix} J_1(i) & P(i)A_d(i) \\ A_d^\top(i)P(i) & -Q + \frac{1}{\eta}E_d^\top(i)E_d(i) \end{pmatrix} < 0 \tag{9.17}
$$

where

$$
J_1(i) = \bar{J}_0(i) + \eta P(i)\left[D_a(i)D_a^\top(i) + D_d(i)D_d^\top(i)\right]P(i) + \frac{1}{\eta}\bar{E}_a^\top(i)\bar{E}_a(i).
$$

Using the Schur complement, it follows that (9.17) is equivalent to

$$
J_1(i) + P(i)A_d(i)\left[Q - \frac{1}{\eta}E_d^\top(i)E_d(i)\right]^{-1} A_d^\top(i)P(i) < 0. \tag{9.18}
$$

Moreover, using the matrix inversion formula (see Appendix B) we have

$$
\left[Q - \frac{1}{\eta}E_d^\top(i)E_d(i)\right]^{-1} = Q^{-1} + Q^{-1}E_d^\top(i)\left[\eta I\right.
$$

$$
\left. - E_d(i)Q^{-1}E_d^\top(i)\right]^{-1} E_d(i)Q^{-1},
$$

which implies that (9.18) is equivalent to (9.16). This completes the proof of Lemma 9.1. $\qquad\square$

With Lemma 9.1, we get the following theorem.

Theorem 9.6 *If there exist symmetric, positive-definite matrices* $X = (X_1, \cdots, X_N) > 0, U > 0$, *matrices* $Y = (Y_1, \cdots, Y_N)$, *and a scalar* $\eta > 0$, *such that the following holds for each* $i \in \mathcal{S}$

$$
\begin{pmatrix} \# & X_i & X_iE_a^\top(i) + Y_i^\top E_b^\top(i) & A_d(i)UE_d^\top(i) & S_i(X) \\ * & -U & 0 & 0 & 0 \\ * & 0 & -\eta I & 0 & 0 \\ * & 0 & 0 & -\eta I + E_d(i)UE_d^\top(i) & 0 \\ * & 0 & 0 & 0 & -\mathcal{X}_i \end{pmatrix} < 0
\tag{9.19}
$$

where

$$
\# = X_iA^\top(i) + A(i)X_i + B(i)Y_i + Y_i^\top B(i) + \lambda_{ii}X_i
$$
$$
+ \eta[D_a(i)D_a^\top(i) + D_d(i)D_d^\top(i)] + A_d(i)UA_d^\top(i),
$$
$$
S_i(X) = \left(\sqrt{\lambda_{i1}}X_i \quad \cdots \quad \sqrt{\lambda_{ii-1}}X_i \ \sqrt{\lambda_{ii+1}}X_i \quad \cdots \quad \sqrt{\lambda_{iN}}X_i\right),
$$

$$\mathcal{X}_i = \text{diag}\{X_1, \cdots, X_{i-1}, X_{i+1}, \cdots, X_N\},$$

then system (9.1) is MSQS under control (8.2) with $K(i) = Y_i X_i^{-1}$.

Proof: In view of Lemma 9.1, to show Theorem 9.6, it suffices to prove that there exist symmetric, positive-definite matrices $P = (P(1), \cdots, P(N)) > 0$ and $Q > 0$, and a scalar $\eta > 0$ satisfying (9.16).

Suppose that matrices $P = (P(1), \cdots, P(N)) > 0$ and $Q > 0$ and scalar $\eta > 0$ satisfy (9.16). Let $X_i = P^{-1}(i), i \in \mathcal{S}$. Pre- and post-multiplying the left-hand side of (9.16) by X_i yields

$$J_2(i) + \eta[D_a(i)D_a^\top(i) + D_d(i)D_d^\top(i)]$$
$$+ \frac{1}{\eta} X_i \bar{E}_a^\top(i)\bar{E}_a(i)X_i + A_d(i)Q^{-1}A_d^\top(i) + A_d(i)Q^{-1}E_d^\top(i)$$
$$\times \left[\eta I - E_d(i)Q^{-1}E_d^\top(i)\right]^{-1} E_d(i)Q^{-1}A_d^\top(i) < 0, \qquad (9.20)$$

where

$$J_2(i) = X_i \bar{A}^\top(i) + \bar{A}(i)X_i + \lambda_{ii}X_i + X_i \left[\sum_{j \neq i} \lambda_{ij}X_j^{-1}\right] X_i + X_i Q X_i.$$

Letting $Q^{-1} = U$ and using the Schur complement, we have that (9.20) holds if and only if

$$\begin{pmatrix} J_3(i) & X_i & X_i\bar{E}_a^\top(i) & A_d(i)UE_d^\top(i) & S_i(X) \\ * & -U & 0 & 0 & 0 \\ * & 0 & -\eta I & 0 & 0 \\ * & 0 & 0 & -\eta I + E_d(i)UE_d^\top(i) & 0 \\ * & 0 & 0 & 0 & -\mathcal{X}_i \end{pmatrix} < 0 \quad (9.21)$$

where

$$J_3(i) = X_i\bar{A}^\top(i) + \bar{A}(i)X_i + \lambda_{ii}X_i + A_d(i)UA_d^\top(i)$$
$$+ \eta[D_a(i)D_a^\top(i) + D_d(i)D_d^\top(i)].$$

Letting $Y_i = K(i)X_i$ in (9.21) leads to (9.19). Thus, if $X_i > 0, Y_i, i \in \mathcal{S}, U > 0$, and a scalar $\eta > 0$ are a set of feasible solutions to (9.19), then $P(i) = X_i^{-1}, Q = U^{-1}$ and $K(i) = Y_i X_i^{-1}$ satisfy (9.16), which means that the closed-loop system is MSQS. This ends the proof of Theorem 9.6. $\quad \square$

Example 9.3 *To illustrate the results of Theorem 9.6, let us consider a system with two modes, and with dynamics described by (9.1). Suppose that* *the system parameters are as follows:*

$$\Lambda = \begin{pmatrix} -3 & 3 \\ 5 & -5 \end{pmatrix}$$

$$A(1) = \begin{pmatrix} -5 & 0 \\ 0 & -12 \end{pmatrix} \qquad A(2) = \begin{pmatrix} 2 & 0.8 \\ 0.2 & -2 \end{pmatrix}$$

$$A_d(1) = \begin{pmatrix} -.3 & 0 \\ 0 & -0.2 \end{pmatrix} \qquad A_d(2) = \begin{pmatrix} -0.5 & 0.1 \\ 0.1 & -0.4 \end{pmatrix}$$

$$B(1) = \begin{pmatrix} 0 \\ -0.1 \end{pmatrix} \qquad B(2) = \begin{pmatrix} 1 \\ 0.1 \end{pmatrix}$$

$$D(1) = \begin{pmatrix} 1 \\ 0.3 \end{pmatrix} \qquad D(2) = \begin{pmatrix} 0 \\ 0.3 \end{pmatrix}$$

$$D_d(1) = \begin{pmatrix} 0.2 \\ 1 \end{pmatrix} \qquad D_d(2) = \begin{pmatrix} 0 \\ 1 \end{pmatrix}$$

$$E_a(1) = \begin{pmatrix} 0 & 0.2 \end{pmatrix} \qquad E_a(2) = \begin{pmatrix} 1 & 0.2 \end{pmatrix}$$

$$E_b(1) = [0.3] \qquad E_b(2) = [0.2]$$
$$E_d(1) = \begin{pmatrix} 0 & 0.1 \end{pmatrix} \qquad E_d(2) = \begin{pmatrix} 0 & -1 \end{pmatrix}.$$

With these data, solving LMIs (9.19) we have the following set of feasible solutions:

$$X_1 = \begin{pmatrix} 75.6794 & -3.6269 \\ -3.6269 & 86.6526 \end{pmatrix} \qquad X_2 = \begin{pmatrix} 65.3928 & 0.2282 \\ 0.2282 & 68.225 \end{pmatrix}$$
$$Y_1 = \begin{pmatrix} 80.4639 & -795.6181 \end{pmatrix} \qquad Y_2 = \begin{pmatrix} -301.6793 & -59.1367 \end{pmatrix}$$

$$U = \begin{pmatrix} 297.6666 & 21.4901 \\ 214.9015 & 114.7958 \end{pmatrix} \qquad \varepsilon = 208.6372.$$

This satisfies the requirements of Theorem 9.6, and therefore we conclude that the system under study is robustly stabilizable in the MSQS sense and

$$K(1) = Y_1 X_1^{-1} = (0.6244 \quad -9.1556)$$
$$K(2) = Y_2 X_2^{-1} = (1.2712 \quad -11.6659)$$

are a set of stabilizing gains.

9.2 Delay-Dependent Robust Stability and Stabilizability

Theorem 9.5 gives a sufficient condition for the stability of system (9.1) and Theorem 9.6 provides a method to design a controller that stabilizes system (9.1). These results are independent of time delay, which means that these results remain valid for any $\tau \in [0, \infty)$. Thus, the results of the previous section are conservative. To reduce such conservatism, this section develops the delay-dependent robust stability and robust stabilizability sufficient conditions for system (9.1). To this end, let us introduce the following assumptions.

Assumption 9.1 *Let the following be satisfied:*

$$\Delta(r_t, t) \equiv \Delta_d(r_t, t), \ D(i) = D_a(i) = D_d(i), \ i \in \mathcal{S},$$

and $x(t), r_t$ are available for feedback.

9.2.1 Delay-Dependent Stability

Since the solution of system (9.1) with $u(t) \equiv 0$ is also a solution of a system described by the dynamics

$$\begin{cases} \dot{x}(t) = [A(r_t, t) + A_d(r_t, t)]x(t) - \int_{t-\tau}^{t} [A_d(r_t, t)A(r_\theta, \theta)x(\theta) \\ \quad + A_d(r_t, t)A_d(r_\theta, \theta)x(\theta - \tau)] \, d\theta \\ x(s) = \phi(s), \ s \in [-2\tau, 0]. \end{cases} \qquad (9.22)$$

Then, to develop the delay-dependent condition for robust stability of system (9.1), it suffices to consider the stability of system (9.22). To this end, we need the following lemma.

Lemma 9.2 *If there exist symmetric, positive-definite matrices $P = (P(1), \cdots, P(N)) > 0, Q > 0, Q_1 > 0$, such that the inequalities*

$$\Theta_0(r_t, t) \triangleq [A(r_t, t) + A_d(r_t, t)]^\top P(r_t)$$

$$+ P(r_t)[A(r_t, t) + A_d(r_t, t)] + \sum_{j=1}^{N} \lambda_{r_t j} P(j)$$

$$+ \tau[A^\top(r_t, t)QA(r_t, t) + Q_1]$$

$$+ 2\tau P(r_t)A_d(r_t, t)Q^{-1}A_d^\top(r_t, t)P(r_t) < 0 \qquad (9.23)$$

$$A_d^\top(r_t, t)QA_d(r_t, t) \le Q_1 \qquad (9.24)$$

hold for every $r_t \in \mathcal{S}$ and all admissible uncertainties, then system (9.1) with $u(t) \equiv 0$ is robust SS.

Remark 9.1 *Let Assumption 9.1 hold. Then, following the same line as in the proof of Theorem 8.3, one can prove that system (9.1) with $u(t) \equiv 0$ is also robustly MES under the conditions given in Lemma 9.2.*

Proof: The proof can be adapted from Lemma 8.1.   □

With Lemma 9.2, a sufficient condition for system (9.1) with $u(t) \equiv 0$ to be SS can be established. This is given by the following theorem.

Theorem 9.7 *If there exist symmetric, positive-definite matrices $X = (X_1, \cdots, X_N) > 0, U > 0, U_1 > 0$ and scalars $\varepsilon_i > 0, i = 1, 2, 3, 4$, such* *that the LMIs*

$$\Phi_1(i) = \begin{pmatrix} J_{11} & J_{12} & J_{13} & S_i(X) \\ J_{12}^\top & -J_{22} & 0 & 0 \\ J_{13}^\top & 0 & -J_{33} & 0 \\ S_i^\top(X) & 0 & 0 & -\mathcal{X}_i \end{pmatrix} < 0 \qquad (9.25)$$

$$\Phi_2(i) = \begin{pmatrix} -U_1 & U_1 A_d^\top(i) & U_1 E_d^\top(i) \\ A_d(i)U_1 & -U_1 + \varepsilon_3 D(i)D^\top(i) & 0 \\ E_d(i)U_1 & 0 & -\varepsilon_3 I \end{pmatrix} < 0 \qquad (9.26)$$

hold for every $i \in \mathcal{S}$, where

$$J_{11} = [A(i) + A_d(i)]X_i + X_i[A(i) + A_d(i)]^\top + \lambda_{ii}X_i$$
$$\qquad + \varepsilon_4 D(i)D^\top(i) + 2\tau A_d(i)U A_d^\top(i) + 2\varepsilon_2 \tau D(i)D^\top(i)$$
$$J_{12} = \begin{pmatrix} X_i A^\top(i) & A_d(i)U E_d^\top(i) \end{pmatrix}$$
$$J_{22} = \begin{pmatrix} \frac{1}{\tau}[U - \varepsilon_1 D(i)D^\top(i)] & 0 \\ 0 & \frac{1}{2\tau}[\varepsilon_2 I - E_d(i)U E_d^\top(i)] \end{pmatrix}$$
$$J_{13} = \begin{pmatrix} X_i & X_i E_a^\top(i) & X_i[E_a(i) + E_d(i)]^\top \end{pmatrix}$$
$$J_{33} = \begin{pmatrix} \frac{1}{\tau}U_1 & 0 & 0 \\ 0 & \frac{\varepsilon}{\varepsilon_1}I & 0 \\ 0 & 0 & \varepsilon_4 I \end{pmatrix},$$

then system (9.1) with $u(t) \equiv 0$ is MSQS.

Proof: In view of Lemma 9.2, to prove Theorem 9.7 it suffices to show that if (9.25) and (9.26) have a set of feasible solutions $P = (P(1), \cdots, P(N)) > 0, Q > 0, Q_1 > 0$, then they satisfy (9.23) and (9.24).

Suppose that $P = (P(1), \cdots, P(N)) > 0, Q > 0, Q_1 > 0$ satisfy (9.23) and (9.24). From Lemma A.1, it follows that for any $\varepsilon_4 > 0$,

$$[A(r_t, t)) + A_d(r_t, t)]^\top P(r_t) + P(r_t)[A(r_t, t)) + A_d(r_t, t)]$$
$$\leq P(r_t)[A(r_t) + A_d(r_t)] + [A(r_t) + A_d(r_t)]^\top P(r_t)$$
$$\qquad + \varepsilon_4 P(r_t)D(r_t)D^\top(r_t)P(r_t)$$
$$\qquad + \frac{1}{\varepsilon_4}[E_a(r_t) + E_d(r_t)]^\top[E_a(r_t) + E_d(r_t)]. \qquad (9.27)$$

Using Lemma A.6, we know that if $Q^{-1} - \varepsilon_1 D(r_t)D^\top(r_t) > 0$ holds for some $\varepsilon_1 > 0$, then

$$A^\top(r_t, t)QA(r_t, t) \leq A^\top(r_t)\left[Q^{-1} - \varepsilon_1 D(r_t)D^\top(r_t)\right]^{-1}A(r_t)$$
$$\qquad + \frac{1}{\varepsilon_1}E_a^\top(r_t)E_a(r_t). \qquad (9.28)$$

Also, if $\varepsilon_2 I - E_d(r_t)Q^{-1}E_d^\top(r_t) > 0$ holds for some $\varepsilon_2 > 0$, then

$$A_d(r_t, t)Q^{-1}A_d^\top(r_t, t) \leq A_d(r_t)Q^{-1}A_d^\top(r_t)$$
$$\qquad + A_d(r_t)Q^{-1}E_d^\top(r_t)\left[\varepsilon_2 I - E_d(r_t)Q^{-1}E_d^\top(r_t)\right]^{-1}$$

$$\times E_d(r_t)Q^{-1}A_d^\top(r_t) + \varepsilon_2 D(r_t)D^\top(r_t). \qquad (9.29)$$

Combining (9.27)-(9.29) yields

$$\begin{aligned}
\Theta_0(r_t, t) \leq{} & P(r_t)[A(r_t) + A_d(r_t)] + [A(r_t) + A_d(r_t)]^\top P(r_t) \\
& + \varepsilon_4 P(r_t)D(r_t)D^\top(r_t)P(r_t) \\
& + \frac{1}{\varepsilon_4}[E_a(r_t) + E_d(r_t)]^\top[E_a(r_t) + E_d(r_t)] + \sum_{j=1}^{N}\lambda_{r_t j}P(j) \\
& + \tau A^\top(r_t)\left[Q^{-1} - \varepsilon_1 D(r_t)D^\top(r_t)\right]^{-1}A(r_t) \\
& + \frac{\tau}{\varepsilon_1}E_a^\top(r_t)E_a(r_t) + \tau Q_1 + 2\tau P(r_t)\left[A_d(r_t)Q^{-1}A_d^\top(r_t)\right. \\
& + A_d(r_t)Q^{-1}E_d^\top(r_t)\left[\varepsilon_2 I - E_d(r_t)Q^{-1}E_d^\top(r_t)\right]^{-1} \\
& \left.\times E_d(r_t)Q^{-1}A_d^\top(r_t) + \varepsilon_2 D(r_t)D^\top(r_t)\right]P(r_t) \overset{\triangle}{=} \Theta_1(r_t).
\end{aligned}$$

Let $X_i = P^{-1}(i), i \in \mathcal{S}, U = Q^{-1}, U_1 = Q_1^{-1}$. Using the Schur complement, we have

$$X_i \Theta_1(i)X_i < 0,$$

which is equivalent to (9.25).

Using Lemma A.6, we know that if there exists $\varepsilon_3 > 0$ such that $Q^{-1} - \varepsilon_3 D(r_t)D^\top(r_t) > 0$ then

$$\begin{aligned}
A_d^\top(r_t, t)QA_d(r_t, t) \leq{} & A_d^\top(r_t)\left[Q^{-1} - \varepsilon_3 D(r_t)D^\top(r_t)\right]^{-1}A_d(r_t) \\
& + \frac{1}{\varepsilon_3}E_d^\top(r_t)E_d(r_t).
\end{aligned}$$

Thus,

$$A_d^\top(r_t)\left[Q^{-1} - \varepsilon_3 D(r_t)D^\top(r_t)\right]^{-1}A_d(r_t) + \frac{1}{\varepsilon_3}E_d^\top(r_t)E_d(r_t) \leq Q_1, \qquad (9.30)$$

which implies in turn (9.24). Pre- and post-multiplying both sides of (9.30) by Q_1^{-1} gives

$$\begin{aligned}
Q_1^{-1}A_d^\top(r_t)&\left[Q^{-1} - \varepsilon_3 D(r_t)D^\top(r_t)\right]^{-1}A_d(r_t)Q_1^{-1} \\
& + \frac{1}{\varepsilon_3}Q_1^{-1}E_d^\top(r_t)E_d(r_t)Q_1^{-1} \leq Q_1^{-1}. \qquad (9.31)
\end{aligned}$$

Letting $U_1 = Q_1^{-1}$ and using the Schur complement, (9.26) is equivalent to (9.31).

Therefore, if matrices $X = (X_1, \cdots, X_N) > 0, U > 0, U_1 > 0$ and scalars $\varepsilon_i > 0, i = 1, 2, 3, 4$, satisfy (9.25) and (9.26), reversing above derivations yields that $P(i) = X_i^{-1}, Q = U^{-1}, Q_1 = U_1^{-1}$ satisfy (9.23) and (9.24). This completes the proof of Theorem 9.7. □

Example 9.4 *Let us consider a system with two modes and dynamics described by (9.1). Suppose that the system has the following parameters:*

$$\Lambda = \begin{pmatrix} -3 & 3 \\ 5 & -5 \end{pmatrix}$$

$$A(1) = \begin{pmatrix} -1 & 0 \\ 0.2 & -2 \end{pmatrix} \qquad A(2) = \begin{pmatrix} -2 & 0.8 \\ 0.2 & -2 \end{pmatrix}$$

$$A_d(1) = \begin{pmatrix} -0.3 & 0 \\ 0 & -0.2 \end{pmatrix} \qquad A_d(2) = \begin{pmatrix} 0.1 & -0.1 \\ 0.1 & -0.1 \end{pmatrix}$$

$$D(1) = \begin{pmatrix} 1 \\ 0.3 \end{pmatrix} \qquad D(2) = \begin{pmatrix} 0 \\ 0.3 \end{pmatrix}$$

$$E_a(1) = \begin{pmatrix} 0 & 0.2 \end{pmatrix} \qquad E_a(2) = \begin{pmatrix} 1 & 0.2 \end{pmatrix}$$

$$E_d(1) = \begin{pmatrix} 0 & 0.1 \end{pmatrix} \qquad E_d(2) = \begin{pmatrix} 0 & -1 \end{pmatrix}.$$

For this set of data, using a one-dimensional search we have that LMIs (9.25) and (9.26) are feasible for any $\tau \in [0, 1.81]$. In case of $\tau = 1.81$, the following are a set of feasible solutions of LMIs (9.25) and (9.26):

$$X_1 = \begin{pmatrix} 591.4871 & 233.3738 \\ 233.3738 & 575.1409 \end{pmatrix} \qquad X_2 = \begin{pmatrix} 591.8618 & 252.6559 \\ 252.6559 & 533.5198 \end{pmatrix}$$

$$U = \begin{pmatrix} 2432.8872 & 723.7556 \\ 723.7556 & 1239.0105 \end{pmatrix} \qquad U_1 = \begin{pmatrix} 6212.7607 & 84.0278 \\ 84.0278 & 467.9090 \end{pmatrix}$$

$$\varepsilon_1 = 780.1672 \; \varepsilon_2 = 0.0047 \qquad \varepsilon_3 = 1105.1064 \; \varepsilon_4 = 280.4517.$$

According to Theorem 9.7, we therefore conclude that the system under study is robustly MSQS for all $\tau \in [0, 1.81]$.

9.2.2   Delay-Dependent Robust Stabilization under a Memoryless Controller

Now, we proceed with the design of a memoryless controller (8.2) that stabilizes system (9.1) in the MSQS sense. Applying controller (8.2) to system (9.1) leads to the closed-loop system described by the following dynamic:

$$\dot{x}(t) = \bar{A}(r_t, t)x(t) + A_d(r_t, t)x(t - \tau), \tag{9.32}$$

with $\bar{A}(r_t, t) = A(r_t, t) + B(r_t, t)K(r_t)$.

To develop delay-dependent stability conditions for the closed-loop system (9.32), we need the following dynamics:

$$\begin{cases} \dot{x}(t) = [\bar{A}(r_t, t) + A_d(r_t, t)]x(t) - \int_{t-\tau}^{t} [A_d(r_t, t)\bar{A}(r_\theta, \theta)x(\theta) \\ \qquad + A_d(r_t, t)A_d(r_\theta, \theta)x(\theta - \tau)] \, d\theta \\ x(s) = \phi(s), \; s \in [-2\tau, 0]. \end{cases} \tag{9.33}$$

For the closed-loop system (9.33) to be robustly stochastically stable, using Theorem 9.7, we need only to check that there exist symmetric, positive-definite matrices $X = (X_1, \cdots, X_N) > 0$, $U > 0, U_1 > 0$, and scalars $\varepsilon_i > 0, i = 1, 2, 3, 4$, such that

$$\bar{\Phi}_1(i) < 0 \tag{9.34}$$

and (9.26) hold, where $\bar{\Phi}_1(i)$ is obtained from $\Phi_1(i)$ by replacing $A(i)$ and $E_a(i)$ with $\bar{A}(i) = A(i) + B(i)K(i)$ and $E_a(i) + E_b(i)K(i)$, respectively. Letting $Y_i = K(i)X_i$ in (9.34) yields the following LMI:

$$\begin{pmatrix} \bar{J}_{11} & \bar{J}_{12} & \bar{J}_{13} & S_i(X) \\ \bar{J}_{12}^\top & -J_{22} & 0 & 0 \\ \bar{J}_{13}^\top & 0 & -J_{33} & 0 \\ S_i^\top(X) & 0 & 0 & -\mathcal{X}_i \end{pmatrix} < 0, \tag{9.35}$$

where J_{22}, J_{33} are defined as in Theorem 9.7 and

$$\begin{aligned} \bar{J}_{11} &= [A(i) + A_d(i)]X_i + X_i[A(i) + A_d(i)]^\top + B(i)Y_i + Y_i^\top B^\top(i) \\ &\quad + \lambda_{ii}X_i + \varepsilon_4 D(i)D^\top(i) + 2\tau A_d(i)UA_d^\top(i) + 2\varepsilon_2\tau D(i)D^\top(i) \\ \bar{J}_{12} &= \left(X_iA^\top(i) + Y_i^\top B^\top(i) \ A_d(i)UE_d^\top(i) \right) \\ \bar{J}_{13} &= \left(I \ X_iE_a^\top(i) + Y_i^\top E_b^\top(i) \ X_i[E_a(i) + E_d(i)]^\top + Y_i^\top E_b^\top(i) \right). \end{aligned}$$

The above derivation shows that if there exist $X = (X_1, \cdots, X_N) > 0, Y = (Y_1, \cdots, Y_N), U > 0, U_1 > 0$ and $\varepsilon_i > 0, i = 1, 2, 3, 4$, satisfying (9.35) and (9.26), then $X_i > 0, K(i) = Y_iX_i^{-1}, U > 0, U_1 > 0$ satisfy (9.34) and (9.26), which means that controller (8.2) with $K(i) = Y_iX_i^{-1}$ stabilizes system (9.22) in the MSQS sense. We therefore get the following theorem.

Theorem 9.8 *If there exist symmetric, positive-definite matrices $X = (X_1, \cdots, X_N) > 0, U > 0, U_1 > 0$, matrices $Y = (Y_1, \cdots, Y_N)$ and scalars $\varepsilon_i > 0, i = 1, 2, 3, 4$, satisfying (9.35) and (9.26) for every $i \in \mathcal{S}$, then controller (8.2) with $K(i) = Y_iX_i^{-1}$ stabilizes system (9.1) in the MSQS sense.*

Theorem 9.8 provides a method of designing a controller that stabilizes system (9.1) in the MSQS sense. To show the usefulness of this method, let us consider the following example.

Example 9.5 *To illustrate the usefulness of the results of Theorem 9.8, let us consider a system with two modes and dynamics described by (9.1)* *with the following parameters:*

$$\Lambda = \begin{pmatrix} -3 & 3 \\ 5 & -5 \end{pmatrix}$$

$$A(1) = \begin{pmatrix} -1 & 0 \\ 0 & -12 \end{pmatrix} \qquad A(2) = \begin{pmatrix} 2 & 0.8 \\ 0.2 & -2 \end{pmatrix}$$

$$A_d(1) = \begin{pmatrix} -0.3 & 0 \\ 0 & -0.2 \end{pmatrix} \qquad A_d(2) = \begin{pmatrix} 0.1 & -0.1 \\ 0.1 & -0.1 \end{pmatrix}$$

$$B(1) = \begin{pmatrix} 0 \\ -0.1 \end{pmatrix} \qquad B(2) = \begin{pmatrix} 1 \\ 0.1 \end{pmatrix}$$

$$D(1) = \begin{pmatrix} 1 \\ 0.3 \end{pmatrix} \qquad D(2) = \begin{pmatrix} 0 \\ 0.3 \end{pmatrix}$$

$$D_d(1) = \begin{pmatrix} 0.2 \\ 1 \end{pmatrix} \qquad D_d(2) = \begin{pmatrix} 0 \\ 1 \end{pmatrix}$$

$$E_a(1) = \begin{pmatrix} 0 & 0.2 \end{pmatrix} \qquad E_a(2) = \begin{pmatrix} 1 & 0.2 \end{pmatrix}$$

$$E_b(1) = [0.3] \qquad E_b(2) = [0.2]$$

$$E_d(1) = \begin{pmatrix} 0 & 0.1 \end{pmatrix} \qquad E_d(2) = \begin{pmatrix} 0 & -1 \end{pmatrix}.$$

For this set of data, utilizing LMI-toolbox of Matlab we obtain that LMIs (9.35) and (9.26) are feasible for any $\tau \in [0, 2.01]$. When $\tau = 2.01$, a set of feasible solutions of LMIs (9.35) and (9.26) are given by

$$X_1 = \begin{pmatrix} 4.8332 & 0.1043 \\ 0.1043 & 0.6408 \end{pmatrix} \qquad X_2 = \begin{pmatrix} 5.4239 & 0.2024 \\ 0.2024 & 0.6645 \end{pmatrix}$$
$$Y_1 = (0.1418 \; -2.1848) \qquad Y_2 = (-18.3297 \; -1.8801)$$

$$U = \begin{pmatrix} 15.3790 & 4.5322 \\ 4.5322 & 12.7887 \end{pmatrix} \qquad U_1 = \begin{pmatrix} 7.9540 & 0.4644 \\ 0.4644 & 3.1446 \end{pmatrix}$$
$$\varepsilon_1 = 3.4572 \quad \varepsilon_2 = 0.1354 \qquad \varepsilon_3 = 24.6373 \quad \varepsilon_4 = 3.3588.$$

In view of Theorem 9.8, the system under study is robustly stabilizable in the MSQS sense for all $\tau \in [0, 2.01]$. Moreover,

$$K(1) = Y_1 X_1^{-1} = \begin{pmatrix} 0.1033 & -3.4262 \end{pmatrix}$$
$$K(2) = Y_2 X_2^{-1} = \begin{pmatrix} 0.1505 & -3.3335 \end{pmatrix}$$

are a set of stabilizing gains.

9.3   Robust Output Stabilization

In the previous sections we addressed state feedback stabilization of the class of jump linear systems with time delay. When the system state is not available for feedback, we can recourse to output feedback control. This section will study the robust stabilization problem of jump linear systems with time delay under output feedback control law.

Consider a dynamical system described by the following dynamics:

$$\begin{cases} \dot{x}(t) = A(r_t, t)x(t) + B(r_t, t)u(t) + A_d(r_t, t)x(t - \tau) \\ x(s) = \phi(s), \ s \in [-\tau, 0] \\ y(t) = C(r_t, t)x(t) + F(r_t, t)u(t) + C_d(r_t, t)x(t - \tau) \end{cases} \quad (9.36)$$

where $x(t) \in \mathbb{R}^n, r_t \in \mathcal{S}, u(t) \in \mathbb{R}^k, y(t) \in \mathbb{R}^m$ denote the state, the mode, the control, and the output of the system, respectively; where

$$A(r_t, t) = A(r_t) + D_1(r_t)\Delta(r_t, t)E_1(r_t)$$
$$B(r_t, t) = B(r_t) + D_1(r_t)\Delta(r_t, t)E_2(r_t)$$
$$A_d(r_t, t) = A_d(r_t) + D_2(r_t)\Delta_d(r_t, t)E_3(r_t)$$
$$C(r_t, t) = C(r_t) + D_3(r_t)\Delta(r_t, t)E_1(r_t)$$
$$F(r_t, t) = F(r_t) + D_3(r_t)\Delta(r_t, t)E_2(r_t)$$
$$C_d(r_t, t) = C_d(r_t) + D_4(r_t)\Delta_d(r_t, t)E_3(r_t)$$

where $\Delta(r_t, t), \Delta_d(r_t, t)$ are unknown time-varying matrices representing the parameter uncertainties and satisfying

$$\begin{cases} \Delta^\top(r_t, t)\Delta(r_t, t) \leq I \\ \Delta_d^\top(r_t, t)\Delta_d(r_t, t) \leq I; \end{cases}$$

and where $A(r_t) \in \mathbb{R}^{n \times n}$, $B(r_t) \in \mathbb{R}^{n \times k}$, $A_d(r_t) \in \mathbb{R}^{n \times n}$, $C(r_t) \in \mathbb{R}^{m \times n}$, $F(r_t) \in \mathbb{R}^{m \times k}$, $C_d(r_t) \in \mathbb{R}^{m \times n}$, $D_l(r_t), 1 \leq l \leq 4$, $E_l(r_t), 1 \leq l \leq 3$ are known matrices of appropriate dimensions. The initial state is assumed to be $(r_0, x(s)) = (r_0, \phi(s))$ for $-\tau \leq s \leq 0$, with $\phi(\cdot) \in \mathcal{L}_2[-\tau, 0]$.

The objective of this section is to design an output controller (8.70) that stabilizes system (9.36).

Applying controller (8.70) to system (9.36) and letting

$$\eta^\top(t) = \begin{pmatrix} x^\top(t) & \xi^\top(t) \end{pmatrix}$$

denote the augmented state vector of the closed-loop system, we have

$$\begin{aligned} \dot{\eta}(t) &= \bar{A}(r_t, t)\eta(t) + \bar{A}_d(r_t, t)I_0\eta(t - \tau) \\ &= \bar{A}(r_t, t)\eta(t) + \bar{A}_d(r_t, t)x(t - \tau), \end{aligned} \quad (9.37)$$

where

$$\bar{A}(r_t, t) = A_0(r_t) + \bar{D}_a(r_t)\Delta(r_t, t)\bar{E}_a(r_t)$$
$$\bar{A}_d(r_t, t) = \bar{A}_d(r_t) + \bar{D}_d(r_t)\Delta(r_t, t)\bar{E}_d(r_t)$$

with

$$\bar{A}_0(r_t) = \begin{pmatrix} A(r_t) & B(r_t)K_C(r_t) \\ K_B(r_t)C(r_t) & K_A(r_t) + K_B(r_t)F(r_t)K_C(r_t) \end{pmatrix}$$

$$\bar{D}_a(r_t) = \begin{pmatrix} D_1(r_t) \\ K_B(r_t)D_3(r_t) \end{pmatrix}$$

$$\bar{E}_a(r_t) = \begin{pmatrix} E_1(r_t) & E_2(r_t)K_C(r_t) \end{pmatrix}$$

$$\bar{A}_d(r_t) = \begin{pmatrix} A_d(r_t) \\ K_B(r_t)C_d(r_t) \end{pmatrix}$$

$$\bar{D}_d(r_t) = \begin{pmatrix} D_2(r_t) \\ K_B(r_t)D_4(r_t) \end{pmatrix}$$

$$\bar{E}_d(r_t) = E_3(r_t).$$

Therefore, to design a controller (8.70) that stabilizes system (9.36) it suffices to find gains $K_A(i), K_B(i), K_C(i), i \in \mathcal{S}$, such that system (9.37) is MSQS.

By the same argument as that of Theorem 9.5, we get the following lemma.

Lemma 9.3 *(i) For a set of given gains $K_A(i), K_B(i), K_C(i), i \in \mathcal{S}$, system (9.37) is MSQS if there exist symmetric, positive-definite matrices $P = (P(1), \cdots, P(N)) > 0$ with $P(i) \in \mathbb{R}^{2n \times 2n}$ and a scalar $\varepsilon > 0$ such that for any $i \in \mathcal{S}$, the following holds for some $Q > 0$:*

$$\begin{pmatrix} J(i) & P(i)\bar{A}_d(i) & P(i)\bar{D}_a(i) & P(i)\bar{D}_d(i) & \bar{E}_a^\top(i) & I \\ * & -J_{22} & 0 & 0 & 0 & 0 \\ * & 0 & -\varepsilon I & 0 & 0 & 0 \\ * & 0 & 0 & -\varepsilon I & 0 & 0 \\ * & 0 & 0 & 0 & -\frac{1}{\varepsilon}I & 0 \\ * & 0 & 0 & 0 & 0 & -Q^{-1} \end{pmatrix} < 0,$$

(9.38)

where

$$J(i) = \bar{A}^\top(i)P(i) + P(i)\bar{A}(i) + \sum_{j=1}^{N} \lambda_{ij}P(j)$$

$$J_{22} = Q - \varepsilon \bar{E}_d^\top(i)\bar{E}_d(i).$$

(ii) If there exist symmetric, positive-definite matrices $P = (P(1), \cdots, P(N)) > 0$ with $P(i) \in \mathbb{R}^{2n \times 2n}$ and a scalar $\varepsilon > 0$ satisfying (9.38) for each $i \in \mathcal{S}$, and if, furthermore, we assume that $\eta(t)$ satisfies (8.2), then the closed-loop system is robustly MES.

Proof: Let $\hat{\eta}(t) = (\eta_s(t), -\tau \le s \le 0) \in \mathbb{C}[-\tau, 0]$ be defined by

$$\hat{\eta}_s(t) = \eta(s + t).$$

Consider the Lyapunov functional candidate

$$V(\hat{\eta}(t), r_t) = \eta^\top(t)P(r_t)\eta(t) + \int_{t-\tau}^{t} x^\top(s)Qx(s)ds.$$

The rest of the proof is similar to that of Theorem 9.5 and is omitted here. □

Now we proceed to cast the problem of finding $K_A(i), K_B(i), K_C(i)$, $P(i), i \in \mathcal{S}$ satisfying (9.38) into the LMI framework. For this purpose, let us partition $P(i)$ as in (8.74). By the same argument as in Lemma 8.2, we can assume without loss of generality that P_{2i} is invertible.

Lemma 9.4 *Suppose that there exist matrices $K_A(i), K_B(i), K_C(i), P(i), i \in$ $\mathcal{S}$ satisfying (9.38). Then, for some given $Q > 0$*

$$
\begin{pmatrix}
\Gamma_{11} & \Gamma_{12} & \Gamma_{13} & \Gamma_{14} & \Gamma_{15} & \Gamma_{16} & \tilde{S}_i(Y) \\
\Gamma_{12}^\top & -\Gamma_{22} & 0 & 0 & 0 & 0 & 0 \\
\Gamma_{13}^\top & 0 & -\varepsilon I & 0 & 0 & 0 & 0 \\
\Gamma_{14}^\top & 0 & 0 & -\varepsilon I & 0 & 0 & 0 \\
\Gamma_{15}^\top & 0 & 0 & 0 & -Q^{-1} & 0 & 0 \\
\Gamma_{16}^\top & 0 & 0 & 0 & 0 & -\frac{1}{\varepsilon}I & 0 \\
\tilde{S}_i^\top(Y) & 0 & 0 & 0 & 0 & 0 & -\mathcal{Y}_i
\end{pmatrix} < 0 \quad (9.39)
$$

$$
\begin{pmatrix} Y_i & I \\ I & X_i \end{pmatrix} > 0 \quad (9.40)
$$

have a set of feasible solutions: $X_i > 0, Y_i > 0, H_i, G_i, M_i, i \in \mathcal{S}$ $\varepsilon > 0$, where

$$
\Gamma_{11} = \begin{pmatrix}
\begin{bmatrix} A(i)Y_i + Y_i A^\top(i) \\ + B(i)G_i + \lambda_{ii}Y_i \\ + G_i^\top B^\top(i) \end{bmatrix} & M_i^\top + A(i) \\
M_i + A^\top(i) & \begin{bmatrix} A^\top(i)X_i + X_i A(i) \\ + H_i C(i) + C^\top(i)H_i^\top \\ + \sum_{j=1}^N \lambda_{ij}X_j \end{bmatrix}
\end{pmatrix},
$$

$$
\Gamma_{12} = \begin{pmatrix} A_d(i) \\ X_i A_d(i) + H_i C_d(i) \end{pmatrix}
$$

$$
\Gamma_{22} = Q - \varepsilon \bar{E}_d^\top(i)\bar{E}_d(i)
$$

$$
\Gamma_{13} = \begin{pmatrix} D_1(i) \\ X_i D_1(i) + H_i D_3(i) \end{pmatrix}
$$

$$
\Gamma_{14} = \begin{pmatrix} D_2(i) \\ X_i D_2(i) + H_i D_4(i) \end{pmatrix}
$$

$$
\Gamma_{15} = \begin{pmatrix} Y_i \\ I \end{pmatrix}
$$

$$
\Gamma_{16} = \begin{pmatrix} Y_i E_1^\top(i) + G_i^\top E_2^\top(i) \\ E_1^\top(i) \end{pmatrix}
$$

$$
\tilde{S}_i(Y) = \begin{pmatrix} \sqrt{\lambda_{i1}}Y_i & \cdots & \sqrt{\lambda_{ii-1}}Y_i & \sqrt{\lambda_{ii+1}}Y_i & \cdots & \sqrt{\lambda_{iN}}Y_i \\ 0 & \cdots & 0 & 0 & \cdots & 0 \end{pmatrix},
$$

$$
\mathcal{Y}_i = \text{diag}\{Y_1, \cdots, Y_{i-1}, Y_{i+1}, \cdots, Y_N\}.
$$

Proof: Let $P(i)$, $K_A(i)$, $K_B(i)$ and $K_C(i)$, $i \in \mathcal{S}$ be a set of feasible solutions of (9.38) and partition $P(i)$ as in (8.74) with P_{2i} invertible. Define matrices

$$
\begin{cases}
X_i = P_{1i} \\
Y_i = \left[P_{1i} - P_{2i} P_{3i}^{-1} P_{2i}^{\top} \right]^{-1} \\
G_i = -K_C(i) P_{3i}^{-1} P_{2i}^{\top} Y_i \\
H_i = P_{2i} K_B(i) \\
M_i = X_i A(i) Y_i + X_i B(i) G_i + H_i C(i) Y_i \\
\quad - P_{2i} [K_A(i) + K_B(i) F(i) K_C(i)] P_{3i}^{-1} P_{2i}^{\top} Y_i \\
\quad + \sum_{j=1}^{N} \lambda_{ij} [P_{1j} - P_{2j} P_{3i}^{-1} P_{2i}^{\top}] Y_i
\end{cases}
$$

and

$$
S_i = \begin{pmatrix} Y_i & I \\ -P_{3i}^{-1} P_{2i}^{\top} Y_i & \mathbf{0} \end{pmatrix}.
$$

Obviously, S_i is invertible. Therefore, we have

$$
0 < S_i^{\top} P(i) S_i = \begin{pmatrix} Y_i & I \\ I & P_{1i} \end{pmatrix} = \begin{pmatrix} Y_i & I \\ I & X_i \end{pmatrix}, \tag{9.41}
$$

yielding (9.40).

Pre- and post-multiplying both sides of (9.38) by

$$
\Xi_i^{\top} \overset{\triangle}{=} \operatorname{diag}\{S_i^{\top}, I, I, I, I, I\}
$$

and Ξ_i yields

$$
\Xi_i^{\top} \Theta(i) \Xi_i < 0, \ \forall i \in S. \tag{9.42}
$$

Direct manipulation gives

$$
S_i^{\top} P(i) \bar{A}(i) S_i = \begin{pmatrix} A(i) Y_i + B(i) G_i & A(i) \\ \# & P_{1i} A(i) + H_i C(i) \end{pmatrix}, \tag{9.43}
$$

where

$$
\begin{aligned}
\# &= [P_{1i} A(i) + P_{2i} K_B(i) C(i)] Y_i - [P_{1i} B(i) K_C(i) \\
&\quad + (P_{2i}(K_A(i) + K_B(i) F(i) K_C(i)))] P_{1i}^{-1} P_{2i}^{-1} Y_i \\
&= P_{1i} A(i) Y_i + H_i C(i) Y_i + X_i B(i) G_i \\
&\quad - P_{2i} [K_A(i) + K_B(i) F(i) K_C(i)] P_{3i}^{-1} P_{2i}^{\top} Y_i.
\end{aligned}
$$

We have also

$$
\begin{aligned}
S_i^{\top} P(j) S_i &= \begin{pmatrix} \begin{bmatrix} Y_i [P_{1j} - P_{2i} P_{3j}^{-1} P_{2j}^{\top} \\ - P_{2j} P_{3i}^{-1} P_{2i}^{\top} \\ + P_{2i} P_{3i}^{-1} P_{3j} P_{3i}^{-1} P_{2i}^{\top}] Y_i \\ P_{1j} Y_i - P_{2j} P_{3i}^{-1} P_{2i}^{\top} Y_i \end{bmatrix} & Y_i P_{1j} - Y_i P_{2i} P_{3i}^{-1} P_{2j}^{\top} \\ & P_{1j} \end{pmatrix} \\
&= \begin{pmatrix} Y_i \left[Y_j^{-1} + [P_{2i} P_{3i}^{-1} P_{3j} - P_{2j}] \\ \times P_{3j}^{-1} [P_{2i} P_{3i}^{-1} P_{3j} - P_{2j}]^{\top} \right] Y_i & Y_i P_{1j} - Y_i P_{2i} P_{3i}^{-1} P_{2j}^{\top} \\ P_{1j} Y_i - P_{2j} P_{3i}^{-1} P_{2i}^{\top} Y_i & P_{1j} \end{pmatrix}. \tag{9.44}
\end{aligned}
$$

Combining (9.43) and (9.44) yields

$$S_i^\top J(i) S_i = \left(\begin{array}{cc} \left[\begin{array}{c} A(i)Y_i + Y_i A^\top(i) \\ +B(i)G_i + \lambda_{ii} Y_i \\ +G_i^\top B^\top(i) \end{array} \right] & M_i^\top + A(i) \\ M_i + A^\top(i) & \left[\begin{array}{c} A^\top(i)P_{1i} + P_{1i}A(i) \\ + H_i C(i) + C^\top(i)H_i^\top \\ + \sum_{j=1}^N \lambda_{ij} P_{1j} \end{array} \right] \end{array} \right)$$

$$+ \sum_{j=1}^N \lambda_{ij} \left(\begin{array}{cc} Y_i[Y_j^{-1} + [P_{2i}P_{3i}^{-1}P_{3j} - P_{2j}] & 0 \\ \times P_{3j}^{-1}[P_{2i}P_{3i}^{-1}P_{3j} - P_{2j}]^\top]Y_i & \\ 0 & 0 \end{array} \right)$$

$$= \Gamma_{11} + \sum_{j=1,j\neq i}^N \lambda_{ij} \left(\begin{array}{cc} Y_i \left[Y_j^{-1} \right] Y_i & 0 \\ 0 & 0 \end{array} \right)$$

$$+ \sum_{j=1,j\neq i}^N \lambda_{ij} \left(\begin{array}{cc} Y_i[P_{2i}P_{3i}^{-1}P_{3j} - P_{2j}]P_{3j}^{-1} & 0 \\ \times [P_{2i}P_{3i}^{-1}P_{3j} - P_{2j}]^\top Y_i & \\ 0 & 0 \end{array} \right).$$

Note that

$$\sum_{j=1,j\neq i}^N \lambda_{ij}[P_{2i}P_{3i}^{-1}P_{3j} - P_{2j}]P_{3j}^{-1}[P_{2i}P_{3i}^{-1}P_{3j} - P_{2j}]^\top \geq 0.$$

Therefore,

$$S_i^\top J(i) S_i \geq \Gamma_{11} + \tilde{S}_i(Y) \mathcal{Y}_i \tilde{S}_i^\top(Y). \tag{9.45}$$

Moreover, simple manipulation yields

$$S_i^\top P(i) = \left(\begin{array}{cc} I & 0 \\ P_{1i} & P_{2i} \end{array} \right)$$

$$S_i^\top P(i) \bar{A}_d(i) = \left(\begin{array}{cc} I & 0 \\ P_{1i} & P_{2i} \end{array} \right) \left(\begin{array}{c} A_d(i) \\ K_B(i)C_d(i) \end{array} \right)$$

$$= \left(\begin{array}{c} A_d(i) \\ X_i A_d(i) + P_{2i}K_B(i)C_d(i) \end{array} \right)$$

$$= \left(\begin{array}{c} A_d(i) \\ X_i A_d(i) + H_i C_d(i) \end{array} \right) = \Gamma_{12} \tag{9.46}$$

$$S_i^\top P(i) \bar{D}_a(i) = \left(\begin{array}{cc} I & 0 \\ P_{1i} & P_{2i} \end{array} \right) \left(\begin{array}{c} D_1(i) \\ K_B(i)D_3(i) \end{array} \right)$$

$$= \left(\begin{array}{c} D_1(i) \\ X_i D_1(i) + P_{2i}K_B(i)D_3(i) \end{array} \right)$$

$$= \left(\begin{array}{c} D_1(i) \\ X_i D_1(i) + H_i D_3(i) \end{array} \right) = \Gamma_{13}(i) \tag{9.47}$$

$$S_i^\top P(i)\bar{D}_d(i) = \begin{pmatrix} I & \mathbf{0} \\ P_{1i} & P_{2i} \end{pmatrix} \begin{pmatrix} D_2(i) \\ K_B(i)D_4(i) \end{pmatrix}$$

$$= \begin{pmatrix} D_2(i) \\ X_i D_2(i) + P_{2i}K_B(i)D_4(i) \end{pmatrix}$$

$$= \begin{pmatrix} D_2(i) \\ X_i D_2(i) + H_i D_4(i) \end{pmatrix} = \Gamma_{14}(i) \qquad (9.48)$$

$$S_i^\top I_0 = \begin{pmatrix} Y_i & -Y_i P_{2i}P_{3i}^{-1} \\ I & \mathbf{0} \end{pmatrix} \begin{pmatrix} I \\ \mathbf{0} \end{pmatrix}$$

$$= \begin{pmatrix} Y_i \\ I \end{pmatrix} = \Gamma_{15}(i) \qquad (9.49)$$

$$S_i^\top \bar{E}_a^\top(i) = \begin{pmatrix} Y_i & -Y_i P_{2i}P_{3i}^{-1} \\ I & \mathbf{0} \end{pmatrix} \begin{pmatrix} E_1^\top(i) \\ K_C^\top(i)E_2^\top(i) \end{pmatrix}$$

$$= \begin{pmatrix} Y_i E_1^\top(i) - Y_i P_{2i}P_{3i}^{-1}K_C^\top(i)E_2^\top(i) \\ E_1^\top(i) \end{pmatrix}$$

$$= \begin{pmatrix} Y_i E_1^\top(i) + G_i^\top E_2^\top(i) \\ E_1^\top(i) \end{pmatrix} = \Gamma_{16}(i). \qquad (9.50)$$

It follows from (9.41)-(9.49) that X_i, Y_i, M_i, H_i, G_i are a set of feasible solutions to (9.39). This completes the proof of Lemma 9.4. □

With Lemma 9.4, we can provide the main result of this subsection.

Theorem 9.9 *There exist symmetric, positive-definite matrices $P = (P(1), \cdots, P(N)) > 0$ and gains $K_A(i), K_B(i), K_C(i), i \in \mathcal{S}$ satisfying (9.38), which implies that system (9.36) is stabilizable in the MSQS sense under output feedback control (8.70) if and only if there exist a matrix $Q > 0$, and a scalar $\varepsilon > 0$ such that the LMIs*

$$\begin{pmatrix} W_1(i) & A_d(i) & W_{13} & W_{14} & S_i(Y) \\ A_d^\top(i) & -W_{22} & & & \\ W_{13}^\top & & -\varepsilon I & & \\ W_{14}^\top & & & -W_{33} & \\ S_i^\top(Y) & & & & -\mathcal{Y}_i \end{pmatrix} < 0 \qquad (9.51)$$

$$\begin{pmatrix} W_2(i) & * & * & * \\ (X_i A_d(i) + H_i C_d(i))^\top & -W_{22} & & \\ (X_i D_1(i) + H_i D_3(i))^\top & & -\varepsilon I & \\ (X_i D_2(i) + H_i D_4(i))^\top & & & -\varepsilon I \end{pmatrix} < 0 \qquad (9.52)$$

$$\begin{pmatrix} Y_i & I \\ I & X_i \end{pmatrix} > 0 \qquad (9.53)$$

are feasible for some $X = (X_1, \cdots, X_N) > 0$, $Y = (Y_1, \cdots, Y_N) > 0$, $G = (G_1, \cdots, G_N)$, $H = (H_1, \cdots, H_N)$, *where*

$$W_1(i) = A(i)Y_i + Y_i A^\top(i) + B(i)G_i + G_i^\top B^\top(i) + \lambda_{ii} Y_i$$

$$W_{13} = \left(\begin{array}{cc} D_1(i) & D_2(i) \end{array} \right)$$

$$W_{22} = Q - \varepsilon E_3^\top(i) E_3(i)$$

$$W_{14} = \left(Y_i \ \ Y_i E_1^\top(i) + G_i^\top E_2^\top(i) \right)$$

$$W_{33} = \text{diag}\left\{ Q^{-1}, \frac{1}{\varepsilon} I \right\}$$

$$W_2(i) = A^\top(i)X_i + X_i A(i) + H_i C(i) + C^\top(i) H_i^\top$$

$$+ \ Q + \sum_{j=1}^{N} \lambda_{ij} X_j + \varepsilon E_1^\top(i) E_1(i).$$

Proof: Necessity. Let $P = (P(1), \cdots, P(N)) > 0$ and gains $K_A(i), K_B(i)$, $K_C(i), i \in \mathcal{S}$ satisfy (9.38). Then from Lemma 9.4, it follows that (9.39) and (9.40) hold for some matrices $X_i > 0, Y_i > 0, H_i, G_i, M_i$ and scalar $\varepsilon > 0$. Using the Schur complement, we have that (9.39) is equivalent to

$$\Gamma_{11} + \Gamma_{12}\Gamma_{22}^{-1}\Gamma_{12}^\top + \frac{1}{\varepsilon}\Gamma_{13}\Gamma_{13}^\top + \frac{1}{\varepsilon}\Gamma_{14}\Gamma_{14}^\top + \Gamma_{15}Q\Gamma_{15}^\top$$

$$+ \ \varepsilon\Gamma_{16}\Gamma_{16}^\top + \tilde{S}_i(Y)\mathcal{Y}_i^{-1}\tilde{S}_i^\top(Y) \triangleq Z(i)$$

$$= \left(\begin{array}{cc} Z_1(i) & Z_2^\top(i) \\ Z_2(i) & Z_3(i) \end{array} \right) < 0. \tag{9.54}$$

Direct manipulation gives

$$Z_1(i) = W_1(i) + A_d(i)\Gamma_{22}^{-1}A_d^\top(i) + \frac{1}{\varepsilon}D_1(i)D_1^\top(i)$$

$$+ \ \frac{1}{\varepsilon}D_2(i)D_2^\top(i) + Y_i \left[\sum_{j\neq i} \lambda_{ij} Y_j^{-1} \right] Y_i$$

$$+ \ Y_i Q Y_i + \varepsilon[Y_i E_1^\top(i) + G_i^\top E_2^\top(i)][E_1(i)Y_i + E_2(i)G_i]$$

$$Z_2(i) = M_i + A^\top(i) + [X_i A_d(i) + H_i C_d(i)]\Gamma_{22}^{-1}A_d^\top(i)$$

$$+ \ \frac{1}{\varepsilon}[X_i D_1(i) + H_i D_3(i)]D_1^\top(i) + \varepsilon E_1^\top(i)[E_1(i)Y_i^\top + E_2(i)G_i]$$

$$+ \ \frac{1}{\varepsilon}[X_i D_2(i) + H_i D_4(i)]D_2^\top(i) + Q Y_i$$

$$Z_3(i) = W_2(i) + [X_i A_d(i) + H_i C_d(i)]\Gamma_{22}^{-1}[X_i A_d(i) + H_i C_d(i)]^\top$$

$$+ \ \frac{1}{\varepsilon}[X_i D_1(i) + H_i D_3(i)][X_i D_1(i) + H_i D_3(i)]^\top$$

$$+ \ \frac{1}{\varepsilon}[X_i D_2(i) + H_i D_4(i)][X_i D_2(i) + H_i D_4(i)]^\top.$$

Obviously, (9.54) implies $i \in \mathcal{S}$

$$Z_1(i) < 0 \qquad (9.55)$$
$$Z_3(i) < 0. \qquad (9.56)$$

Using the Schur complement, we have that (9.55) and (9.56) are equivalent to (9.51) and (9.52), respectively. This completes the proof of the necessity.

Sufficiency: Let $X_i, Y_i, G_i, H_i, i \in \mathcal{S}$ satisfy (9.51)-(9.53). Then define matrices

$$P(i) = \begin{pmatrix} X_i & Y_i^{-1} - X_i \\ Y_i^{-1} - X_i & X_i - Y_i^{-1} \end{pmatrix}, \quad i \in \mathcal{S}, \qquad (9.57)$$

which are obviously positive-definite from the Schur complement and (9.53). Define the gain matrices as follows:

$$\begin{cases}
K_A(i) = [X_i - Y_i^{-1}]^{-1} \Big[X_i A(i) Y_i + X_i B(i) G_i + H_i C(i) Y_i \\
\qquad + A^\top(i) + [X_i A_d(i) + H_i C_d(i)][Q - \varepsilon E_d^\top(i) E_d(i)]^{-1} A_d^\top(i) \\
\qquad + \sum_{j=1}^{N} \lambda_{ij} Y_j^{-1} + \frac{1}{\varepsilon}[X_i D_1(i) + H_i D_3(i)] D_1^\top(i) Y_i \\
\qquad + \frac{1}{\varepsilon}[X_i D_2(i) + H_i D_4(i)] D_2^\top(i) Y_i \\
\qquad + Q Y_i + \varepsilon E_1^\top(i)[E_1(i) Y_i^\top + E_2(i) G_i] - H_i F_i G_i \Big] Y_i^{-1} \\
K_B(i) = [Y_i^{-1} - X_i]^{-1} H_i \\
K_C(i) = G_i Y_i^{-1}.
\end{cases} \qquad (9.58)$$

Let

$$T_i = \begin{pmatrix} Y_i & Y_i \\ I & 0 \end{pmatrix}.$$

Substituting $K_A(i), K_B(i), K_C(i)$ and $P(i), i \in \mathcal{S}$ into (9.38), pre- and post-multiplying both sides of (9.38) by $\mathrm{diag}\{T_i^\top, I, I, I, I\}$ and $\mathrm{diag}\{T_i, I, I, I, I\}$, respectively, and using the same argument as in the proof of necessity, we get that (9.38) is equivalent to

$$\begin{pmatrix} Z_1(i) & \mathbf{0} \\ \mathbf{0} & Z_3(i) \end{pmatrix} < 0,$$

which follows from (9.51) and (9.52). This concludes the proof of Theorem 9.9. $\qquad \Box$

Remark 9.2 *Theorem 9.9 establishes a necessary and sufficient condition for system (9.36) to be robustly stabilizable in the MSQS sense under controller (8.70).*

Example 9.6 *Let us consider a system with two modes and let the dynamics of the continuous variables $x(t)$ be described by (9.36) with the following* *parameters:*

$$\Lambda = \begin{pmatrix} -3 & 3 \\ 5 & -5 \end{pmatrix}$$

$$A(1) = \begin{pmatrix} -1 & 0 \\ 0.4 & -10 \end{pmatrix} \qquad A(2) = \begin{pmatrix} -1 & 0.8 \\ 0.2 & -2 \end{pmatrix}$$

$$A_d(1) = \begin{pmatrix} -0.3 & 0 \\ 0 & -0.2 \end{pmatrix} \qquad A_d(2) = \begin{pmatrix} 0.1 & -0.1 \\ 0.1 & -0.1 \end{pmatrix}$$

$$B(1) = \begin{pmatrix} 1 & 0 \\ 0 & -0.1 \end{pmatrix} \qquad B(2) = \begin{pmatrix} 0 & 1 \\ 0 & 0.1 \end{pmatrix}$$

$$D_1(1) = \begin{pmatrix} 1 & 0 \\ 0 & 0.3 \end{pmatrix} \qquad D_1(2) = \begin{pmatrix} 0 & 1 \\ 0 & 0.3 \end{pmatrix}$$

$$D_2(1) = \begin{pmatrix} 0 & 0.2 \\ 0 & 1 \end{pmatrix} \qquad D_2(2) = \begin{pmatrix} 1 & 0 \\ 0 & 1 \end{pmatrix}$$

$$D_3(1) = \begin{pmatrix} 0 & 0 \\ 1 & 1 \end{pmatrix} \qquad D_3(2) = \begin{pmatrix} 1 & 0 \\ 0 & 1 \end{pmatrix}$$

$$D_4(1) = \begin{pmatrix} 0 & 0.2 \\ 0 & 0.1 \end{pmatrix} \qquad D_4(2) = \begin{pmatrix} 1 & 0 \\ 1 & 1 \end{pmatrix}$$

$$E_1(1) = \begin{pmatrix} 1 & 0 \\ 0 & 0.2 \end{pmatrix} \qquad E_1(2) = \begin{pmatrix} 1 & 0 \\ 0 & 0.2 \end{pmatrix}$$

$$E_2(1) = \begin{pmatrix} 0 & 0.3 \\ 0 & 1 \end{pmatrix} \qquad E_2(2) = \begin{pmatrix} 0.2 & 0 \\ 1 & 0 \end{pmatrix}$$

$$E_3(1) = \begin{pmatrix} 0.2 & 0 \\ 0.2 & 0.1 \end{pmatrix} \qquad E_3(2) = \begin{pmatrix} -.1 & 0 \\ 0 & -0.2 \end{pmatrix}$$

$$C(1) = \begin{pmatrix} 1 & 0 \\ 0 & 1 \end{pmatrix} \qquad C(2) = \begin{pmatrix} 0 & 1 \\ 1 & 0 \end{pmatrix}$$

$$F(1) = \begin{pmatrix} 1 & 0 \\ 0 & 1 \end{pmatrix} \qquad F(2) = \begin{pmatrix} 1 & 0.5 \\ 1 & 0 \end{pmatrix}$$

$$C_d(1) = \begin{pmatrix} 1 & 1 \\ 0 & 1 \end{pmatrix} \qquad C_d(2) = \begin{pmatrix} 0 & 0.4 \\ 1 & 0 \end{pmatrix}.$$

Let $\varepsilon = 0.81$ *and* $Q = \begin{pmatrix} 1.5 & 0 \\ 0 & 1.5 \end{pmatrix}$. *With these parameters, solving LMIs (9.51)-(9.53) yields a set of feasible solutions given by*

$$X_1 = \begin{pmatrix} 1.9389 & -0.0196 \\ -0.0196 & 1.0505 \end{pmatrix} \qquad X_2 = \begin{pmatrix} 1.8113 & -0.0085 \\ -0.0085 & 1.7354 \end{pmatrix}$$

$$Y_1 = \begin{pmatrix} 3.3824 & 0.0207 \\ 0.0207 & 4.1103 \end{pmatrix} \qquad Y_2 = \begin{pmatrix} 3.0896 & 0.1136 \\ 0.1136 & 2.4577 \end{pmatrix}$$

$$G_1 = \begin{pmatrix} -21.9703 & -0.6127 \\ -0.9347 & -0.8837 \end{pmatrix} \qquad G_2 = \begin{pmatrix} -0.6160 & -0.4945 \\ -19.6889 & -3.6419 \end{pmatrix}$$

$$H_1 = \begin{pmatrix} -0.1785 & -0.5898 \\ 0.0159 & -0.2356 \end{pmatrix} \qquad H_2 = \begin{pmatrix} -0.4507 & -1.0334 \\ 0.1253 & -0.7237 \end{pmatrix}.$$

Therefore, according to Theorem 9.9 the system under consideration is robustly stabilizable in the MSQS sense using dynamical output feedback controller (8.70). Substituting the above obtained data into (9.58) yields a set of stabilizing gains as follows:

$$K_A(1) = \begin{pmatrix} -7.4745 & -0.4978 \\ 0.2876 & -12.7435 \end{pmatrix} \quad K_A(2) = \begin{pmatrix} -8.5683 & -1.7432 \\ -1.4299 & -1.5528 \end{pmatrix}$$

$$K_B(1) = \begin{pmatrix} -0.1084 & -0.3622 \\ 0.0173 & -0.2999 \end{pmatrix} \quad K_B(2) = \begin{pmatrix} -0.3039 & -0.6899 \\ 0.0976 & -0.5377 \end{pmatrix}$$

$$K_C(1) = \begin{pmatrix} -6.4948 & -0.1163 \\ -0.2750 & -0.2136 \end{pmatrix} \quad K_C(2) = \begin{pmatrix} -0.1923 & -0.1923 \\ -6.3289 & -1.1892 \end{pmatrix}$$

9.4 Robust Stability and Stabilization of JLS with Discrete and Distributed Time Delays

This section deals with stability and stabilization of jump linear uncertain systems with discrete and distributed time delays, with the following dynamics

$$\begin{cases} \dot{x}(t) = A(r_t, t)x(t) + A_1(r_t, t)x(t - \tau_1) \\ \quad + A_2(r_t, t) \int_{t-\tau_2}^{t} x(s)ds + B(r_t, t)u(t) \\ x(s) = \phi(s), \quad s \in [-\max\{\tau_1, \tau_2\}, 0] \end{cases} \quad (9.59)$$

where $x(t) \in \mathbb{R}^n, u(t) \in \mathbb{R}^k$ are the system state and the control input; where τ_1, τ_2 are constants representing the discrete and the distributed time delays in the system, respectively; where

$$A(r_t, t) = A(r_t) + D(r_t)\Delta(r_t, t)E(r_t)$$
$$A_1(r_t, t) = A_1(r_t) + D_1(r_t)\Delta_1(r_t, t)E_1(r_t)$$
$$A_2(r_t, t) = A_2(r_t) + D(r_t)\Delta_2(r_t, t)E_2(r_t)$$
$$B(r_t, t) = B(r_t) + D(r_t)\Delta(r_t, t)E_b(r_t)$$

where $A(r_t), A_1(r_t), A_2(r_t) \in \mathbb{R}^{n \times n}$, $B(r_t) \in \mathbb{R}^{n \times k}$, $D(r_t), E(r_t), D_1(r_t)$, $E_1(r_t), E_2(r_t), E_b(r_t)$ are constant matrices of appropriate dimensions; and where $\Delta(r_t, t), \Delta_1(r_t, t), \Delta_2(r_t, t)$ are time-varying unknown matrices with appropriate dimensions representing the uncertainties in the system parameters.

Remark 9.3 *If the evolution of $x(t)$ depends on the history in the interval $[t - \tau_4, t - \tau_3]$ with $\tau_3 < \tau_4$ being constants, that is,*

$$\dot{x}(t) = A(r_t, t)x(t) + A_1(r_t, t)x(t - \tau_1)$$
$$+ A_2(r_t, t) \int_{t-\tau_2}^{t} x(s)ds + C(r_t, t) \int_{t-\tau_4}^{t-\tau_3} x(s)ds,$$

then (9.59) can be extended to include this case by considering

$$\dot{x}(t) = A(r_t, t)x(t) + A_1(r_t, t)x(t - \tau_1)$$
$$+ A_2(r_t, t) \int_{t-\tau_2}^{t} x(s)ds + C(r_t, t) \int_{t-\tau_4}^{t} x(s)ds$$
$$- C(r_t, t) \int_{t-\tau_3}^{t} x(s)ds.$$

The goal of this section is to study the robust stability and stabilization of system (9.59). To this end, let us assume that there exists a constant symmetric matrix M such that Assumption (8.1) holds for all $s \in [t - \max\{\tau_1, \tau_2\}, t]$ and also suppose that the system state $x(t)$ and mode r_t are available for feedback.

9.4.1 Robust Stability of JLS with Discrete and Distributed Time Delays

Let us begin with the robust stability of system (9.59) with $u(t) \equiv 0$.

Lemma 9.5 *If there exist symmetric, positive-definite matrices* $P = (P(1), \cdots, P(N)) > 0, Q_0 > 0$ *and* $Q_1 > 0$ *such that*

$$\Theta(r_t, t) \triangleq \begin{pmatrix} J_1(r_t, t) & P(r_t)A_1(r_t, t) & P(r_t)A_2(r_t, t) \\ A_1^\top(r_t, t)P(r_t) & -Q_0 & 0 \\ A_2^\top(r_t, t)P(r_t) & 0 & -\frac{1}{\tau_2}Q_1 \end{pmatrix} < 0 \tag{9.60}$$

holds for all $r_t \in S$ *and for all admissible uncertainties, where*

$$J_1(r_t, t) = A^\top(r_t, t)P(r_t) + P(r_t)A(r_t, t) + \sum_{j=1}^{N} \lambda_{r_t j}P(j) + \tau_2 Q_1 + Q_0,$$

then the closed-loop system (9.59) is robustly MES.

Proof: Let $\mathbf{x}(t) \in \mathbb{C}[-\tau, 0]$ with $\mathbf{x}_s(t) = x(t + s)$ and let $\mathcal{A}$ be the weak infinitesimal generator of $\{(\mathbf{x}(t), r_t), t \geq 0\}$. Let us consider the following Lyapunov functional candidate:

$$V(\mathbf{x}(t), r_t) = x^\top(t)P(r_t)x(t) + V_1(\mathbf{x}(t), r_t), \tag{9.61}$$

where

$$V_1(\mathbf{x}(t), r_t) = \int_{t-\tau_1}^{t} x^\top(s)Q_0 x(s)ds + \int_{-\tau_2}^{0} \int_{t+\theta}^{t} x^\top(s)Q_1 x(s)dsd\theta.$$

Then, direct manipulation gives

$$\mathcal{A}[x^\top(t)P(r_t)x(t)] = x^\top(t)\left[A^\top(r_t, t)P(r_t) + P(r_t)A(r_t, t)\right.$$

$$+ \sum_{j=1}^{N} \lambda_{r_t j} P(j) \Big] x(t)$$

$$+ 2x^\top(t) P(r_t) A_d(r_t, t) x(t - \tau_1) + h(t) \quad (9.62)$$

where $h(t) = 2x^\top(t) P(r_t) A_2(r_t, t) \int_{t-\tau_2}^{t} x(s) ds$.

$$\mathcal{A}V_1(\mathbf{x}(t), r_t) = x^\top(t) Q_0 x(t) - x^\top(t - \tau_1) Q_0 x(t - \tau_1)$$

$$+ \tau_2 x^\top(t) Q_1 x(t) - \int_{t-\tau_2}^{t} x^\top(s) Q_1 x(s) ds. \quad (9.63)$$

In view of Lemma A.1, we have

$$h(t) \leq \tau_2 x^\top(t) P(r_t) A_2(r_t, t) Q_1^{-1} A_2^\top(r_t, t) P(r_t) x(t)$$

$$+ \int_{t-\tau_2}^{t} x^\top(s) Q_1 x(s) ds. \quad (9.64)$$

Combining (9.62)-(9.64) yields

$$\mathcal{A}V(\mathbf{x}(t), r_t) \leq x^\top(t) \Big[\bar{A}^\top(r_t, t) P(r_t) + P(r_t) \bar{A}(r_t, t)$$

$$+ \sum_{j=1}^{m} \lambda_{r_t j} P(j) \Big] x(t) + 2x^\top(t) P(r_t) A_1(r_t, t) x(t - \tau_1)$$

$$+ \tau_2 x^\top(t) P(r_t) A_2(r_t, t) Q_1^{-1} A_2^\top(r_t, t) P(r_t) x(t)$$

$$+ \tau_2 x^\top(t) Q_1 x(t) + x^\top(t) Q_0 x(t) - x^\top(t - \tau_1) Q_0 x(t - \tau_1)$$

$$= \left(x^\top(t) \ x^\top(t - \tau_1) \right) \Theta_1(r_t, t) \left(\begin{array}{c} x(t) \\ x(t - \tau_1) \end{array} \right) \quad (9.65)$$

where

$$\Theta_1(r_t, t) = \left(\begin{array}{cc} J_0(r_t, t) & P(r_t) A_1(r_t, t) \\ * & -Q_0 \end{array} \right)$$

with

$$J_0(r_t, t) = \bar{A}^\top(r_t, t) P(r_t) + P(r_t) \bar{A}(r_t, t) + \sum_{j=1}^{N} \lambda_{r_t j} P(j)$$

$$+ \tau_2 P(r_t) A_2(r_t, t) Q_1^{-1} A_2^\top(r_t, t) P(r_t) + \tau_2 Q_1 + Q_0.$$

Using the Schur complement, we get that (9.60) holds if and only if

$$\Theta_1(r_t, t) < 0$$

holds for all admissible uncertainties. This means that (9.65) is negative-definite. The rest of the proof of this lemma is similar to that of Theorem 8.3 and is omitted here.   □

Note that

$$
\Theta(r_t, t) = \begin{pmatrix} J_2(r_t) & P(r_t)A_d(r_t) & P(r_t)A_2(r_t) \\ * & -Q_0 & 0 \\ * & 0 & -\frac{1}{\tau_2}Q_1 \end{pmatrix}
$$
$$
+ \begin{pmatrix} P(r_t)D(r_t) & P(r_t)D_1(r_t) & P(r_t)D(r_t) \\ 0 & 0 & 0 \\ 0 & 0 & 0 \end{pmatrix}
$$
$$
\times \tilde{\Delta}(r_t, t) \begin{pmatrix} E(r_t) & 0 & 0 \\ 0 & E_1(r_t) & 0 \\ 0 & 0 & E_2(r_t) \end{pmatrix}
$$
$$
+ \begin{pmatrix} E^{\top}(r_t) & 0 & 0 \\ 0 & E_1^{\top}(r_t) & 0 \\ 0 & 0 & E_2^{\top}(r_t) \end{pmatrix}
$$
$$
\times \tilde{\Delta}^{\top}(r_t, t) \begin{pmatrix} D^{\top}(r_t)P(r_t) & 0 & 0 \\ D_1^{\top}(r_t)P(r_t) & 0 & 0 \\ D^{\top}(r_t)P(r_t) & 0 & 0 \end{pmatrix}, \tag{9.66}
$$

where

$$
J_2(r_t) = A(r_t)P(r_t) + P(r_t)A(r_t) + \sum_{j=1}^{N} \lambda_{r_t j} P(j) + \tau_2 Q_1 + Q_0,
$$
$$
\tilde{\Delta}(r_t, t) = \begin{pmatrix} \Delta_1(r_t, t) & 0 & 0 \\ 0 & \Delta_2(r_t, t) & 0 \\ 0 & 0 & \Delta_1(r_t, t) \end{pmatrix}.
$$

Combining (9.66) and Lemma A.5 yields that (9.60) holds for all admissible uncertainties if there exists $\varepsilon > 0$ such that for all $i \in \mathcal{S}$,

$$
\begin{pmatrix} J_2(i) & P(i)A_1(i) & P(i)A_2(i) \\ * & -Q_0 & 0 \\ * & 0 & -\frac{1}{\tau_2}Q_1 \end{pmatrix}
$$
$$
+ \varepsilon \begin{pmatrix} P(i)D(i) & P(i)D_1(i) & P(i)D(i) \\ 0 & 0 & 0 \\ 0 & 0 & 0 \end{pmatrix} \begin{pmatrix} D^{\top}(i)P(i) & 0 & 0 \\ D_1^{\top}(i)P(i) & 0 & 0 \\ D^{\top}(i)P(i) & 0 & 0 \end{pmatrix}
$$
$$
+ \frac{1}{\varepsilon} \begin{pmatrix} E^{\top}(i) & 0 & 0 \\ 0 & E_1^{\top}(i) & 0 \\ 0 & 0 & E_2^{\top}(i) \end{pmatrix} \begin{pmatrix} E(i) & 0 & 0 \\ 0 & E_1(i) & 0 \\ 0 & 0 & E_2(i) \end{pmatrix} < 0
$$

$$
\tag{9.67}
$$

holds.

Let $X_i = P^{-1}(i)$. Pre- and post-multiplying both sides of (9.67) by $\text{diag}\{X_i, I, I\}$ yields

$$\begin{pmatrix} J_3(i) & A_1(i) & A_2(i) \\ A_1^\top(i) & -Q_0 + \frac{1}{\varepsilon}E_1^\top(i)E_1(i) & 0 \\ A_2^\top(i) & 0 & -\frac{1}{\tau_2}Q_1 + \frac{1}{\varepsilon}E_2^\top(i)E_2(i) \end{pmatrix} < 0, \quad (9.68)$$

where

$$J_3(i) = X_i A(i) + A(i)X_i + \lambda_{ii}X_i + X_i\left[\sum_{j \neq i}\lambda_{ij}X_j^{-1}\right]X_i$$
$$+ X_i[\tau_2 Q_1 + Q_0]X_i + 2\varepsilon D(i)D^\top(i) + \varepsilon D_1(i)D_1^\top(i)$$
$$+ \frac{1}{\varepsilon}X_i E^\top(i)E(i)X_i.$$

Using the Schur complement, (9.68) implies that

$$J_3(i) + A_1(i)\left[Q_0 - \frac{1}{\varepsilon}E_1^\top(i)E_1(i)\right]^{-1}A_1^\top(i)$$
$$+ A_2(i)\left[\frac{1}{\tau_2}Q_1 - \frac{1}{\varepsilon}E_2^\top(i)E_2(i)\right]^{-1}A_2^\top(i) < 0. \quad (9.69)$$

Using matrix inversion formula we have

$$\left[Q_0 - \frac{1}{\varepsilon}E_1^\top(i)E_1(i)\right]^{-1} = Q_0^{-1} + Q_0^{-1}E_1^\top(i)\left[\varepsilon I\right.$$
$$\left. - E_1(i)Q_0^{-1}E_1^\top(i)\right]^{-1}E_1(i)Q_0^{-1},$$

from which it follows that the left-hand side of (9.69) equals

$$J_3(i) + A_1(i)Q_0^{-1}A_1^\top(i)$$
$$+ A_1(i)Q_0^{-1}E_1^\top(i)\left[\varepsilon I - E_1(i)Q_0^{-1}E_1^\top(i)\right]^{-1}E_1(i)Q_0^{-1}A_1^\top(i)$$
$$+ A_2(i)\left[\frac{Q_1}{\tau_2} - \frac{1}{\varepsilon}E_2^\top(i)E_2(i)\right]^{-1}A_2^\top(i) < 0. \quad (9.70)$$

Using the Schur complement, (9.70) is equivalent to

$$\begin{pmatrix} J_4(i) & J_{12}(i) & A_2(i)Q_1^{-1} \\ * & -J_{22}(i) & 0 \\ * & 0 & -\frac{1}{\tau_2}Q_1^{-1} + \frac{1}{\varepsilon}Q_1^{-1}E_2^\top(i)E_2(i)Q_1^{-1} \end{pmatrix} < 0,$$

where

$$J_4(i) = X_i A^\top(i) + A(i)X_i + \lambda_{ii}X_i + X_i\left[\sum_{j \neq i}\lambda_{ij}X_j^{-1}\right]X_i$$
$$+ 2\varepsilon D(i)D^\top(i) + \varepsilon D_1(i)D_1^\top(i) + A_1(i)Q_0^{-1}A_1^\top(i)$$

$$J_{12}(i) = \begin{pmatrix} X_i & X_i & X_i E(i) & A_1(i)Q_0^{-1}E_1^\top(i) \end{pmatrix}$$

$$J_{22}(i) = \begin{pmatrix} \frac{1}{\tau_2}Q_1^{-1} & & & \\ & Q_0^{-1} & & \\ & & \varepsilon I & \\ & & & \varepsilon I - E_1(i)Q_0^{-1}E_1^\top(i) \end{pmatrix}$$

yielding

$$\begin{pmatrix} J_5(i) & J_{12}(i) & A_2(i) & 0 & S_i(X) \\ J_{12}^\top(i) & -J_{22}(i) & 0 & 0 & 0 \\ A_2^\top(i) & 0 & -\frac{1}{\tau_2}Q_1 & E_2^\top(i) & 0 \\ 0 & 0 & E_2(i) & -\varepsilon I & 0 \\ S_i^\top(X) & 0 & 0 & 0 & -\mathcal{X}_i \end{pmatrix} < 0, \qquad (9.71)$$

where

$$\begin{aligned} J_5(i) &= X_i A^\top(i) + A(i)X_i + \lambda_{ii}X_i + 2\varepsilon D(i)D^\top(i) \\ &\quad + \varepsilon D_1(i)D_1^\top(i) + A_1(i)Q_0^{-1}A_1^\top(i). \end{aligned}$$

Letting $Q_0^{-1} = U_0$ in (9.71) yields

$$\begin{pmatrix} J_6(i) & \tilde{J}_{12}(i) & A_2(i) & 0 & S_i(X) \\ \tilde{J}_{12}(i) & -\tilde{J}_{22}(i) & 0 & 0 & 0 \\ A_2^\top(i) & 0 & -\frac{1}{\tau_2}Q_1 & E_2^\top(i) & 0 \\ 0 & 0 & E_2(i) & -\varepsilon I & 0 \\ S_i^\top(X) & 0 & 0 & 0 & -\mathcal{X}_i \end{pmatrix} < 0, \qquad (9.72)$$

where

$$\begin{aligned} J_6(i) &= A(i)X_i + X_i A^\top(i) + \lambda_{ii}X_i + 2\varepsilon D(i)D^\top(i) \\ &\quad + \varepsilon D_1(i)D_1^\top(i) + A_1(i)U_0 A_1^\top(i) \\ \tilde{J}_{12}(i) &= \begin{pmatrix} X_i & X_i & X_i E^\top(i) & A_1(i)U_0 E_1^\top(i) \end{pmatrix} \\ \tilde{J}_{22}(i) &= \begin{pmatrix} Q_1^{-1} & & & \\ & U_0 & & \\ & & \varepsilon I & \\ & & & \varepsilon I - E_1(i)U_0 E_1^\top(i) \end{pmatrix}. \end{aligned}$$

From the above argument, we therefore get the following theorem.

Theorem 9.10 *If there exist symmetric, positive matrices $X = (X_1, \cdots, X_N) > 0$, $U_0 > 0$ and scalar $\varepsilon > 0$ such that (9.72) holds for all $i \in S$ and for some $Q_1 > 0$, then system (9.1) with $u(t) \equiv 0$ is robustly stable in the MSQS sense.*

Remark 9.4 *The stability condition provided by Theorem 9.10 depends on the distributed delay τ_2.*

Example 9.7 *Consider a dynamical time delay system with two modes and assume that the dynamic is described by (9.59) with $u(t) \equiv 0$ with the* *following parameters:*

$$\Lambda = \begin{pmatrix} -2 & 2 \\ 3 & -3 \end{pmatrix} \qquad \tau_2 = 1$$

$$A(1) = \begin{pmatrix} -10 & 0 \\ 0 & -2 \end{pmatrix} \qquad A(2) = \begin{pmatrix} -2 & 0.8 \\ 0.2 & -2 \end{pmatrix}$$

$$A_1(1) = \begin{pmatrix} -0.3 & 0 \\ 0 & -0.2 \end{pmatrix} \qquad A_1(2) = \begin{pmatrix} -0.1 & 0 \\ 0.1 & -1 \end{pmatrix}$$

$$A_2(1) = \begin{pmatrix} -3 & 0 \\ 0 & -0.2 \end{pmatrix} \qquad A_2(2) = \begin{pmatrix} -1 & 0.1 \\ 0.1 & -0.1 \end{pmatrix}$$

$$D(1) = \begin{pmatrix} 1 \\ 0.3 \end{pmatrix} \qquad D(2) = \begin{pmatrix} 0 \\ 0.3 \end{pmatrix}$$

$$D_1(1) = \begin{pmatrix} 0.2 \\ 0.1 \end{pmatrix} \qquad D_1(2) = \begin{pmatrix} 0 \\ 0.31 \end{pmatrix}$$

$$E(1) = \begin{pmatrix} 0 & 0.2 \end{pmatrix} \qquad E(2) = \begin{pmatrix} 0 & 0.2 \end{pmatrix}$$

$$E_b(1) = [0.02] \qquad E_b(2) = -[0.1]$$

$$E_1(1) = \begin{pmatrix} 0 & 0.1 \end{pmatrix} \qquad E_1(2) = \begin{pmatrix} 0 & -0.51 \end{pmatrix}$$

$$E_2(1) = \begin{pmatrix} -0.21 & -0.1 \end{pmatrix} \quad E_2(2) = \begin{pmatrix} 0.1 & 0.4 \end{pmatrix}.$$

With these data and $Q_1 = \begin{pmatrix} 0.5 & 0 \\ 0 & 0.5 \end{pmatrix}$, *(9.72) has the following set of feasible solutions:*

$$X_1 = \begin{pmatrix} 12.6139 & 0.0176 \\ 0.0176 & 2.6297 \end{pmatrix} \quad X_2 = \begin{pmatrix} 6.1247 & -0.2972 \\ -0.2972 & 2.0893 \end{pmatrix}$$

$$U = \begin{pmatrix} 99.1691 & 7.6995 \\ 7.6995 & 3.2407 \end{pmatrix} \quad \varepsilon = 5.9719.$$

According to Theorem 9.10, the system under study is stochastically stable.

Since LMI (9.72) is delay-dependent, a one-dimensional search yields that LMI (9.72) remains feasible for $\tau_2 \in [0, 5.335]$. *At* $\tau_2 = 5.335$, *the following is a set of feasible solutions:*

$$X_1 = \begin{pmatrix} 3.1953 & -0.1078 \\ -0.1078 & 0.5495 \end{pmatrix} \quad X_2 = \begin{pmatrix} 1.3411 & -0.0801 \\ -0.0801 & 0.4534 \end{pmatrix}$$

$$U = \begin{pmatrix} 32.7419 & 1.9348 \\ 1.9348 & 0.6673 \end{pmatrix} \quad \varepsilon = 1.1221.$$

According to Theorem 9.10, the system under study is stochastically stable for all $\tau_2 \in [0, 5.335]$.

9.4.2 Robust Stabilization of JLS with Discrete and Distributed Time Delays

Now, we proceed with the design of a memory controller, that is,

$$u(t) = K_1(r_t)x(t) + K_2(r_t) \int_{t-\tau_2}^{t} x(s)ds \tag{9.73}$$

which stabilizes system (9.59) in the stochastic sense.

Substituting (9.73) into (9.59) yields that the closed-loop system is described by the dynamic

$$\dot{x}(t) = \bar{A}(r_t, t)x(t) + A_1(r_t, t)x(t - \tau_1) + \bar{A}_2(r_t, t) \int_{t-\tau_2}^{t} x(s)ds \tag{9.74}$$

where

$$\bar{A}(r_t, t) = \bar{A}(r_t) + D(r_t)\Delta_1(r_t, t)\bar{E}(r_t)$$
$$\bar{A}_2(r_t, t) = \bar{A}_2(r_t) + D(r_t)\Delta_2(r_t, t)\bar{E}_2(r_t)$$

with $\bar{A}(r_t) = A(r_t) + B(r_t)K_1(r_t)$, $\bar{E}(r_t) = E(r_t) + E_b(r_t)K_1(r_t)$, $\bar{A}_2(r_t) = A_2(r_t) + B(r_t)K_2(r_t)$, $\bar{E}_2(r_t) = E_2(r_t) + E_b(r_t)K_2(r_t)$.

In view of Theorem 9.10, we find that for a given set of gains $K_1(i)$, $K_2(i)$, $i \in \mathcal{S}$, the closed-loop system (9.74) is stochastically stable if there exist symmetric, positive-definite matrices $X = (X_1, \cdots, X_N) > 0$, $U_0 > 0$ and scalar $\varepsilon > 0$ such that the following holds for some matrix $Q_1 > 0$:

$$\begin{pmatrix} \bar{J}_6(i) & \bar{J}_{12}(i) & \bar{A}_2(i) & 0 & S_i(X) \\ * & -\bar{J}_{22}(i) & 0 & 0 & 0 \\ * & 0 & -\frac{1}{\tau_2}Q_1 & \bar{E}_2^\top(i) & 0 \\ * & 0 & \bar{E}_2(i) & -\varepsilon I & 0 \\ * & 0 & 0 & 0 & -\mathcal{X}_i \end{pmatrix} < 0, \tag{9.75}$$

where

$$\bar{J}_6(i) = A(i)X_i + X_i A^\top(i) + \lambda_{ii} X_i + 2\varepsilon D(i)D^\top(i) \\ + \varepsilon D_1(i)D_1^\top(i) + A_1(i)U_0 A_1^\top(i),$$

$$\bar{J}_{12}(i) = \begin{pmatrix} X_i & X_i & X_i \bar{E}^\top(i) & A_1(i)U_0 E_1^\top(i) \end{pmatrix},$$

$$\bar{J}_{22}(i) = \begin{pmatrix} \frac{1}{\tau_2}Q_1^{-1} & & & \\ & U_0 & & \\ & & \varepsilon I & \\ & & & \varepsilon I - E_1(i)U_0 E_1^\top(i) \end{pmatrix}.$$

Letting $Y_i = K_1(i)X_i$, $i \in \mathcal{S}$ in (9.75) yields

$$\begin{pmatrix} \hat{J}_6(i) & \hat{J}_{12}(i) & \bar{A}_2(i) & 0 & S_i(X) \\ * & -\hat{J}_{22}(i) & 0 & 0 & 0 \\ * & 0 & -Q_1 & \bar{E}_2^\top(i) & 0 \\ * & 0 & \bar{E}_2(i) & -\varepsilon I & 0 \\ * & 0 & 0 & 0 & -\mathcal{X}_i \end{pmatrix} < 0, \tag{9.76}$$

where

$$
\begin{aligned}
\hat{J}_6(i) &= \bar{A}(i)X_i + X_i\bar{A}^\top(i) + B(i)Y_i + Y_i^\top B^\top(i) + \lambda_{ii}X_i \\
&\quad + 2\varepsilon D(i)D^\top(i) + \varepsilon D_1(i)D_1^\top(i) + A_1(i)U_0 A_1^\top(i) \\
\hat{J}_{12}(i) &= \left(\begin{array}{cccc} X_i & X_i & X_iE^\top(i) + Y_i^\top B^\top(i) & A_1(i)U_0E_1^\top(i) \end{array}\right) \\
\bar{J}_{22}(i) &= \begin{pmatrix} \frac{1}{\tau_2}Q_1^{-1} & & & \\ & U_0 & & \\ & & \varepsilon I & \\ & & & \varepsilon I - E_1(i)U_0E_1^\top(i) \end{pmatrix}.
\end{aligned}
$$

Therefore, we get the following theorem.

Theorem 9.11 *If there exist symmetric, positive-definite matrices $P = (P(1), \cdots, P(N)) > 0, Q_0 > 0$, and matrices $Y = (Y_1, \cdots, Y_N)$, $K_2 = (K_2(1), \cdots, K_2(N))$ and a scalar $\varepsilon > 0$ such that (9.76) holds for every $i \in S$ for some given $Q_1 > 0$, then controller (9.73) with $K_1(i) = Y_iX_i^{-1}, K_2(i)$ stabilizes system (9.59) in the stochastic sense.*

Remark 9.5 *Note that the left-hand side of (9.76) is increasing with respect to τ_2 in the positive-definite sense, i.e., if $\tau_2' < \tau_2''$ then if (9.76) holds for τ_2'', then it holds also for τ_2'. Therefore, we can find the upper bound $\bar{\tau}_2$ of τ_2 at which (9.76) is feasible and get a stabilizing controller with $\tau_2 = \bar{\tau}_2$. This controller will robustly stabilize system (9.59) for all $\tau_2 \in [0, \bar{\tau}_2]$.*

Example 9.8 *Consider a system described by (9.59) with the following parameters:*

$$
\Lambda = \begin{pmatrix} -2 & 2 \\ 3 & -3 \end{pmatrix} \qquad \tau_2 = 0.4
$$

$$
A(1) = \begin{pmatrix} -1 & 0 \\ 0 & -2 \end{pmatrix} \qquad A(2) = \begin{pmatrix} -2 & 0.8 \\ 0.2 & -2 \end{pmatrix}
$$

$$
A_1(1) = \begin{pmatrix} -0.3 & 0 \\ 0 & -0.2 \end{pmatrix} \qquad A_1(2) = \begin{pmatrix} -0.1 & 0 \\ 0.1 & -1 \end{pmatrix}
$$

$$
A_2(1) = \begin{pmatrix} -3 & 0 \\ 0 & -0.2 \end{pmatrix} \qquad A_2(2) = \begin{pmatrix} -1 & 0.1 \\ 0.1 & -0.1 \end{pmatrix}
$$

$$
D(1) = \begin{pmatrix} 1 \\ 0.3 \end{pmatrix} \qquad D(2) = \begin{pmatrix} 0 \\ 0.3 \end{pmatrix}
$$

$$
B(1) = \begin{pmatrix} 1 \\ 1 \end{pmatrix} \qquad B(2) = \begin{pmatrix} 1 \\ 0 \end{pmatrix}
$$

$$
D_1(1) = \begin{pmatrix} 0.2 \\ 0.1 \end{pmatrix} \qquad D_1(2) = \begin{pmatrix} 0 \\ 0.31 \end{pmatrix}
$$

$$E(1) = \begin{pmatrix} 0 & 0.2 \end{pmatrix} \qquad E(2) = \begin{pmatrix} 0 & 0.2 \end{pmatrix}$$

$$E_b(1) = [0.02] \qquad\qquad E_b(2) = [-0.1]$$

$$E_1(1) = \begin{pmatrix} 0 & 0.1 \end{pmatrix} \qquad E_1(2) = \begin{pmatrix} 0 & -0.51 \end{pmatrix}$$

$$E_2(1) = \begin{pmatrix} -0.21 & -0.1 \end{pmatrix} \quad E_2(2) = \begin{pmatrix} 0.1 & 0.4 \end{pmatrix}.$$

With these data and $Q_1 = \begin{pmatrix} 0.5 & 0 \\ 0 & 0.5 \end{pmatrix}$, solving (9.76) yields

$$X_1 = \begin{pmatrix} 64.9664 & 58.4192 \\ 58.4192 & 64.7918 \end{pmatrix} \qquad X_2 = \begin{pmatrix} 42.6526 & 2.9984 \\ 2.9984 & 10.2222 \end{pmatrix}$$

$$U = \begin{pmatrix} 29.9763 & -0.4814 \\ -0.4814 & 10.0665 \end{pmatrix} \qquad \varepsilon = 6.8435$$

$$Y_1 = (-2256.1218 \ -2317.2746) \quad Y_2 = (-716.7147 \ 16.7026)$$

$$K_2(1) = (10.1195 \ -0.01388) \qquad K_2(2) = (1.1365 \ 0.4859).$$

Thus, we can conclude in view of Theorem 9.11 that the system under study is stochastically stable under controller (9.73) and

$$K_1(1) = Y_1 X_1^{-1} = (-13.5658 \ -23.5334)$$
$$K_1(2) = Y_2 X_2^{-1} = (-17.2746 \ 6.7011)$$

and $K_2(1), K_2(2)$ are a set of stabilizing gains.

Using a one-dimensional search, we find that (9.76) remains feasible for all $\tau_2 \in [0, 0.5224]$. For $\tau_2 = 0.5224$, a set of feasible solutions is given by

$$X_1 = \begin{pmatrix} 100.0152 & 94.4756 \\ 94.4756 & 99.2027 \end{pmatrix} \qquad X_2 = \begin{pmatrix} 30.0724 & 2.0397 \\ 2.0397 & 8.0698 \end{pmatrix}$$

$$U = \begin{pmatrix} 15.7195 & -2.1523 \\ -2.1523 & 6.2955 \end{pmatrix} \qquad \varepsilon = 3.4319$$

$$Y_1 = (-9607.6075 \ -9654.9580) \quad Y_2 = (-339.2928 \ 16.1274)$$

$$K_2(1) = (10.0548 \ -0.09302) \qquad K_2(2) = (1.2730 \ -0.3477).$$

Therefore, controller (9.73) with

$$K_1(1) = Y_1 X_1^{-1} = (-41.1015 \ -58.1825)$$
$$K_1(2) = Y_2 X_2^{-1} = (-11.6172 \ 4.9348)$$

and $K_2(1), K_2(2)$ stabilizes the system under study for all $\tau_2 \in [0, 0.5224]$.

9.5   Robust Stability and Stabilization for JLS with Mode-Dependent Time Delays

This section considers the stability of JLS with uncertain parameters and time delays, which are assumed to be dependent on the mode of the system. For this purpose, let us consider a system described by the following dynamics:

$$\begin{cases} \dot{x}(t) = A(r_t,t)x(t) + A_d(r_t,t)x(t-\tau_{r_t}) + B(r_t,t)u(t) \\ x(s) = \phi(s), \ [-\bar{\tau},0] \\ y(t) = C(r_t,t)x(t), \end{cases} \tag{9.77}$$

where $x(t), A(r_t,t), A_d(r_t,t), B(r_t,t), u(t)$ are as in (9.1) and τ_{r_t} denotes the time delay in the system when the mode is in $r_t, \bar{\tau} = \max\{\tau_j, j \in \mathcal{S}\}$, $y(t) \in \mathbb{R}^k$ is the measurement output, and

$$C(r_t,t) = C(r_t) + D_c(r_t)\Delta(r_t,t)E_a(r_t) \in \mathbb{R}^{k\times n}$$

with $C(r_t), D_c(r_t), E_c(r_t)$ being known matrices of appropriate dimensions. Let $\varrho = [1 + (\bar{\tau} - \underline{\tau})\bar{\lambda}]$, $\underline{\tau} = \min\{\tau_i, i \in \mathcal{S}\}$, $\bar{\lambda} = \max\{|\lambda_{ii}|, i \in \mathcal{S}\}$.

9.5.1   Robust SS of JLS with Mode-Dependent Time Delays

First, let us establish some sufficient conditions for system (9.77) to be robustly stochastically stable(SS). The following theorem gives such results.

Theorem 9.12 *If there exist symmetric, positive-definite matrices $P = (P(1), \cdots, P(N)) > 0, Q > 0$ such that*

$$A^\top(r_t,t)P(r_t) + P(r_t)A(r_t,t) + \sum_{j=1}^N \lambda_{r_tj}P(j) + \varrho Q + P(r_t) < 0 \tag{9.78}$$

$$A_d^\top(r_t,t)P(r_t)A_d(r_t,t) < Q \tag{9.79}$$

holds for every $r_t \in \mathcal{S}$ and for all admissible uncertainties, then system (9.77) with $u(t) \equiv 0$ is SS.

Proof: The proof is the same as that of Theorem 8.16.   □

From the above theorem, the following theorem follows directly.

Theorem 9.13 *If there exist symmetric, positive-definite matrices $P = (P(1), \cdots, P(N)) > 0, Q > 0$, and scalars $\varepsilon_{1i} > 0, \varepsilon_{2i} > 0$ satisfying the LMIs for every $i \in \mathcal{S}$*

$$\begin{pmatrix} J(i) + \varepsilon_{1i}E_a^\top(i)E_a(i) & P(i)D_a(i) \\ D_a^\top(i)P(i) & -\varepsilon_{1i}I \end{pmatrix} < 0 \tag{9.80}$$

$$\begin{pmatrix} -Q + \varepsilon_{2i}E_d^\top(i)E_d(i) & A_d^\top(i)P(i) & 0 \\ P(i)A_d^\top(i) & -P(i) & P(i)D_d(i) \\ 0 & D_d^\top(i)P(i) & -\varepsilon_{2i}I \end{pmatrix} < 0, \tag{9.81}$$

then system (9.77) with $u(t) \equiv 0$ is robustly SS, where $J(i) = A^\top(i)P(i) + P(i)A(i) + \sum_{j=1}^{N} \lambda_{ij}P(j) + \varrho Q + P(i)$.

Proof: Let the matrices $P = (P(1), \cdots, P(N)) > 0$ and $Q > 0$ be a set of feasible solutions of LMIs (9.80) and (9.81). By virtue of Theorem 9.12, to prove this it suffices to prove that the matrices $P = (P(1), \cdots, P(N)) > 0, Q > 0$ satisfy (9.78) and (9.79). The left-hand side of (9.78) can be rewritten as

$$J(r_t) + P(r_t)D_a(r_t)\Delta(r_t, t)E_a(r_t) + E_a^\top(r_t)\Delta^\top(r_t, t)D_a^\top(r_t, t)P(r_t).$$

Using Lemma A.5, we obtain that (9.78) holds if and only if there exist scalars $\varepsilon_{1i} > 0$ satisfying

$$J(i) + \varepsilon_{1i}E_a^\top(i)E_a(i) + \frac{1}{\varepsilon_{1i}}P(i)D_a(i)D_a^\top(i)P(i) < 0,$$

which is equivalent to (9.80).

Using the Schur complement, (9.79) equivalent to

$$\begin{pmatrix} -Q & A_d^\top(r_t, t)P(r_t) \\ P(r_t)A_d(r_t, t) & -P(r_t) \end{pmatrix} < 0 \qquad (9.82)$$

holds for all admissible uncertainties. Noting that the left-hand side of the above inequality can be rewritten as

$$\begin{pmatrix} -Q & A_d^\top(r_t)P(r_t) \\ P(r_t)A_d(r_t) & -P(r_t) \end{pmatrix} + \begin{pmatrix} 0 \\ P(r_t)D_a(r_t) \end{pmatrix} \Delta_d(r_t, t) \begin{pmatrix} E_d(r_t) & 0 \end{pmatrix}$$
$$+ \begin{pmatrix} E_d^\top(r_t) \\ 0 \end{pmatrix} \Delta_d^\top(r_t, t) \begin{pmatrix} 0 & D_a^\top(r_t)P(r_t) \end{pmatrix}.$$

Therefore, using Lemma A.5, (9.82) holds for all admissible uncertainties if and only if there exist scalars $\varepsilon_{2i} > 0$ such that

$$\begin{pmatrix} -Q & A_d^\top(r_t)P(r_t) \\ P(r_t)A_d(r_t) & -P(r_t) \end{pmatrix} + \varepsilon_{2i} \begin{pmatrix} E_d^\top(i)E_d(i) & 0 \\ 0 & 0 \end{pmatrix}$$
$$+ \frac{1}{\varepsilon_{2i}} \begin{pmatrix} 0 \\ P(i)D_a(i) \end{pmatrix} \begin{pmatrix} 0 & D_a^\top(i)P(i) \end{pmatrix} < 0, \qquad (9.83)$$

which is equivalent to (9.81). This completes the proof of Theorem 9.13. $\square$

Now we proceed with the design of a state feedback memoryless controller in the form of (8.2) that stabilizes (9.77) in the SS sense. To this end, let us give the following lemma.

Lemma 9.6 *System (9.77) with $u(t) \equiv 0$ is robustly SS if the following LMIs are feasible for every $i \in \mathcal{S}$ with respect to $X_i > 0, U > 0, \varepsilon_{1i} >$*

$0, \varepsilon_{2i} > 0$:

$$
\begin{pmatrix}
\# & X_i & X_i E_a^\top(i) & S_i(X) \\
X_i & -\frac{1}{\varrho}U & 0 & 0 \\
E_a(i)X_i & 0 & -\varepsilon_{1i}I & 0 \\
S_i^\top(X) & 0 & 0 & -\mathcal{X}_i
\end{pmatrix} < 0 \tag{9.84}
$$

$$
\begin{pmatrix}
-U & UA_d^\top(i) & UE_d^\top(i) \\
A_d(i)U & -X_i + \varepsilon_{2i}D_a(i)D_a^\top(i) & 0 \\
E_d(i)U & 0 & -\varepsilon_{2i}I
\end{pmatrix} < 0, \tag{9.85}
$$

where

$$
\# = X_i A^\top(i) + A(i)X_i + \lambda_{ii}X_i + X_i + \varepsilon_{1i}D_a(i)D_a(i),
$$

$S_i(X)$ and $\mathcal{X}_i$ are defined as in (8.11) and (8.12), respectively.

Proof: From the proof of Theorem 9.12, we find that (9.78) holds for all admissible uncertainties if and only if there exist scalars $\varepsilon_{1i} > 0$ such that

$$
J(i) + \varepsilon_{1i}P(i)D_a(i)D_a^\top(i)P(i) + \frac{1}{\varepsilon_{1i}}E_a^\top(i)E_a(i) < 0.
$$

Letting $X_i = P^{-1}(i), U = Q^{-1}$, and multiplying both sides of the above inequality by X_i yields

$$
X_i J(i)X_i + \varepsilon_{1i}D_a(i)D_a^\top(i) + \frac{1}{\varepsilon_{1i}}X_i E_a^\top(i)E_a(i)X_i < 0.
$$

Letting $U = P^{-1}$ and noting that

$$
X_i J(i)X_i = X_i A^\top(i) + A(i)X_i + \lambda_{ii}X_i + X_i \left[\sum_{j\neq i}\lambda_{ij}X_j^{-1}\right]X_i
$$
$$
+ \varrho X_i Q X_i + X_i,
$$

we obtain (9.84) using the Schur complement.

Using (9.83), we obtain that (9.79) holds for all admissible uncertainties if and only if there exist scalars $\varepsilon_{2i} > 0$, $i = 1, \ldots, N$ such that

$$
\begin{pmatrix}
-Q & A_d^\top(i)P(i) \\
P(i)A_d(i) & -P(i) + \varepsilon_{2i}P(i)D_a(i)D_a^\top(i)P(i)
\end{pmatrix}
$$
$$
+ \frac{1}{\varepsilon_{2i}}\begin{pmatrix} E_d^\top(i) \\ 0 \end{pmatrix}(E_d(i)\ 0) < 0,
$$

which is equivalent to

$$
\begin{pmatrix}
-Q & A_d^\top(i)P(i) & E_d^\top(i) \\
P(i)A_d(i) & -P(i) + \varepsilon_{2i}P(i)D_a(i)D_a^\top(i)P(i) & 0 \\
E_d^\top(i)U & 0 & -\varepsilon_{2i}I
\end{pmatrix} < 0.
$$

Pre- and post-multiplying the above inequality by $\text{diag}\{U, X_i, I\}$ leads to (9.85). This completes the proof of Lemma 9.6. $\qquad\square$

Substituting (8.2) into (9.77) yields the dynamics of the closed-loop system as

$$\dot{x}(t) = [\bar{A}(r_t) + D_a(r_t)\Delta(r_t,t)\bar{E}_a(r_t)]x(t) + A_d(r_t,t)x(t - \tau_{r_t}),$$

where $\bar{A}(r_t) = A(r_t) + B(r_t)K(r_t)$, $\bar{E}_a(r_t) = E_a(r_t) + E_b(r_t)K(r_t)$. Using Lemma 9.6, we obtain that if there exist symmetric, positive-definite matrices $X_i > 0, U > 0$, and scalars $\varepsilon_{1i} > 0, \varepsilon_{2i} > 0$ satisfying (9.85) and the following:

$$\begin{pmatrix} \#_1 & X_i & X_i\bar{E}_a^\top(i) & S_i(X) \\ X_i & -\frac{1}{\varrho}U & 0 & 0 \\ \bar{E}_a(i)X_i & 0 & -\varepsilon_{1i}I & 0 \\ S_i^\top(X) & 0 & 0 & -\mathcal{X}_i \end{pmatrix} < 0,$$

then controller (8.2) stabilizes system (9.77), where $\#_1 = X_i\bar{A}^\top(i) + \bar{A}(i)X_i + \lambda_{ii}X_i + X_i + \varepsilon_{1i}D_a(i)D_a(i)$. Letting $Y_i = K(i)X_i$ in previous inequality leads to the following theorem.

Theorem 9.14 *If there exist symmetric, positive-definite matrices $X_i > 0, Y_i, U > 0$, and scalars $\varepsilon_{1i}, \varepsilon_{2i} > 0$ satisfying (9.85) and*

$$\begin{pmatrix} \#_2 & X_i & X_iE_a^\top(i) + Y_i^\top E_b^\top(i) & S_i(X) \\ X_i & -\frac{1}{\varrho}U & 0 & 0 \\ E_a(i)X_i + E_b(i)Y_i & 0 & -\varepsilon_{1i}I & 0 \\ S_i^\top(X) & 0 & 0 & -\mathcal{X}_i \end{pmatrix} < 0$$

for every $i \in \mathcal{S}$, where $\#_2 = X_i\bar{A}^\top(i) + \bar{A}(i)X_i + Y_i^\top B^\top(i) + B(i)Y_i + \lambda_{ii}X_i + X_i + \varepsilon_{1i}D_a(i)D_a(i)$, then controller (8.2) with $K(i) = Y_iX_i^{-1}$ stabilizes system (9.77).

Now, we come to develop some sufficient conditions for system (9.77) to be robustly MES. For this purpose, let us give the following definition.

Definition 9.4 *System (9.77) is said to be robustly MSQS if there exist symmetric, positive-definite matrices $P = (P(1), \cdots, P(N)) > 0, Q > 0$ such that the following LMI holds for all $r_t \in \mathcal{S}$ and for all admissible uncertainties,*

$$\Theta(r_t,t) \triangleq \begin{pmatrix} \# & P(r_t)A_d(r_t,t) \\ A_d^\top(r_t,t)P(r_t) & -Q \end{pmatrix} < 0, \tag{9.86}$$

where $\# = A^\top(r_t,t)P(r_t) + P(r_t)A(r_t,t) + \sum_{j=1}^N \lambda_{r_t j}P(j) + \varrho Q$.

Theorem 9.15 *Under Assumption 8.3, if system (9.77) with $u(t) \equiv 0$ is robustly MSQS, then it is robustly MES.*

Proof: The proof is the same as that of Theorem 8.17 and is omitted here. □

Theorem 9.16 *System (9.77) with $u(t) \equiv 0$ is robust MSQS if there exist symmetric, positive-definite matrices $P = (P(1), \cdots, P(N)) > 0, Q > 0$ and a scalar $\varepsilon > 0$ such that the following hold for any $i \in S$*

$$
\begin{pmatrix}
J_0(i) & P(i)A_d(i) & P(i)D(i) & P(i)D_a(i) \\
* & -Q + \varepsilon E_d^\top(i)E_d(i) & 0 & 0 \\
* & 0 & -\varepsilon I & 0 \\
* & 0 & 0 & -\varepsilon I
\end{pmatrix} < 0, \quad (9.87)
$$

where $J_0(i) = A^\top(i)P(i) + P(i)A(i) + \sum_{j=1}^{N} \lambda_{ij}P(j) + \varrho Q + \varepsilon E_a^\top(i)E_a(i)$.

Proof: Note that

$$
\Theta(r_t, t) = \begin{pmatrix} J_1(r_t) & P(r_t)A_d(r_t) \\ A_d^\top(r_t)P(r_t) & -Q \end{pmatrix} + \begin{pmatrix} P(r_t)D_a(r_t) & P(r_t)D_d(r_t) \\ 0 & 0 \end{pmatrix}
$$
$$
\begin{pmatrix} \Delta(r_t, t) & 0 \\ 0 & \Delta_d(r_t, t) \end{pmatrix} \begin{pmatrix} E_a(r_t) & 0 \\ 0 & E_d(r_t) \end{pmatrix}
$$
$$
+ \left[\begin{pmatrix} P(r_t)D_a(r_t) & P(r_t)D_d(r_t) \\ 0 & 0 \end{pmatrix} \begin{pmatrix} \Delta(r_t, t) & 0 \\ 0 & \Delta_d(r_t, t) \end{pmatrix} \right.
$$
$$
\left. \times \begin{pmatrix} E_a(r_t) & 0 \\ 0 & E_d(r_t) \end{pmatrix} \right]^\top, \quad (9.88)
$$

where $J_1(r_t) = A^\top(r_t)P(r_t) + P(r_t)A(r_t) + \sum_{j=1}^{N} \lambda_{r_t j}P(j) + \varrho Q$.

Thus, using Lemma A.5 we have $\Theta(r_t, t) < 0$ for all admissible uncertainties if there exists $\varepsilon > 0$ such that the following holds:

$$
\begin{pmatrix} J_1(i) & P(i)A_d(i) \\ A_d^\top(i)P(i) & -Q \end{pmatrix} + \varepsilon \begin{pmatrix} E_a^\top(i)E_a(i) & 0 \\ 0 & E_d^\top(i)E_d(i) \end{pmatrix}
$$
$$
+ \frac{1}{\varepsilon} \begin{pmatrix} P(i)D_a(i) & P(i)D_d(i) \\ 0 & 0 \end{pmatrix} \begin{pmatrix} D_a^\top(i)P(i) & 0 \\ D_d^\top(i)P(i) & 0 \end{pmatrix}
$$
$$
= \begin{pmatrix} J_0(i) & P(i)A_d(i) \\ A_d^\top(i)P(i) & -Q + \varepsilon E_d^\top(i)E_d(i) \end{pmatrix}
$$
$$
+ \frac{1}{\varepsilon} \begin{pmatrix} P(i)D_a(i) & P(i)D_d(i) \\ 0 & 0 \end{pmatrix} \begin{pmatrix} D_a^\top(i)P(i) & 0 \\ D_d^\top(i)P(i) & 0 \end{pmatrix} < 0, \quad (9.89)
$$

which is equivalent to (9.87) using the Schur complement. This completes the proof of Theorem 9.16. □

Example 9.9 *Let us consider a system described by (9.77) with $u(t) \equiv 0$ and system parameters are as follows: The system has two modes, i.e., $S = \{1, 2\}$. The Markov process that governs mode switching takes values in S and has generator*

$$
\begin{pmatrix} -3 & 3 \\ 7 & -7 \end{pmatrix}.
$$

Associated with modes 1 and 2, the system has time delay $\tau_1 = 0.2$ and $\tau_2 = 1.28$, respectively.

$$A(1) = \begin{pmatrix} -1 & 0 \\ 0 & -2 \end{pmatrix} \qquad A(2) = \begin{pmatrix} -5 & 0.5 \\ 0.2 & -1 \end{pmatrix}$$

$$A_d(1) = \begin{pmatrix} -0.3 & 0 \\ 0 & -0.2 \end{pmatrix} \qquad A_d(2) = \begin{pmatrix} 0.1 & -0.1 \\ 0.1 & -0.1 \end{pmatrix}$$

$$D(1) = \begin{pmatrix} 0.8 \\ 0.3 \end{pmatrix} \qquad D(2) = \begin{pmatrix} 0 \\ 0.3 \end{pmatrix}$$

$$D_d(1) = \begin{pmatrix} 0.2 \\ 0.1 \end{pmatrix} \qquad D_d(2) = \begin{pmatrix} 0 \\ 1 \end{pmatrix}$$

$$E_a(1) = \begin{pmatrix} 0 & 0.2 \end{pmatrix} \qquad E_a(2) = \begin{pmatrix} 0.2 & 0.2 \end{pmatrix}.$$

With these data, solving (9.87) yields

$$P(1) = \begin{pmatrix} 4.8549 & -0.8713 \\ -0.8713 & 6.6263 \end{pmatrix} \qquad P(2) = \begin{pmatrix} 2.6863 & -0.6071 \\ -0.6071 & 9.2340 \end{pmatrix}$$

$$Q = \begin{pmatrix} 0.5810 & -0.2749 \\ -0.2749 & 1.0569 \end{pmatrix} \qquad \varepsilon = 5.1530.$$

According to Theorem 9.16, we can conclude that the system under study is stochastically stable.

Let us now switch to the stabilization problem of the class of systems we are considering in this section. This will be the subject of the next subsection.

9.5.2 Robust Stabilization of JLS with Mode-Dependent Time Delays

Throughout this subsection, we assume that the system state vector $x(t)$ and mode r_t are available for feedback. This section deals with the design of a memoryless state feedback controller (8.2) that stabilizes system (9.77) in the MSQS sense. To this end, let us introduce the following lemma.

Lemma 9.7 *System (9.77) with $u(t) \equiv 0$ is robustly MSQS if there exist symmetric, positive-definite matrices $X = (X_1, \cdots, X_N) > 0, U > 0$ and a scalar $\eta > 0$ such that the following hold for all $i \in \mathcal{S}$:*

$$\begin{pmatrix} J(i) & X_i & X_i E_a^\top(i) & A_d(i)UE_d^\top(i) & S_i(X) \\ * & -\dfrac{U}{\varrho} & 0 & 0 & 0 \\ * & 0 & -\eta I & 0 & 0 \\ * & 0 & 0 & -\eta I + E_d(i)UE_d^\top(i) & 0 \\ * & 0 & 0 & 0 & -\mathcal{X}_i \end{pmatrix} < 0, \qquad (9.90)$$

where

$$J(i) = X_i A^\top(i) + A(i)X_i + \lambda_{ii}X_i + \eta D_a(i)D_a^\top(i)$$

$$+ \eta D_d(i) D_d^\top(i) + A_d(i) U A_d^\top(i).$$

Proof: In view of Theorem 9.16, system (9.77) with $u(t) \equiv 0$ is robustly MSQS if there exist symmetric, positive-definite matrices $P = (P(1), \cdots, P(N)) > 0, Q > 0$ and a scalar $\varepsilon > 0$ satisfying (9.87), which is equivalent to

$$\Xi(i) = A^\top(i) P(i) + P(i) A(i) + \sum_{j=1}^{N} \lambda_{ij} P(j) + \varrho Q + \varepsilon E_a^\top(i) E_a(i)$$

$$+ \frac{1}{\varepsilon} P(i) D_a(i) D_a^\top(i) P(i) + \frac{1}{\varepsilon} P(i) D_d(i) D_d^\top(i) P(i)$$

$$+ P(i) A_d(i) \Big[Q - \varepsilon E_d^\top(i) E_d(i) \Big]^{-1} A_d^\top(i) P(i) < 0.$$

Using the matrix inversion formula, we have

$$\Big[Q - \varepsilon E_d^\top(i) E_d(i) \Big]^{-1}$$

$$= Q^{-1} + Q^{-1} E_d^\top(i) \Big(\frac{1}{\varepsilon} I - E_d(i) Q^{-1} E_d^\top(i) \Big)^{-1} E_d(i) Q^{-1}, \quad (9.91)$$

yielding in turn that $\Xi(i)$ can be rewritten as follows:

$$\Xi(i) = A^\top(i) P(i) + P(i) A(i) + \sum_{j=1}^{N} \lambda_{ij} P(j)$$

$$+ \varrho Q + \varepsilon E_a^\top(i) E_a(i) + \frac{1}{\varepsilon} P(i) D_a(i) D_a^\top(i) P(i)$$

$$+ \frac{1}{\varepsilon} P(i) D_d(i) D_d^\top(i) P(i) + P(i) A_d(i) \Big[Q^{-1}$$

$$+ Q^{-1} E_d^\top(i) \Big(\frac{1}{\varepsilon} I - E_d(i) Q^{-1} E_d^\top(i) \Big)^{-1} E_d(i) Q^{-1} \Big] A_d^\top(i) P(i).$$

$$(9.92)$$

Let $X_i = P^{-1}(i), U = Q^{-1}, \eta = \frac{1}{\varepsilon}$. Pre- and post-multiplying both sides of (9.92) by X_i yields

$$X_i \Xi(i) X_i = X_i A^\top(i) + A(i) X_i + \lambda_{ii} X_i$$

$$+ X_i \Big[\sum_{j \neq i} \lambda_{ij} X_j^{-1} \Big] X_i + \varrho X_i U^{-1} X_i$$

$$+ \frac{1}{\eta} X_i E_a^\top(i) E_a(i) X_i + \eta D_a(i) D_a^\top(i) + \eta D_d(i) D_d^\top(i)$$

$$+ A_d(i) \Big[U + U E_d^\top(i) \big(\eta I - E_d(i) U E_d^\top(i) \big)^{-1} E_d(i) U \Big] A_d^\top(i) < 0,$$

which is equivalent to (9.90) using the Schur complement. This terminates the proof of Lemma 9.7 $\square$

For a given set of gains $K = (K(1), \cdots, K(N))$, substituting (8.2) into (9.77) gives the following closed-loop system dynamic

$$\dot{x}(t) = [\bar{A}(r_t) + D_a(r_t)\Delta(r_t, t)\bar{E}_a(r_t)]x(t) + A_d(r_t, t)x(t - \tau_{r_t}) \quad (9.93)$$

where

$$\bar{A}(r_t) = A(r_t) + B(r_t)K(r_t)$$
$$\bar{E}_a(r_t) = E_a(r_t) + E_b(r_t)K(r_t).$$

In view of Lemma 9.7, we obtain that controller (8.2) robustly stabilizes system (9.77) in the MSQS sense if there exist symmetric, positive-definite matrices $X = (X_1, \cdots, X_N) > 0$, $U > 0$ and a scalar $\eta > 0$ satisfying

$$\begin{pmatrix} \bar{J}(i) & X_i & X_i\bar{E}_a^\top(i) & A_d(i)UE_d^\top(i) & S_i(X) \\ * & -\frac{U}{\varrho} & 0 & 0 & 0 \\ * & 0 & -\eta I & 0 & 0 \\ * & 0 & 0 & -\eta I + E_d(i)UE_d^\top(i) & 0 \\ * & 0 & 0 & 0 & -\mathcal{X}_i \end{pmatrix} < 0, \quad (9.94)$$

where

$$\bar{J}(i) = X_i\bar{A}^\top(i) + \bar{A}(i)X_i + \lambda_{ii}X_i + \eta D_a(i)D_a^\top(i)$$
$$+ \eta D_d(i)D_d^\top(i) + A_d(i)UA_d^\top(i).$$

Letting $Y_i = K(i)X_i$ in (9.94) yields

$$\begin{pmatrix} \hat{J}(i) & J_{12}(i) & S_i(X) \\ J_{12}^\top(i) & -J_{22}(i) & 0 \\ S_i^\top(X) & 0 & -\mathcal{X}_i \end{pmatrix} < 0 \quad (9.95)$$

where

$$\hat{J}(i) = A(i)X_i + X_iA^\top(i) + B(i)Y_i + Y_i^\top B^\top(i) + \lambda_{ii}X_i$$
$$+ \eta D_a(i)D_a^\top(i) + \eta D_d(i)D_d^\top(i) + A_d(i)UA_d^\top(i)$$
$$J_{12}(i) = \begin{pmatrix} X_i & X_iE_a^\top(i) + Y_i^\top E_b^\top(i) & A_d(i)UE_d^\top(i) \end{pmatrix}$$
$$J_{22}(i) = \begin{pmatrix} \frac{U}{\varrho} & & \\ & \eta I & \\ & & \eta I - E_d(i)UE_d^\top(i) \end{pmatrix}.$$

We therefore get the following theorem.

Theorem 9.17 *If there exist symmetric, positive-definite matrices $X = (X_1, \cdots, X_N) > 0, U > 0$, matrices $Y = (Y_1, \cdots, Y_N)$ and a scalar $\eta > 0$ satisfying (9.95) for every $i \in \mathcal{S}$, then controller (8.2) with $K(i) = Y_iX_i^{-1}, i \in \mathcal{S}$ robustly stabilizes system (9.77) in the MSQS sense.*

Example 9.10 *Let us consider a system described by (9.77) with parameters as follows. The Markov process that represents the system mode switching takes values in* $S = \{1, 2\}$ *and has the generator*

$$\Lambda = \begin{pmatrix} -3 & 3 \\ 5 & -5 \end{pmatrix}.$$

The system state contains time delays $\tau_1 = 0.2$ *and* $\tau_2 = 1.80$ *when the system is in mode* $r_t = 1$ *and* $r_t = 2$, *respectively.*

$$A(1) = \begin{pmatrix} -1 & 0 \\ 0 & -12 \end{pmatrix} \qquad A(2) = \begin{pmatrix} 2 & 0.8 \\ 0.2 & -2 \end{pmatrix}$$

$$A_d(1) = \begin{pmatrix} -0.3 & 0 \\ 0 & -0.2 \end{pmatrix} \qquad A_d(2) = \begin{pmatrix} 0.1 & -0.1 \\ 0.1 & -0.1 \end{pmatrix}$$

$$B(1) = \begin{pmatrix} 0 \\ -0.1 \end{pmatrix} \qquad B(2) = \begin{pmatrix} 1 \\ 0.1 \end{pmatrix}$$

$$D(1) = \begin{pmatrix} 1 \\ 0.3 \end{pmatrix} \qquad D(2) = \begin{pmatrix} 0 \\ 0.3 \end{pmatrix}$$

$$D_d(1) = \begin{pmatrix} 0.2 \\ 1 \end{pmatrix} \qquad D_d(2) = \begin{pmatrix} 0 \\ 1 \end{pmatrix}$$

$$E_a(1) = \begin{pmatrix} 0 & 0.2 \end{pmatrix} \qquad E_a(2) = \begin{pmatrix} 1 & 0.2 \end{pmatrix}$$

$$E_b(1) = [0.3] \qquad E_b(2) = [0.2]$$

$$E_d(1) = \begin{pmatrix} 0 & 0.1 \end{pmatrix} \qquad E_d(2) = \begin{pmatrix} 0 & -1 \end{pmatrix}.$$

For these data, LMI (9.95) has the following set of feasible solutions:

$$X_1 = \begin{pmatrix} 1.800 & 0.0597 \\ 0.0597 & 1.3300 \end{pmatrix} \qquad X_2 = \begin{pmatrix} 2.7627 & 0.1704 \\ 0.1704 & 0.5362 \end{pmatrix}$$
$$Y_1 = \begin{pmatrix} -0.0835 & -3.4856 \end{pmatrix} \qquad Y_2 = \begin{pmatrix} -14.1173 & -1.4098 \end{pmatrix}$$
$$U = \begin{pmatrix} 17.2703 & 0.4314 \\ 0.4314 & 1.5392 \end{pmatrix} \qquad \varepsilon = 1.7443$$

In view of Theorem 9.17, the system under study is robust stabilizable under controller (8.2). Moreover,

$$K(1) = Y_1 X_1^{-1} = \begin{pmatrix} 0.0405 & -2.6225 \end{pmatrix}$$
$$K(2) = Y_2 X_2^{-1} = \begin{pmatrix} -5.0467 & -1.025 \end{pmatrix}$$

are a set of stabilizing gains.

9.5.3 Robust Output Feedback Stabilization

This section deals with the design of an output feedback controller (8.70) that stabilizes system (9.77) in the MSQS sense. Substitute (8.70) into

(9.77) and let $\eta^\top(t) = \left(x^\top(t)\,\xi^\top(t)\right)$ be the augmented state, then we have

$$
\begin{cases}
\dot{\eta}(t) & = \bar{A}(r_t,t)\eta(r_t) + \bar{A}_d(r_t,t)I_0\eta(t - \tau_{r_t}) \\
& = \bar{A}(r_t,t)\eta(r_t) + \bar{A}_d(r_t,t)x(t - \tau_{r_t}) \\
\eta(s) & = \left(\ \phi(s), 0\ \right)
\end{cases}
\tag{9.96}
$$

where

$$
\bar{A}(r_t,t) = \left(
\begin{array}{cc}
A(r_t,t) & B(r_t,t)K_C(r_t) \\
K_B(r_t)C(r_t,t) & K_A(r_t)
\end{array}
\right),
$$

$$
\bar{A}_d(r_t,t) = \left(
\begin{array}{c}
A_d(r_t,t) \\
0
\end{array}
\right)\quad I_0 = (I\ 0).
$$

To design a stabilizing controller in the form of (8.70), let us introduce some notation:

$$
\bar{A}(i) = \left(
\begin{array}{cc}
A(i) & B(i)K_C(i) \\
K_B(i)C(i) & K_A(i)
\end{array}
\right)\quad
\bar{D}_a(i) = \left(
\begin{array}{cc}
D_a(i) & D_a(i) \\
K_B(i)D_c(i) & 0
\end{array}
\right)
$$

$$
\bar{E}_a(i) = \left(
\begin{array}{cc}
E_a(i) & 0 \\
0 & E_b(i)K_C(i)
\end{array}
\right)\quad
\bar{A}_d(i) = \left(
\begin{array}{c}
A_d(i) \\
0
\end{array}
\right)
$$

$$
\bar{D}_d(i) = \left(
\begin{array}{c}
D_d(i) \\
0
\end{array}
\right)\quad\quad\quad\quad\quad
\bar{E}_d(i) = E_d(i).
$$

With this notation, we have

$$
\bar{A}(r_t,t) = \bar{A}(r_t) + \bar{D}_a(r_t)\bar{\Delta}(r_t,t)\bar{E}_a(r_t)
\tag{9.97}
$$

$$
\bar{A}_d(r_t,t) = \bar{A}_d(r_t) + \bar{D}_d(r_t)\Delta(r_t,t)\bar{E}_d(r_t),
\tag{9.98}
$$

where $\bar{\Delta}(r_t,t) = \left(
\begin{array}{cc}
\Delta(r_t,t) & 0 \\
0 & \Delta(r_t,t)
\end{array}
\right).$

Using Theorem 9.16, we have that system (9.96) is robustly stochastically stable if

$$
\left(
\begin{array}{cccc}
\# & * & * & * \\
\bar{A}_d^\top(i)P(i) & -Q + \varepsilon E_d^\top(i)E_d(i) & 0 & 0 \\
\bar{D}_a^\top(i)P(i) & 0 & -\varepsilon I & 0 \\
\bar{D}_d^\top(i)P(i) & 0 & 0 & -\varepsilon I
\end{array}
\right) < 0,
\tag{9.99}
$$

where

$$
\# = \bar{A}^\top(i)P(i) + P(i)\bar{A}(i) + \sum_{j=1}^{N}\lambda_{ij}P(j) + \varrho I_0^\top QI_0 + \varepsilon\bar{E}_a^\top(i)\bar{E}_a(i),
$$

from which using the Schur complement, the following lemma follows.

Lemma 9.8 *System (9.96) is robustly MSQS if there exist matrices $P = (P(1),\cdots,P(N)) > 0$ with $P(i) \in \mathbb{R}^{2n\times 2n}$ such that for any $i \in \mathcal{S}$ the* *following LMI holds for some given matrix $Q > 0$ and scalar $\varepsilon > 0$*

$$\begin{pmatrix} J(i) & * & * & * & * & * \\ \bar{A}_d^\top(i)P(i) & -Q + \varepsilon E_d^\top(i)E_d(i) & 0 & 0 & 0 & 0 \\ \bar{D}_a^\top(i)P(i) & 0 & -\varepsilon I & 0 & 0 & 0 \\ \bar{D}_d^\top(i)P(i) & 0 & 0 & -\varepsilon I & 0 & 0 \\ \bar{E}_a(i) & 0 & 0 & 0 & -\frac{1}{\varepsilon}I & 0 \\ I_0 & 0 & 0 & 0 & 0 & -\frac{Q^{-1}}{\varrho} \end{pmatrix} < 0,$$

(9.100)

where $J(i) = \bar{A}^\top(i)P(i) + P(i)\bar{A}(i) + \sum_{j=1}^{N} \lambda_{ij}P(j)$, $I_0 = (I\ \mathbf{0})$.

For a given set of gains $K_A(i), K_B(i), K_C(i)$, Lemma 9.8 provides a sufficient condition for the closed-loop system to be MSQS. Unfortunately, (9.100) is obviously nonlinear with respect to $K_A(i), K_B(i), K_C(i)$ and $P(i)$. Thus, this lemma cannot be used to synthesize a stabilizing controller. To cast the control synthesis problem into the LMI framework, let us partition $P(i)$ as in (8.74). Without loss of generality, we can assume that P_{2i} is invertible.

Lemma 9.9 *If there exist symmetric, positive-definite matrices $K_A(i)$, $K_B(i), K_C(i)$ and $P(i) > 0$ satisfying (9.100), then the following LMIs are feasible for some $X_i > 0, Y_i > 0, G_i, H_i, M_i, i \in \mathcal{S}$:*

$$\begin{pmatrix} \Gamma_{11}(i) & * & * & * & * & * & * \\ \Gamma_{12}^\top(i) & -\Gamma_{22}(i) & 0 & 0 & 0 & 0 & 0 \\ \Gamma_{13}^\top(i) & 0 & -\varepsilon I & 0 & 0 & 0 & 0 \\ \Gamma_{14}^\top(i) & 0 & 0 & -\varepsilon I & 0 & 0 & 0 \\ \Gamma_{15}^\top(i) & 0 & 0 & 0 & -\frac{1}{\varepsilon}I & 0 & 0 \\ \Gamma_{16}^\top(i) & 0 & 0 & 0 & 0 & -\frac{Q^{-1}}{\varrho} & 0 \\ \tilde{S}_i^\top(Y) & 0 & 0 & 0 & 0 & 0 & -\mathcal{Y}_i \end{pmatrix} < 0$$

(9.101)

$$\begin{pmatrix} Y_i & I \\ I & X_i \end{pmatrix} > 0$$

(9.102)

where

$$\Gamma_{11}(i) = \begin{pmatrix} W_{11}(i) & M_i^\top + A(i) \\ M_i + A^\top(i) & W_{22}(i) \end{pmatrix}$$

with

$$W_{11}(i) = A(i)Y_i + Y_i A^\top(i) + B(i)G_i + G_i^\top B^\top(i) + \lambda_{ii}Y_i$$

$$W_{22}(i) = A^\top(i)X_i + X_i A(i) + H_i C(i) + C^\top(i)H_i^\top + \sum_{j=1}^{N} \lambda_{ij}X_j$$

$$\Gamma_{22}(i) = Q - \varepsilon E_d^\top(i)E_d(i)$$

$$\Gamma_{12}(i) = \begin{pmatrix} A_d(i) \\ X_i A_d(i) \end{pmatrix}$$

$$\Gamma_{13}(i) = \begin{pmatrix} D_a(i) & D_a(i) \\ X_i D_a(i) + H_i D_c(i) & X_i D_a(i) \end{pmatrix}$$

$$\Gamma_{14}(i) = \begin{pmatrix} D_d(i) \\ X_i D_d(i) \end{pmatrix}$$

$$\Gamma_{15}(i) = \begin{pmatrix} Y_i E_a^\top(i) & G_i^\top E_b^\top(i) \\ E_a^\top(i) & 0 \end{pmatrix}$$

$$\Gamma_{16}(i) = \begin{pmatrix} Y_i \\ I \end{pmatrix}$$

$$\tilde{S}_i(Y) = \begin{pmatrix} \sqrt{\lambda_{i1}} Y_i & \cdots & \sqrt{\lambda_{ii-1}} Y_i & \sqrt{\lambda_{ii+1}} Y_i & \cdots & \sqrt{\lambda_{iN}} Y_i \\ 0 & \cdots & 0 & 0 & \cdots & 0 \end{pmatrix}$$

$$\mathcal{Y}_i = \text{diag}\{Y_1, \cdots, Y_{i-1}, Y_{i+1}, \cdots, Y_N\}.$$

Proof: The proof can be adapted from Lemma 8.3 and is omitted here. $\square$

In view of the Schur complement, LMI (9.101) is feasible for some $X_i > 0, Y_i > 0, G_i, H_i, M_i, i \in S$ if and only if

$$\Gamma_{11}(i) + \Gamma_{12}(i)\Gamma_{22}^{-1}(i)\Gamma_{12}^\top(i) + \frac{1}{\varepsilon}\Gamma_{13}(i)\Gamma_{13}^\top(i) + \frac{1}{\varepsilon}\Gamma_{14}(i)\Gamma_{14}^\top(i)$$
$$+ \varepsilon\Gamma_{15}(i)\Gamma_{15}^\top(i) + \varrho\Gamma_{16}(i)Q\Gamma_{16}^\top(i)$$
$$+ \tilde{S}_i(Y)\mathcal{Y}_i\tilde{S}_i^\top(Y) \triangleq Z_i = \begin{pmatrix} Z_{1i} & Z_{2i}^\top \\ Z_{2i} & Z_{3i} \end{pmatrix} < 0. \tag{9.103}$$

The last inequality implies in turn that

$$Z_{1i} < 0 \tag{9.104}$$
$$Z_{3i} < 0. \tag{9.105}$$

Direct computation gives

$$Z_{1i} = A(i)Y_i + Y_i A^\top(i) + B(i)G_i + G_i^\top B^\top(i) + \lambda_{ii} Y_i$$
$$+ A_d(i)[Q - \varepsilon E_d^\top(i)E_d(i)]^{-1} A_d^\top(i) + \frac{2}{\varepsilon} D_a(i)D_a^\top(i)$$
$$+ \frac{1}{\varepsilon} D_d(i)D_d^\top(i) + \varrho Y_i Q Y_i + \varepsilon Y_i E_a^\top(i)E_a(i)Y_i$$
$$+ G_i^\top E_b^\top(i)E_b(i)G_i + Y_i\left[\sum_{j\neq i}\lambda_{ij}Y_j^{-1}\right]Y_i,$$

$$Z_{3i} = A^\top(i)X_i + X_i A(i) + H_i C(i) + C^\top(i)H_i^\top + \sum_{j=1}^{N}\lambda_{ij}X_j$$
$$+ X_i A_d(i)[Q - \varepsilon E_d^\top(i)E_d(i)]^{-1} A_d^\top(i)X_i + \frac{1}{\varepsilon} X_i D_a(i)D_a^\top(i)X_i$$

$$+ \frac{1}{\varepsilon}[X_i D_a(i) + H_i D_c(i)][X_i D_a(i) + H_i D_c(i)]^\top$$

$$+ \frac{1}{\varepsilon}X_i D_d(i)D_d^\top(i)X_i + \varrho Q + \varepsilon E_a^\top(i)E_a(i)$$

$$Z_{2i} = M_i + A^\top(i) + X_i A_d(i)[Q - \varepsilon E_d^\top(i)E_d(i)]^{-1}A_d^\top(i)$$

$$+ \frac{1}{\varepsilon}[X_i D_a(i) + H_i D_c(i)]D_a^\top(i) + X_i D_a(i)D_a^\top(i)$$

$$+ \frac{1}{\varepsilon}X_i D_d(i)D_d^\top(i) + \varrho Q Y_i + \varepsilon E_a^\top(i)E_a(i)Y_i.$$

Using matrix inversion formula, we have

$$[Q - \varepsilon E_d^\top(i)E_d(i)]^{-1}$$
$$= Q^{-1} + Q^{-1}E_d^\top(i)[\frac{1}{\varepsilon}I - E_d(i)Q^{-1}E_d^\top(i)]^{-1}E_d(i)Q^{-1},$$

from which it follows that

$$Z_{1i} = A(i)Y_i + Y_i A^\top(i) + B(i)G_i + G_i^\top B^\top(i) + \lambda_{ii}Y_i$$

$$+ A_d(i)Q^{-1}E_d^\top(i)[\frac{1}{\varepsilon}I - E_d(i)Q^{-1}E_d^\top(i)]^{-1}E_d(i)Q^{-1}A_d^\top(i)$$

$$+ \frac{2}{\varepsilon}D_a(i)D_a^\top(i) + A_d(i)Q^{-1}A_d^\top(i) + \frac{1}{\varepsilon}D_d(i)D_d^\top(i)$$

$$+ \varrho Y_i Q Y_i + \varepsilon Y_i E_a^\top(i)E_a(i)Y_i$$

$$+ G_i^\top E_b^\top(i)E_b(i)G_i + Y_i\left[\sum_{j\neq i}\lambda_{ij}Y_j^{-1}\right]Y_i.$$

Using the Schur complement, we have that $Z_{1i} < 0$ and $Z_{3i} < 0$ are equivalent to the following two LMIs, respectively:

$$\left(\begin{array}{cccc}
W_1(i) & A_d(i)Q^{-1}E_d^\top(i) & Y_i \\
E_d(i)Q^{-1}A_d^\top(i) & -\frac{1}{\varepsilon} + E_d(i)Q^{-1}E_d^\top(i) & \\
Y_i & 0 & -\frac{Q^{-1}}{\varrho} \\
E_a^\top(i)Y_i & 0 & 0 \\
E_b(i)G_i & 0 & 0 \\
S_i^\top(Y) & 0 & 0
\end{array}\right.$$

$$\left.\begin{array}{ccc}
Y_i E_a(i) & G_i^\top E_b^\top(i) & S_i(Y) \\
0 & 0 & 0 \\
0 & 0 & 0 \\
-\frac{1}{\varepsilon}I & 0 & 0 \\
0 & -I & 0 \\
0 & 0 & -\mathcal{Y}_i
\end{array}\right) < 0 \qquad (9.106)$$

$$
\begin{pmatrix}
W_2(i) & * & * & * & * \\
A_d^\top(i)X_i & -Q + \varepsilon E_d^\top(i)E_d(i) & 0 & 0 & 0 \\
D_a^\top(i)X_i + D_c^\top(i)H_i^\top & 0 & -\varepsilon I & 0 & 0 \\
D_a^\top(i)X_i & 0 & 0 & -\varepsilon I & 0 \\
D_d^\top(i)X_i & 0 & 0 & 0 & -\varepsilon I
\end{pmatrix} < 0,
$$

$$(9.107)$$

where

$$
\begin{aligned}
W_1(i) ={}& A(i)Y_i + Y_iA^\top(i) + B(i)G_i + G_i^\top B^\top(i) + \lambda_{ii}Y_i \\
& + A_d(i)Q^{-1}A_d^\top(i) + \frac{2}{\varepsilon}D_a(i)D_a^\top(i) + \frac{1}{\varepsilon}D_d(i)D_d^\top(i), \\
W_2(i) ={}& A^\top(i)X_i + X_iA(i) + H_iC(i) + C^\top(i)H_i^\top \\
& + \sum_{j=1}^{N}\lambda_{ij}X_j + \varrho Q + \varepsilon E_a^\top(i)E_a(i).
\end{aligned}
$$

On the other hand, let $X_i, Y_i, G_i, H_i, i \in \mathcal{S}$ be a set of feasible solutions to LMIs (9.106), (9.107), and (9.102) and let us define matrices $P(i)$ as in (8.91) and gains

$$
\begin{cases}
K_A(i) = \left[X_i - Y_i^{-1}\right]^{-1}\Big[[A^\top(i) + X_iA_d(i)[Q \\
\quad - \varepsilon E_d^\top(i)E_d(i)]^{-1}A_d^\top(i) + \frac{1}{\varepsilon}[X_iD_a(i) + H_iD_c(i)]D_a^\top(i) \\
\quad + X_iD_a(i)D_a^\top(i) + \frac{1}{\varepsilon}X_iD_d(i)D_d^\top(i) + X_iB(i)G_i]Y_i^{-1} \\
\quad + \varrho Q + \varepsilon E_a^\top(i)E_a(i) \\
\quad + X_iA(i) + H_iC(i) + \sum_{j=1}^{N}\lambda_{ij}Y_j^{-1}\Big] \\
K_B(i) = \left[X_i - Y_i^{-1}\right]^{-1}H_i \\
K_C(i) = G_iY_i^{-1}.
\end{cases}
\tag{9.108}
$$

From (9.102) and the Schur complement, we can conclude that the $P(i)$ are positive-definite. Direct manipulation yields that $P(i)$, $K_A(i)$, $K_B(i)$, $K_C(i)$, $i \in \mathcal{S}$ defined above satisfy (9.100). Therefore, we get the following lemma.

Lemma 9.10 *There exist symmetric, positive-definite matrices $P = (P(1), \cdots, P(N)) > 0$ and $K_A(i)$, $K_B(i)$, $K_C(i)$, $i \in \mathcal{S}$ satisfying (9.100) for some given $Q > 0$ and a scalar $\varepsilon > 0$ if and only if LMIs (9.102), (9.106) and (9.107) are feasible for some $X = (X_1, \cdots, X_N) > 0$, $Y = (Y_1, \cdots, Y_N) > 0$, $G = (G_1, \cdots, G_N)$ and $H = (H_1, \cdots, H_N)$.*

Combining Lemma 9.8 and Lemma 9.10, we get the following theorem.

Theorem 9.18 *If there exist matrices $X_i > 0$, $Y_i > 0$, G_i, H_i, $i \in \mathcal{S}$ satisfying (9.106), (9.107) and (9.102), controller (8.70) with gains defined by (9.108) robustly stabilizes system (9.96) in the MSQS sense.*

Example 9.11 *Consider a system described by (9.77) with the following parameters*

$$S = \{1, 2\} \qquad \tau_1 = 0.3$$
$$\tau_2 = 1.23 \qquad \Lambda = \begin{pmatrix} -2 & 2 \\ 5 & -5 \end{pmatrix}$$

$$A(1) = \begin{pmatrix} -1 & 0 \\ 1.4 & -10 \end{pmatrix} \qquad A(2) = \begin{pmatrix} -3 & -0.8 \\ 0.2 & -2 \end{pmatrix}$$

$$A_d(1) = \begin{pmatrix} -0.3 & 0 \\ 0 & -0.2 \end{pmatrix} \qquad A_d(2) = \begin{pmatrix} 0.1 & -0.1 \\ 0.1 & -0.1 \end{pmatrix}$$

$$B(1) = \begin{pmatrix} 1 & 0 \\ 0 & -0.1 \end{pmatrix} \qquad B(2) = \begin{pmatrix} 0 & 1 \\ 0 & 0.1 \end{pmatrix}$$

$$D_a(1) = \begin{pmatrix} 1 & 0 \\ 0 & 0.3 \end{pmatrix}, \qquad D_a(2) = \begin{pmatrix} 0 & 1 \\ 0 & -0.3 \end{pmatrix}$$

$$D_d(1) = \begin{pmatrix} 0 & 0.2 \\ 0 & 1 \end{pmatrix} \qquad D_d(2) = \begin{pmatrix} -1 & 0 \\ 0 & 1 \end{pmatrix}$$

$$D_c(1) = \begin{pmatrix} 0 & 1 \\ 1 & 0 \end{pmatrix} \qquad D_c(2) = \begin{pmatrix} 1 & 0 \\ 01 & \end{pmatrix}$$

$$E_a(1) = \begin{pmatrix} 1 & 0 \\ 0 & 0 \end{pmatrix} \qquad E_a(2) = \begin{pmatrix} -1 & 0 \\ 0 & 0.2 \end{pmatrix}$$

$$E_d(1) = \begin{pmatrix} 0 & 0.3 \\ 0 & -1 \end{pmatrix} \qquad E_d(2) = \begin{pmatrix} -0.2 & 0 \\ 1 & 0 \end{pmatrix}$$

$$E_b(1) = \begin{pmatrix} 0.2 & 0 \\ 0.2 & 0.1 \end{pmatrix} \qquad E_b(2) = \begin{pmatrix} -0.1 & 0 \\ 0 & -0.2 \end{pmatrix}$$

$$C(1) = \begin{pmatrix} 1 & 0 \\ 0 & 1 \end{pmatrix} \qquad C(2) = \begin{pmatrix} 0 & 1 \\ 1 & 0 \end{pmatrix}.$$

Let $\varepsilon_1 = 0.81$ and $Q = \begin{pmatrix} 1.5 & 0 \\ 0 & 1.5 \end{pmatrix}$. With the above data solving (9.106), (9.107), and (9.102) gives the following set of feasible solutions:

$$X_1 = \begin{pmatrix} 108.1086 & -9.6097 \\ -9.6097 & 105.4823 \end{pmatrix} \qquad X_2 = \begin{pmatrix} 79.5152 & 8.6522 \\ 8.6522 & 107.4812 \end{pmatrix}$$

$$Y_1 = \begin{pmatrix} 1.34026 & 0.05909 \\ 0.05909 & 1.5968 \end{pmatrix} \qquad Y_2 = \begin{pmatrix} 1.41075 & 0.1322 \\ 0.1322 & 0.4128 \end{pmatrix}$$

$$G_1 = \begin{pmatrix} -24.9798 & -4.9911 \\ 49.95056 & 19.9728 \end{pmatrix} \qquad G_2 = \begin{pmatrix} 0 & 0 \\ -24.9623 & -2.4945 \end{pmatrix}$$

$$H_1 = \begin{pmatrix} -1490.6285 & -151.1869 \\ -151.2168 & -1399.9929 \end{pmatrix}$$

$$H_2 = \begin{pmatrix} 98.2781 & -1491.5182 \\ -1483.7903 & 98.2767 \end{pmatrix}.$$

Thus in view of Theorem 9.18 system (9.77) is robustly stabilizable in the MSQS sense. Moreover, substituting $X_i, Y_i, G_i, H_i, i \in \mathcal{S}$ into (9.108) yields a set of stabilizing gains as follows:

$$K_A(1) = \begin{pmatrix} -168.727 & -47.9006 \\ -140.5607 & -143.6264 \end{pmatrix}$$

$$K_A(2) = \begin{pmatrix} -398.1964 & 245.5875 \\ 31.6228 & -161.7669 \end{pmatrix}$$

$$K_B(1) = \begin{pmatrix} -140.1137 & -13.4342 \\ -14.24626 & -134.7447 \end{pmatrix}$$

$$K_B(2) = \begin{pmatrix} 17.3545 & -191.2477 \\ -142.8038 & 17.1240 \end{pmatrix}$$

$$K_C(1) = \begin{pmatrix} -18.5305 & -2.4398 \\ 36.7779 & 11.1467 \end{pmatrix}$$

$$K_C(2) = \begin{pmatrix} 0 & 0 \\ -17.6579 & -0.3877 \end{pmatrix}.$$

9.6 Notes

The class of deterministic systems with time delay has been extensively studied over the last two decades. Many results have been reported to the literature; among them we quote the following references [77, 118, 149] for constant delay; [77, 105] for time-varying delay; [137, 175] for constant delay and uncertainties; [49, 110, 108] for time-varying delay and parameter uncertainties; [49, 108, 110, 206, 207, 215] for delay-independent stabilization; [118, 128, 155, 189, 190] for delay-dependent stabilization. For the time-varying delay system, delay-dependent stability is dependent on the maximum value of the time derivative of the time-varying delay. The stability with discrete and distributed time delays was addressed by [118].

For a Markov jump system with time delay the results are few. For instance, delay-independent stabilization was addressed in [10]. The robust stabilization of discrete-time jump uncertain system was studied by [18]. This chapter gives most of the well-known results for this class of systems.

The results presented in this chapter can be extended to the case of time-varying time delay, $\tau(t)$, which satisfies the conditions

$$\dot{\tau}(t) \leq a, \ \tau(t) \leq \tau^*$$

where a, τ^* are known constants.

System (9.1) has polytopic uncertainties, that is, $A(r_t, t), A_d(r_t, t), B(r_t, t)$ represented as follows:

$$(A(r_t, t), A_d(r_t, t), B(r_t, t)) = \sum_{l=1}^{\ell} f_l(t)(A_l(r_t), A_{dl}(r_t), B_l(r_t)),$$

where $f_l(t) \geq 0, \sum_{l=1}^{\ell} f_l(t) = 1$, $A_l(r_t), A_{dl}(r_t), B_l(r_t), 1 \leq l \leq l$ are known matrices. Delay-dependent robust stability of jump systems with polytopic uncertainties was studied by [26]. The results of this chapter (except output feedback stabilization) can be extended to polytopic uncertain systems without difficulty.

10

$\mathcal{H}_\infty$ Control and Filtering Problems for Markov Jump Systems with Time Delay

Practical systems are always subject to exogenous disturbance input which can cause performances degradations. The tools we developed are unfortunately inefficient for eliminating the effect of the disturbance. One alternative is to use the $\mathcal{H}_\infty$ control, which entails searching for a controller that stabilizes the system, and at the same time assures disturbance rejection at a given level. The $\mathcal{H}_\infty$ control problem has received considerable attention and thus numerous achievements have been reported to the literature. In the linear time-invariant context, early $\mathcal{H}_\infty$ control theory emerged as a frequency domain design technique [83]. An important breakthrough in this line of research was the derivation of state-space solutions to the standard $\mathcal{H}_\infty$ control problem in terms of the solutions of two algebraic Riccati equations [64]. For more information we refer the reader to [87, 106, 113, 130, 139, 154, 170, 174]. The objective of this chapter is to address the delay-independent and delay-dependent $\mathcal{H}_\infty$ control problems for Markov jump systems with time delay and $\mathcal{H}_\infty$-filtering problems. Benjelloun, Boukas, and Yang [10] were the first to consider the $\mathcal{H}_\infty$ control problem for the class of systems we are considering in this chapter. Other results have been established by Boukas and his coauthors. The rest of this chapter is organized as follows. Section 10.1 deals with the $\mathcal{H}_\infty$ control problem. Delay-independent conditions are established to test if a given system is stable and satisfies the desired disturbance rejection level. Section 10.2 overcomes the conservatism of Section 10.1 by establishing a delay-dependent test for $\mathcal{H}_\infty$. In Section 10.3, we deal with output feedback control. Section 10.4 covers the $\mathcal{H}_\infty$-filtering problem.

10.1 $\mathcal{H}_\infty$ Control

The objective of this section is to study $\mathcal{H}_\infty$ control for the class of Markov jump linear systems with constant time delay. For this purpose, let $\{r_t, t \geq 0\}$ be a Markov process that describes the system mode taking values in a finite state space $\mathcal{S} = \{1, 2, \cdots, N\}$ with generator $\Lambda = (\lambda_{ij}), i, j \in \mathcal{S}$. Consider the following dynamical system:

$$
\begin{cases}
\dot{x}(t) = A(r_t)x(t) + A_d(r_t)x(t-\tau) + B(r_t)u(t) + B_w(r_t)w(t) \\
x(s) = \phi(s), \ s \in [-\tau, 0] \\
z(t) = C(r_t)x(t) + C_d(r_t)x(t-\tau) + C_u(r_t)u(t) + C_w(r_t)w(t) \\
y(t) = G(r_t)x(t) + G_d(r_t)x(t-\tau) + G_u(r_t)u(t) + G_w(r_t)w(t)
\end{cases}
\tag{10.1}
$$

where $x(t) \in \mathbb{R}^n$ is the state vector; $u(t) \in \mathbb{R}^m$ is the control input; $z(t) \in \mathbb{R}^p$ is the controlled output, $y(t) \in \mathbb{R}^r$ is the measured output; $w(t) \in \mathbb{R}^q$ is the exogenous disturbance input, which is assumed to be in $\mathcal{L}_2[0, \infty)$; $\tau > 0$ is the time delay in the system; r_t denotes the system mode; $\phi(\cdot) \in \mathbb{C}[-\tau, 0]$ is the initial condition; $A(r_t) \in R^{n \times n}$, $A_d(r_t) \in R^{n \times n}$, $B(r_t) \in \mathbb{R}^{n \times m}$, $B_w(r_t) \in \mathbb{R}^{n \times q}$, $C(r_t) \in \mathbb{R}^{p \times n}$, $C_d(r_t) \in \mathbb{R}^{p \times n}$, $C_u(r_t) \in \mathbb{R}^{p \times m}$, $C_w(r_t) \in \mathbb{R}^{p \times q}$, $G(r_t) \in \mathbb{R}^{p \times n}$, $G_u(r_t) \in \mathbb{R}^{r \times n}$, $G_w(r_t) \in \mathbb{R}^{r \times q}$ are given matrices of appropriate dimensions.

Remark 10.1 *The time delay can be chosen as time-varying, but here we will restrict ourselves to constant time delay. The extension of the results of this chapter to dynamical systems with time-varying time delay is straightforward.*

Definition 10.1 *Let $\gamma > 0$ be a given positive constant. System (10.1) with $u(t) \equiv 0$ is said to be stochastically stable with γ-disturbance attenuation if there exists a constant $M(r_0, \phi(\cdot))$ with $M(r_0, 0) = 0$, for all $r_0 \in \mathcal{S}$, such that the following holds:*

$$
\|z\|_2 \overset{\Delta}{=} \left[\mathbb{E} \int_0^\infty z^\top(t)z(t)dt | (r_0, \phi(\cdot)) \right]^{1/2} \leq \gamma \left[\|w\|_2^2 + M(r_0, \phi(\cdot)) \right]^{1/2}. \tag{10.2}
$$

Definition 10.2 *System (10.1) with $u(t) \equiv 0$ is said to be internally mean square quadratically stable (MSQS) if there exist symmetric, positive-definite matrices $P = (P(1), \cdots, P(N)) > 0$ and $Q > 0$ satisfying the following for every $i \in \mathcal{S}$:*

$$
\Theta(i) = \begin{pmatrix} J(i) & P(i)A_d(i) \\ A_d^\top(i)P(i) & -Q \end{pmatrix} < 0, \ \forall i \in \mathcal{S} \tag{10.3}
$$

where $J(i) = A^\top(i)P(i) + P(i)A(i) + \sum_{j=1}^N \lambda_{ij}P(j) + Q$.

By virtue of Definition 8.3, it is obvious that internal MSQS means that system (10.1) is MSQS in the case of $w(t) \equiv 0$, i.e., system (10.1) is free of input disturbance. Likewise, we can give the following definition.

Definition 10.3 *System (10.1) with $u(t) \equiv 0$ is said to be internally SS (MES) if it is SS (MES) in the case of $w(t) \equiv 0$*

Definition 10.4 *System (10.1) is said to be stabilizable with γ-disturbance in the SS (MES, MSQS) sense if there exist matrices $K(i), i \in \mathcal{S}$, such that the closed-loop system under control*

$$u(t) = K(r_t)x(t) \tag{10.4}$$

is SS (MES, MSQS) and satisfies (10.2).

10.1.1 $\mathcal{H}_\infty$-Performance

Let us develop sufficient conditions for system (10.1) with $u(t) \equiv 0$ to be stochastically stable and to satisfy (10.2). The following theorem shows that in the case of $w(t) \not\equiv 0$, internal MSQS implies stochastic stability.

Theorem 10.1 *If system (10.1) with $u(t) \equiv 0$ is internally MSQS, then it is stochastically stable.*

Proof: Let us define a stochastic process $\{\mathbf{x}_t, t \geq 0\}$ with $\mathbf{x}_t \in \mathbb{C}[-\tau, 0]$ given by

$$\mathbf{x}_t(s) = x(t + s), \ s \in [-\tau, 0].$$

Then, $\{(\mathbf{x}_t, r_t), t \geq 0\}$ is a strong Markov process taking values in the state space $\mathbb{C}[-\tau, 0] \times \mathcal{S}$ with initial condition $(\mathbf{x}_0, r_0) = (\phi(\cdot), r_0)$. Suppose that $P(i) > 0, i \in \mathcal{S}, Q > 0$ are a set of feasible solutions of LMIs (10.3). Let us consider Lyapunov functional candidate as follows:

$$V(\mathbf{x}_t, r_t) = x^\top(t)P(r_t)x(t) + \int_{t-\tau}^t x^\top(s)Qx(s)ds.$$

Direct manipulation gives

$$\mathcal{A}V(\mathbf{x}_t, r_t) = x^\top(t)\left[A^\top(r_t)P(r_t) + P(r_t)A(r_t) + \sum_{j=1}^N \lambda_{r_t j}P(j) + Q\right]x(t)$$
$$+ 2x^\top(t)P(r_t)A_d(r_t)x(t-\tau) - x^\top(t-\tau)Qx(t-\tau)$$
$$+ 2x^\top(t)P(r_t)B_w(r_t)w(t) \tag{10.5}$$

Since $P(i)B_w(i)B_w^\top(i)P(i) \geq 0$, there must exist constant $\varepsilon_0 > 0$ such that

$$\tilde{\Theta}(i) = \begin{pmatrix} J(i) + \varepsilon_0 P(i)B_w(i)B_w^\top(i)P(i) & P(i)A_d(i) \\ A_d^\top(i)P(i) & -Q \end{pmatrix} < 0$$

holds for all $i \in \mathcal{S}$.

Using Lemma A.1, we get the following for any $\varepsilon > 0$:

$$2x^\top(t)P(r_t)B_w(r_t)w(t) \leq \varepsilon x^\top(t)P(r_t)B_w(r_t)B_w^\top(r_t)P(r_t)x(t)$$
$$+ \frac{1}{\varepsilon}w^\top(t)w(t).$$

Combining this with (10.5) yields

$$\mathcal{A}V(\mathbf{x}_t, r_t) \leq x^\top(t)\left[A^\top(r_t)P(r_t) + P(r_t)A(r_t) + \sum_{j=1}^N \lambda_{r_t j}P(j) + Q\right]x(t)$$
$$+ 2x^\top(t)P(r_t)A_d(r_t)x(t-\tau) - x^\top(t-\tau)Qx(t-\tau)$$
$$+ \varepsilon x^\top(t)P(r_t)B_w(r_t)B_w^\top(r_t)P(r_t)x(t) + \frac{1}{\varepsilon}w^\top(t)w(t)$$
$$= \xi_t^\top \tilde{\Theta}(r_t)\xi_t + \frac{1}{\varepsilon}w^\top(t)w(t), \qquad (10.6)$$

where $\xi_t = \begin{pmatrix} x(t) \\ x(t-\tau) \end{pmatrix}$.

Using Dynkin's formula, we have

$$\mathbb{E}[V(\mathbf{x}_t, r_t)] - V(\mathbf{x}_0, r_0)] = \mathbb{E}\left[\int_0^t \mathcal{A}V(\mathbf{x}_s, r_s)ds\right],$$

which combined with (10.6) yields

$$\mathbb{E}[V(\mathbf{x}_t, r_t)] - V(\mathbf{x}_0, r_0)] \leq \mathbb{E}\left[\int_0^t \xi_s^\top \tilde{\Theta}(r_s)\xi_s ds\right]$$
$$+ \frac{1}{\varepsilon}\int_0^t w^\top(s)w(s)ds. \qquad (10.7)$$

Since $V(\mathbf{x}_t, r_t)$ is nonnegative, (10.7) implies

$$\mathbb{E}[V(\mathbf{x}_t, r_t)] + \mathbb{E}\left[\int_0^t \xi_s^\top[-\tilde{\Theta}(r_s)]\xi_s ds\right] \leq V(\mathbf{x}_0, r_0)] + \frac{1}{\varepsilon}\int_0^t w^\top(s)w(s)ds,$$

which yields

$$\min_{i\in\mathcal{S}}\{\lambda_{\min}(-\tilde{\Theta}(i))\}\mathbb{E}\left[\int_0^t x^\top(s)x(s)ds\right] \leq \mathbb{E}\left[\int_0^t \xi_s^\top[-\tilde{\Theta}(r_s)]\xi_s ds\right]$$
$$\leq V(\mathbf{x}_0, r_0) + \frac{1}{\varepsilon}\int_0^\infty w^\top(s)w(s)ds.$$

This proves that system (10.1) is stochastically stable. □

Theorem 10.2 *Let γ be a given positive constant. If there exist symmetric, positive-definite matrices $P = (P(1), \cdots, P(N))$ and Q such that the following LMI holds for every $i \in \mathcal{S}$,*

$$\Theta_h(i) = \left(\begin{array}{cc} \left[\begin{array}{c} A^\top(i)P(i) + P(i)A(i) \\ + \sum_{j=1}^N \lambda_{ij}P(j) \\ + Q + C^\top(i)C(i) \end{array}\right] & P(i)A_d(i) + C^\top(i)C_d(i) \\ A_d^\top(i)P(i) + C_d^\top(i)C(i) & -Q + C_d^\top(i)C_d(i) \\ C_w^\top(i)C(i) + B_w^\top(i)P(i) & C_w^\top(i)C_d(i) \end{array}\right.$$

$$\left.\begin{array}{c} C^\top(i)C_w(i) + P(i)B_w(i) \\ C_d^\top(i)C_w(i) \\ C_w^\top(i)C_w(i) - \gamma^2 I \end{array}\right) < 0, \quad (10.8)$$

then system (10.1) with $u(t) \equiv 0$ is stochastically stable and satisfies

$$\|z\| \le \left[\gamma^2\|w\|_2^2 + x_0^\top P(r_0)x_0 + \int_{-\tau}^0 \phi^\top(s)Q\phi(s)ds\right]^{1/2}, \quad (10.9)$$

where $x_0 = \phi(0)$, which means that the system is stochastically stable with γ-disturbance attenuation.

Proof: Obviously, (10.8) implies the following inequality

$$\left(\begin{array}{cc} \left[\begin{array}{c} A^\top(i)P(i) + P(i)A(i) \\ + \sum_{j=1}^N \lambda_{ij}P(j) \\ + Q + C^\top(i)C(i) \end{array}\right] & P(i)A_d(i) + C^\top(i)C_d(i) \\ A_d^\top(i)P(i) + C_d^\top(i)C(i) & -Q + C_d^\top(i)C_d(i) \end{array}\right) < 0. (10.10)$$

Noting that the left-hand side of (10.10) can be rewritten as

$$\left(\begin{array}{cc} \left[\begin{array}{c} A^\top(i)P(i) + P(i)A(i) \\ + \sum_{j=1}^N \lambda_{ij}P(j) + Q \end{array}\right] & P(i)A_d(i) \\ A_d^\top(i)P(i) & -Q \end{array}\right)$$
$$+ \left(\begin{array}{c} C^\top(i) \\ C_d^\top(i) \end{array}\right)(C(i) \; C_d(i)),$$

which combined with (10.10) yields

$$\left(\begin{array}{cc} \left[\begin{array}{c} A^\top(i)P(i) + P(i)A(i) \\ + \sum_{j=1}^N \lambda_{ij}P(j) + Q \end{array}\right] & P(i)A_d(i) \\ A_d^\top(i)P(i) & -Q \end{array}\right) < 0.$$

This proves that the system under study is internally MSQS. Using now Theorem 10.1, we conclude that system (10.1) with $u(t) \equiv 0$ is stochastically stable.

Next, we proceed to prove that (10.9) is satisfied. To this end, let us define the performance function

$$J_T = \mathbb{E}\left[\int_0^T [z^\top(t)z(t) - \gamma^2 w^\top(t)w(t)]dt\right].$$

To prove (10.9), it suffices to establish

$$J_\infty \le V(\mathbf{x}_0, r_0) = x_0^\top P(r_0)x_0 + \int_{-\tau}^0 \phi(s)Q\phi(s)ds.$$

Using Dynkin's formula, we have

$$\mathbb{E}\left[\int_0^T \mathcal{A}V(\mathbf{x}_t, r_t)]dt\right] = \mathbb{E}[V(\mathbf{x}_T, r_T)] - \mathbb{E}[V(\mathbf{x}_0, r_0)].$$

Noting (10.5), direct computation gives

$$z^\top(t)z(t) - \gamma^2 w^\top(t)w(t) + \mathcal{A}V(\mathbf{x}_t, r_t) = \eta^\top(t)\Theta_h(r_t)\eta(t),$$

where $\eta^\top(t) = \left(x^\top(t)\ x^\top(t-\tau)\ w^\top(t)\right)$. Therefore,

$$\begin{aligned}
J_T &= \mathbb{E}\left[\int_0^T [z^\top(t)z(t) - \gamma^2 w^\top(t)w(t) + \mathcal{A}V(\mathbf{x}_t, r_t)]dt\right] \\
&\quad - \mathbb{E}\left[\int_0^T \mathcal{A}V(\mathbf{x}_t, r_t)]dt\right] \\
&= \mathbb{E}\left[\int_0^T \eta^\top(t)\Theta_h(r_t)\eta(t)dt\right] - \mathbb{E}[V(\mathbf{x}_T, r_T)] + V(\mathbf{x}_0, r_0). \quad (10.11)
\end{aligned}$$

Since $\Theta_h(i) < 0$ and $\mathbb{E}[V(\mathbf{x}_T, r_T)] \geq 0$, (10.11) implies

$$J_T \leq V(\mathbf{x}_0, r_0),$$

yielding

$$J_\infty \leq V(\mathbf{x}_0, r_0),$$

i.e.,

$$\|z\|_2^2 - \gamma^2\|w\|_2^2 \leq x_0^\top P(r_0)x_0 + \int_{-\tau}^0 \phi^\top(s)Q\phi(s)ds.$$

This yields

$$\|z\|_2^2 \leq \gamma^2\|w\|_2^2 + x_0^\top P(r_0)x_0 + \int_{-\tau}^0 \phi^\top(s)Q\phi(s)ds,$$

which gives (10.9).

This completes the proof of Theorem 10.2.   □

Remark 10.2 *When the initial conditions are null, the condition (10.9) becomes* $\|z(t)\|_2 \leq \gamma\|w(t)\|_2$, *which is the standard form used in many papers (see for example [9]).*

The disturbance attenuation level γ was assumed to be a given positive constant. In practice, we are interested in determining the minimum $\gamma > 0$ disturbance rejection level a system can have. For a given system, the following optimization problem can be used to compute this minimum disturbance attenuation level γ:

$$\mathcal{P}_1: \quad \min_{\mu>0, P(i)>0, i\in\mathcal{S}, Q>0} \mu$$

$$\text{s.t. } \Theta_\mu(i) < 0$$

where $\Theta_\mu(i)$ is obtained from Θ_h by replacing γ^2 with μ.

The following example shows the usefulness of the results of the previous theorem.

Example 10.1 *Let us consider a hybrid system with two modes denoted by $\mathcal{S} = \{1, 2\}$. The Markov process which governs the mode switching of the system takes values in $\mathcal{S}$ and has the infinitesimal generator*

$$\Lambda = \begin{pmatrix} -5 & 5 \\ 3 & -3 \end{pmatrix}.$$

Let the system be described by (10.1) with $u(t) \equiv 0$ and suppose that the system parameters are as follows:

$$A(1) = \begin{pmatrix} -2 & 0.1 \\ 0 & -4 \end{pmatrix} \qquad A(2) = \begin{pmatrix} -1 & 0.2 \\ -1 & -3 \end{pmatrix}$$

$$A_d(1) = \begin{pmatrix} 0.8 & 0 \\ 0.3 & 0.2 \end{pmatrix} \qquad A_d(2) = \begin{pmatrix} 0.4 & 0 \\ 0.3 & 0.2 \end{pmatrix}$$

$$C(1) = \begin{pmatrix} 1 & 1 \\ 1 & 0 \end{pmatrix} \qquad C(2) = \begin{pmatrix} 0 & 1 \\ 1 & 0 \end{pmatrix}$$

$$B_w(1) = \begin{pmatrix} 1 & 0 \\ 0 & -1 \end{pmatrix} \qquad B_w(2) = \begin{pmatrix} -1 & 1 \\ 1 & 0 \end{pmatrix}$$

$$C_w(1) = \begin{pmatrix} 1 & 0 \\ 1 & -1 \end{pmatrix} \qquad C_w(2) = \begin{pmatrix} 0 & 0.2 \\ 0 & 1 \end{pmatrix}$$

$$C_d(1) = \begin{pmatrix} 1 & 0 \\ 0 & 1 \end{pmatrix} \qquad C_d(2) = \begin{pmatrix} 1 & 1 \\ 0 & 1 \end{pmatrix}$$

With the above data, solving $\mathcal{P}_1$ yields the following solutions:

$$P(1) = \begin{pmatrix} 383.5752 & 32.5218 \\ 32.5218 & 229.8900 \end{pmatrix} \qquad P(2) = \begin{pmatrix} 454.6091 & 17.9084 \\ 17.9084 & 262.5447 \end{pmatrix}$$

$$Q = \begin{pmatrix} 460.2800 & 98.7244 \\ 98.7244 & 776.6260 \end{pmatrix} \qquad \mu = 1063.5.$$

According to Theorem 10.2, the system under study is stochastically stable with disturbance attenuation level $\gamma = 32.6118$.

When we cannot reach the desired disturbance rejection a design of an appropriate controller is required. The next section will address this problem.

10.1.2 Synthesis of Memoryless Controller

Our goal here is to design a controller of the form (10.4) that stochastically stabilizes system (10.1) and guarantees a given disturbance attenuation level γ, and to establish how to determine the corresponding optimal disturbance attenuation level. For this purpose we will assume complete access to the state vector and the system mode at each time t.

Substituting controller (10.4) into (10.1) yields the dynamics

$$\begin{cases} \dot{x}(t) = \bar{A}(r_t)x(t) + A_d(r_t)x(t-\tau) + B_w(r_t)w(t) \\ x(s) = \phi(s), \ s \in [-\tau, 0] \\ z(t) = \bar{C}(r_t)x(t) + C_d(r_t)x(t-\tau) + C_w(r_t)w(t), \end{cases} \tag{10.12}$$

where $\bar{A}(r_t) = A(r_t) + B(r_t)K(r_t)$ and $\bar{C}(r_t) = C(r_t) + C_u(r_t)K(r_t)$.

Applying Theorem 10.2 to system (10.12) gives the following proposition.

Proposition 10.1 *Let $\gamma > 0$ be a given constant. If there exist symmetric, positive-definite matrices $P = (P(1), \cdots, P(N)) > 0$ and Q such that the LMI*

$$\bar{\Theta}_h(i) = \left(\begin{array}{ccc} \left[\begin{array}{c} \bar{A}^\top(i)P(i) + P(i)\bar{A}(i) \\ + \sum_{j=1}^N \lambda_{ij}P(j) \\ + Q + \bar{C}^\top(i)\bar{C}(i) \end{array} \right] & P(i)A_d(i) + \bar{C}^\top(i)C_d(i) \\ A_d^\top(i)P(i) + C_d^\top(i)\bar{C}(i) & -Q + C_d^\top(i)C_d(i) \\ C_w^\top(i)\bar{C}(i) + B_w^\top(i)P(i) & C_w^\top(i)C_d(i) \end{array} \right.$$
$$\left. \begin{array}{c} \bar{C}^\top(i)C_w(i) + P(i)B_w(i) \\ C_d^\top(i)C_w(i) \\ C_w^\top(i)C_w(i) - \gamma^2 I \end{array} \right) < 0 \tag{10.13}$$

holds for every $i \in \mathcal{S}$, then system (10.1) under controller (10.4) is stochastically stable and satisfies

$$\|z\| \leq \left[\gamma^2 \|w\|_2^2 + x_0^\top P(r_0)x_0 + \int_{-\tau}^0 \phi^\top(s)Q\phi(s)ds \right]^{1/2}. \tag{10.14}$$

Note that

$$\bar{\Theta}_h(i) = \left(\begin{array}{ccc} \bar{J}(i) & P(i)A_d(i) & P(i)B_w(i) \\ A_d^\top(i)P(i) & -Q & 0 \\ B_w^\top(i)P(i) & 0 & -\gamma^2 I \end{array} \right)$$
$$+ \left(\begin{array}{c} \bar{C}^\top(i) \\ C_d^\top(i) \\ C_w^\top(i) \end{array} \right) \left(\bar{C}(i) \ C_d(i) \ C_w(i) \right)$$

with $\bar{J}(i) = \bar{A}^\top(i)P(i) + P(i)\bar{A}(i) + \sum_{j=1}^N \lambda_{ij}P(j) + Q$.

Using the Schur complement, (10.13) is equivalent to

$$\left(\begin{array}{cccc} \bar{J}(i) & P(i)A_d(i) & P(i)B_w(i) & \bar{C}^\top(i) \\ A_d^\top(i)P(i) & -Q & 0 & C_d^\top(i) \\ B_w^\top(i)P(i) & 0 & -\gamma^2 I & C_w^\top(i) \\ \bar{C}(i) & C_d(i) & C_w(i) & -I \end{array} \right) < 0. \tag{10.15}$$

Let $X_i = P^{-1}(i)$. Pre- and post-multiplying both sides of (10.15) by diag$\{X_i, I, I, I\}$ yields

$$
\begin{pmatrix}
X_i \bar{J}(i) X_i & A_d(i) & B_w(i) & X_i \bar{C}^\top(i) \\
A_d^\top(i) & -Q & 0 & C_d^\top(i) \\
B_w^\top(i) & 0 & -\gamma^2 I & C_w^\top(i) \\
\bar{C}(i) X_i & C_d(i) & C_w(i) & -I
\end{pmatrix} < 0. \tag{10.16}
$$

Notice that

$$
X_i \bar{J}(i) X_i = X_i \bar{A}^\top(i) + \bar{A}(i) X_i + \lambda_{ii} X_i + X_i \left[\sum_{j \neq i} \lambda_{ij} X_j^{-1} \right] X_i + X_i Q X_i
$$

Let $\mathcal{X}_i$ and $S_i(X)$ be defined by (8.12) and (8.11), respectively. Then we get

$$
X_i \left[\sum_{j \neq i} \lambda_{ij} X_j^{-1} \right] X_i = S_i(X) \mathcal{X}_i S_i^\top(X).
$$

Using now the Schur complement, (10.16) is equivalent to

$$
\begin{pmatrix}
\tilde{J}(i) & A_d(i) & B_w(i) & X_i \bar{C}^\top(i) & X_i & S_i(X) \\
A_d^\top(i) & -Q & 0 & C_d^\top(i) & 0 & 0 \\
B_w^\top(i) & 0 & -\gamma^2 I & C_w^\top(i) & 0 & 0 \\
\bar{C}(i) X_i & C_d(i) & C_w(i) & -I & 0 & 0 \\
X_i & 0 & 0 & 0 & -Q^{-1} & 0 \\
S_i^\top(X) & 0 & 0 & 0 & 0 & -\mathcal{X}_i
\end{pmatrix} < 0, \tag{10.17}
$$

where $\tilde{J}(i) = X_i \bar{A}^\top(i) + \bar{A}(i) X_i + \lambda_{ii} X_i$.

From above discussion, we get the following theorem.

Theorem 10.3 *If there exist symmetric, positive-definite matrices $X = (X_1, \cdots, X_N) > 0, U > 0$ and $Y = (Y_1, \cdots, Y_N)$, such that the LMIs*

$$
\left(
\begin{array}{ccc}
\hat{J}(i) & A_d(i) U & B_w(i) \\
U A_d^\top(i) & -U & 0 \\
B_w^\top(i) & 0 & -\gamma^2 I \\
C(i) X_i + C_u Y_i & C_d(i) U & C_w(i) \\
X_i & 0 & 0 \\
S_i^\top(X) & 0 & 0
\end{array}
\right.
$$

$$
\left.
\begin{array}{ccc}
X_i C^\top(i) + Y_i^\top C_u^\top(i) & X_i & S_i(X) \\
U C_d^\top(i) & 0 & 0 \\
C_w^\top(i) & 0 & 0 \\
-I & 0 & 0 \\
0 & -U & 0 \\
0 & 0 & -\mathcal{X}_i
\end{array}
\right) < 0, \tag{10.18}
$$

are feasible for every $i \in \mathcal{S}$, *with* $\hat{J}(i) = X_i A^\top(i) + Y_i^\top B^\top(i) + A(i)X_i + B(i)Y_i + \lambda_{ii}X_i$, *then controller (10.4) with* $K(i) = Y_i X_i^{-1}, i \in \mathcal{S}$, *stochastically stabilizes system (10.1), and moreover, the closed-loop system verifies the γ-disturbance attenuation level.*

Proof: Let $U = Q^{-1}$. Pre- and post-multiplying both sides of (10.17) by $\mathrm{diag}\{I, U, I, I, I, I\}$ yields

$$\begin{pmatrix} \tilde{J}(i) & A_d(i)U & B_w(i) & X_i\bar{C}^\top(i) & X_i & S_i(X) \\ UA_d^\top(i) & -U & 0 & UC_d^\top(i) & 0 & 0 \\ B_w^\top(i) & 0 & -\gamma^2 I & C_w^\top(i) & 0 & 0 \\ \bar{C}(i)X_i & C_d(i)U & C_w(i) & -I & 0 & 0 \\ X_i & 0 & 0 & 0 & -U & 0 \\ S_i^\top(X) & 0 & 0 & 0 & 0 & -\mathcal{X}_i \end{pmatrix} < 0. \quad (10.19)$$

Letting $Y_i = K(i)X_i$ in (10.19) leads to the proof of Theorem 10.3. □

Theorem 10.3 provides a design method of a controller that stochastically stabilizes system (10.1) and guarantees the given level disturbance attenuation. Evidently, a problem of interest is to design a controller that stabilizes system (10.1) and minimizes the disturbance attenuation level. This problem can be cast into a linear optimization problem and thus solved using LMI toolbox easily. Let us replace γ^2 by μ in (10.18) and consider the following optimization problem:

$$\mathcal{P}_2: \quad \begin{matrix} \min \\ X = (X_1, \ldots, X_N) > 0, U > 0, \\ Y = (Y_1, \ldots, Y_N), \mu > 0 \end{matrix} \quad \mu$$

s.t. (10.18) with γ^2 replaced by μ

Therefore, we get the following theorem.

Theorem 10.4 *Let* $X = (X_1, \cdots, X_N) > 0, Y = (Y_1, \cdots, Y_N), U > 0, \mu_0 > 0$ *be the solutions to the optimization problem* $\mathcal{P}_2$. *Then, controller (10.4) stochastically stabilizes the system (10.1) and the closed-loop system attains the optimal disturbance attenuation level* $\sqrt{\mu_0}$.

To show the usefulness of the results of this theorem, let us consider the following numerical example.

Example 10.2 *Let us consider a system described by (10.1) with the following parameters:* $\mathcal{S} = \{1, 2\}$ *and the Markov process that governs the mode-switching has infinitesimal generator* $\Lambda = \begin{pmatrix} -5 & 5 \\ 3 & -3 \end{pmatrix}$. *Let the rest* *of data be given by*

$$A(1) = \begin{pmatrix} 2 & 0.1 \\ 0 & -4 \end{pmatrix} \qquad A(2) = \begin{pmatrix} -1 & 0.2 \\ -1 & 3 \end{pmatrix}$$

$$A_d(1) = \begin{pmatrix} -0.8 & 0 \\ 0.3 & 0.2 \end{pmatrix} \qquad A_d(2) = \begin{pmatrix} -0.4 & 0 \\ 0.3 & 0.2 \end{pmatrix}$$

$$C(1) = \begin{pmatrix} 1 & 1 \\ 1 & 0 \end{pmatrix} \qquad C(2) = \begin{pmatrix} 0 & 1 \\ 1 & 0 \end{pmatrix},$$

$$B(1) = \begin{pmatrix} 1 & 0 \\ 0 & -1 \end{pmatrix} \qquad B(2) = \begin{pmatrix} -1 & 0.1 \\ 1 & 0 \end{pmatrix}$$

$$B_w(1) = \begin{pmatrix} 1 & 0 \\ 0 & -1 \end{pmatrix} \qquad B_w(2) = \begin{pmatrix} -1 & 1 \\ 1 & 0 \end{pmatrix}$$

$$C_w(1) = \begin{pmatrix} 1 & 0 \\ 0 & -1 \end{pmatrix} \qquad C_w(2) = \begin{pmatrix} 2 & 0 \\ 0 & 1 \end{pmatrix}$$

$$C_u(1) = \begin{pmatrix} 1 & 0 \\ 1 & -1 \end{pmatrix} \qquad C_u(2) = \begin{pmatrix} 0 & 0.2 \\ 0 & 1 \end{pmatrix}$$

$$C_d(1) = \begin{pmatrix} 1 & 0 \\ 0 & 1 \end{pmatrix} \qquad C_d(2) = \begin{pmatrix} 1 & 1 \\ 0 & 1 \end{pmatrix}$$

With this set of parameters, solving $\mathcal{P}_2$ gives the following solutions:

$$X_1 = \begin{pmatrix} 0.0394 & -0.0276 \\ -0.0276 & 0.0522 \end{pmatrix} \quad X_2 = \begin{pmatrix} 0.0808 & -0.0468 \\ -0.0468 & 0.0422 \end{pmatrix}$$

$$Y_1 = \begin{pmatrix} -0.7846 & 0.7343 \\ -0.5725 & 1.2122 \end{pmatrix} \quad Y_2 = \begin{pmatrix} 1643.6 & -1643.8 \\ -0.1 & -0.1 \end{pmatrix}$$

$$U = \begin{pmatrix} 0.1243 & -0.0953 \\ -0.0953 & 0.4569 \end{pmatrix} \quad \mu = 6406.5.$$

According to Theorem 10.4, controller (10.4) with

$$K(1) = \begin{pmatrix} -15.9793 & 5.6180 \\ 2.7364 & 24.6573 \end{pmatrix} \quad K(2) = \begin{pmatrix} -6241 & -45891 \\ -7 & -10 \end{pmatrix}$$

stochastically stabilizes the system under study and the closed-loop system verifies disturbance attenuation level $\gamma = 80.0408$.

The conditions we developed so far are delay-independent which make them conservative. In the next section we will develop delay-dependent conditions.

10.2 Delay-Dependent $\mathcal{H}_\infty$ Control

The results developed in the previous section are independent of the time delay of the system, which means that these results hold for any $\tau \in [0, \infty)$, and thus are conservative. This section establishes the delay-dependent stability and $\mathcal{H}_\infty$ control for system (10.1).

10.2.1 Delay-Dependent $\mathcal{H}_\infty$ Performance

The goal of the subsection is

- to develop sufficient conditions for system (10.1) with $u(t) \equiv 0$ to be stochastically stable with a given disturbance attenuation level;

- to address the problem of minimizing a disturbance attenuation level.

For this purpose, let us assume the following.

Proposition 10.2 *If there exist symmetric, positive-definite matrices $X = (X_1, \cdots, X_N) > 0$, $Q > 0, Q_1 > 0$ satisfying the LMIs*

$$\begin{pmatrix} J_1(i) & X_i & X_i A^\top(i) & A_d(i)Q_1 & S_i(X) \\ X_i & -\frac{1}{\tau}Q & 0 & 0 & 0 \\ A(i)X_i & 0 & -\frac{1}{\tau}Q_1 & 0 & 0 \\ Q_1 A_d^\top(i) & 0 & 0 & -\frac{1}{2\tau}Q_1 & 0 \\ S_i^\top(X) & 0 & 0 & 0 & -\mathcal{X}_i \end{pmatrix} \triangleq \Gamma(i) < 0 \quad (10.20)$$

$$\begin{pmatrix} Q & QA_d^\top(i) \\ A_d(i)Q & Q_1 \end{pmatrix} > 0 \quad\quad (10.21)$$

for every $i \in \mathcal{S}$, where

$$J_1(i) = X_i[A(i) + A_d(i)]^\top + [A(i) + A_d(i)]X_i + \lambda_{ii}X_i,$$

and $S_i(X), \mathcal{X}_i$ are defined as in (8.11) and (8.12), respectively, then system (10.1) is SS.

Proof: Suppose that $w(s) = 0, -2\tau \leq s \leq 0$. Noting that for a given $\tau > 0$, we have

$$x(t - \tau) = x(t) - \int_{-\tau}^{0} \dot{x}(t + \theta)d\theta = x(t) - \int_{-\tau}^{0} [A(r_{t+\theta})x(t + \theta) \\ + A_d(r_{t+\theta})x(t + \theta - \tau) + B_w(r_{t+\theta})w(t + \theta)]d\theta.$$

Substituting the above expression into (10.1), we get

$$\begin{cases} \dot{x}(t) = [A(r_t) + A_d(r_t)]x(t) - \int_{-\tau}^{0} A_d(r_t)\Big[A(r_{t+\theta})x(t + \theta) \\ \qquad + A_d(r_{t+\theta})x(t + \theta - \tau) + B_w(r_{t+\theta})w(t + \theta)\Big]d\theta \quad (10.22) \\ \qquad + B_w(r_t)w(t) \\ x(t) = \phi(t), \ r_t = r_0, \ w(t) = 0, \ t \in [-2\tau, 0]. \end{cases}$$

Suppose that $X_i > 0, i \in \mathcal{S}, Q > 0$, and $Q_1 > 0$ satisfy (10.20) and (10.21). Then, there exists a scalar $\varepsilon > 0$ satisfying

$$\Gamma(i) + \varepsilon \begin{pmatrix} I + \tau A_d^\top(i)A_d(i) & 0 \\ 0 & 0 \end{pmatrix} < 0. \quad\quad (10.23)$$

Let $P(i) = X_i^{-1}$. Combining (10.20) and (10.23), it is easy to show that

$$\Theta(r_t) + \varepsilon\left[P^2(r_t) + \tau P(r_t)A_d(r_t)A_1^\top(r_t)P(r_t)\right] < 0 \qquad (10.24)$$

holds for every $r_t \in \mathcal{S}$, where

$$\Theta(r_t) = \tau[A^\top(r_t)Q_1^{-1}A(r_t) + Q^{-1}] + [A(r_t) + A_d(r_t)]^\top P(r_t)$$

$$+ P(r_t)[A(r_t) + A_d(r_t)] + \sum_{j=1}^N \lambda_{r_t j}P(j)$$

$$+ 2\tau P(r_t)A_d(r_t)Q_1 A_d^\top(r_t)P(r_t).$$

Let us consider a Lyapunov functional candidate as follows:

$$V(\mathbf{x}_t, \mathbf{r}_t) = x(t)^\top P(r_t)x(t) + V_1(\mathbf{x}_t, \mathbf{r}_t),$$

where $\mathbf{r}_t = (r_s, t - 2\tau \le s \le t)$ and

$$V_1(\mathbf{x}_t, \mathbf{r}_t) = \int_{-\tau}^0 \int_{t+\theta}^t x^\top(s)Q_0(r_s)x(s)dsd\theta$$

$$+ \int_{-\tau}^0 \int_{t-\tau+\theta}^t x^\top(s)Q^{-1}x(s)dsd\theta$$

$$+ \frac{1}{\varepsilon}\int_{-\tau}^0 \int_{t+\theta}^t w^\top(s)B_w^\top(r_s)B_w(r_s)w(s)dsd\theta,$$

where $\Theta_0(r_t) = A^\top(r_t)Q_1^{-1}A(r_t)$

Let $\mathcal{A}$ be the weak generator of $\{(\mathbf{x}_t, \mathbf{r}_t), t \ge 0\}$. Then,

$$\mathcal{A}V(\mathbf{x}_t, \mathbf{r}_t) = \mathcal{A}V_1(\mathbf{x}_t, \mathbf{r}_t) + \mathcal{A}(x(t)^\top P(r_t)x(t))$$

$$= \tau x(t)^\top[Q_0(r_t) + Q^{-1}]x(t)$$

$$- \int_{-\tau}^0 x^\top(t+\theta)Q_0(r_{t+\theta})x(t+\theta)d\theta$$

$$- \frac{1}{\varepsilon}\int_{-\tau}^0 w^\top(t+\theta)B_w^\top(r_{t+\theta})B_w(r_{t+\theta})w(t+\theta)d\theta$$

$$- \int_{-\tau}^0 x^\top(t-\tau+\theta)Q^{-1}x(t-\tau+\theta)d\theta$$

$$+ x(t)^\top\Big[[A(r_t) + A_d(r_t)]^\top P(r_t) + P(r_t)[A(r_t) + A_d(r_t)]$$

$$+ \sum_{j=1}^N \lambda_{r_t j}P(j)\Big]x(t) + \frac{\tau}{\varepsilon}w^\top(t)B_w^\top(r_t)B_w(r_t)w(t)$$

$$+ 2x^\top(t)P(r_t)B_w(r_t)w(t) + h_1(t), \qquad (10.25)$$

where

$$h_1(t) = -2\int_{-\tau}^0 x(t)^\top P(r_t)A_d(r_t)\Big[A(r_{t+\theta})x(t+\theta)$$

$$+ B_w(r_{t+\theta})w(t+\theta) + A_d(r_{t+\theta})x(t+\theta-\tau)\Big]d\theta.$$

Notice that

$$-\int_{-\tau}^0 x^\top(t+\theta)Q_0(r_{t+\theta})x(t+\theta)d\theta$$

$$-2\int_{-\tau}^0 x(t)^\top P(r_t)A_d(r_t)A(r_{t+\theta})x(t+\theta)d\theta$$

$$\begin{aligned}
=-\int_{-\tau}^0 & [Q_1 A_d^\top(r_t)P(r_t)x(t) + A(r_{t+\theta})x(t+\theta)]Q_1^{-1} \\
& \times [Q_1 A_d^\top(r_t)P(r_t)x(t) + A(r_{t+\theta})x(t+\theta)]d\theta \\
& + \tau x^\top(t)P(r_t)A_d(r_t)Q_1 A_d^\top(r_t)P(r_t)x(t)
\end{aligned} \tag{10.26}$$

$$\begin{aligned}
-2\int_{-\tau}^0 & x(t)^\top P(r_t)A_d(r_t)B_w(r_{t+\theta})w(t+\theta)d\theta \\
& \leq \ \varepsilon\tau x^\top(t)P(r_t)A_d(r_t)A_d^\top(r_t)P(r_t)x(t) \\
& + \frac{1}{\varepsilon}\int_{-\tau}^0 w^\top(t+\theta)B_w^\top(r_{t+\theta})B_w(r_{t+\theta})w(t+\theta)d\theta
\end{aligned} \tag{10.27}$$

$$\begin{aligned}
-2\int_{-\tau}^0 & x^\top(t)P(r_t)A_d(r_t)A_d(r_{t+\theta})x(t+\theta-\tau)d\theta \\
& \leq \ \tau x^\top(t)P(r_t)A_d(r_t)Q_1 A_d^\top(r_t)P(r_t)x(t) \\
& + \int_{-\tau}^0 x^\top(t+\theta-\tau)A_d^\top(r_{t+\theta})Q_1^{-1}A_d(r_{t+\theta})x(t+\theta-\tau)d\theta \\
& \leq \ \tau x^\top(t)P(r_t)A_d(r_t)Q_1 A_d^\top(r_t)P(r_t)x(t) \\
& + \int_{-\tau}^0 x^\top(t+\theta-\tau)Q^{-1}x(t+\theta-\tau)d\theta.
\end{aligned} \tag{10.28}$$

The last inequality follows from inequality

$$A_d^\top(r_{t+\theta})Q_1^{-1}A_d(r_{t+\theta}) \leq Q^{-1},$$

which is implied by (10.21). Moreover, using Lemma A.1 yields

$$\begin{aligned}
2x^\top(t)P(r_t)B_w(r_t)w(t) & \leq \varepsilon x^\top(t)P(r_t)P(r_t)x(t) \\
& + \frac{1}{\varepsilon}w^\top(t)B_w^\top(r_t)B_w(r_t)w(t).
\end{aligned} \tag{10.29}$$

Combining now (10.25)-(10.29) yields

$$\begin{aligned}
\mathcal{A}V(\mathbf{x}_t,\mathbf{r}_t) \leq \ & x^\top(t)\Theta(r_t)x(t) + \frac{\tau+1}{\varepsilon}w^\top(t)B_w^\top(r_t)B_w(r_t)w(t) \\
& + \varepsilon\tau x^\top(t)P(r_t)A_d(r_t)A_d^\top(r_t)P(r_t)x(t) \\
& + \varepsilon x^\top(t)P(r_t)P(r_t)x(t)
\end{aligned}$$

$$= x^\top(t)\Xi(r_t)x(t) + \frac{\tau+1}{\varepsilon}w^\top(t)B_w^\top(r_t)B_w(r_t)w(t),$$

where

$$\Xi(r_t) = \Theta(r_t) + \varepsilon\tau P(r_t)A_d(r_t)A_d^\top(r_t)P(r_t) + \varepsilon P(r_t)P(r_t) < 0.$$

In view of (10.24), applying Dynkin's formula, we have

$$\mathbb{E}[V(\mathbf{x}_t, r_t)] - \mathbb{E}[V(\mathbf{x}_0, r_0)] = \mathbb{E}\left[\int_0^T AV(\mathbf{x}_s, r_s)ds\right]$$

$$\leq \mathbb{E}\left[\int_0^t x^\top(s)\Xi(r_s)x(s)ds\right]$$

$$+ \int_0^t \frac{\tau+1}{\varepsilon}w^\top(s)B_w^\top(r_s)B_w(r_s)w(s)ds,$$

which yields

$$\mathbb{E}[V(\mathbf{x}_t, r_t)] + \mathbb{E}\left[\int_0^t \lambda_{\min}(-\Xi(r_s))x^\top(s)x(s)ds\right]$$

$$\leq \mathbb{E}[V(\mathbf{x}_0, r_0)] + \frac{\tau+1}{\varepsilon}\int_0^t w^\top(s)B_w^\top(r_s)B_w(r_s)w(s)ds$$

$$\leq \mathbb{E}[V(\mathbf{x}_0, r_0)] + \frac{\tau+1}{\varepsilon}\max\{\lambda_{\max}(B_w^\top(i)B_w(i))\}\|w\|_2^2. \quad (10.30)$$

Noting that $\mathbb{E}[V(\mathbf{x}_t, r_t)] \geq 0$ and using (10.30), we have

$$\min_{i\in\mathcal{S}}[\lambda_{\min}(-\Xi(i))]\int_0^\infty \mathbb{E}[x^\top(t)x(t)]dt$$

$$\leq \mathbb{E}[v(\mathbf{x}_0, r_0)] + \frac{\tau+1}{\varepsilon}\max\{\lambda_{\max}(B_w^\top(i)B_w(i))\}\|w\|_2^2,$$

which completes the proof of Proposition 10.2. $\qquad\square$

With Proposition 10.2, we get the following theorem.

Theorem 10.5 *If there exist symmetric, positive-definite matrices* $X = (X_1, \cdots, X_N) > 0$, $Q > 0$, $Q_1 > 0$ *such that*

$$\begin{pmatrix} J_1(i) & B_w(i) & X_iC^\top(i) & J_{14}(i) & S_i(X) \\ B_w^\top(i) & -\gamma^2 I & C_w^\top(i) & 0 & 0 \\ C(i)X_i & C_w(i) & -I & [\ B_w^\top(i)\quad 0\ ] & 0 \\ J_{14}^\top(i) & 0 & \begin{bmatrix} B_w(i) \\ 0 \end{bmatrix} & -\frac{1}{\tau}Q & 0 \\ S_i^\top(X) & 0 & 0 & 0 & -\mathcal{X}_i \end{pmatrix} < 0 \quad (10.31)$$

$$\begin{pmatrix} Q & QA_d^\top(i) \\ A_d(i)Q & Q_1 \end{pmatrix} > 0 \quad (10.32)$$

hold for every $i \in \mathcal{S}$, where

$$J_{14}(i) = \begin{pmatrix} \mathbf{0} & X_i A^\top(i) & X_i & A_d(i)Q_1 \end{pmatrix}$$

$$\mathcal{Q} = \mathrm{diag}\{Q_1, Q_1, Q, \frac{1}{3}Q_1\},$$

and $S_i(X)$ and $\mathcal{X}_i$ are defined as in (8.11)and (8.12), respectively, then system (10.1) with $u(t) \equiv 0$ is SS and

$$\|z\|_2 \le \gamma \left[\|w\|_2^2 + \lambda \left(\int_{-\tau}^0 \int_\theta^0 \phi^\top(s)\phi(s)dsd\theta \right. \right.$$

$$\left. \left. + \int_{-\tau}^0 \int_{\theta-\tau}^0 \phi^\top(s)\phi(s)dsd\theta \right)^{1/2} \right], \qquad (10.33)$$

where $\lambda = \max\{\max_i[\lambda_{\max}(A^\top(i)Q_1^{-1}A(i))], \lambda_{\max}(Q^{-1})\}$.

Proof: Suppose that X_i, Q, Q_1 satisfy (10.31) and (10.32). In view of the Schur complement, it follows from (10.31) that the following hold for all $i \in \mathcal{S}$:

$$\begin{pmatrix} \tilde{J}_1(i) & B_w(i) & X_i C^\top(i) & \mathbf{0} \\ B_w^\top(i) & -\gamma^2 I & C_w^\top(i) & B_w^\top(i) \\ C(i)X_i & C_w(i) & -I & \mathbf{0} \\ \mathbf{0} & B_w(i) & \mathbf{0} & -\frac{1}{\tau}Q_1 \end{pmatrix} < 0, \qquad (10.34)$$

where

$$\tilde{J}_1(i) = J_1(i) + S_i^\top(X)\mathcal{X}_i^{-1}S_i(X) + \tau X_i A^\top(i)Q_1^{-1}A(i)X_i$$
$$+ \tau X_i Q^{-1} X_i + 3\tau A_d(i)Q_1 A_d^\top(i).$$

From (10.34), we have

$$J_1(i) + S_i^\top(X)\mathcal{X}_i^{-1}S_i(X) + \tau X_i A^\top(i)Q_1^{-1}A(i)X_i$$
$$+ \tau X_i Q^{-1} X_i + 2\tau A_d(i)Q_1 A_d^\top(i) < 0.$$

It implies that X_i, Q, Q_1 satisfy LMI (10.20). Therefore, from Proposition 10.2, we conclude that system (10.1) is stochastically stable.

Next, we proceed to prove (10.33). Let X_i, Q, Q_1 be a set of solutions to (10.20), (10.21), and consider the Lyapunov functional candidate as follows:

$$V_h(\mathbf{x}_t, \mathbf{r}_t) = x^\top(t)P(r_t)x(t) + V_1(\mathbf{x}_t, \mathbf{r}_t),$$

where

$$V_1(\mathbf{x}_t, \mathbf{r}_t) = \int_{-\tau}^0 \int_{t+\theta}^t x^\top(s)Q_0(r_s)x(s)dsd\theta$$

$$+ \int_{-\tau}^0 \int_{t-\tau+\theta}^t x^\top(s)Q^{-1}x(s)dsd\theta$$

$$+ \int_{-\tau}^0 \int_{t+\theta}^t w^\top(s)B_w^\top(r_s)Q_1^{-1}B_w(r_s)w(s)dsd\theta.$$

Then, noting that $w(t) = 0, t \in [-2\tau, 0]$ we have

$$\mathbb{E}[V_1(\mathbf{x}_0, \mathbf{r}_0)] = \int_{-\tau}^{0} \int_{\theta}^{0} \phi^\top(s) Q_0(r_0) \phi(s) \, ds \, d\theta$$
$$+ \int_{-\tau}^{0} \int_{\theta-\tau}^{0} \phi^\top(s) Q^{-1} \phi(s) \, ds \, d\theta$$
$$\leq \lambda \left[\int_{-\tau}^{0} \int_{\theta}^{0} \phi^\top(s) \phi(s) \, ds \, d\theta \right.$$
$$\left. + \int_{-\tau}^{0} \int_{\theta-\tau}^{0} \phi^\top(s) \phi(s) \, ds \, d\theta \right]. \tag{10.35}$$

Moreover,

$$\mathcal{A}V_h(\mathbf{x}_t, \mathbf{r}_t) \leq x^\top(t) \Theta(r_t) x(t)$$
$$+ \tau w^\top(t) B_w^\top(r_t) Q_1^{-1} B_w(r_t) w(t)$$
$$- \int_{-\tau}^{0} w^\top(t+\theta) B_w^\top(r_{t+\theta}) Q_1^{-1} B_w(r_{t+\theta}) w(t+\theta) \, d\theta$$
$$- 2 \int_{-\tau}^{0} x^\top(t) P(r_t) A_d(r_t) B_w(r_{t+\theta}) w(t+\theta) \, d\theta$$
$$+ 2 x^\top(t) P(r_t) B_w(r_t) w(t). \tag{10.36}$$

Using Lemma A.1 yields

$$-2 \int_{-\tau}^{0} x^\top(t) P(r_t) A_d(r_t) B_w(r_{t+\theta}) w(t+\theta) \, d\theta$$
$$\leq \tau x^\top(t) P(r_t) A_d(r_t) Q_1 A_d^\top(r_t) P(r_t) x(t)$$
$$+ \int_{-\tau}^{0} w^\top(t+\theta) B_w^\top(r_{t+\theta}) Q_1^{-1} B_w(r_{t+\theta}) w(t+\theta) \, d\theta.$$

Combining this with (10.36) leads to

$$\mathcal{A}V_h(\mathbf{x}_t, \mathbf{r}_t) \leq x^\top(t) \tilde{\Theta}(r_t) x(t) + \tau w^\top(t) B_w^\top(r_t) Q_1^{-1} B_w(r_t) w(t)$$
$$+ 2 x^\top(t) P(r_t) B_w(r_t) w(t), \tag{10.37}$$

where

$$\tilde{\Theta}(r_t) = \tau [A^\top(r_t) Q_1^{-1} A(r_t) + Q^{-1}] + [A(r_t) + A_d(r_t)]^\top P(r_t)$$
$$+ P(r_t)[A(r_t) + A_d(r_t)] + \sum_{j=1}^{N} \lambda_{r_t j} P(j)$$
$$+ 3\tau P(r_t) A_d(r_t) Q_1 A_d^\top(r_t) P(r_t).$$

Let us define performance $J_T, T \geq 0$ as follows:

$$J_T = \mathbb{E} \left[\int_{0}^{T} \left(z^\top(t) z(t) - \gamma^2 w^\top(t) w(t) \right) dt \right], \quad \forall T > 0.$$

Using Dynkin's formula, we obtain

$$\mathbb{E}\left[\int_0^T \mathcal{A}V_h(\mathbf{x}_t, r_t)dt | r_0, \phi(\cdot)\right] = \mathbb{E}[V_h(\mathbf{x}_T, r_T)] - \mathbb{E}[V_h(\mathbf{x}_0, r_0)].$$

In view of (10.35), for any $T > 0$, J_T can be rewritten as

$$J_T = \mathbb{E}\left[\int_0^T z^\top(t)z(t) - \gamma^2 w^\top(t)w(t) + \mathcal{A}V_h(\mathbf{x}_t, r_t)dt\right]$$
$$- \mathbb{E}\left[\int_0^T \mathcal{A}V_h(\mathbf{x}_t, \mathbf{r}_t)dt\right]$$
$$= \mathbb{E}\left[\int_0^T \left(\ x^\top(t)\ \ w^\top(t)\ \right)\Xi(r_t)\left(\begin{array}{c} x(t) \\ w(t) \end{array}\right)dt\right]$$
$$+ \mathbb{E}[V_h(\mathbf{x}_0, \mathbf{r}_0)] - \mathbb{E}[V_h(\mathbf{x}_T, \mathbf{r}_T)]$$
$$\leq \mathbb{E}\left[\int_0^T \left(\ x^\top(t)\ \ w^\top(t)\ \right)\Xi(r_t)\left(\begin{array}{c} x(t) \\ w(t) \end{array}\right)dt\right]$$
$$+ \mathbb{E}[V_h(\mathbf{x}_0, \mathbf{r}_0)], \tag{10.38}$$

where

$$\Xi(r_t) = \left(\begin{array}{cc} \tilde{\Theta}(r_t) + C^\top(r_t)C(r_t) & C^\top(r_t)C_w(r_t) + P(r_t)B_w(r_t) \\ C_w^\top(r_t)C(r_t) + B_w^\top(r_t)P(r_t) & \left[\begin{array}{c} -\gamma^2 I + C_w^\top(r_t)C_w(r_t) \\ +\tau B_w^\top(r_t)Q_1^{-1}B_w(r_t) \end{array}\right] \end{array}\right).$$

Using now the Schur complement one can check that (10.31) implies the inequality

$$\Xi(r_t) < 0.$$

Combining this with (10.38) and (10.35) yields

$$J_\infty \leq \mathbb{E}[V_h(\mathbf{x}_0, \mathbf{r}_0)]$$
$$\leq \lambda\left[\int_{-\tau}^0 \int_\theta^0 \phi^\top(s)\phi(s)dsd\theta + \int_{-\tau}^0 \int_{\theta-\tau}^0 \phi^\top(s)\phi(s)dsd\theta\right]$$

i.e.,

$$\mathbb{E}\left[\int_0^\infty z^\top(t)z(t) - \gamma^2 w^\top(t)w(t)dt\right]$$
$$\leq \lambda\left[\int_{-\tau}^0 \int_\theta^0 \phi^\top(s)\phi(s)dsd\theta + \int_{-\tau}^0 \int_{\theta-\tau}^0 \phi^\top(s)\phi(s)dsd\theta\right].$$

This ends the proof of Theorem 10.5. $\qquad\square$

For a given system, the optimal disturbance attenuation level can be obtained by solving an optimization problem. The following corollary summaries such results.

Corollary 10.1 *Let μ_0 be the solutions of the linear optimization problem*

$$\min_{\substack{\mu > 0, X = (X_1, \cdots, X_N) > 0, \\ Q > 0, Q_1 > 0,}} \mu$$

s.t. (10.32) and (10.31)
with ρ^2 replaced by μ,

then system (10.1) with $u(t) \equiv 0$ is stochastically stable with optimal disturbance attenuation level $\sqrt{\mu_0}$.

10.2.2 Synthesis of Stabilizing Controller

Now we proceed with the design of a controller with structure defined by (10.4) that stabilizes system (10.1) using the delay-dependent stability developed in the previous subsection. Substituting (10.4) into (10.1) yields

$$\begin{cases} \dot{x}(t) = [A(r_t) + B(r_t)K(r_t)]x(t) + A_d(r_t)x(t-\tau) + B_w(r_t)w(t) \\ x(t) = \phi(t), \ t \in [-\tau, 0) \\ z(t) = [C(r_t) + C_d(r_t)K(r_t)]x(t) + C_w(r_t)w(t). \end{cases} \quad (10.39)$$

Replacing $A(i)$ and $C(i)$ with $A(i) + B(i)K(i)$ and $C(i) + C_u(i)K(i)$ in (10.31) and letting $Y_i = K(i)X_i$, we get the following theorem from Theorem 10.5.

Theorem 10.6 *Let $\gamma > 0$ be a given constant. If there exist a set of matrices $X = (X_1, \cdots, X_N) > 0, Y = (Y_1, \cdots, Y_N)$, such that*

$$\begin{pmatrix} \tilde{J}_{11}(i) & B_w(i) & \tilde{J}_{13}(i) & \tilde{J}_{14}(i) & S_i(X) \\ B_w^\top(i) & -\gamma^2 I & C_w^\top(i) & [\ B_w^\top(i) \ \ 0\ ] & 0 \\ \tilde{J}_{13}^\top(i) & C_w(i) & -I & 0 & 0 \\ \tilde{J}_{14}^\top(i) & \begin{bmatrix} B_w(i) \\ 0 \end{bmatrix} & 0 & -\frac{1}{\tau}\mathcal{Q} & 0 \\ S_i^\top(X) & 0 & 0 & 0 & -\mathcal{X}_i \end{pmatrix} < 0 \quad (10.40)$$

$$\begin{pmatrix} Q & QA_d^\top(i) \\ A_d(i)Q & Q_1 \end{pmatrix} > 0 \quad (10.41)$$

hold for all $i \in \mathcal{S}$, where

$$\begin{aligned} \tilde{J}_{11}(i) &= [A(i) + A_d(i)]X_i + B(i)Y_i + X_i[A(i) + A_d(i)]^\top \\ &\quad + Y_i^\top B^\top(i) + \lambda_{ii}X_i, \\ \tilde{J}_{13}(i) &= X_i C^\top(i) + Y_i^\top C_d^\top(i), \\ \tilde{J}_{14}(i) &= (\ 0 \quad X_i A^\top(i) + Y_i^\top B^\top(i) \quad X_i \quad A_d(i)Q_1\ ), \\ \mathcal{Q} &= \text{diag}\{Q_1, Q_1, Q, \frac{1}{3}Q_1\}, \end{aligned}$$

then controller (10.4) with $K_i = Y_i X_i^{-1}$ stochastically stabilizes system (10.1) and the closed-loop system verifies noise attenuation level γ.

In the above theorem τ is assumed to be known and constant. When τ is an unknown constant, its upper bound can be obtained by solving a GEVP. Let $v = 1/\tau, \gamma^2 = \delta$. Then, (10.40) is equivalent to

$$
\begin{pmatrix}
\tilde{J}_{11}(i) & B_w(i) & \tilde{J}_{13}(i) & \tilde{J}_{14}(i) & S_i(X) \\
B_w^\top(i) & -\delta I & C_w^\top(i) & \mathbf{0} & \mathbf{0} \\
\tilde{J}_{13}^\top(i) & C_w(i) & -I & [\ B_w^\top(i)\ \ \mathbf{0}\ ] & \mathbf{0} \\
\tilde{J}_{14}^\top(i) & \mathbf{0} & \begin{bmatrix} B_w(i) \\ \mathbf{0} \end{bmatrix} & -\tilde{\Gamma} & \mathbf{0} \\
S_i^\top(X) & \mathbf{0} & \mathbf{0} & \mathbf{0} & -\mathcal{X}_i
\end{pmatrix} < 0 \quad (10.42)
$$

$$
\begin{pmatrix} \Gamma & \mathbf{0} \\ \mathbf{0} & \Gamma_1 \end{pmatrix} < v \begin{pmatrix} Q & \mathbf{0} \\ \mathbf{0} & Q_1 \end{pmatrix}, \tag{10.43}
$$

where $\tilde{\Gamma} = \text{diag}\{\Gamma_1, \Gamma_1, \Gamma, \frac{1}{3}\Gamma_1\}$. The optimization can be formulated as follows:

$$
\mathcal{P}_3: \quad \underset{\substack{v > 0, (X_1, \ldots, X_N) > 0, \\ (Y_1, \ldots, Y_N), \\ Q > 0, Q_1 > 0, \Gamma > 0, \Gamma_1 > 0}}{\min} \quad v
$$

$$
\text{s.t. } (10.41), (10.42), (10.43)
$$

From above discussion, we get the following theorem.

Theorem 10.7 *Let v_0, δ_0 be the solutions of $\mathcal{P}_3$, then controller (10.4) with $K(i) = Y_i X_i^{-1}, i \in \mathcal{S}$ stabilizes system (10.1) in the stochastic sense for all $\tau \in [0, \frac{1}{v_0}]$, and the closed-loop system attains the optimal noise attenuation level $\sqrt{\delta_0}$.*

Example 10.3 *Let us consider an MJLS described by dynamics (10.1) with two modes $\mathcal{S} = \{1, 2\}$ and suppose that the Markov process that describes the mode switching has the infinitesimal generator matrix $\Lambda = \begin{pmatrix} -2 & 2 \\ 3 & -3 \end{pmatrix}$.*
Suppose that the system has the following parameters:

$$
A(1) = \begin{pmatrix} 2 & 0 \\ 0 & -2 \end{pmatrix} \qquad A(2) = \begin{pmatrix} -10 & 1 \\ 0 & 1 \end{pmatrix}
$$

$$
A_d(1) = \begin{pmatrix} 0 & 0 \\ 0.1 & -0.2 \end{pmatrix} \qquad A_d(2) = \begin{pmatrix} 0.1 & 0 \\ 0.2 & -0.2 \end{pmatrix}
$$

$$
C(1) = \begin{pmatrix} 0.1 & 0 \\ 0.1 & 0 \end{pmatrix} \qquad C(2) = \begin{pmatrix} 0.1 & 0 \\ 0.1 & 0.1 \end{pmatrix}
$$

$$
B(1) = \begin{pmatrix} 0.1 & 0 \\ 0 & -0.1 \end{pmatrix} \qquad B(2) = \begin{pmatrix} -0.1 & 0 \\ 0.1 & 0 \end{pmatrix}
$$

$$B_w(1) = \begin{pmatrix} 0.1 & 0 \\ 0 & -0.1 \end{pmatrix} \quad B_w(2) = \begin{pmatrix} -0.1 & 0.1 \\ 0.4 & 0 \end{pmatrix}$$

$$C_w(1) = \begin{pmatrix} 0.1 & 0 \\ 0 & -0.3 \end{pmatrix} \quad C_w(2) = \begin{pmatrix} 0.2 & 0 \\ 0 & 0.5 \end{pmatrix}$$

$$C_u(1) = \begin{pmatrix} 0.1 & 0 \\ 0.1 & -1 \end{pmatrix} \quad C_u(2) = \begin{pmatrix} 0 & 0.2 \\ 0 & 0.1 \end{pmatrix}$$

$$C_d(1) = \begin{pmatrix} 0.1 & 0 \\ 0 & 0.1 \end{pmatrix} \quad C_d(2) = \begin{pmatrix} 0.1 & 0 \\ 0 & 0.1 \end{pmatrix}$$

With these data, solving LMIs (10.40) and (10.41) yields the following solutions

$$X_1 = \begin{pmatrix} 22.0156 & 14.5834 \\ 14.5834 & 10.5306 \end{pmatrix} \quad X_2 = \begin{pmatrix} 22.5991 & 14.9805 \\ 14.9805 & 10.7201 \end{pmatrix}$$

$$Y_1 = \begin{pmatrix} -1172.6 & -743.4 \\ 157.2 & 483.0 \end{pmatrix} \quad Y_2 = \begin{pmatrix} -909.0255 & -833.8469 \\ -37.6279 & -25.6481 \end{pmatrix}$$

$$Q = \begin{pmatrix} 306.6142 & 34.6194 \\ 34.6194 & 282.4461 \end{pmatrix} \quad Q_1 = \begin{pmatrix} 348.5520 & 242.7110 \\ 242.7110 & 284.4395 \end{pmatrix}$$

$$\delta = 477.6905.$$

According to Theorem 10.6, the system under study is stochastically stable under controller (10.4) with gains

$$K(1) = \begin{pmatrix} -78.6491 & 38.3232 \\ -281.1867 & 435.2658 \end{pmatrix}$$

$$K(2) = \begin{pmatrix} 153.8818 & -292.8215 \\ -1.0731 & -0.8930 \end{pmatrix}$$

and the closed-loop system satisfies the disturbance attenuation level $\gamma = 21.8561$. The upper bound for the time delay is $\tau = 2.5$.

The results presented so far are based on the assumption of complete access to the state vector and to the mode system. When this assumption is not satisfied, output feedback control is an alternative to control the system in order to get the desired performances. This will be the subject of the next section.

10.3 Output Feedback $\mathcal{H}_\infty$ Control Problem

In the previous section under the assumption of the complete access to the state and the mode system, a method to design a state feedback memoryless controller that stochastically stabilizes system (10.1) and minimizes the disturbance attenuation level of the closed-loop system was provided. When the system state $x(t)$ is not available for feedback, we can recourse to the dynamical output feedback control. This section addresses the stabilization problem under a dynamical output feedback controller.

The object of this section is to design a dynamical output feedback controller of the form

$$\begin{cases} \dot{\xi}(t) = K_A(r_t)\xi(t) + K_B(r_t)y(t) \\ u(t) = K_C(r_t)\xi(t), \ t \geq 0, \end{cases} \tag{10.44}$$

that stabilizes system (10.1) in the SS sense and such that the closed-loop system verifies a given noise attenuation level.

Substituting (10.44) into (10.1) yields the dynamics of the closed-loop system as follows:

$$\begin{cases} \dot{\eta}(t) = \bar{A}(r_t)\eta(t) + \bar{A}_d(r_t)x(t-\tau) + \bar{B}_w(r_t)w(t) \\ z(t) = \bar{C}(r_t)\eta(t) + C_d(r_t)x(t-\tau) + C_w(r_t)w(t) \end{cases} \tag{10.45}$$

where $\eta(t) = \begin{pmatrix} x(t) \\ \xi(t) \end{pmatrix}$,

$$\bar{A}(r_t) = \begin{pmatrix} A(r_t) & B(r_t)K_C(r_t) \\ K_B(r_t)G(r_t) & K_A(r_t) + K_B(r_t)G_u(r_t)K_C(r_t) \end{pmatrix}$$

$$\bar{A}_d(r_t) = \begin{pmatrix} A_d(r_t) \\ K_B(r_t)G_d(r_t) \end{pmatrix} \quad \bar{B}_w(r_t) = \begin{pmatrix} B_w(r_t) \\ K_B(r_t)G_w(r_t) \end{pmatrix}$$

$$\bar{C}(r_t) = (C(r_t)\ C_u(r_t)K_C(r_t)).$$

By the same argument as in the proof of Proposition 10.1, we have the following lemma.

Lemma 10.1 *If there exist symmetric, positive-definite matrices $Q > 0$, $(P(1), \cdots, P(N)) > 0$ with $P(i) \in \mathbb{R}^{2n \times 2n}$, $Q \in \mathbb{R}^{n \times n}$, such that*

$$\begin{pmatrix} \bar{J}(i) & P(i)\bar{A}_d(i) & P(i)B_w(i) & \bar{C}^\top(i) & I_0 \\ \bar{A}_d^\top(i)P(i) & -Q & 0 & C_d^\top(i) & 0 \\ B_w^\top(i)P(i) & 0 & -\gamma^2 I & C_w^\top(i) & 0 \\ \bar{C}(i) & C_d(i) & C_w(i) & -I & 0 \\ I_0^\top & 0 & 0 & 0 & -Q^{-1} \end{pmatrix} < 0 \tag{10.46}$$

holds for every $i \in \mathcal{S}$, where $\bar{J}(i) = \bar{A}^\top(i)P(i) + P(i)\bar{A}(i) + \sum_{j=1}^N \lambda_{ij}P(j)$ with $I_0 = \begin{pmatrix} I \\ 0 \end{pmatrix}$, then system (10.45) is stochastically stable with γ-disturbance attenuation level.

With Lemma 10.1, we now come to design gains $K_A(i), K_B(i), K_C(i), i \in \mathcal{S}$ using LMI framework. For this purpose, let us partition $P(i)$ as in (8.74). Without loss of generality, we assume that P_{2i} are invertible.

Lemma 10.2 *If there exist symmetric, positive-definite matrices $P = (P(1), \cdots, P(N)) > 0$, matrices $K_A(i), K_B(i), K_C(i), i \in \mathcal{S}$ satisfying*

(10.46) for some given symmetric, positive-definite matrix $Q > 0$, which means that the closed-loop system under controller (10.44) is stochastically stable with γ-disturbance attenuation, then the LMIs

$$
\begin{pmatrix}
\Gamma_{11}(i) & \Gamma_{12}(i) & \Gamma_{13}(i) & \Gamma_{14}(i) & \Gamma_{15}(i) & \tilde{S}_i(Y) \\
\Gamma_{12}^\top(i) & -Q & 0 & C_d^\top(i) & 0 & 0 \\
\Gamma_{13}^\top(i) & 0 & -\gamma^2 I & C_w^\top(i) & 0 & 0 \\
\Gamma_{14}^\top(i) & C_d(i) & C_w(i) & -I & 0 & 0 \\
\Gamma_{15}^\top(i) & 0 & 0 & 0 & -Q^{-1} & 0 \\
\tilde{S}_i^\top(Y) & 0 & 0 & 0 & 0 & -\mathcal{Y}_i
\end{pmatrix} < 0 \quad (10.47)
$$

$$
\begin{pmatrix} Y_i & I \\ I & X_i \end{pmatrix} > 0 \qquad (10.48)
$$

have a set of feasible solutions $X_i > 0, Y_i > 0, F_i, H_i, M_i, i \in \mathcal{S}$ where

$$
\Gamma_{11} = \begin{pmatrix}
\begin{bmatrix} A(i)Y_i + Y_i A^\top(i) \\ +B(i)F(i) + \lambda_{ii}Y_i \\ +F_i^\top B^\top(i) \end{bmatrix} & M_i^\top + A(i) \\
M_i + A^\top(i) & \begin{bmatrix} A^\top(i)X_i + X_i A(i) \\ + H_i G(i) + G^\top(i)H_i^\top \\ + \sum_{j=1}^N \lambda_{ij} X_j \end{bmatrix}
\end{pmatrix}
$$

$$
\Gamma_{12} = \begin{pmatrix} A_d(i) \\ X_i A_d(i) + H_i G_d(i) \end{pmatrix}
$$

$$
\Gamma_{13} = \begin{pmatrix} B_w(i) \\ X_i B_w(i) + H_i B_w(i) \end{pmatrix}
$$

$$
\Gamma_{14} = \begin{pmatrix} Y_i C^\top(i) + F_i^\top C_u^\top(i) \\ C^\top(i) \end{pmatrix}
$$

$$
\Gamma_{15} = \begin{pmatrix} Y_i \\ I \end{pmatrix}
$$

$$
\tilde{S}_i(Y) = \begin{pmatrix} \sqrt{\lambda_{i1}}Y_i & \cdots & \sqrt{\lambda_{ii-1}}Y_i & \sqrt{\lambda_{ii+1}}Y_i & \cdots & \sqrt{\lambda_{iN}}Y_i \\ 0 & \cdots & 0 & 0 & \cdots & 0 \end{pmatrix}
$$

$$
\mathcal{Y}_i = \mathrm{diag}\{Y_1, \cdots, Y_{i-1}, Y_{i+1}, \cdots, Y_N\}.
$$

Proof: Let $P(i)$, $K_A(i)$, $K_B(i)$ and $K_C(i)$, $i \in \mathcal{S}$ be a set of feasible solutions of (10.46). Partition $P(i)$ as in (8.74) with P_{2i} invertible, and define matrices $X_i, Y_i, F_i, H_i, M_i, i \in \mathcal{S}$ as follows:

$$\begin{cases} X_i = P_{1i} \\ Y_i = \left[P_{1i} - P_{2i}P_{3i}^{-1}P_{2i}^\top \right]^{-1} \\ F_i = -K_C(i)P_{3i}^{-1}P_{2i}^\top Y_i \\ H_i = P_{2i}K_B(i) \\ M_i = X_i A(i)Y_i + X_i B(i)F_i + H_i G_i Y_i \\ \quad - P_{2i}[K_A(i) + K_B(i)G_u(i)K_C(i)]P_{3i}^{-1}P_{2i}^\top Y_i \\ \quad + \sum_{j=1}^{N} \lambda_{ij}[P_{1j} - P_{2j}P_{3i}^{-1}P_{2i}^\top]Y_i \end{cases} \qquad (10.49)$$

and

$$S_i = \begin{pmatrix} Y_i & I \\ -P_{3i}^{-1}P_{2i}^\top Y_i & \mathbf{0} \end{pmatrix}.$$

Obviously, S_i are invertible. Therefore, we have

$$0 < S_i^\top P(i) S_i = \begin{pmatrix} Y_i & I \\ I & P_{1i} \end{pmatrix} = \begin{pmatrix} Y_i & I \\ I & X_i \end{pmatrix}, \qquad (10.50)$$

yielding (10.47).

Pre- and post-multiplying both sides of (10.47) by

$$\Xi_i^\top \triangleq \operatorname{diag}\{S_i^\top, I, I, I, I\}$$

and Ξ_i yields

$$\Xi_i^\top \Theta(i)\Xi_i < 0, \ \forall i \in \mathcal{S}. \qquad (10.51)$$

Direct manipulation gives

$$S_i^\top P(i)\bar{A}(i)S_i = \begin{pmatrix} A(i)Y_i + B(i)F_i & A(i) \\ \# & P_{1i}A(i) + H_i G(i) \end{pmatrix}, \qquad (10.52)$$

where

$$\# = P_{1i}A(i)Y_i + H_i G(i)Y_i + X_i B(i)F_i - P_{2i}[K_A(i) \\ \quad + K_B(i)G_u(i)K_C(i)]P_{3i}^{-1}P_{2i}^\top Y_i.$$

$$\begin{aligned} S_i^\top P(j) S_i &= \begin{pmatrix} \begin{bmatrix} Y_i[P_{1j} - P_{2i}P_{3j}^{-1}P_{2j}^\top \\ - P_{2j}P_{3i}^{-1}P_{2i}^\top \\ + P_{2i}P_{3i}^{-1}P_{3j}P_{3i}^{-1}P_{2i}^\top]Y_i \\ P_{1j}Y_i - P_{2j}P_{3i}^{-1}P_{2i}^\top Y_i \end{bmatrix} & Y_i P_{1j} - Y_i P_{2i}P_{3i}^{-1}P_{2j}^\top \\ Y_j \end{pmatrix} \\ &= \begin{pmatrix} Y_i \left[Y_j^{-1} + [P_{2i}P_{3i}^{-1}P_{3j} - P_{2j}] \\ \times P_{3j}^{-1}[P_{2i}P_{3i}^{-1}P_{3j} - P_{2j}]^\top \right] Y_i & Y_i P_{1j} - Y_i P_{2i}P_{3i}^{-1}P_{2j}^\top \\ P_{1j}Y_i - P_{2j}P_{3i}^{-1}P_{2i}^\top Y_i & Y_j \end{pmatrix}. \end{aligned}$$

$$(10.53)$$

Combining (10.52) and (10.53) yields

$$
S_i^\top J(i) S_i = \left(
\begin{array}{cc}
\begin{bmatrix} A(i)Y_i + Y_i A^\top(i) \\ + B(i)F_i + \lambda_{ii} Y_i \\ + F_i^\top B^\top(i) \end{bmatrix} & M_i^\top + A(i) \\[2em]
M_i + A^\top(i) & \begin{bmatrix} A^\top(i) P_{1i} + P_{1i} A(i) \\ + H_i G(i) + G^\top(i) H_i^\top \\ + \sum_{j=1}^N \lambda_{ij} Y_j \end{bmatrix}
\end{array}
\right)
$$

$$
+ \sum_{j=1}^N \lambda_{ij} \left(
\begin{array}{cc}
Y_i[Y_j^{-1} + [P_{2i}P_{3i}^{-1}P_{3j} - P_{2j}] & 0 \\
\times P_{3j}^{-1}[P_{2i}P_{3i}^{-1}P_{3j} - P_{2j}]^\top] & \\
0 & 0
\end{array}
\right)
$$

$$
= \Gamma_{11} + \sum_{j=1,j\neq i}^N \lambda_{ij} \left(
\begin{array}{cc}
Y_i [Y_j^{-1}] Y_i & 0 \\
0 & 0
\end{array}
\right)
$$

$$
+ \sum_{j=1,j\neq i}^N \lambda_{ij} \left(
\begin{array}{cc}
Y_i[P_{2i}P_{3i}^{-1}P_{3j} - P_{2j}]P_{3j}^{-1} & 0 \\
\times [P_{2i}P_{3i}^{-1}P_{3j} - P_{2j}]^\top Y_i & \\
0 & 0
\end{array}
\right). \qquad (10.54)
$$

Moreover, simple manipulation yields

$$
S_i^\top P(i) \bar{A}_d(i) = \left(
\begin{array}{c}
A_d(i) \\
X_i A_d(i) + P_{2i} K_B(i) G_d(i)
\end{array}
\right) = \Gamma_{12} \qquad (10.55)
$$

$$
S_i^\top P(i) \bar{B}_w(i) = \left(
\begin{array}{c}
B_w(i) \\
P_{1i} B_w(i) + P_{2i} K_B(i) B_w(i)
\end{array}
\right) = \Gamma_{13}(i) \quad (10.56)
$$

$$
S_i^\top \bar{C}^\top(i) = \left(
\begin{array}{cc}
Y_i & -Y_i P_{2i} P_{3i}^{-1} \\
I & 0
\end{array}
\right) \left(
\begin{array}{c}
C^\top(i) \\
K_C^\top(i) C_u^\top(i)
\end{array}
\right)
$$

$$
= \left(
\begin{array}{c}
Y_i C^\top(i) + F_i^\top C_u^\top(i) \\
C^\top(i)
\end{array}
\right) = \Gamma_{14}(i) \qquad (10.57)
$$

$$
S_i^\top I_0^\top = \left(
\begin{array}{c}
Y_i \\
I
\end{array}
\right) = \Gamma_{15}(i). \qquad (10.58)
$$

Noting that

$$
\sum_{j=1,j\neq i}^N \lambda_{ij}[P_{2i}P_{3i}^{-1}P_{3j} - P_{2j}]P_{3j}^{-1}[P_{2i}P_{3i}^{-1}P_{3j} - P_{2j}]^\top \geq 0,
$$

it follows from (10.50)-(10.58) that X_i, Y_i, M_i, H_i, F_i are a set of feasible solutions to (10.47)-(10.48). This completes the proof of Lemma 10.1. $\quad\square$

Let $\Gamma_{1j}(i), i \in \mathcal{S}, 1 \leq j \leq 5$ be defined in (10.47) and let us introduce some notations.

$$
\Gamma_{11}(i) = \left(
\begin{array}{cc}
\Gamma_{1111}(i) & \Gamma_{1112}(i) \\
\Gamma_{1112}^\top(i) & \Gamma_{1122}(i)
\end{array}
\right)
$$

$$\alpha_{12}(i) = \begin{pmatrix} \Gamma_{12}(i) & \Gamma_{13}(i) & \Gamma_{14}(i) & \Gamma_{15}(i) & \tilde{S}_i(Y) \end{pmatrix} \triangleq \begin{pmatrix} \alpha_{121}(i) \\ \alpha_{122}(i) \end{pmatrix}$$

$$\alpha_{22}(i) = \begin{pmatrix} -Q & \mathbf{0} & C_d^\top(i) & \mathbf{0} & \mathbf{0} \\ \mathbf{0} & -\gamma^2 I & C_w^\top(i) & \mathbf{0} & \mathbf{0} \\ C_d(i) & C_w(i) & -I & \mathbf{0} & \mathbf{0} \\ \mathbf{0} & \mathbf{0} & \mathbf{0} & -Q^{-1} & \mathbf{0} \\ \mathbf{0} & \mathbf{0} & \mathbf{0} & \mathbf{0} & -\mathcal{Y}_i \end{pmatrix},$$

that is,

$$\Gamma_{1111}(i) = A(i)Y_i + Y_i A^\top(i) + B(i)F(i) + \lambda_{ii}Y_i + F_i^\top B^\top(i)$$
$$\Gamma_{1112}(i) = M_i^\top + A(i)$$
$$\Gamma_{1122}(i) = A^\top(i)X_i + X_i A(i) + H_i G(i) + G^\top(i)H_i^\top + \sum_{j=1}^{N} \lambda_{ij} X_j$$
$$\alpha_{121}(i) = \begin{pmatrix} A_d(i) & B_w(i) & Y_i C^\top(i) + F^\top(i)C_u^\top(i) & Y_i & S_i(Y) \end{pmatrix}$$
$$\alpha_{122}(i) = \begin{pmatrix} X_i A_d(i) + H_i G_d(i) & X_i B_w(i) + H_i B_w(i) & C^\top(i) & I & \mathbf{0} \end{pmatrix}.$$

Theorem 10.8 *(i) If there exist matrices $K_A(i), K_B(i), K_C(i), P(i), i \in \mathcal{S}$ satisfying (10.46) for some given symmetric and positive-definite matrix $Q > 0$, which means that the closed-loop system under controller (10.44) is stochastically stable with γ-disturbance attenuation, then the following LMIs have a set of feasible solutions $X_i > 0, Y_i > 0, F_i, H_i, i \in \mathcal{S}$*

$$\begin{pmatrix} \Gamma_{1111}(i) & \alpha_{121}(i) \\ \Gamma_{121}^\top(i) & \alpha_{22}(i) \end{pmatrix} < 0 \tag{10.59}$$

$$\begin{pmatrix} \Gamma_{1122}(i) & \alpha_{122}(i) \\ \Gamma_{122}^\top(i) & \tilde{\alpha}_{22}(i) \end{pmatrix} < 0 \tag{10.60}$$

$$\begin{pmatrix} Y_i & I \\ I & X_i \end{pmatrix} > 0, \tag{10.61}$$

where

$$\tilde{\alpha}_{22}(i) = \begin{pmatrix} -Q & \mathbf{0} & C_d^\top(i) & \mathbf{0} \\ \mathbf{0} & -\gamma^2 I & C_w^\top(i) & \mathbf{0} \\ C_d(i) & C_w(i) & -I & \mathbf{0} \\ \mathbf{0} & \mathbf{0} & \mathbf{0} & -Q^{-1} \end{pmatrix}$$

(ii) On the other hand, if LMIs (10.59)-(10.61) have a set of feasible solutions $X_i > 0, Y_i > 0, F_i, H_i, i \in \mathcal{S}$, then controller (10.44) with gains

given by

$$
\begin{cases}
K_A(i) = [X_i - Y_i^{-1}]^{-1}\Big[X_i A(i) + X_i B(i) F_i Y_i^{-1} + H_i G_i \\
\qquad + \sum_{j=1}^N \lambda_{ij} Y_j^{-1} + [A^\top(i) - \alpha_{122}(i)\alpha_{22}^{-1}(i)\alpha_{121}^\top(i)]Y_i^{-1}\Big] \quad (10.62) \\
\qquad + [X_i - Y_i^{-1}]^{-1} H_i G_u(i) F_i Y_i^{-1} \\
K_B(i) = [Y_i^{-1} - X_i]^{-1} H_i \\
K_C(i) = F_i Y_i^{-1}
\end{cases}
$$

stabilizes system (10.1), and the closed-loop system has γ-disturbance level.

Proof: (i). Using the Schur complement, (10.47) holds if and only if the following holds

$$
\Gamma(i) - \begin{pmatrix} \alpha_{121}(i) \\ \alpha_{122}(i) \end{pmatrix} \alpha_{22}^{-1}(i) \begin{pmatrix} \alpha_{121}^\top(i) & \alpha_{122}(i) \end{pmatrix} < 0, \; i \in \mathcal{S},
$$

implying

$$
\Gamma_{1111}(i) - \alpha_{121}(i)\alpha_{22}^{-1}(i)\alpha_{121}^\top(i) < 0 \qquad (10.63)
$$
$$
\Gamma_{1122}(i) - \alpha_{122}(i)\alpha_{22}^{-1}(i)\alpha_{122}^\top(i) < 0. \qquad (10.64)
$$

Note that (10.63) is equivalent to (10.59), and (10.64) is equivalent to

$$
\begin{pmatrix} \Gamma_{1122}(i) & \alpha_{122}(i) \\ \alpha_{122}^\top(i) & \alpha_{22}(i) \end{pmatrix} < 0,
$$

which holds if and only if (10.60) holds since $-\mathcal{Y}_i$ are negative-definite. This proves (i).

Proof of (ii). Let $X_i > 0, Y_i > 0, F_i, H_i, i \in \mathcal{S}$ be a set of feasible solutions to LMIs (10.59)-(10.61). Define the gains matrices as in (10.62) and define

$$
P(i) = \begin{pmatrix} X_i & Y_i^{-1} - X_i \\ Y_i^{-1} - X_i & X_i - Y_i^{-1} \end{pmatrix}, \; i \in \mathcal{S}.
$$

Then, it is easy to check that $K_A(i), K_B(i), K_C(i)$, and $P(i)$ defined above satisfy (10.46) if and only if

$$
\begin{pmatrix} \Gamma_{1111}(i) - \alpha_{121}(i)\alpha_{22}^{-1}(i)\alpha_{121}^\top(i) & 0 \\ 0 & \Gamma_{1122}(i) - \alpha_{122}(i)\alpha_{22}^{-1}(i)\alpha_{122}^\top(i) \end{pmatrix} < 0,
$$

which is implied by (10.59) and (10.60). This completes the proof of (ii) and concludes the proof of Theorem 10.8. $\qquad \square$

To show the usefulness of the design algorithm for the output feedback controller, let us work out the following numerical example.

Example 10.4 *Consider a two-mode system described by (10.1) with the following data:*

$$
\Lambda = \begin{pmatrix} -2 & 2 \\ 3 & -3 \end{pmatrix} \quad \mathcal{S} = \{1, 2\}
$$

$$A(1) = \begin{pmatrix} 2 & 0 \\ 0.1 & -2 \end{pmatrix} \qquad A(2) = \begin{pmatrix} -10 & 1 \\ 0 & 1 \end{pmatrix}$$

$$A_d(1) = \begin{pmatrix} 0 & 0 \\ 0.1 & -0.2 \end{pmatrix} \qquad A_d(2) = \begin{pmatrix} 0.1 & 0 \\ 0.2 & 0.2 \end{pmatrix}$$

$$C(1) = \begin{pmatrix} 0.1 & 0 \\ 0.1 & 0 \end{pmatrix} \qquad C(2) = \begin{pmatrix} 0.1 & 0 \\ 0.1 & 0.1 \end{pmatrix}$$

$$B(1) = \begin{pmatrix} 0.1 & 0 \\ 0 & -0.1 \end{pmatrix} \qquad B(2) = \begin{pmatrix} -0.1 & 0 \\ 0.1 & 0 \end{pmatrix}$$

$$B_w(1) = \begin{pmatrix} 0.1 & 0 \\ 0 & -0.1 \end{pmatrix} \qquad B_w(2) = \begin{pmatrix} -0.1 & 0.1 \\ 0.4 & 0 \end{pmatrix}$$

$$C_w(1) = \begin{pmatrix} 0.1 & 0 \\ 0 & -0.3 \end{pmatrix} \qquad C_w(2) = \begin{pmatrix} 0.2 & 0 \\ 0 & 0.5 \end{pmatrix}$$

$$C_u(1) = \begin{pmatrix} 0.1 & 0 \\ 0.1 & -1 \end{pmatrix} \qquad C_u(2) = \begin{pmatrix} 0 & 0.2 \\ 0 & 0.1 \end{pmatrix}$$

$$C_d(1) = \begin{pmatrix} 0.1 & 0 \\ 0 & 0.1 \end{pmatrix} \qquad C_d(2) = \begin{pmatrix} 0.1 & 0 \\ 0 & 0.1 \end{pmatrix}$$

$$G(1) = \begin{pmatrix} 0.1 & 0 \\ 0.1 & 0 \end{pmatrix} \qquad G(2) = \begin{pmatrix} 0.1 & 0 \\ 0.1 & 0.1 \end{pmatrix}$$

$$G_w(1) = \begin{pmatrix} 0.1 & 0 \\ 0 & -0.3 \end{pmatrix} \qquad G_w(2) = \begin{pmatrix} 0.2 & 0 \\ 0 & 0.5 \end{pmatrix}$$

$$G_u(1) = \begin{pmatrix} 0.1 & 0 \\ 0.1 & -1 \end{pmatrix} \qquad G_u(2) = \begin{pmatrix} 0 & 0.2 \\ 0 & 0.1 \end{pmatrix}$$

$$G_d(1) = \begin{pmatrix} 0.1 & 0 \\ 0 & 0.1 \end{pmatrix} \qquad C_d(2) = \begin{pmatrix} 0.1 & 0 \\ 0 & 0.1 \end{pmatrix}.$$

With these data and letting $Q = I$, solving LMIs (10.59)-(10.61) gives the following set of solutions:

$$X_1 = \begin{pmatrix} 0.1833 & -0.4448 \\ -0.4448 & 13.2116 \end{pmatrix} \qquad X_2 = \begin{pmatrix} 0.2991 & -1.3192 \\ -1.3192 & 33.1910 \end{pmatrix}$$

$$Y_1 = \begin{pmatrix} 9.7414 & 0.2640 \\ 0.2640 & 8.5941 \end{pmatrix} \qquad Y_2 = \begin{pmatrix} 9.1177 & -0.8968 \\ -0.8968 & 6.5997 \end{pmatrix}$$

$$H_1 = \begin{pmatrix} -9.7548 & -10.8168 \\ -8.8589 & 26.4602 \end{pmatrix} \qquad H_2 = \begin{pmatrix} -13.7649 & -10.8814 \\ -50.6916 & -64.7808 \end{pmatrix}$$

$$F_1 = \begin{pmatrix} -925.1819 & -81.2833 \\ 81.1931 & 536.9898 \end{pmatrix} \qquad F_2 = \begin{pmatrix} -308.6855 & -461.6766 \\ 0 & 0 \end{pmatrix}$$

$$\mu = 86.3734.$$

According to Theorem 10.8, the system is stochastically stable under controller (10.44) with

$$K_A(1) = \begin{pmatrix} -3918.6 & -9611.1 \\ -106.2 & -205.6 \end{pmatrix}$$

$$K_A(2) = \begin{pmatrix} -33.7465 & 4.4424 \\ -5.6202 & -6.9038 \end{pmatrix}$$

$$K_B(1) = \begin{pmatrix} -153.0787 & -151.1113 \\ -5.8393 & -3.0759 \end{pmatrix}$$

$$K_B(2) = \begin{pmatrix} -117.9713 & -100.7039 \\ -6.2991 & -6.0282 \end{pmatrix}$$

$$K_C(1) = \begin{pmatrix} -94.7967 & -6.5455 \\ 6.6467 & 62.2796 \end{pmatrix}$$

$$K_C(2) = \begin{pmatrix} -41.2883 & -75.5647 \\ 0 & 0 \end{pmatrix}$$

and the closed-loop system satisfies the disturbance attenuation level $\gamma = \sqrt{\mu} = 9.2937$.

In practice, we are sometimes interested in estimating the state vector and then using it to compute the control law that stabilizes the system and assures the desired performances. This can be done by designing an appropriate filter. The Kalman filter is the most popular one. Recently, with the development of the $\mathcal{H}_\infty$ control we have seen the introduction of the $\mathcal{H}_\infty$-filter, which is mainly based of the $\mathcal{H}_\infty$ theory. The next section will address this problem for the class of systems with Markov jumps and time delay.

10.4 Filtering

This section deals with the $\mathcal{H}_\infty$-filtering problems for jump linear systems. For this purpose, let us consider the dynamics

$$\begin{cases} \dot{x}(t) = A(r_t)x(t) + A_d(r_t)x(t - \tau) + B_w(r_t)w(t) \\ y(t) = G(r_t)x(t) + G_w(r_t)w(t) \\ z(t) = C(r_t)x(t) + C_w(r_t)w(t) \\ x(s) = \phi(s), \ s \in [-\tau, 0] \end{cases} \tag{10.65}$$

where $\{r_t, t \geq 0\}$ is a Markov process as defined before, which specifies the mode-switching of the system; $x(t) \in \mathbb{R}^n$ is the state vector; $y(t) \in \mathbb{R}^s$ is the measurement vector; $w(\cdot) \in \mathcal{L}_2[0, \infty)$ is the noise disturbance; $z(t) \in \mathbb{R}^p$ is the signal to be estimated; τ is the time delay in the system; $\phi(\cdot)$ is a given initial function which is continuous on $[-\tau, 0)$; and $A(r_t)$, $A_d(r_t)$, $B_w(r_t)$, $G(r_t)$, $G_w(r_t)$, $C(r_t)$, $C_w(r_t)$ are matrices with appropriate dimensions.

The goal of this section is to design a linear filter of order n of the following form

$$\begin{cases} \dot{\hat{x}}(t) = K_A(r_t)\hat{x}(t) + K_B(r_t)y(t) \\ \hat{x}(0) = x_0 \stackrel{\Delta}{=} \phi(0), \ \hat{x}(s) = 0, \ s \in [-\tau, 0] \\ \hat{z}(t) = K_C(r_t)\hat{x}(t) \end{cases} \qquad (10.66)$$

which gives an estimate of the state vector $\hat{x}(t)$ at time t and can ensure that the extended system $\{(x^\top(t), \hat{x}(t)), t \geq 0\}$ is SS and the estimation error $e(t) = \hat{z}(t) - z(t)$ satisfies the condition

$$\|e\|_2 \leq \gamma \|w\|_2, \qquad (10.67)$$

where γ is a given positive scalar.

If (10.67) is satisfied, the error system of the filter (10.66) is said to verify the noise attenuation level γ.

Remark 10.3 $K_A(i), K_B(i)$ and $K_C(i)$ in (10.66) are the constant matrices that must be designed.

We proceed with the development of LMI-based conditions for the extended system with a filter of the form (10.66) to verify (10.67). To this end, let us consider the augmented state $\tilde{x}(t) = (x_t^\top, \hat{x}_t^\top)^\top$. Obviously, from (10.1) and (10.66), it follows that $\tilde{x}$ satisfies the following dynamics

$$\begin{cases} \dot{\tilde{x}}(t) = \tilde{A}(r_t)\tilde{x}(t) + \tilde{A}_d(r_t)I_0\tilde{x}(t - \tau) + \tilde{B}_w(r_t)w(t) \\ \quad = \tilde{A}(r_t)x(t) + \tilde{A}_d(r_t)I_0x(t - \tau) + \tilde{B}_w(r_t)w(t) \\ \tilde{x}(0) = (x_0^\top, x_0^\top)^\top, \end{cases} \qquad (10.68)$$

where

$$x(0) = \phi(0) \quad \tilde{A}(i) = \begin{pmatrix} A(i) & 0 \\ K_B(i)G(i) & K_A(i) \end{pmatrix}$$

$$\tilde{A}_d(i) = \begin{pmatrix} A_d(i) \\ 0 \end{pmatrix} \quad \tilde{B}_w(i) = \begin{pmatrix} B_w(i) \\ K_B(i)G_w(i) \end{pmatrix}$$

$$I_0 = \begin{pmatrix} I & 0 \end{pmatrix}.$$

Then, the filtering error $e(t) = z(t) - \hat{z}(t)$ satisfies

$$e(t) = \tilde{C}(r_t)\tilde{x}(t) + \tilde{C}_w(r_t)w(t), \qquad (10.69)$$

where

$$\tilde{C}(i) = \begin{pmatrix} C(i) & -K_C(i) \end{pmatrix} \quad \tilde{C}_w(i) = C_w(i).$$

Let $P(i) > 0, Q > 0, i \in \mathcal{S}$ be a set of matrices and define

$$\Theta_0(i) = \begin{pmatrix} J_1(i) & P(i)\tilde{A}_d(i) & P(i)\tilde{B}_w(i) & \tilde{C}^\top(i) & I_0^\top \\ \tilde{A}_d^\top(i)P(i) & -Q & 0 & 0 & 0 \\ \tilde{B}_w^\top(i)P(i) & 0 & -\gamma^2 I & \tilde{C}_w^\top(i) & 0 \\ \tilde{C}(i) & 0 & \tilde{C}_w(i) & -I & 0 \\ I_0 & 0 & 0 & 0 & -Q^{-1} \end{pmatrix},$$

$$J_1(i) = \tilde{A}^\top(i)P(i) + P(i)\tilde{A}(i) + \sum_{s=1}^{N} \lambda_{is} P(s).$$

Proposition 10.3 *Let* γ *be a given positive constant. If there exist symmetric, positive-definite matrices* $P = (P(1), \cdots, P(N)) > 0$ *such that*

$$\Theta_0(i) < 0 \tag{10.70}$$

$$(\,I \quad I\,)\, P(r_0) \begin{pmatrix} I \\ I \end{pmatrix} < \gamma^2 R \tag{10.71}$$

$$Q < \gamma^2 R_1 \tag{10.72}$$

hold for some symmetric, positive-definite matrices Q, *then the extended system (10.68) is SS and, moreover,*

$$\|\hat{z}(t) - z(t)\|_2 \leq \left[\gamma^2 \|w\|_2^2 + x_0^\top R x_0 + \int_{-\tau}^{0} \phi^\top(t) R_1 \phi(t) dt\right]^{1/2} \tag{10.73}$$

holds, where $R, R_1 \in \mathbb{R}^{n \times n}$ *are symmetric, nonnegative-definite matrices representing the weighting of the initial conditions.*

Proof: We first prove that under the condition (10.70) the extended system (10.68) is internal MSQS. Obviously, it follows from (10.70) that

$$\begin{pmatrix} J_{11}(i) & P(i)\tilde{A}_d(i) & P(i)\tilde{B}_w(i) & \tilde{C}^\top(i) \\ \tilde{A}_d^\top(i)P(i) & -Q & 0 & 0 \\ \tilde{B}_w^\top(i)P(i) & 0 & -\gamma^2 I & \tilde{C}_w^\top(i) \\ \tilde{C}(i) & 0 & \tilde{C}_w(i) & -I \end{pmatrix} < 0, \tag{10.74}$$

where $J_{11}(i) = J_1(i) + I_0^\top Q I_0$. (10.74) in turn implies

$$\begin{pmatrix} J_{11}(i) & P(i)\tilde{A}_d(i) \\ \tilde{A}_d(i)^\top P(i) & -Q \end{pmatrix} < 0.$$

(10.75) means that $\{\tilde{x}_t\}$ is internal MSQS and thus is SS.

Now, we come to prove (10.73). To this end, let us define the following $\mathcal{H}_\infty$ performance

$$J_T = \mathbb{E}\left[\int_0^T \left[e^\top(t)e(t) - \gamma^2 w^\top(t)w(t)\right]dt\right], \forall T > 0$$

and define $\tilde{\mathbf{x}}(t) = (\tilde{x}(s), t - \tau \leq s \leq t)$ and let $\hat{\mathcal{A}}$ be the generator of $\{(\tilde{\mathbf{x}}(t), r_t), t \geq 0\}$.

Consider the following Lyapunov functional candidate as follows:

$$V(\tilde{\mathbf{x}}_t, r_t) = \tilde{x}^\top(t)P(r_t)\tilde{x}(t) + \int_{-\tau}^{0} x^\top(t+\theta)Qx(t+\theta)d\theta, \forall r_t \in \mathcal{S}. \tag{10.75}$$

Then

$$J_T = \mathbb{E}\left[\int_0^T \left[e^\top(t)e(t) - \gamma^2 w^\top(t)w(t) + \hat{\mathcal{A}}V(\tilde{\mathbf{x}}_t, r_t)\right]dt\right.$$

$$- \mathbb{E}\left[\int_0^T \hat{A}V(\tilde{\mathbf{x}}_t, r_t)dt\right]. \tag{10.76}$$

Using Dynkin's formula, we have

$$\mathbb{E}\left[\int_0^T \hat{A}V(\tilde{\mathbf{x}}_t, r_t)dt\right] = \mathbb{E}[V(\tilde{\mathbf{x}}_T, r_T)] - \mathbb{E}[V(\tilde{\mathbf{x}}_0, r_0)].$$

From (10.75), we have

$$\mathbb{E}[V(\tilde{x}_0, r_0)] = \left\{\mathbb{E}[\tilde{x}^\top(0)P(r_0)\tilde{x}(0)] + \int_{-\tau}^0 \phi^\top(s)Q\phi(s)ds\right\}. \tag{10.77}$$

Note that $\tilde{x}^\top(0) = \left(x^\top(0)\ x^\top(0)\right)^\top$ and $\tilde{x}^\top(t) = \left(\phi^\top(t)\ 0\right)^\top$, $t \in [-\tau, 0)$. In view of (10.71), (10.72) and (10.77), we have

$$\mathbb{E}[V(\tilde{x}(0), r(0))] \leq \left\{\mathbb{E}\left[x^\top(0)\left(I\ \ I\right)P(r_0)\begin{pmatrix}I\\I\end{pmatrix}x(0)\right]\right.$$
$$\left. + \gamma^2 \int_{-\tau}^0 \phi^\top(s)R_1\phi(s)ds\right\}$$
$$\leq \gamma^2 \mathbb{E}[x^\top(0)Rx(0)] + \gamma^2 \int_{-\tau}^0 \phi^\top(s)R_1\phi(s)ds.$$

Simple manipulation gives

$$J_T = \mathbb{E}\left[\int_0^T \tilde{\xi}_t^\top \Theta_2(r_t)\tilde{\xi}_t dt\right] + \mathbb{E}[V(\tilde{x}(0), r_0)] - \mathbb{E}[V(\tilde{x}(T), r_T)]$$
$$\leq \mathbb{E}\left[\int_0^T \tilde{\xi}_t^\top \Theta_2(r_t)\tilde{\xi}_t dt\right] + \mathbb{E}[V(\tilde{x}(0), r_0)] \tag{10.78}$$

where

$$\begin{pmatrix} J_{11}(i) + \tilde{C}^\top(i)\tilde{C}(i) & P(i)\tilde{A}_d(i) & P(i)\tilde{B}_w(i) + \tilde{C}^\top(i)\tilde{C}_w(i) \\ \tilde{A}_d^\top(i)P(i) & -Q & 0 \\ \tilde{B}_w^\top(i)P(i) + \tilde{C}_w^\top(i)\tilde{C}(i) & 0 & -\gamma^2 I + \tilde{C}_w^\top(i)\tilde{C}_w(i) \end{pmatrix}$$
$$\stackrel{\triangle}{=} \Theta_2(i)$$

and $\tilde{\xi}_t = (\tilde{x}^\top(t), x^\top(t-\tau), w^\top(t))^\top$. Using the Schur complement, (10.74) implies in turn that

$$\Theta_2(i) < 0.$$

Combining this with (10.78) implies that the following holds for all $T > 0$:

$$J_T \leq \mathbb{E}[V(\tilde{x}(0), r_0)] \leq \gamma^2 x^\top(0)Rx(0) + \gamma^2 \int_{-\tau}^0 \phi^\top(s)R_1\phi(s)ds.$$

Therefore, we get

$$J_\infty = \mathbb{E}\left[\int_0^\infty [e^\top(t)e(t) - \gamma^2 w^\top(t)w(t)]dt\right]$$

$$\leq \gamma^2 x^\top(0)Rx(0) + \gamma^2 \int_{-\tau}^0 \phi^\top(t)R_1\phi(t)dt.$$

This gives

$$\|e\|_2^2 \leq \left[\gamma^2\|w\|_2^2 + x^\top(0)Rx(0) + \int_{-\tau}^0 \phi^\top(t)R_1\phi(t)dt\right]$$

and this completes the proof of Proposition 10.3.   □

In case of filtering problems where the effect of the initial state is ignored, i.e., initial condition $\phi(\cdot)$ is set to zero, then the following corollary follows immediately from Proposition 10.3.

Corollary 10.2 *Let the initial conditions of system (10.1) be zero, i.e., $\phi(t) \equiv 0, t \in [-\tau, 0]$. If there exists a set of matrices $P = (P(1), \cdots, P(N)) > 0$, such that*

$$\Theta_0(i) < 0, \ \forall i \in \mathcal{S}$$

holds for some given $Q > 0$, then

$$\|e\|_2 \leq \gamma\|w\|_2.$$

In Proposition 10.3, the minimal noise attenuation level γ that can be verified by the filter of the form of (10.66) can be obtained by solving the optimization problem

$$\mathcal{P}_0: \quad \min_{v > 0, P = (P(1), \cdots, P(N)) > 0,} \quad v,$$

$$s.t. \begin{cases} \Theta_v(i) < 0, \\ \begin{pmatrix} I & I \end{pmatrix} P(r_0) \begin{pmatrix} I \\ I \end{pmatrix} \leq vR, \\ Q \leq vR_1, \end{cases}$$

where $\Theta_v(i)$ is obtained from $\Theta_0(i)$ by replacing γ^2 by v. Thus, if the linear programming problem $\mathcal{P}_0$ has a solutions v, then by using Proposition 10.3, the corresponding error of the filter (10.66) is stable with noise attenuation level $\sqrt{v}$.

Proposition 10.3 provides an LMI-based sufficient condition for the filtering error system to be stable with noise attenuation level γ. However, when $K_A(i), K_B(i), K_C(i)$ are unknown, inequality (10.70) is nonlinear with respect to $K_A(i), K_B(i), K_C(i), P(i)$ and thus Proposition 10.3 does not solve the filtering design problem. To cast the filtering design problem into the LMI framework, a technique of change of variables is used.

The discussion above shows that to ensure that the filter given by (10.66) satisfies the given noise attenuation level, we need the following condition to be satisfied: $\Theta_0(i) < 0$, for which we will develop some sufficient and necessary conditions.

Lemma 10.3 *Let γ be a given positive constant. For system (10.1), if there exist symmetric, positive-definite matrices $P = (P(1), \cdots, P(N)) > 0$, matrices $K_A(i)$, $K_B(i)$, $K_C(i)$, $i \in \mathcal{S}$ such that (10.70) holds, then the LMIs*

$$\begin{pmatrix} \Gamma_{11}(i) & \Gamma_{12}(i) & \Gamma_{13}(i) & \Gamma_{14}(i) & \Gamma_{15}(i) & T_i(Y) \\ \Gamma_{12}^\top(i) & -Q & 0 & 0 & 0 & 0 \\ \Gamma_{13}^\top(i) & 0 & -\gamma^2 I & \tilde{C}_w^\top(i) & 0 & 0 \\ \Gamma_{14}^\top(i) & 0 & \tilde{C}_w(i) & -I & 0 & 0 \\ \Gamma_{15}^\top(i) & 0 & 0 & 0 & -Q^{-1} & 0 \\ T_i^\top(Y) & 0 & 0 & 0 & 0 & -\mathcal{Y}_i \end{pmatrix} < 0 \quad (10.79)$$

$$\begin{pmatrix} Y_i & I \\ I & X_i \end{pmatrix} > 0 \quad (10.80)$$

have a set of feasible solutions $X_i > 0, Y_i > 0, M_i, H_i, L_i$, for some given $Q > 0$, where

$$\Gamma_{11}(i) = \begin{pmatrix} W_{1i} & M_i^\top + A(i) \\ M_i + A^\top(i) & W_{2i} \end{pmatrix}$$

$$W_{1i} = A(i)Y_i + Y_i A^\top(i) + \lambda_{ii} Y_i$$

$$W_{2i} = A^\top(i)X_i + X_i A(i) + H_i G(i) + G^\top(i)H_i^\top + \sum_{m=1}^N \lambda_{im} X_m$$

$$\Gamma_{12}(i) = \begin{pmatrix} A_d(i) \\ X_i A_d(i) \end{pmatrix}$$

$$\Gamma_{13}(i) = \begin{pmatrix} B_w(i) \\ X_i B_w(i) + H_i G_w(i) \end{pmatrix}$$

$$\Gamma_{14}(i) = \begin{pmatrix} Y_i C^\top(i) + L_i^\top \\ C^\top(i) \end{pmatrix}$$

$$\Gamma_{15}(i) = \begin{pmatrix} Y_i \\ I \end{pmatrix}$$

$$T_i(Y) = \begin{pmatrix} \sqrt{\lambda_{i1}}Y_i & \cdots & \sqrt{\lambda_{ii-1}}Y_i & \sqrt{\lambda_{ii+1}}Y_i & \cdots & \sqrt{\lambda_{iN}}Y_i \\ 0 & \cdots & 0 & 0 & \cdots & 0 \end{pmatrix}$$

$$\mathcal{Y}_i = \text{diag}\{Y_1, \cdots, Y_{i-1}, Y_{i+1}, \cdots, Y_N\}.$$

Proof: Let $P(i)$, $K_A(i)$, $K_B(i)$, $K_C(i)$, $i \in \mathcal{S}$ be a set of feasible solutions of (10.70) and partition $P(i)$ as follows:

$$P(i) = \begin{pmatrix} P_{1i} & P_{2i} \\ P_{2i}^\top & P_{3i} \end{pmatrix}.$$

Due to the strict nature of (10.70) and using the same argument as in Lemma 8.2 of [48], we may assume that P_{2i} is invertible without loss of generality. Then, define

$$
\begin{cases}
X_i = P_{1i} \\
Y_i = \left[P_{1i} - P_{2i}P_{3i}^{-1}P_{2i}^{\top}\right]^{-1} \\
L_i = K_C(i)P_{3i}^{-1}P_{2i}^{\top}Y_i \\
H_i = P_{2i}K_B(i) \\
M_i = X_iA(i)Y_i + H_iG(i)Y_i \\
\qquad - P_{2i}K_A(i)P_{3i}^{-1}P_{2i}^{\top}Y_i + \sum_{m=1}^{N}\lambda_{im}[P_{1m} - P_{2m}P_{3i}^{-1}P_{2i}^{\top}]Y_i
\end{cases}
\tag{10.81}
$$

and

$$
S_i = \begin{pmatrix} Y_i & I \\ -P_{3i}^{-1}P_{2i}^{\top}Y_i & \mathbf{0} \end{pmatrix}.
$$

Obviously, S_i is invertible. Thus we have

$$
S_i^{\top}P(i)S_i = \begin{pmatrix} Y_i & I \\ I & P_{1i} \end{pmatrix} = \begin{pmatrix} Y_i & I \\ I & X_i \end{pmatrix} > 0
\tag{10.82}
$$

yielding (10.80).

Pre- and post-multiplying $\Theta_0(i,j)$ by $\Xi_i^{\top} \triangleq \mathrm{diag}\{S_i^{\top}, I, I, I, I\}$ and Ξ_i yield

$$
\Xi_i^{\top}\Theta_0(i)\Xi_i < 0, \ \forall i \in \mathcal{S}.
\tag{10.83}
$$

Direct manipulation gives

$$
S_i^{\top}P(i)\tilde{A}(i)S_i = \begin{pmatrix} A(i)Y_i & A(i) \\ \begin{bmatrix} P_{1i}A(i)Y_i + H_iG(i)Y_i \\ -P_{2i}K_A(i)P_{3i}^{-1}P_{2i}^{\top}Y_i \end{bmatrix} & P_{1i}A(i) + H_iG(i) \end{pmatrix}
\tag{10.84}
$$

$$
S_i^{\top}P(j)S_i = \begin{pmatrix} \begin{bmatrix} Y_i[P_{1j} - P_{2i}P_{3j}^{-1}P_{2j}^{\top} \\ -P_{2j}P_{3i}^{-1}P_{2i}^{\top} \\ +P_{2i}P_{3i}^{-1}P_{3j}P_{3i}^{-1}P_{2i}^{\top}]Y_i \\ P_{1j}Y_i - P_{2j}P_{3i}^{-1}P_{2i}^{\top}Y_i \end{bmatrix} & Y_iP_{1j} - Y_iP_{2i}P_{3i}^{-1}P_{2j}^{\top} \\[2pt] & X_j \end{pmatrix}
$$

$$
= \begin{pmatrix} \begin{bmatrix} Y_i[Y_j^{-1} + [P_{2i}P_{3i}^{-1}P_{3j} \\ -P_{2j}]P_{3j}^{-1}[P_{2i}P_{3i}^{-1}P_{3j} \\ -P_{2j}]^{\top}]Y_i \\ P_{1j}Y_i - P_{2j}P_{3i}^{-1}P_{2i}^{\top}Y_i \end{bmatrix} & Y_iP_{1j} - Y_iP_{2i}P_{3i}^{-1}P_{2j}^{\top} \\[2pt] & X_j \end{pmatrix}.
\tag{10.85}
$$

Combining (10.84) and (10.85) yields

$$
S_i^{\top}J_1(i)S_i = \begin{pmatrix} A(i)Y_i + Y_iA^{\top}(i) + \lambda_{ii}Y_i & M_i^{\top} + A(i) \\[4pt] M_i + A^{\top}(i) & \begin{bmatrix} A^{\top}(i)X_i + X_iA(i) \\ +H_iG(i) + G^{\top}(i)H_i^{\top} \\ +\sum_{m=1}^{N}\lambda_{im}X_m \end{bmatrix} \end{pmatrix}
$$

$$+ \sum_{m=1}^{N} \lambda_{im} \left(\begin{bmatrix} Y_i[Y_j^{-1} + [P_{2i}P_{3i}^{-1}P_{3j} - P_{2j}]P_{3j}^{-1} \\ [P_{2i}P_{3i}^{-1}P_{3j} - P_{2j}]^\top \end{bmatrix} & \mathbf{0} \\ \mathbf{0} & \mathbf{0} \end{bmatrix} \right)$$

$$= \Gamma_{11}(i) + \sum_{m=1,m\neq i}^{N} \lambda_{im} \begin{pmatrix} Y_i[Y_m^{-1}]Y_i & \mathbf{0} \\ \mathbf{0} & \mathbf{0} \end{pmatrix}$$

$$+ \sum_{m=1,m\neq i}^{N} \lambda_{im} \left(\begin{bmatrix} Y_i[P_{2i}P_{3i}^{-1}P_{3m} - P_{2m}]P_{3m}^{-1} \\ \times [P_{2i}P_{3i}^{-1}P_{3m} - P_{2m}]^\top Y_i \end{bmatrix} & \mathbf{0} \\ \mathbf{0} & \mathbf{0} \end{bmatrix} \right). \quad (10.86)$$

Moreover, simple manipulation yields

$$S_i^\top P(i) = \begin{pmatrix} I & \mathbf{0} \\ P_{1i} & P_{2i} \end{pmatrix}$$

$$S_i^\top P(i)A(i) = \begin{pmatrix} I & \mathbf{0} \\ P_{1i} & P_{2i} \end{pmatrix} \begin{pmatrix} A(i) \\ \mathbf{0} \end{pmatrix}$$

$$= \begin{pmatrix} A_d(i) \\ X_i A_d(i) \end{pmatrix} = \Gamma_{12}(i) \quad (10.87)$$

$$S_i^\top P(i)\tilde{B}_w(i) = \begin{pmatrix} I & \mathbf{0} \\ P_{1i} & P_{2i} \end{pmatrix} \begin{pmatrix} B_w(i) \\ K_B(i)G_w(i) \end{pmatrix}$$

$$= \begin{pmatrix} B_w(i) \\ P_{1i}B_w(i) + P_{2i}K_B(i)G_w(i) \end{pmatrix}$$

$$\begin{pmatrix} B_w(i) \\ X_i B_w(i) + H_i G_w(i) \end{pmatrix} = \Gamma_{13}(i), \quad (10.88)$$

$$S_i^\top \tilde{C}(i) = \begin{pmatrix} Y_i & -Y_i P_{2i}P_{3i}^{-1} \\ I & \mathbf{0} \end{pmatrix} \begin{pmatrix} C^\top(i) \\ -K_C^\top(i) \end{pmatrix}$$

$$= \begin{pmatrix} Y_i C^\top(i) + Y_i P_{2i}P_{3i}^{-1}K_C^\top(i) \\ C^\top(i) \end{pmatrix} = \Gamma_{14}(i), \quad (10.89)$$

$$S_i^\top I_0^\top = \begin{pmatrix} Y_i & -Y_i P_{2i}P_{3i}^{-1} \\ I & \mathbf{0} \end{pmatrix} \begin{pmatrix} I \\ \mathbf{0} \end{pmatrix} = \begin{pmatrix} Y_i \\ I \end{pmatrix} = \Gamma_{15}(i). \quad (10.90)$$

Noting that

$$\sum_{m=1,m\neq i}^{N} \lambda_{im}[P_{2i}P_{3i}^{-1}P_{3m} - P_{2m}]P_{3m}^{-1}[P_{2i}P_{3i}^{-1}P_{3m} - P_{2m}]^\top \geq 0,$$

it follows from (10.83), (10.86), (10.87), and (10.88) that (10.79) is feasible for each i and each j with respect to X_i, Y_i, M_i, H_i, L_i. This completes the proof of Lemma 10.3.   □

From the proof of Lemma 10.3, we get the following theorem.

Theorem 10.9 *There exist matrices* $K_A(i), K_B(i), K_C(i), i \in \mathcal{S}, P(i) > 0, i \in \mathcal{S}$ *satisfying (10.70) for some given matrix* $Q > 0$ *if and only if there exist a set of matrices* $X_i > 0, Y_i > 0, H_i, L_i$ *such that the following hold:*

$$
\left(
\begin{array}{cccc}
\left[\begin{array}{c} A(i)Y_i + Y_i A^\top(i) \\ +\lambda_{ii} Y_i \\ +A_d(i)Q^{-1}A_d^\top(i) \end{array}\right] & \left[\begin{array}{c} Y_i C^\top(i) C_w(i) \\ +L_i^\top C_w(i) + B_w(i) \end{array}\right] & Y_i & S_i(Y) \\
\left[\begin{array}{c} C_w^\top(i)C(i)Y_i \\ +C_w^\top(i)L_i \\ +B_w^\top(i) \end{array}\right] & -\gamma^2 I + C_w^\top(i)C_w(i) & 0 & 0 \\
Y_i & 0 & -Q^{-1} & 0 \\
S_i^\top(Y) & 0 & 0 & -\mathcal{Y}_i
\end{array}
\right) < 0
$$
(10.91)

$$
\left(
\begin{array}{ccc}
W_{2i} + Q & \left[\begin{array}{c} C^\top(i)C_w(i) + \\ X_i B_w(i) \\ +H_i G_w(i) \end{array}\right] & X_i A_d(i) \\
\left[\begin{array}{c} C_w^\top(i)C(i) + B_w^\top(i)X_i \\ +G_w^\top(i)H_i^\top \end{array}\right] & -\gamma^2 I + C_w^\top(i)C_w(i) & 0 \\
A_d^\top(i)X_i & 0 & -Q
\end{array}
\right) < 0
$$
(10.92)

$$
\left(\begin{array}{cc} Y_i & I \\ I & X_i \end{array}\right) > 0,
$$
(10.93)

where

$$
S_i(Y) = \left(\begin{array}{ccccc} \sqrt{\lambda_{i1}}Y_i, & \cdots & \sqrt{\lambda_{ii-1}}Y_i & \sqrt{\lambda_{ii+1}}Y_i & \cdots & \sqrt{\lambda_{iN}}Y_i \end{array}\right)
$$

$$
\mathcal{Y}_i = \text{diag}\{Y_1, \cdots, Y_{i-1}, Y_{i+1}, \cdots, Y_N\}.
$$

Proof: *Necessity*. In view of Lemma 10.3, to prove (i) it suffices that (10.70) implies (10.91) and (10.92). Let $X_i, Y_i, H_i, L_i, M_i, i \in \mathcal{S}$ be a set of feasible solutions to (10.79) and (10.80). Using the Schur complement, we find that (10.79) is equivalent to

$$
\begin{aligned}
Z(i) &\triangleq \left(\begin{array}{cc} Z_{11}(i) & Z_{12}(i) \\ Z_{12}^\top(i) & Z_{22}(i) \end{array}\right) \\
&= \Gamma_{11}(i) + \left(\begin{array}{cc} T_i(Y)\mathcal{Y}_i T_i^\top(Y) & 0 \\ 0 & 0 \end{array}\right) + \left(\begin{array}{c} Y_i \\ I \end{array}\right) Q \left(\begin{array}{cc} Y_i & I \end{array}\right) \\
&\quad + \Gamma_{12}(i)Q^{-1}\Gamma_{12}^\top(i) + [\Gamma_{13}(i) + \Gamma_{14}(i)C_w(i)] \\
&\quad \times [\gamma^2 I - C_w^\top(i)C_w(i)]^{-1}[\Gamma_{13}(i) + \Gamma_{14}(i)C_w(i)]^\top < 0. \quad (10.94)
\end{aligned}
$$

Direct manipulation gives

$$Z_{11}(i) = A(i)Y_i + Y_i A^\top(i) + \lambda_{ii} Y_i + Y_i \left[\sum_{j \neq i} Y_j^{-1} \right] Y_i$$

$$+ [Y_i C^\top(i) C_w(i) + L_i^\top C_w(i) + B_w(i)][\gamma^2 I - C_w^\top(i) C_w(i)]^{-1}$$
$$\times [Y_i C^\top(i) C_w(i) + L_i^\top C_w(i) + B_w(i)]^\top$$
$$+ A_d(i) Q^{-1} A_d^\top(i) + Y_i Q Y_i,$$

$$Z_{12}^\top(i) = A^\top(i) + X_i A(i) Y_i + H_i G^\top(i) Y_i$$

$$+ [Y_i^{-1} - X_i] K_A(i) Y_i + \sum_{m=1}^N Y_m^{-1} Y_i$$
$$+ [C^\top(i) C_w(i) + X_i B_w(i) + H_i G_w(i)][\gamma^2 I - C_w^\top(i) C_w(i)]^{-1}$$
$$\times [Y_i C^\top(i) C_w(i) + L_i^\top C_w(i) + B_w(i)]^\top$$
$$+ X_i A_d(i) Q^{-1} A_d^\top(i) + Q Y_i$$

$$Z_{22}(i) = A^\top(i) X_i + X_i A(i) X_i + H_i G(i) + G^\top(i) H_i^\top$$

$$+ \sum_{m=1}^N \lambda_{im} X_m + Q + X_i A_d(i) Q^{-1} A_d^\top(i)$$
$$+ [C^\top(i) C_w(i) + X_i B_w(i) + H_i D_i G_w(i)][\gamma^2 I - C_w^\top(i) C_w(i)]^{-1}$$
$$\times [C^\top(i) C_w(i) + X_i B_w(i) + H_i G_w(i)]^\top.$$

Therefore, (10.94) implies that

$$Z_{11}(i) < 0, \ \forall i \in \mathcal{S}$$
$$Z_{22}(i) < 0, \ \forall i \in \mathcal{S}$$

which are equivalent to (10.91) and (10.92), respectively. This completes the proof of necessity.

Sufficiency. Let $X_i > 0, Y_i > 0$, H_i and L_i be a set of feasible solutions to (10.91)-(10.93). Then, for every $i \in \mathcal{S}$ define the matrices $P(i), K_A(i), K_B(i), K_C(i), i \in \mathcal{S}$ as follows:

$$P(i) = \begin{pmatrix} X_i & Y_i^{-1} - X_i \\ Y_i^{-1} - X_i & X_i - Y_i^{-1} \end{pmatrix}, \tag{10.95}$$

$$\begin{cases} K_A(i) = [X_i - Y_i^{-1}]^{-1} \\ \qquad \times \left[\mathcal{W}_i Y_i^{-1} + X_i A(i) + H_i G(i) + \sum_{m=1}^N \lambda_{im} Y_m^{-1} \right] \\ K_B(i) = [Y_i^{-1} - X_i]^{-1} H_i \\ K_C(i) = L_i Y_i^{-1} \end{cases} \tag{10.96}$$

where

$$\mathcal{W}_i = A^\top(i) + [C^\top(i) C_w(i) + X_i B_i + H_i D_i G_w(i)][\gamma^2 I - C_w^\top(i) C_w(i)]^{-1}$$

$$\times\; [Y_iC^\top(i)C_w(i) + L_i^\top C_w(i) + B_i]^\top + X_iA_d(i)Q^{-1}A_d^\top(i) + QY_i.$$

Obviously, $P(i)$ is symmetric, it is easy to check that $P(i)$ are positive-definite. Letting

$$S_i = \begin{pmatrix} Y_i & I \\ Y_i & \mathbf{0} \end{pmatrix}$$

and substituting $K_A(i), K_B(i), K_C(i), P(i)$ into $\Theta_0(i)$, it is easy to check that

$$\tilde{S}_i^\top \Theta_0(i)\tilde{S}_i < 0, i \in \mathcal{S}$$

with $\tilde{S}_i = \text{diag}\{S_i, I, \cdots, I\}$, which is equivalent to

$$\begin{pmatrix} Z_{11}(i) & \mathbf{0} \\ \mathbf{0} & Z_{22}(i) \end{pmatrix} < 0, \; i \in \mathcal{S}$$

which is implied by (10.91) and (10.92). This ends the proof of sufficiency, which completes the proof of the theorem. □

From Theorem 10.9, we get the following corollary.

Corollary 10.3 *Let $X_i > 0, Y_i > 0, H_i, L_i$ be a set of matrices that satisfy (10.91)-(10.93) and*

$$\begin{pmatrix} \gamma^2 R & I \\ I & Y_{r_0} \end{pmatrix} > 0 \tag{10.97}$$

$$Q_l < \gamma^2 R_1, \; 1 \le l \le \nu, \tag{10.98}$$

then the filtering error with parameters defined by (10.96) verifies (10.73).

Proof: In view of Theorem 10.9, (10.91)-(10.93) implies (10.70). Let $P(i)$ be defined by (10.95). Then we have

$$\begin{pmatrix} I & I \end{pmatrix} P(r_0) \begin{pmatrix} I \\ I \end{pmatrix} = Y_{r_0}^{-1}.$$

Consequently, (10.97) implies (10.71) and the proof follows from Proposition 10.3. □

The filter parameters given by (10.96) that attain the optimal noise attenuation level can be obtained by solving the following optimization problem.

$$\mathcal{P}_1: \quad \begin{array}{c} \min \\ (X_1, \ldots, X_N) > 0, \\ (Y_1, \ldots, Y_N) > 0, \\ (H_1, \ldots, H_N), \\ (L_1, \ldots, L_N), v > 0 \end{array} \quad v$$

$$\text{s.t. (10.91)-(10.93), (10.97), (10.98)}$$

with γ^2 replaced by v

Theorem 10.9 shows that it is very difficult to design a filter using general $P(i), i \in \mathcal{S}$. However, using block diagonal $P(i), i \in \mathcal{S}$ we can get a very simple design method as follows.

Theorem 10.10 *If there exist symmetric, positive-definite matrices $X = (X_1, \cdots, X_N) > 0, Y = (Y_1, \cdots, Y_N) > 0, Q > 0$ and $U_i, V_i, W_i, i \in \mathcal{S}$ of appropriate dimensions such that*

$$\Lambda_0(i) < 0 \tag{10.99}$$
$$X_{r_0} + Y_{r_0} < \gamma^2 R \tag{10.100}$$
$$Q < \gamma^2 R_1 \tag{10.101}$$

where

$$\Lambda_0(i) = \begin{pmatrix} \theta_{11}(i) & G^\top(i)V_i^\top & X_i A_d(i) & X_i B_w(i) & C^\top(i) \\ V_i G(i) & \theta_{22}(i) & 0 & V_i G_w(i) & -W_i^\top \\ A_d^\top(i)X_i & 0 & -Q & 0 & 0 \\ B_w^\top(i)X_i & G_w^\top(i)V_i & 0 & -\gamma^2 I & C_w^\top(i) \\ C(i) & -W_i & 0 & C_w(i) & -I \end{pmatrix}$$

$$\theta_{11}(i) = A^\top(i)X_i + X_i A(i) + \sum_{m=1}^N \lambda_{im} X_m + Q$$

$$\theta_{22}(i) = U_i + U_i^\top + \sum_{m=1}^N \lambda_{im} Y_m,$$

then the estimation error of the filter of the form of (10.66) with

$$\begin{cases} K_A(i) = Y_i^{-1} U_i \\ K_B(i) = Y_i^{-1} V_i \\ K_C(i) = W_i \end{cases} \tag{10.102}$$

is stochastically stable with noise attenuation level γ.

Proof: Let the symmetric, positive-definite matrices X_i, Y_i, Q and the matrices $U_i, V_i, W_i, i \in \mathcal{S}$ be a set of feasible solutions of LMIs (10.99)-(10.101). Let us define $K_A(i), K_B(i), K_C(i)$ as in (10.102) and let $P(i) = \begin{pmatrix} X_i & 0 \\ 0 & Y_i \end{pmatrix}$. It is easy to check that $P(i), Q$ are a set of feasible solutions (10.70)-(10.72). Therefore, the proof of Theorem 10.10 follows from Theorem 10.3. $\square$

The filter design and noise attenuation level optimization problem can be formulated as follows.

Corollary 10.4 *Let ν_0 be the optimal solutions of the following optimization problem*

$$\mathcal{P}_2: \qquad \begin{array}{c} \min \\ v > 0, (X_1, \ldots, X_N) > 0, (Y_1, \ldots, Y_N) > 0, \\ (U_1, \ldots, U_N), (V_1, \ldots, V_N), (W_1, \ldots, W_N), Q \end{array} \qquad v$$

$$s.t. \begin{cases} \Lambda_v(i) < 0, \forall i \in \mathcal{S} \\ X_i + Y_i < vR, \\ Q < vR_1, \end{cases}$$

where $\Lambda_v(i)$ is obtained from $\Lambda_0(i)$ by replacing γ^2 with v. Then, the filter with parameters given by (10.102) stabilizes system (10.1) and attains the optimal attenuation level $\sqrt{v_0}$.

Let us now work out the following numerical example to show the usefulness of the previous results.

Example 10.5 *Consider a dynamical time delay jump linear system with two modes, i.e., $\mathcal{S} = \{1, 2\}$. Consider the Markov process describing the mode-switching has the generator $\Lambda = \begin{pmatrix} -7 & 7 \\ 6 & -6 \end{pmatrix}$.*

The system parameters are as follows:

$$A(1) = \begin{pmatrix} -2 & 0.1 \\ 0 & -1 \end{pmatrix} \qquad A(2) = \begin{pmatrix} -2.5 & 0 \\ 0.1 & -2.1 \end{pmatrix}$$

$$A_d(1) = \begin{pmatrix} -1 & 0 \\ 0 & -0.1 \end{pmatrix} \qquad A_d(2) = \begin{pmatrix} -1 & 0 \\ 0.1 & -1 \end{pmatrix}$$

$$B_w(1) = \begin{pmatrix} 1 & 0 \\ 0.1 & 0.2 \end{pmatrix} \qquad B_w(2) = \begin{pmatrix} 1 & 0 \\ 0 & 0 \end{pmatrix}$$

$$G(1) = \begin{pmatrix} 0.1 & 0 \\ 0.3 & 0 \end{pmatrix} \qquad G(2) = \begin{pmatrix} 0.1 & 0 \\ 0.3 & 0 \end{pmatrix}$$

$$G_w(1) = \begin{pmatrix} 1.4 & 0 \\ 0 & 0.3 \end{pmatrix} \qquad G_w(2) = \begin{pmatrix} 1.4 & 0 \\ 0 & 0.3 \end{pmatrix}$$

$$C(1) = \begin{pmatrix} 1.3 & 0 \\ 0 & 0.1 \end{pmatrix} \qquad C(2) = \begin{pmatrix} 1.3 & 0 \\ 0 & 0.1 \end{pmatrix}$$

$$C_w(1) = \begin{pmatrix} 1 & 0 \\ 0 & 0 \end{pmatrix} \qquad C_w(2) = \begin{pmatrix} 1 & 0 \\ 0 & 0 \end{pmatrix}.$$

Using Theorem 10.10, the solutions of $\mathcal{P}_1$ gives

$$K_A(1) = \begin{pmatrix} -4.4703 & 0.2589 \\ -0.3418 & -2.7151 \end{pmatrix} \qquad K_A(2) = \begin{pmatrix} -5.1013 & 0.3915 \\ 0.4753 & -4.9469 \end{pmatrix}$$

$$K_B(1) = \begin{pmatrix} 1.1985 & 1.2731 \\ 0.0006 & 1.4746 \end{pmatrix} \qquad K_B(2) = \begin{pmatrix} 1.2432 & 1.5763 \\ -0.1156 & 0.3112 \end{pmatrix}$$

$$K_C(1) = \begin{pmatrix} -1.2755 & -0.0497 \\ 0.0051 & -1.2038 \end{pmatrix} \qquad K_C(2) = \begin{pmatrix} -1.2614 & 0.0191 \\ -0.0042 & -1.0636 \end{pmatrix}$$

and noise attenuation level $\gamma = 1.0$

   The solutions of $\mathcal{P}_2$ gives another filter with parameters:

$$K_A(1) = \begin{pmatrix} 2.1921 & 0.2006 \\ -0.2276 & -2.1639 \end{pmatrix} \quad K_A(2) = \begin{pmatrix} 0.4016 & 0 \\ 0 & -0.399 \end{pmatrix}$$

$$K_B(1) = \begin{pmatrix} 0.3678 & -0.0192 \\ -0.00226 & 0.5552 \end{pmatrix} \quad K_B(2) = \begin{pmatrix} 0.1269 & 0.4637 \\ 0.0002 & -0.0007 \end{pmatrix}$$

$$K_C(1) = \begin{pmatrix} 0.6752 & 0.2691 \\ 0.0053 & 0.0027 \end{pmatrix} \quad K_C(2) = \begin{pmatrix} 1.3315 & -0.001 \\ 0.0004 & 1.2681 \end{pmatrix}$$

which verifies noise attenuation level $\gamma = 2.7835$.

This example shows that the design method in Theorem 10.9 is less conservative than that of Theorem 10.10.

10.5 Notes

The $\mathcal{H}_\infty$ control problem for jump linear system is studied by [63, 180]. Mixed $\mathcal{H}_2/\mathcal{H}_\infty$ control of discrete time Markov jump linear systems is addressed by [59]. For jump linear systems with time delay, Benjelloun et al. [9] and Cao and Lam [41] deal with the $\mathcal{H}_\infty$ control problem. Delay-dependent $\mathcal{H}_\infty$ control is addressed by [26]. For jump linear system without time delay, [143] studies the output feedback stabilization, [62] studied the output feedback $\mathcal{H}_\infty, \mathcal{H}_2$-control and the filtering problem using LMI-toolbox. The output feedback $\mathcal{H}_\infty$ for jump linear systems with time delay is reported first here. Reference [63] looks at robust $\mathcal{H}_\infty$-filtering for uncertain Markov jump linear systems without time delay. Robust $\mathcal{H}_\infty$ filtering for Markov jump linear systems with time delay and polytopic uncertainties is studied in [23].

11
Robust $\mathcal{H}_\infty$ and Guaranteed Cost Control for Jump Linear Systems with Time Delay

The linear Markov jump model that is usually used in the analysis and design phases is an approximation of a real nonlinear system with Markov jumps in the neighborhood of the operating point. For many reasons well known in the control community, the systems parameters and the operating point change with time and therefore, the fixed linear Markov jump model is not adequate to guarantee robustness of system performance. Besides this, the system can be affected by exogenous disturbances, which will make the degradation worse. To overcome this surprise, the control engineer should take care of these uncertainties and exogenous disturbances during the analysis and design phases to guarantee the required stability and other system performance, despite the presence of uncertainties in the system. The results presented so far for the class of linear systems with Markov jumps and time delay are not adequate to guarantee the robustness of the desired performance. The robust $\mathcal{H}_\infty$ control problem was developed to maintain robustness of stability and performance when it known algorithms lack robustness, that is, the system parameters have uncertainties. This chapter deals with the robust $\mathcal{H}_\infty$ control and the guaranteed cost control problems for jump linear systems with norm-bounded uncertainties and time delay. The rest of the chapter is organized as follows. In Section 11.1, we deal with the robust $\mathcal{H}_\infty$ control problem. Section 11.2 considers the guaranteed cost control problem. In Section 11.3, we cover the output feedback guaranteed cost control problem.

11.1 $\mathcal{H}_\infty$ Control

Let $\{r_t, t \geq 0\}$ be a continuous-time Markov process taking values in the finite state space $\mathcal{S} = \{1, 2, \cdots, N\}$ and suppose that the Markov process $\{r_t, t \geq 0\}$ has the generator matrix $\Lambda = (\lambda_{ij}), i, j \in \mathcal{S}$. Let us now consider a hybrid system with N modes and assume that its mode switching in time is governed by the Markov process $\{r_t, t \geq 0\}$. The system is described by the following dynamics:

$$
\begin{cases}
\dot{x}(t) = A(r_t, t)x(t) + A_d(r_t, t)x(t - \tau(r_t)) \\
\qquad + B(r_t, t)u(t) + B_w(r_t)w(t) \\
x(s) = \phi(s), \; [-\bar{\tau}, 0] \\
y(t) = C(r_t, t)x(t) + F(r_t, t)u(t) + C_d(r_t, t)x(t - \tau(r_t)) \\
\qquad + B_y(r_t)w(t) \\
z(t) = C_z(r_t, t)x(t) + F_z(r_t, t)u(t) + G_z(r_t)w(t),
\end{cases}
\tag{11.1}
$$

where $x(t) \in \mathbb{R}^{n_1}$, $u(t) \in R^{n_2}$, $y(t) \in R^{n_3}$, $z(t) \in R^{n_4}$, $w(t) \in \mathbb{R}^{n_5}$ denote, respectively, the state, the control input, the measured output, the controlled output, and the system disturbance, $\tau(r_t)$ denotes the time delay in the system when the mode is in r_t, $\bar{\tau} = \max\{\tau(j), j \in \mathcal{S}\}$, $\phi(\cdot)$ is the initial function, and

$$
\begin{aligned}
A(r_t, t) &= A(r_t) + D_a(r_t)\Delta(r_t, t)E_a(r_t) \\
B(r_t, t) &= B(r_t) + D_a(r_t)\Delta(r_t, t)E_b(r_t) \\
A_d(r_t, t) &= A_d(r_t) + D_d(r_t)\Delta_d(r_t, t)E_d(r_t) \\
C(r_t, t) &= C(r_t) + D_c(r_t)\Delta(r_t, t)E_c(r_t) \\
C_z(r_t, t) &= C_z(r_t) + D_{CZ}\Delta(r_t, t)E_{CZ}(r_t) \\
C_d(r_t, t) &= C_d(r_t) + D_{cd}\Delta(r_t, t)E_{cd}(r_t) \\
F(r_t, t) &= F(r_t) + D_{CZ}(r_t)\Delta(r_t, t)E_f(r_t) \\
F_z(r_t, t) &= F_z(r_t) + D_{CZ}(r_t)\Delta(r_t, t)E_f(r_t)
\end{aligned}
$$

with $A(r_t)$, $B(r_t)$, $A_d(r_t)$, $B_w(r_t)$, $B_y(r_t)$, $C(r_t)$, $C_z(r_t)$, $C_d(r_t)$, $D_{cd}(r_t)$, $E_{cd}(r_t)$, $F(r_t)$, $F_z(r_t)$, $G_z(r_t)$, $D_a(r_t)$, $E_a(r_t)$, $E_b(r_t)$, $D_d(r_t)$, $E_d(r_t)$, $D_w(r_t)$, $E_w(r_t)$, $D_c(r_t)$, $E_c(r_t)$, $D_{cz}(r_t)$, $E_{cz}(r_t)$, $E_f(r_t)$, $D_g(r_t)$, $E_g(r_t)$ being known matrices of appropriate dimensions, $\Delta(r_t, t)$, $\Delta_d(r_t, t)$ being time-varying unknown matrices satisfying

$$
\begin{aligned}
\Delta^\top(r_t, t)\Delta(r_t, t) &\leq I \\
\Delta_d^\top(r_t, t)\Delta_d(r_t, t) &\leq I.
\end{aligned}
$$

The system disturbance $w(t)$ is assumed to belong to $\mathcal{L}_2[0, \infty)$.

This section addresses stability and stabilization and $\mathcal{H}_\infty$ control of system (11.1).

11.1.1 $\mathcal{H}_\infty$ Performance

This subsection develops sufficient conditions to guarantee that system (11.1) with $u(t) \equiv 0$ is robustly stochastically stable and verifies the noise attenuation level γ. For this purpose, let us introduce the following definition.

Definition 11.1 *System (11.1) with $u(t) \equiv 0$ is said to be internally MSQS if there exist matrices $P = (P(1), \cdots, P(N)) > 0, Q > 0$ such that the following holds for all admissible uncertainties*

$$\Theta(r_t, t) \triangleq \begin{pmatrix} \# & P(r_t)A_d(r_t, t) \\ A_d^\top(r_t, t)P(r_t) & -Q \end{pmatrix} < 0, \quad \forall r_t \in \mathcal{S} \quad (11.2)$$

where

$$\# = A^\top(r_t, t)P(r_t) + P(r_t)A(r_t, t) + \sum_{j=1}^{N} \lambda_{ij} P(j) + \varrho Q$$

with $\varrho = 1 + (\bar{\tau} - \underline{\tau})\bar{\lambda}$, $\underline{\tau} = \min\{\tau(i), i \in \mathcal{S}\}, \bar{\tau} = \max\{\tau(i), i \in \mathcal{S}\}$, $\bar{\lambda} = \max\{|\lambda_{ii}|, i \in \mathcal{S}\}$.

By the same proof as in Theorem 9.5, we have the following proposition.

Proposition 11.1 *If there exist symmetric, positive-definite matrices $P = (P(1), \cdots, P(N)) > 0, Q > 0$ and a scalar $\varepsilon > 0$ such that for every $i \in \mathcal{S}$,*

$$\begin{pmatrix} \# & P(i)A_d(i) & P(i)D_a(i) & P(i)D_d(i) \\ * & -Q + \varepsilon E_d^\top(i)E_d(i) & 0 & 0 \\ * & 0 & -\varepsilon I & 0 \\ * & 0 & 0 & -\varepsilon I \end{pmatrix} < 0 \quad (11.3)$$

holds, where

$$\# = A^\top(i)P(i) + P(i)A(i) + \sum_{j=1}^{N} \lambda_{ij} P(j) + \varrho Q + \varepsilon E_a^\top(i)E_a(i),$$

then system (11.1) is internally MSQS.

By definition, *internal MSQS* means that system (11.1) with $u(t) \equiv 0$ and zero noise disturbance is MSQS. In general, under the assumption that $w(\cdot) \in \mathcal{L}_2[0, \infty)$, the following proposition shows that internal MSQS implies robust stochastic stability.

Proposition 11.2 *Let $w(\cdot) \in \mathcal{L}_2[0, \infty)$. If system (11.1) with $u(t) \equiv 0$ is internal MSQS, then it is robustly stochastically stable.*

Proof: The proof can be adapted from Theorem 10.1.   □

The following theorem establishes the stochastic stability of system (11.1) with $u(t) \equiv 0$.

Theorem 11.1 *If there exist symmetric, positive matrices $Q > 0$, $P = (P(1), \cdots, P(N)) > 0$ such that for all admissible uncertainties and for every $r_t \in \mathcal{S}$,*

$$
\begin{pmatrix}
\mathcal{N}_1(r_t, t) & P(r_t)A_d(r_t, t) & \begin{bmatrix} P(r_t)B_w(r_t) \\ +C_z^\top(r_t, t)G_z(r_t) \end{bmatrix} \\
A_d^\top(r_t, t)P(r_t) & -Q & 0 \\
\begin{bmatrix} B_w^\top(r_t)P(r_t) \\ +G_z^\top(r_t)C_z(r_t, t) \end{bmatrix} & 0 & -\gamma^2 I + G_z^\top(r_t)G_z(r_t)
\end{pmatrix}
$$

$$
\triangleq \mathcal{H}(r_t, t) < 0 \quad (11.4)
$$

holds, where

$$
\mathcal{N}_1(r_t, t) = A^\top(r_t, t)P(r_t) + P(r_t)A(r_t, t) + \sum_{j=1}^N \lambda_{r_t j} P(j)
$$

$$
+ C_z^\top(r_t, t)C_z(r_t, t) + \varrho Q,
$$

then system (11.1) with $u(t) \equiv 0$ is robustly stochastically stable and satisfies noise attenuation level γ, i.e.,

$$
\|z\| \leq \Bigg[\gamma^2 \|w\|_2^2 + x_0^\top P(r_0)x_0 + \int_{-\tau(r_0)}^0 \phi^\top(s)Q\phi(s)ds
$$

$$
+ \bar{\lambda} \int_{-\bar{\tau}}^{-\underline{\tau}} \int_\theta^0 \phi(s)Q\phi(s)ds d\theta \Bigg]^{1/2}, \quad (11.5)
$$

where $x_0 = \phi(0)$.

Proof: Note that (11.4) implies that

$$
\begin{pmatrix}
\mathcal{N}_1(r_t, t) & P(r_t)A_d(r_t, t) \\
A_d^\top(r_t, t)P(r_t) & -Q
\end{pmatrix} < 0
$$

holds for all admissible uncertainties and for all $r_t \in \mathcal{S}$. Since $C_z^\top(r_t, t) C_z(r_t, t) \geq 0$, this yields

$$
\begin{pmatrix}
\mathcal{N}_2(r_t, t) & P(r_t)A_d(r_t, t) \\
A_d^\top(r_t, t)P(r_t) & -Q
\end{pmatrix} < 0 \quad (11.6)
$$

where

$$
\mathcal{N}_2(r_t, t) = A^\top(r_t, t)P(r_t) + P(r_t)A(r_t, t) + \sum_{j=1}^N \lambda_{r_t j} P(j) + \varrho Q.
$$

Following Definition 11.1, this means that system (11.1) with $u(t) \equiv 0$ is internally stable. Using now Proposition 11.2, we conclude that system (11.1) with $u(t) \equiv 0$ is robustly stochastically stable.

Let us now prove that system (11.1) with $u(t) \equiv 0$ satisfies the noise attenuation level γ.

For this purpose, let us define $\mathbf{x}_s(t) = x(s+t), t-\tau \le s \le t$ and consider a Lyapunov functional candidate as follows:

$$V(\mathbf{x}(t), r_t) = x^\top(t)P(r_t)x(t) + V_1(\mathbf{x}(t), r_t) + V_2(\mathbf{x}(t), r_t) \qquad (11.7)$$

where

$$V_1(\mathbf{x}(t), r_t) = \int_{t-\tau(r_t)}^t x^\top(s)Qx(s)ds$$

$$V_2(\mathbf{x}(t), r_t) = \int_{-\bar{\tau}}^{-\underline{\tau}} \int_{t+\theta}^t x^\top(s)\bar{\lambda}Qx(s)dsd\theta.$$

Let $\mathcal{A}$ be the infinitesimal generator of the Markov process $\{(\mathbf{x}(t), r_t), t \ge 0\}$. By the same procedure as in (8.134), it is easy to check that

$$\mathcal{A}V(\mathbf{x}(t), r_t) \le x^\top(t)\left[A^\top(r_t, t)P(r_t) + P(r_t)A(r_t, t) + \sum_{j=1}^N \lambda_{r_t j}P(j)\right]x(t)$$

$$+ 2x^\top(t)P(r_t)A_d(r_t, t)x(t - \tau(r_t))$$
$$+ x^\top(t)Qx(t) - x^\top(t - \tau(r_t))Qx(t - \tau(r_t))$$
$$+ 2x^\top(t)P(r_t)B_w(r_t)w(t) + \bar{\lambda}(\bar{\tau} - \underline{\tau})x^\top(t)Qx(t). \qquad (11.8)$$

Let us define a performance function J_T by

$$J_T = \mathbb{E}\left[\int_0^T [z^\top(t)z(t) - \gamma^2 w^\top(t)w(t)]dt\right].$$

Then, to establish (11.5), it suffices to check that

$$J_\infty \le \mathbb{E}[V(\mathbf{x}(0), r_0)]. \qquad (11.9)$$

Note that

$$J_T = \mathbb{E}\left[\int_0^T \left(z^\top(t)z(t) - \gamma^2 w^\top(t)w(t) + \mathcal{A}V(\mathbf{x}(t), r_t)\right) dt\right]$$

$$- \mathbb{E}\left[\int_0^T \mathcal{A}V(\mathbf{x}(t), r_t)dt\right]. \qquad (11.10)$$

From Dynkin's formula, it follows that

$$\mathbb{E}\left[\int_0^T \mathcal{A}V(\mathbf{x}(t), r_t)dt\right] = \mathbb{E}[V(\mathbf{x}(T), r_T)] - \mathbb{E}[V(\mathbf{x}(0), r_0)],$$

which combined with (11.10) yields

$$J_T = \mathbb{E}\left[\int_0^T \left(z^\top(t)z(t) - \gamma^2 w^\top(t)w(t) + \mathcal{A}V(\mathbf{x}(t), r_t)\right) dt\right]$$

$$+ \mathbb{E}[V(\mathbf{x}(0), r_0)] - \mathbb{E}[V(\mathbf{x}(T), r_T)]$$

$$\leq \mathbb{E}\left[\int_0^T z^\top(t)z(t) - \gamma^2 w^\top(t)w(t) + \mathcal{A}V(\mathbf{x}(t), r_t)dt\right]$$
$$+ \mathbb{E}[V(\mathbf{x}(0), r_0)] \qquad (11.11)$$

Using now (11.8), we have

$$z^\top(t)z(t) - \gamma^2 w^\top(t)w(t) + \mathcal{A}V(\mathbf{x}(t), r_t)$$
$$\leq [C_z(r_t, t)x(t) + G_z(r_t)w(t)]^\top [C_z(r_t, t)x(t) + G_z(r_t)w(t)]$$
$$+ x^\top(t)\left[A^\top(r_t, t)P(r_t) + P(r_t)A(r_t, t) + \sum_{j=1}^N \lambda_{r_t, j}P(j)\right]x(t)$$
$$+ 2x^\top(t)P(r_t)A_d(r_t, t)x(t - \tau(r_t)) + x^\top(t)Qx(t)$$
$$- x^\top(t - \tau(r_t))Qx(t - \tau(r_t)) + 2x^\top(t)P(r_t)B_w(r_t)w(t)$$
$$+ \bar{\lambda}(\bar{\tau} - \underline{\tau})x^\top(t)Qx(t) - \gamma^2 w^\top(t)w(t)$$
$$= \xi^\top(t)\mathcal{H}(r_t, t)\xi(t), \qquad (11.12)$$

where

$$\xi(t) = \left(\begin{array}{ccc} x^\top(t) & x^\top(t - \tau(r_t)) & w^\top(t) \end{array}\right).$$

Based on (11.4), we get

$$z^\top(t)z(t) - \gamma^2 w^\top(t)w(t) + \mathcal{A}V(\mathbf{x}(t), r_t) < 0$$

Combining (11.12) and (11.11) yields that

$$J_T \leq V(\mathbf{x}(0), r_0), \ \forall T > 0,$$

which implies (11.9). This completes the proof of Theorem 11.1. $\qquad \square$

Theorem 11.1 provides a sufficient condition for system (11.1) with $u(t) \equiv 0$ to be stochastically stable and verifies the noise attenuation level γ. However, since $\mathcal{H}(r_t, t)$ contains the uncertainties of the system, it is difficult to verify condition (11.4). Next, we will develop the LMI-based condition that guarantees (11.4). For this purpose, note that

$$\mathcal{H}(r_t, t) = \left(\begin{array}{ccc} \mathcal{N}_2(r_t, t) & P(r_t)A_d(r_t, t) & P(r_t)B_w(r_t) \\ A_d^\top(r_t, t)P(r_t) & -Q & 0 \\ B_w^\top(r_t)P(r_t) & 0 & -\gamma^2 I \end{array}\right)$$
$$+ \left(\begin{array}{c} C_z^\top(r_t, t) \\ 0 \\ G_z^\top(r_t) \end{array}\right)\left(\begin{array}{ccc} C_z(r_t, t) & 0 & G_z(r_t) \end{array}\right),$$

where $\mathcal{N}_2(r_t, t) = A^\top(r_t, t)P(r_t) + P(r_t)A(r_t, t) + \sum_{j=1}^N \lambda_{r_t j}P(j) + \varrho Q.$

Therefore, using the Schur complement (11.4) holds if and only if

$$
\begin{pmatrix}
\mathcal{N}_2(r_t,t) & P(r_t)A_d(r_t,t) & P(r_t)B_w(r_t) & C_z^\top(r_t,t) \\
A_d^\top(r_t,t)P(r_t) & -Q & \mathbf{0} & \mathbf{0} \\
B_w^\top(r_t)P(r_t) & \mathbf{0} & -\gamma^2 I & G_z^\top(r_t) \\
C_z(r_t,t) & \mathbf{0} & G_z(r_t) & -I
\end{pmatrix}
$$

$$
\triangleq \mathcal{H}_1(r_t,t) < 0.
$$

Furthermore, notice that we can rewrite $\mathcal{H}_1(r_t,t)$ as follows:

$$
\mathcal{H}_1(r_t,t) =
$$
$$
\begin{pmatrix}
\mathcal{N}_2(r_t) & P(r_t)A_d(r_t) & P(r_t)B_w(r_t) & C_z^\top(r_t) \\
A_d^\top(r_t)P(r_t) & -Q & \mathbf{0} & \mathbf{0} \\
B_w^\top(r_t)P(r_t) & \mathbf{0} & -\gamma^2 I & G_z^\top(r_t) \\
C_z(r_t,t) & \mathbf{0} & G_z(r_t) & -I
\end{pmatrix}
$$
$$
+
\begin{pmatrix}
P(r_t)D_a(r_t) & P(r_t)D_d(r_t) & \mathbf{0} & E_{CZ}^\top(r_t) \\
\mathbf{0} & \mathbf{0} & \mathbf{0} & \mathbf{0} \\
\mathbf{0} & \mathbf{0} & \mathbf{0} & \mathbf{0} \\
\mathbf{0} & \mathbf{0} & \mathbf{0} & \mathbf{0}
\end{pmatrix}
\tilde{\Delta}(r_t,t)
$$
$$
\times
\begin{pmatrix}
E_a(r_t) & & & \\
& E_d(r_t) & & \\
& & \mathbf{0} & \\
& & & D_{CZ}^\top(r_t)
\end{pmatrix}
$$
$$
+
\begin{pmatrix}
E_a^\top(r_t) & & & \\
& E_d^\top(r_t) & & \\
& & \mathbf{0} & \\
& & & D_{CZ}(r_t)
\end{pmatrix}
$$
$$
\times \tilde{\Delta}^\top(r_t,t)
\begin{pmatrix}
D_a^\top(r_t)P(r_t) & \mathbf{0} & \mathbf{0} & \mathbf{0} \\
D_d^\top(r_t)P(r_t) & \mathbf{0} & \mathbf{0} & \mathbf{0} \\
\mathbf{0} & \mathbf{0} & \mathbf{0} & \mathbf{0} \\
E_{CZ}(r_t) & \mathbf{0} & \mathbf{0} & \mathbf{0}
\end{pmatrix},
$$

where

$$
\mathcal{N}_2(r_t) = A^\top(r_t)P(r_t) + P(r_t)A(r_t) + \sum_{j=1}^N \lambda_{r_t j} + \varrho Q,
$$

$$
\tilde{\Delta}(r_t,t) =
\begin{pmatrix}
\Delta(r_t,t) & & & \\
& \Delta_d(r_t,t) & & \\
& & \Delta(r_t,t) & \\
& & & \Delta^\top(r_t,t)
\end{pmatrix}.
$$

Using now Lemma A.5, we have that for any $\varepsilon > 0$,

$$\mathcal{H}_1(r_t, t) \le$$

$$\begin{pmatrix} \mathcal{N}_2(r_t) & P(r_t)A_d(r_t) & P(r_t)B_w(r_t) & C_z^\top(r_t) \\ A_d^\top(r_t)P(r_t) & -Q & 0 & 0 \\ B_w^\top(r_t)P(r_t) & 0 & -\gamma^2 I & G_z^\top(r_t) \\ C_z(r_t, t) & 0 & G_z(r_t) & -I \end{pmatrix}$$

$$+ \frac{1}{\varepsilon} \begin{pmatrix} P(r_t)D_a(r_t) & P(r_t)D_d(r_t) & 0 & E_{CZ}^\top(r_t) \\ 0 & 0 & 0 & 0 \\ 0 & 0 & 0 & 0 \\ 0 & 0 & 0 & 0 \end{pmatrix}$$

$$\times \begin{pmatrix} D_a^\top(r_t)P(r_t) & 0 & 0 & 0 \\ D_d^\top(r_t)P(r_t) & 0 & 0 & 0 \\ 0 & 0 & 0 & 0 \\ E_{CZ}(r_t) & 0 & 0 & 0 \end{pmatrix}$$

$$+ \varepsilon \begin{pmatrix} E_a^\top(r_t)E_a(r_t) & & & \\ & E_d^\top(r_t)E_d(r_t) & & \\ & & 0 & \\ & & & D_{CZ}(r_t)D_{CZ}^\top(r_t) \end{pmatrix}$$

$$= \begin{pmatrix} \mathcal{N}_3(r_t) & P(r_t)A_d(r_t) \\ A_d^\top(r_t)P(r_t) & -Q + \varepsilon E_d^\top(r_t)E_d(r_t) \\ B_w^\top(r_t)P(r_t) & 0 \\ C_z(r_t, t) & 0 \end{pmatrix}$$

$$\begin{pmatrix} P(r_t)B_w(r_t) & C_z^\top(r_t) \\ 0 & 0 \\ -\gamma^2 I & G_z^\top(r_t) \\ G_z(r_t) & -I + \varepsilon D_{CZ}(r_t)D_{CZ}^\top(r_t) \end{pmatrix}$$

$$+ \frac{1}{\varepsilon} \begin{pmatrix} P(r_t)D_a(r_t) & P(r_t)D_d(r_t) & E_{CZ}^\top(r_t) \\ 0 & 0 & 0 \\ 0 & 0 & 0 \\ 0 & 0 & 0 \end{pmatrix}$$

$$\times \begin{pmatrix} D_a^\top(r_t)P(r_t) & 0 & 0 & 0 \\ D_d^\top(r_t)P(r_t) & 0 & 0 & 0 \\ E_{CZ}(r_t) & 0 & 0 & 0 \end{pmatrix} \triangleq \mathcal{H}_2(r_t) \qquad (11.13)$$

where

$$\mathcal{N}_3(r_t) = A^\top(r_t)P(r_t) + P(r_t)A(r_t) + \sum_{j=1}^{N} \lambda_{r_t j} P(j)$$

$$+ \varrho Q + \varepsilon E_a^\top(r_t)E_a(r_t).$$

Using again the Schur complement, $\mathcal{H}_2(r_t) < 0$ if and only if

$$
\begin{pmatrix}
\mathcal{N}_3(i) & P(i)A_d(i) & P(i)B_w(i) \\
A_d^\top(i)P(i) & -Q + \varepsilon E_d^\top(i)E_d(i) & 0 \\
B_w^\top(i)P(i) & 0 & -\gamma^2 I \\
C_z(i) & 0 & G_z(i) \\
\mathcal{H}_{15}^\top(i) & 0 & 0
\end{pmatrix}
$$

$$
\begin{pmatrix}
C_z^\top(i) & \mathcal{H}_{15}(i) \\
0 & 0 \\
G_z^\top(i) & 0 \\
-I + \varepsilon D_{CZ}(i)D_{CZ}^\top(i) & 0 \\
0 & -\varepsilon I
\end{pmatrix} < 0, \qquad (11.14)
$$

where $\mathcal{H}_{15}(i) = \begin{pmatrix} P(i)D_a(i) & P(i)D_d(i) & E_{cz}^\top(i) \end{pmatrix}$.
From above derivation, we get the following theorem.

Theorem 11.2 *Let γ be a given positive constant. If there exist symmetric, positive-definite matrices $P = (P(1), \cdots, P(N)) > 0$, $Q > 0$ and a positive constant $\varepsilon > 0$ satisfying (11.14), then system (11.1) with $u(t) \equiv 0$ is robust stochastically stable and verifies the noise attenuation level γ.*

In Theorem 11.2, the attenuation level γ is assumed to be a given constant. For a given set of parameters, if system (11.1) with $u(t) \equiv 0$ is stochastically stable, the optimal noise attenuation level that the given system can have, can be obtained by solving the following optimization problem:

$$
\mathcal{P}_0 : \qquad \min_{\substack{\rho > 0, \ P = (P(1), \dots, P(N)) > 0 \\ \Gamma(i) < 0, \forall i \in \mathcal{S}}} \rho
$$

where $\Gamma(i)$ is obtained from the left-hand matrix of (11.14) by replacing γ^2 with ρ.

Let μ_0 be the optimal solution on $\mathcal{P}_0$. Then system (11.1) with $u(t) \equiv 0$ verifies the noise attenuation level $\sqrt{\mu_0}$.

In some practical systems, it can happen that the time delay is constant and doesn't depend of the mode system, that is, $\tau(r_t) = \tau$ and $\bar{\tau} = \underline{\tau} = \tau$. From Theorem 11.2, we directly obtain for this case the following corollary.

Corollary 11.1 *Let the time delay in system (11.1) be independent of the system mode, i.e., $\tau(r_t) \equiv \tau > 0$. Then, if there exist symmetric, positive-definite matrices $P = (P(1), \cdots, P(N)) > 0, Q > 0$ and a scalar $\varepsilon > 0$ such* that for any $i \in \mathcal{S}$,

$$
\left(
\begin{array}{ccc}
\# & P(r_t)A_d(r_t) & P(r_t)B_w(r_t) \\
A_d^\top(r_t)P(r_t) & -Q + \varepsilon E_d^\top(r_t)E_d(r_t) & 0 \\
B_w^\top(r_t)P(r_t) & 0 & -\gamma^2 I \\
C_z(r_t) & 0 & G_z(r_t) \\
\mathcal{H}_{15}^\top(r_t) & 0 & 0
\end{array}
\right.
$$
$$
\left.
\begin{array}{cc}
C_z^\top(r_t) & \mathcal{H}_{15}(r_t) \\
0 & 0 \\
G_z^\top(r_t) & 0 \\
-I + \varepsilon D_{CZ}(r_t)D_{CZ}^\top(r_t) & 0 \\
0 & -\varepsilon I
\end{array}
\right) < 0 \qquad (11.15)
$$

holds, where $\mathcal{H}_{15}(r_t)$ is defined as in (11.14) and

$$
\# = A^\top(r_t)P(r_t) + P(r_t)A(r_t) + \sum_{j=1}^{N} \lambda_{r_t j} P(j) + Q
$$
$$
+ \varepsilon E_a^\top(r_t)E_a(r_t),
$$

then system (11.1) with $u(t) \equiv 0$ is robustly stochastically stable and verifies the noise attenuation level γ.

Proof: The proof can be obtained by following the same lines as in the derivation of Theorem 11.2 and considering the following Lyapunov functional candidate:

$$
V(\mathbf{x}_t, r_t) = x^\top(t)Px(t) + \int_{t-\tau}^{t} x^\top(s)Qx(s)ds.
$$

The details are omitted.   □

Remark 11.1 *The time-varying mode-independent time delay case is studied by Benjelloun et al. [9]. LMI sufficient conditions have been developed to guarantee robust stochastic stability and stabilizability.*

Example 11.1 *To show the usefulness of the results of the previous theorem, let us consider a system with dynamics described by (11.1) with $u(t) \equiv 0$ and suppose that the system has two modes, i.e., $\mathcal{S} = \{1, 2\}$, and the parameters are as follows:*

$$
\Lambda = \left(
\begin{array}{cc}
-3 & 3 \\
2 & -2
\end{array}
\right)
$$

with $\tau_1 = 0.5, \tau_2 = 1$

$$A(1) = \begin{pmatrix} -2 & 0 \\ 0 & -0.9 \end{pmatrix} \qquad A(2) = \begin{pmatrix} -2.1 & -0.1 \\ -0.1 & -1 \end{pmatrix}$$

$$B(1) = \begin{pmatrix} 2 & 0 \\ 0 & 1 \end{pmatrix} \qquad B(2) = \begin{pmatrix} 1 & 0 \\ 0 & 0 \end{pmatrix}$$

$$A_d(1) = \begin{pmatrix} -1 & 0 \\ -1 & -1 \end{pmatrix} \qquad A_d(2) = \begin{pmatrix} -1 & 0 \\ -1 & -1 \end{pmatrix}$$

$$B_w(1) = \begin{pmatrix} 1 & 0 \\ 0 & 0 \end{pmatrix} \qquad B_w(2) = \begin{pmatrix} 0.8 & 0 \\ 0 & 0 \end{pmatrix}$$

$$C(1) = \begin{pmatrix} 2 & 1 \\ 0 & 1 \end{pmatrix} \qquad C(2) = \begin{pmatrix} 0.2 & -1 \\ 1 & 0 \end{pmatrix}$$

$$C_z(1) = \begin{pmatrix} 1 & 1 \\ 0 & 0 \end{pmatrix} \qquad C_z(2) = \begin{pmatrix} 2 & -1 \\ 1 & 0 \end{pmatrix}$$

$$F_z(1) = \begin{pmatrix} -0.1 & 1 \\ 0 & 0 \end{pmatrix} \qquad F_z(2) = \begin{pmatrix} -0.2 & 1 \\ 1 & 0 \end{pmatrix}$$

$$D_a(1) = \begin{pmatrix} 1 \\ 0.3 \end{pmatrix} \qquad D_a(2) = \begin{pmatrix} 0 \\ 0.3 \end{pmatrix}$$

$$D_d(1) = \begin{pmatrix} 0.2 \\ 1 \end{pmatrix} \qquad D_d(2) = \begin{pmatrix} 0 \\ 1 \end{pmatrix}$$

$$D_c(1) = \begin{pmatrix} 0.2 \\ 1 \end{pmatrix} \qquad D_c(2) = \begin{pmatrix} 0 \\ 1 \end{pmatrix}$$

$$D_{CZ}(1) = \begin{pmatrix} -0.2 \\ 1 \end{pmatrix} \qquad D_{CZ}(2) = \begin{pmatrix} 1 \\ 1 \end{pmatrix}$$

$$D_f(1) = \begin{pmatrix} 0.2 \\ 1 \end{pmatrix} \qquad D_f(2) = \begin{pmatrix} 0 \\ 1 \end{pmatrix}$$

$$
\begin{array}{ll}
E_a(1) = \begin{pmatrix} 0 & 0.2 \end{pmatrix} & E_a(2) = \begin{pmatrix} 1 & 0.2 \end{pmatrix} \\
E_b(1) = \begin{pmatrix} 1 & 1 \end{pmatrix} & E_b(2) = \begin{pmatrix} 0 & -1 \end{pmatrix} \\
E_c(1) = \begin{pmatrix} 1 & -1 \end{pmatrix} & E_c(2) = \begin{pmatrix} 0 & 1 \end{pmatrix} \\
E_{CZ}(1) = \begin{pmatrix} 1 & 0.1 \end{pmatrix} & E_{CZ}(2) = \begin{pmatrix} -0.1 & 1 \end{pmatrix} \\
E_d(1) = \begin{pmatrix} 0 & 0.1 \end{pmatrix} & E_d(2) = \begin{pmatrix} 0 & -1 \end{pmatrix} \\
E_f(1) = \begin{pmatrix} 1 & 0 \end{pmatrix} & E_f(2) = \begin{pmatrix} 0 & 1 \end{pmatrix}.
\end{array}
$$

With these data, solving $\mathcal{P}_0$ yields the following set of solutions.

$$\mu_0 = 141.3702 \qquad P(1) = \begin{pmatrix} 5.6761 & -1.2398 \\ -1.2398 & 6.0142 \end{pmatrix}$$

$$P(2) = \begin{pmatrix} 17.3735 & 0.4549 \\ 0.4549 & 4.8569 \end{pmatrix} \qquad Q = \begin{pmatrix} 222.4522 & 11.7423 \\ 11.7423 & 216.3246 \end{pmatrix}$$

$$\varepsilon = 0.4452.$$

Therefore, according to Theorem 11.2 the system under study is stochastically stable and verifies the noise attenuation level $\sqrt{\mu_0} = 11.889$.

11.1.2   Synthesis of Robust Controller

This subsection considers the problem of designing a memoryless state feedback controller of the form of (8.2) which guarantees that the closed-loop system is stochastically stable and verifies a given noise attenuation level γ. To this end, let us first of all develop another stability condition for system (11.1) with $u(t) \equiv 0$ that can be used as a basis for the results of this subsection. We will assume complete access to the state vector and to the mode system at each time t.

Let us begin with Theorem 11.1. Using the Schur complement, (11.4) holds if and only if the following holds for all admissible uncertainties and for all $r_t \in \mathcal{S}$

$$\mathcal{N}_1(r_t, t) + P(r_t)A_d(r_t, t)Q^{-1}A_d^\top(r_t, t)P(r_t) + [P(r_t)B_w(r_t)$$
$$+ C_z^\top(r_t, t)G_z(r_t)][\gamma^2 I - G_z^\top(r_t)G_z(r_t)]^{-1}$$
$$\times [P(r_t)B_w(r_t) + C_z^\top(r_t, t)G_z(r_t)]^\top \triangleq \mathcal{H}_3(r_t, t) < 0. \qquad (11.16)$$

Since

$$A^\top(r_t, t)P(r_t) + P(r_t)A(r_t, t) = A^\top(r_t)P(r_t) + P(r_t)A(r_t)$$
$$+ P(r_t)D_a(r_t)\Delta(r_t, t)E_a(r_t)$$
$$+ E_a^\top(r_t)\Delta^\top(r_t, t)D_a^\top(r_t)P(r_t),$$

using Lemma A.5, we have for any $\varepsilon_1 > 0$,

$$A^\top(r_t, t)P(r_t) + P(r_t)A(r_t, t) \leq A^\top(r_t)P(r_t) + P(r_t)A(r_t)$$
$$+ \varepsilon_1 P(r_t)D_a(r_t)D_a^\top(r_t)P(r_t) + \frac{1}{\varepsilon_1}E_a^\top(r_t)E_a(r_t). \qquad (11.17)$$

By virtue of Lemma A.6, if $\varepsilon_2 I - E_d(r_t)Q^{-1}E_d^\top(r_t) > 0$, then

$$A_d(r_t, t)Q^{-1}A_d^\top(r_t, t) \leq A_d(r_t)Q^{-1}A_d^\top(r_t) + A_d(r_t)Q^{-1}E_d^\top(r_t)$$
$$\times [\varepsilon_2 I - E_d(r_t)Q^{-1}E_d^\top(r_t)]^{-1}E_d(r_t)Q^{-1}A_d^\top(r_t)$$
$$+ \varepsilon_2 D_d(r_t)D_d^\top(r_t). \qquad (11.18)$$

Combining now (11.17) and (11.18) yields

$$\mathcal{H}_3(r_t, t) \leq \mathcal{N}_4(r_t) + C_z^\top(r_t, t)C_z(r_t, t)$$
$$+ [P(r_t)B_w(r_t) + C_z^\top(r_t, t)G_z(r_t)][\gamma^2 I - G_z^\top(r_t)G_z(r_t)]^{-1}$$
$$\times [P(r_t)B_w(r_t) + C_z^\top(r_t, t)G_z(r_t)]^\top \triangleq \mathcal{H}_4(r_t, t),$$

where

$$\mathcal{N}_4(r_t) = A^\top(r_t)P(r_t) + P(r_t)A(r_t) + \varepsilon_1 P(r_t)D_a(r_t)D_a^\top(r_t)P(r_t)$$
$$+ \frac{1}{\varepsilon_1}E_a^\top(r_t)E_a(r_t) + \sum_{j=1}^{N}\lambda_{r_tj}P(j) + \varrho Q$$
$$+ P(r_t)A_d(r_t)Q^{-1}A_d^\top(r_t)P(r_t)$$

$$+ \varepsilon_2 P(r_t) D_d(r_t) D_d^\top(r_t) P(r_t) + P(r_t) A_d(r_t) Q^{-1} E_d^\top(r_t)$$
$$\times [\varepsilon_2 I - E_d(r_t) Q^{-1} E_d^\top(r_t)]^{-1} E_d(r_t) Q^{-1} A_d^\top(r_t) P(r_t).$$

Using the Schur complement, $\mathcal{H}_4(r_t, t) < 0$ if and only if

$$\begin{pmatrix} \mathcal{N}_4(r_t) + C_z^\top(r_t, t) C_z(r_t, t) & P(r_t) B_w(r_t) + C_z^\top(r_t, t) G_z(r_t) \\ B_w^\top(r_t) P(r_t) + G_z^\top(r_t) C_z(r_t, t) & -\gamma^2 I + G_z^\top(r_t) G_z(r_t) \end{pmatrix} < 0. \tag{11.19}$$

Noting that

$$\begin{pmatrix} \mathcal{N}_4(r_t) + C_z^\top(r_t, t) C_z(r_t, t) & P(r_t) B_w(r_t) + C_z^\top(r_t, t) G_z(r_t) \\ B_w^\top(r_t) P(r_t) + G_z^\top(r_t) C_z(r_t, t) & -\gamma^2 I + G_z^\top(r_t) G_z(r_t) \end{pmatrix}$$
$$= \begin{pmatrix} \mathcal{N}_4(r_t) & P(r_t) B_w(r_t) \\ B_w^\top(r_t) P(r_t) & -\gamma^2 I \end{pmatrix}$$
$$+ \begin{pmatrix} C_z^\top(r_t, t) \\ G_z^\top(r_t) \end{pmatrix} \begin{pmatrix} C_z(r_t, t) & G_z(r_t) \end{pmatrix}.$$

Therefore, (11.19) is equivalent to

$$\begin{pmatrix} \mathcal{N}_4(r_t) & P(r_t) B_w(r_t) & C_z^\top(r_t, t) \\ B_w^\top(r_t) P(r_t) & -\gamma^2 I & G_z^\top(r_t) \\ C_z(r_t, t) & G_z(r_t) & -I \end{pmatrix} < 0. \tag{11.20}$$

Moreover,

$$\begin{pmatrix} \mathcal{N}_4(r_t) & P(r_t) B_w(r_t) & C_z^\top(r_t, t) \\ B_w^\top(r_t) P(r_t) & -\gamma^2 I & G_z^\top(r_t) \\ C_z(r_t, t) & G_z(r_t) & -I \end{pmatrix}$$
$$= \begin{pmatrix} \mathcal{N}_4(r_t) & P(r_t) B_w(r_t) & C_z^\top(r_t) \\ B_w^\top(r_t) P(r_t) & -\gamma^2 I & G_z^\top(r_t) \\ C_z(r_t) & G_z(r_t) & -I \end{pmatrix}$$
$$+ \begin{pmatrix} \mathbf{0} \\ \mathbf{0} \\ D_{CZ}(r_t) \end{pmatrix} \Delta(r_t, t) \begin{pmatrix} E_{CZ}(r_t) & \mathbf{0} & \mathbf{0} \end{pmatrix}$$
$$+ \begin{pmatrix} E_{CZ}^\top(r_t) \\ \mathbf{0} \\ \mathbf{0} \end{pmatrix} \Delta^\top(r_t, t) \begin{pmatrix} \mathbf{0} & \mathbf{0} & D_{CZ}(r_t) \end{pmatrix}.$$

In view of Lemma A.5, we have the following for any $\varepsilon_3 > 0$,

$$\begin{pmatrix} \mathcal{N}_4(r_t) & P(r_t) B_w(r_t) & C_z^\top(r_t, t) \\ B_w^\top(r_t) P(r_t) & -\gamma^2 I & G_z^\top(r_t) \\ C_z(r_t, t) & G_z(r_t) & -I \end{pmatrix}$$
$$\leq \begin{pmatrix} \mathcal{N}_4(r_t) + \frac{1}{\varepsilon_3} E_{CZ}^\top(r_t) E_{CZ}(r_t) & P(r_t) B_w(r_t) & C_z^\top(r_t) \\ B_w^\top(r_t) P(r_t) & -\gamma^2 I & G_z^\top(r_t) \\ C_z(r_t) & G_z(r_t) & -I + \varepsilon_3 D_{CZ}(r_t) D_{CZ}^\top(r_t) \end{pmatrix}$$
$$\triangleq \mathcal{H}_5(r_t). \tag{11.21}$$

Let $X_i = P^{-1}(i), i \in \mathcal{S}$. Pre- and post-multiplying $\mathcal{H}_5(i)$ by $\Xi_i = \text{diag}\{X_i, I, I\}$ yields

$$\Xi_i \mathcal{H}_5(i)\Xi_i =$$
$$\begin{pmatrix} \begin{bmatrix} X_i\mathcal{N}_4(i)X_i \\ + \frac{1}{\varepsilon_3}X_i E_{CZ}^\top(i)E_{CZ}(i)X_i \end{bmatrix} & B_w(i) & X_iC_z^\top(i) \\ B_w^\top(i) & -\gamma^2 I & G_z^\top(i) \\ C_z(i)X_i & G_z(i) & -I + \varepsilon_3 D_{CZ}(i)D_{CZ}^\top(i) \end{pmatrix},$$

where

$$X_i\mathcal{N}_4(i)X_i = X_iA^\top(i) + A(i)X_i + \varepsilon_1 D_a(i)D_a^\top(i)$$
$$+ \frac{1}{\varepsilon_1}X_i E_a^\top(i)E_a(i)X_i + \lambda_{ii}X_i + X_i\left[\sum_{j=1,j\neq i}^N \lambda_{ij}X_j^{-1}\right]X_i$$
$$+ \varrho X_iQX_i + A_d(i)Q^{-1}A_d^\top(i) + A_d(r_t)Q^{-1}E_d^\top(i)$$
$$\times [\varepsilon_2 I - E_d(i)Q^{-1}E_d^\top(i)]^{-1}E_d(i)Q^{-1}A_d^\top(i)$$
$$+ \varepsilon_2 D_d(i)D_d^\top(i). \tag{11.22}$$

Let $S_i(X)$ and $\mathcal{X}_i$ be defined as in (8.11) and (8.12), respectively. Noting that

$$S_i(X)\mathcal{X}_i S_i^\top(X) = X_i\left[\sum_{j=1,j\neq i}^N \lambda_{ij}X_j^{-1}\right]X_i$$

and using the Schur complement, we get that

$$\Xi_i \mathcal{H}_5(i)\Xi_i < 0 \tag{11.23}$$

holds if and only if

$$\begin{pmatrix} T_{11}(i) & B_w(i) & X_iC_z^\top(i) & T_{14}(i) & S_i(X) \\ B_w^\top(i) & -\gamma^2 I & G_z^\top(i) & 0 & 0 \\ C_z(i)X_i & G_z(i) & -I + \varepsilon_3 D_{CZ}(i)D_{CZ}^\top(i) & 0 & 0 \\ T_{14}^\top(i) & 0 & 0 & -T_{44}(i) & 0 \\ S_i^\top(X) & 0 & 0 & 0 & -\mathcal{X}_i \end{pmatrix} < 0,$$

where

$$T_{11}(i) = X_iA^\top(i) + A(i)X_i + \varepsilon_1 D_a(i)D_a^\top(i) + \lambda_{ii}X_i$$
$$+ A_d(i)Q^{-1}A_d^\top(i) + \varepsilon_2 D_d(i)D_d^\top(i)$$
$$T_{14}(i) = \begin{pmatrix} X_iE_{CZ}^\top(i) & X_iE_a^\top(i) & X_i & A_d(i)Q^{-1}E_d^\top(i) \end{pmatrix}$$
$$T_{44}(i) = \begin{pmatrix} \varepsilon_3 I & & & \\ & \varepsilon_1 I & & \\ & & \frac{1}{\varrho}Q^{-1} & \\ & & & \varepsilon_2 I - E_d(i)Q^{-1}E_d^\top(i) \end{pmatrix}.$$

Letting $U = Q^{-1}$, we obtain the following lemma.

Lemma 11.1 *Let γ be a given positive constant. If there exist symmetric, positive-definite matrices $X = (X_1, \cdots, X_N) > 0$, $U > 0$ and scalars $\varepsilon_j > 0, 1 \le j \le 3$, such that for all $i \in \mathcal{S}$*

$$\begin{pmatrix} \mathcal{T}_{11}(i) & B_w(i) & X_i C_z^\top(i) & \mathcal{T}_{14}(i) & S_i(X) \\ B_w^\top(i) & -\gamma^2 I & G_z^\top(i) & 0 & 0 \\ C_z(i)X_i & G_z(i) & -I + \varepsilon_3 D_{CZ}(i)D_{CZ}^\top(i) & 0 & 0 \\ \mathcal{T}_{14}^\top(i) & 0 & 0 & -\mathcal{T}_{44}(i) & 0 \\ S_i^\top(X) & 0 & 0 & 0 & -\mathcal{X}_i \end{pmatrix} < 0$$

(11.24)

holds, then system (11.1) with $u(t) \equiv 0$ is robustly stochastically stable and verifies the noise attenuation level γ, where

$$\mathcal{T}_{11}(i) = X_i A^\top(i) + A(i)X_i + \varepsilon_1 D_a(i)D_a^\top(i) + \lambda_{ii}X_i$$
$$+ A_d(i)U A_d^\top(i) + \varepsilon_2 D_d(i)D_d^\top(i),$$
$$\mathcal{T}_{14}(i) = \begin{pmatrix} X_i E_{CZ}^\top(i) & X_i E_a^\top(i) & X_i & A_d(i)U E_d^\top(i) \end{pmatrix}$$

$$\mathcal{T}_{44}(i) = \begin{pmatrix} \varepsilon_3 I & & & \\ & \varepsilon_1 I & & \\ & & \frac{1}{\varrho}U & \\ & & & \varepsilon_2 I - E_d(i)U E_d^\top(i) \end{pmatrix}.$$

With Lemma 11.1, we can consider the problem of synthesizing a stabilizing memoryless controller of the form of (8.2). Substituting (8.2) into (11.1) yields the dynamics of the closed-loop system as follows:

$$\begin{cases} \dot{x}(t) = \bar{A}(r_t, t)x(t) + \bar{A}_d(r_t, t)x(t - \tau(r_t)) + B_w(r_t)w(t) \\ z(t) = \bar{C}_z(r_t, t)x(t) + G_z(r_t)w(t), \end{cases}$$

(11.25)

where

$$\bar{A}(r_t, t) = \bar{A}(r_t) + D_a(r_t)\Delta(r_t, t)\bar{E}_a(r_t)$$
$$\bar{C}_z(r_t, t) = \bar{C}_z(r_t) + D_{CZ}(r_t)\Delta(r_t, t)\bar{E}_{CZ}(r_t),$$

with

$$\bar{A}(r_t) = A(r_t) + B(r_t)K(r_t) \qquad \bar{E}_a(r_t) = E_a(r_t) + E_b(r_t)K(r_t)$$
$$\bar{C}_z(r_t) = C_z(r_t) + F_z(r_t)K(r_t) \quad \bar{E}_{CZ}(r_t) = E_{CZ}(r_t) + E_{fz}(r_t)K(r_t).$$

From Lemma 11.1, it follows that the closed-loop system (11.25) is robustly stochastically stable and verifies the noise attenuation level γ if there exist symmetric, positive-definite matrices $X = (X_1, \cdots, X_N) > 0$, $U > 0$ and scalars $\varepsilon_j > 0, 1 \le j \le 3$, such that for all $i \in \mathcal{S}$

$$\begin{pmatrix} \bar{\mathcal{T}}_{11}(i) & B_w(i) & X_i \bar{C}_z^\top(i) & \bar{\mathcal{T}}_{14}(i) & S_i(X) \\ B_w^\top(i) & -\gamma^2 I & G_z^\top(i) & 0 & 0 \\ \bar{C}_z(i)X_i & G_z(i) & -I + \varepsilon_3 D_{CZ}(i)D_{CZ}^\top(i) & 0 & 0 \\ \bar{\mathcal{T}}_{14}^\top(i) & 0 & 0 & -\mathcal{T}_{44}(i) & 0 \\ S_i^\top(X) & 0 & 0 & 0 & -\mathcal{X}_i \end{pmatrix} < 0$$

$$(11.26)$$

holds, then system (11.1) with $u(t) \equiv 0$ is robustly stochastically stable and verifies the noise attenuation level γ, where

$$\bar{T}_{11}(i) = X_i \bar{A}^\top(i) + \bar{A}(i)X_i + \varepsilon_1 D_a(i)D_a^\top(i) + \lambda_{ii}X_i$$
$$+ A_d(i)UA_d^\top(i) + \varepsilon_2 D_d(i)D_d^\top(i)$$
$$\bar{T}_{14}(i) = \left(X_i \bar{E}_{CZ}^\top(i) \quad X_i \bar{E}_a^\top(i) \quad X_i \quad A_d(i)UE_d^\top(i) \right).$$

From the above derivation, we get the following theorem.

Theorem 11.3 *Let γ be a given positive constant. If there exist symmetric, positive-definite matrices $X = (X_1, \cdots, X_N) > 0$, $U > 0$, matrices $Y = (Y_1, \cdots, Y_N)$ and scalars $\varepsilon_j > 0, 1 \le j \le 3$, such that for any $i \in \mathcal{S}$,*

$$\begin{pmatrix} \mathcal{H}_{11}(i) & B_w(i) & X_i C_z^\top(i) + Y_i^\top F_z^\top(i) & \mathcal{H}_{14}(i) & S_i(X) \\ B_w^\top(i) & -\gamma^2 I & G_z^\top(i) & 0 & 0 \\ \begin{bmatrix} C_z(i)X_i \\ +F_z(i)Y_i \end{bmatrix} & G_z(i) & -I + \varepsilon_3 D_{CZ}(i)D_{CZ}^\top(i) & 0 & 0 \\ \mathcal{H}_{14}^\top(i) & 0 & 0 & -\mathcal{H}_{44}(i) & 0 \\ S_i^\top(X) & 0 & 0 & 0 & -\mathcal{X}_i \end{pmatrix}$$
$$< 0 \quad (11.27)$$

holds, then the closed-loop system (11.1) with controller (8.2) with $K(i) = Y_i X_i^{-1}$, $i \in \mathcal{S}$, is robustly stochastically stable and verifies the noise attenuation level γ, where

$$\mathcal{H}_{11}(i) = X_i A^\top(i) + A(i)X_i + B(i)Y_i + Y_i^\top B^\top(i) + \varepsilon_1 D_a(i)D_a^\top(i)$$
$$+ \lambda_{ii}X_i + A_d(i)UA_d^\top(i) + \varepsilon_2 D_d(i)D_d^\top(i)$$
$$\mathcal{H}_{14}(i) = \left(X_i E_{CZ}^\top(i) + Y_i^\top E_{fz}^\top(i) \quad X_i E_a^\top(i) + Y_i^\top E_b^\top(i) \right.$$
$$\left. X_i \quad A_d(i)UE_d^\top(i) \right)$$
$$\mathcal{H}_{44}(i) = \begin{pmatrix} \varepsilon_3 I & & & \\ & \varepsilon_1 I & & \\ & & \frac{1}{\varrho}U & \\ & & & \varepsilon_2 I - E_d(i)UE_d^\top(i) \end{pmatrix}.$$

Proof: Letting $Y_i = K(i)X_i$ in (11.26) leads to Theorem 11.3.   □

In Theorem 11.3, the noise attenuation level γ is assumed to be a given constant. In case of γ being an unknown constant, the optimization problem of finding the gain matrices $K = (K(1), \cdots, K(N))$ such that the closed-loop system is stochastically stable and minimizes the noise attenuation level γ can be cast into a linear optimization problem. For this purpose, let us consider the following LMI:

$$
\begin{pmatrix}
\mathcal{H}_{11}(i) & B_w(i) & X_i C_z^\top(i) + Y_i^\top F_z^\top(i) & \mathcal{H}_{14}(i) & S_i(X) \\
B_w^\top(i) & -\rho I & G_z^\top(i) & 0 & 0 \\
\begin{bmatrix} C_z(i)X_i \\ + F_z(i)Y_i \end{bmatrix} & G_z(i) & -I + \varepsilon_3 D_{CZ}(i)D_{CZ}^\top(i) & 0 & 0 \\
\mathcal{H}_{14}^\top(i) & 0 & 0 & -\mathcal{H}_{44}(i) & 0 \\
S_i^\top(X) & 0 & 0 & 0 & -\mathcal{X}_i
\end{pmatrix}
< 0. \qquad (11.28)
$$

Let $X = (X_1, \cdots, X_N) > 0$, $Y = (Y_1, \cdots, Y_N)$, $U > 0$, scalars $\varepsilon_j > 0, 1 \leq j \leq 3$, and let μ_0 be the solution to the optimization problem

$$
\mathcal{P}_1: \quad \begin{array}{c} \min \\ \mu > 0, (X_1, \cdots, X_N) > 0, \\ U > 0, (Y_1, \cdots, Y_N), \varepsilon_j > 0 \end{array} \quad \mu
$$

s.t. (11.28)

then controller (8.2) with $K(i) = Y_i X_i^{-1}$ stabilizes system (11.1) and the closed-loop system verifies the noise attenuation level $\sqrt{\mu_0}$.

As we did previously, if the time delay doesn't depend on the mode system, from Theorem 11.3, we obtain the following corollary.

Corollary 11.2 *If the time delay in system (11.1) satisfies $\tau(r_t) \equiv \tau$ and there exist symmetric, positive-definite matrices $X = (X_1, \cdots, X_N) > 0$, $U > 0$, matrices $Y = (Y_1, \cdots, Y_N)$ and scalars $\varepsilon_j > 0, 1 \leq j \leq 3$, such that for all $i \in \mathcal{S}$,*

$$
\begin{pmatrix}
\mathcal{H}_{11}(i) & B_w(i) & X_i C_z^\top(i) + Y_i^\top F_z^\top(i) & \mathcal{H}_{14}(i) & S_i(X) \\
B_w^\top(i) & -\gamma^2 I & G_z^\top(i) & 0 & 0 \\
\begin{bmatrix} C_z(i)X_i \\ +F_z(i)Y_i \end{bmatrix} & G_z(i) & -I + \varepsilon_3 D_{CZ}(i)D_{CZ}^\top(i) & 0 & 0 \\
\mathcal{H}_{14}^\top(i) & 0 & 0 & -\hat{\mathcal{H}}_{44}(i) & 0 \\
S_i^\top(X) & 0 & 0 & 0 & -\mathcal{X}_i
\end{pmatrix}
< 0 \qquad (11.29)
$$

holds, where $\mathcal{H}_{11}(i)$, $\mathcal{H}_{14}(i)$ are defined in (11.27) and

$$
\hat{\mathcal{H}}_{44}(i) = \begin{pmatrix} \varepsilon_3 I & & & \\ & \varepsilon_1 I & & \\ & & U & \\ & & & \varepsilon_2 I - E_d(i) U E_d^\top(i) \end{pmatrix},
$$

then the closed-loop system (11.1) under controller (8.2) with $K(i) = Y_i X_i^{-1}, i \in \mathcal{S}$, is robustly stochastically stable and verifies the noise attenuation level γ.

Example 11.2 *Let us consider a system with dynamics described by (11.1) and suppose that the system has two modes, that is, $\mathcal{S} = \{1, 2\}$, with the* *infinitesimal generator*

$$\Lambda = \begin{pmatrix} -3 & 3 \\ 2 & -2 \end{pmatrix}$$

with $\tau_1 = 0.2$, $\tau_2 = 1$.

Let us also suppose that the system parameters are as follows:

$$A(1) = \begin{pmatrix} 2 & 0 \\ 0 & -0.9 \end{pmatrix} \qquad A(2) = \begin{pmatrix} 2.1 & -0.1 \\ 0.1 & 1 \end{pmatrix}$$

$$B(1) = \begin{pmatrix} 0 & 1 \\ 0 & 1 \end{pmatrix} \qquad B(2) = \begin{pmatrix} 1 & 0 \\ 0 & 1 \end{pmatrix}$$

$$A_d(1) = \begin{pmatrix} -0.1 & 0 \\ -0.1 & -0.1 \end{pmatrix} \qquad A_d(2) = \begin{pmatrix} -0.1 & 0 \\ -0.1 & -0.1 \end{pmatrix}$$

$$B_w(1) = \begin{pmatrix} 1 & 0 \\ 0 & 0 \end{pmatrix} \qquad B_w(2) = \begin{pmatrix} 0.8 & 0 \\ 0 & 0 \end{pmatrix}$$

$$C(1) = \begin{pmatrix} 0.3 & 0.1 \\ 0 & 1 \end{pmatrix} \qquad C(2) = \begin{pmatrix} 0.2 & -0.1 \\ 0.1 & 0 \end{pmatrix}$$

$$C_z(1) = \begin{pmatrix} 0.1 & 0.1 \\ 0 & 0 \end{pmatrix} \qquad C_z(2) = \begin{pmatrix} 0.2 & -1 \\ 1 & 0 \end{pmatrix}$$

$$F_z(1) = \begin{pmatrix} -0.1 & 1 \\ 0 & 0 \end{pmatrix} \qquad F_z(2) = \begin{pmatrix} -0.2 & 1 \\ 1 & 0 \end{pmatrix}$$

$$G_z(1) = \begin{pmatrix} 0.1 & 0.1 \\ 0 & 0 \end{pmatrix} \qquad G_z(2) = \begin{pmatrix} 0.2 & -1 \\ 0.2 & 0 \end{pmatrix}$$

$$D_a(1) = \begin{pmatrix} 1 \\ 0.3 \end{pmatrix} \qquad D_a(2) = \begin{pmatrix} 0 \\ 0.3 \end{pmatrix}$$

$$D_d(1) = \begin{pmatrix} 0.2 \\ 1 \end{pmatrix} \qquad D_d(2) = \begin{pmatrix} 0 \\ 1 \end{pmatrix}$$

$$D_c(1) = \begin{pmatrix} 0.2 \\ 1 \end{pmatrix} \qquad D_c(2) = \begin{pmatrix} 0 \\ 1 \end{pmatrix}$$

$$D_{CZ}(1) = \begin{pmatrix} -0.2 \\ 1 \end{pmatrix} \qquad D_{CZ}(2) = \begin{pmatrix} 1 \\ 1 \end{pmatrix}$$

$$D_f(1) = \begin{pmatrix} 0.2 \\ 1 \end{pmatrix} \qquad D_f(2) = \begin{pmatrix} 0 \\ 1 \end{pmatrix}$$

$$E_a(1) = \begin{pmatrix} 0 & 0.2 \end{pmatrix}, \qquad E_a(2) = \begin{pmatrix} 1 & 0.2 \end{pmatrix}$$
$$E_b(1) = \begin{pmatrix} 1 & 1 \end{pmatrix} \qquad E_b(2) = \begin{pmatrix} 0 & -1 \end{pmatrix}$$
$$E_c(1) = \begin{pmatrix} 1 & -1 \end{pmatrix} \qquad E_c(2) = \begin{pmatrix} 0 & 1 \end{pmatrix}$$
$$E_{CZ}(1) = \begin{pmatrix} 1 & 0.1 \end{pmatrix} \qquad E_{CZ}(2) = \begin{pmatrix} -0.1 & 1 \end{pmatrix}$$
$$E_d(1) = \begin{pmatrix} 0 & 0.1 \end{pmatrix} \qquad E_d(2) = \begin{pmatrix} 0 & -1 \end{pmatrix}$$
$$E_f(1) = \begin{pmatrix} 0.1 & 0 \end{pmatrix} \qquad E_f(2) = \begin{pmatrix} 0 & 0.1 \end{pmatrix}$$

$$F(1) = \begin{pmatrix} 0.1 & 0 \\ 0 & 0.1 \end{pmatrix} \qquad F(2) = \begin{pmatrix} 0.1 & 0 \\ 0 & -1 \end{pmatrix}$$

$$E_{fz}(1) = \begin{pmatrix} 0.1 & 0 \end{pmatrix} \qquad E_{fz}(2) = \begin{pmatrix} 0 & 0.1 \end{pmatrix}$$

With these data, solving $\mathcal{P}_1$ yields the following set of solutions:

$$\mu_0 = 197.1969 \qquad\qquad X_1 = \begin{pmatrix} 0.0343 & 0.2188 \\ 0.2188 & 3.6392 \end{pmatrix}$$

$$X_2 = \begin{pmatrix} 0.1616 & -0.0292 \\ -0.0292 & 0.0189 \end{pmatrix} \quad U = \begin{pmatrix} 12.1185 & -9.7491 \\ -9.7491 & 18.8334e \end{pmatrix}$$

$$\varepsilon_1 = 0.3856 \; \varepsilon_2 = 0.0214 \qquad \varepsilon_3 = 0.2174.$$

According to Theorem 11.3, the system under study is robustly stabilizable and a set of stabilizing gains can be given by

$$K(1) = \begin{pmatrix} 32.6459 & -1.9238 \\ -34.8656 & 1.8209 \end{pmatrix}$$

$$K(2) = \begin{pmatrix} -7.4822 & -8.7867 \\ -2.5825 & -17.5115 \end{pmatrix}$$

which verifies the noise attenuation level $\gamma = \sqrt{\mu_0} = 14.0427$.

11.2 Guaranteed Cost Control

The problem we will address in this section is the guaranteed cost control(GCC) problem. It consists of designing a controller that robustly stabilizes systems (11.1) and at the same guarantees a given upper bound for the cost function. Mathematically, it can be formulated as follows: Given a dynamical system with the appropriate assumptions, find a controller $u(.)$ that robustly stabilizes the systems and at the same time guarantees that the cost (11.31) is bounded for all admissible uncertainties $\Delta(t)$ and $\Delta_d(t)$.

To address the guaranteed cost control problem let us use the system dynamics (11.1) with $w(t) \equiv 0$, i.e.,

$$\begin{cases} \dot{x}(t) = A(r_t, t)x(t) + A_d(r_t, t)x(t - \tau(r_t)) \\ \qquad + B(r_t, t)u(t) \;\; x(s) = \phi(s), \; [-\bar{\tau}, 0] \\ y(t) = C(r_t, t)x(t) + F(r_t, t)u(t) + C_d(r_t, t)x(t - \tau(r_t)) \\ z(t) = C_z(r_t, t)x(t) + F_z(r_t, t)u(t). \end{cases} \quad (11.30)$$

Let us consider the cost function defined by

$$J(r_0, \phi(\cdot)) = \mathbb{E}\left[\int_0^\infty [x^\top(t)R_1(r_t)x(t) + u^\top(t)R_2(r_t)u(t)]dt | r_0, \phi(\cdot) \right]$$

$$(11.31)$$

where $R_1(i), R_2(i) \geq 0, i \in \mathcal{S}$ are a set of semipositive matrices.

This section deals with the design of a controller $u(\cdot)$ that stochastically stabilizes system (11.30) and guarantees a certain upper bound of cost index J defined by (11.31). For this purpose, let us first compute the guaranteed cost performance for system (11.30) with $u(t) \equiv 0$.

11.2.1 Guaranteed Cost Performance

Let us in this subsection establish the guaranteed performance for the unforced system. Plugging in $u(t) \equiv 0$, the cost function becomes

$$J_0(r_0, \phi(\cdot)) = \mathbb{E}\left[\int_0^\infty [x^\top(t)R_1(r_t)x(t)]dt | r_0, \phi(\cdot)\right], \tag{11.32}$$

The following theorem gives our first result on guaranteed cost.

Theorem 11.4 *If there exist matrices* $P = (P(1), \cdots, P(N)) > 0, Q > 0$ *and a scalar* $\varepsilon > 0$ *such that for every* $i \in \mathcal{S}$

$$\begin{pmatrix} \# & P(i)A_d(i) & P(i)D_a(i) & P(i)D_d(i) \\ * & -Q + \varepsilon E_d^\top(i)E_d(i) & 0 & 0 \\ * & 0 & -\varepsilon I & 0 \\ * & 0 & 0 & -\varepsilon I \end{pmatrix} < 0 \tag{11.33}$$

holds, where

$$\# = A^\top(i)P(i) + P(i)A(i) + \sum_{j=1}^N \lambda_{ij}P(j)$$

$$+ \varrho Q + \varepsilon E_a^\top(i)E_a(i) + R_1(i),$$

then system (11.30) with $u(t) \equiv 0$ *is stochastically stable and the cost defined by (11.31) satisfies the following for all admissible uncertainties:*

$$J \leq \mathrm{tr}[P(r_t)x_0x_0^\top] + \mathrm{tr}\left[Q\int_{-\tau}^0 \phi(s)\phi^\top(s)ds\right]$$

$$+ \bar{\lambda}\mathrm{tr}\left[\int_{-\bar{\tau}}^{-\underline{\tau}} Q\phi(s)\phi^\top(s)ds\right]. \tag{11.34}$$

Proof: Since $R_1(i) > 0$, (11.33) implies (11.3), which means by Proposition 11.1 that the system under study is stochastically stable and thus the cost (11.31) is well defined. By the same argument as in Proposition 11.1, we obtain that for all admissible uncertainties

$$\Theta(r_t, t) + \begin{pmatrix} R_1(r_t) & 0 \\ 0 & 0 \end{pmatrix} \triangleq \hat{\Theta}(r_t, t) < 0, \ r_t \in \mathcal{S} \tag{11.35}$$

holds, where $\Theta(r_t, t)$ is defined by (11.2).

Let us consider a Lyapunov functional candidate as follows:

$$V(\mathbf{x}(t), r_t) = x^\top(t)P(r_t)x(t) + V_1(\mathbf{x}(t), r_t) + V_2(\mathbf{x}(t), r_t)$$

where

$$V_1(\mathbf{x}(t), r_t) = \int_{t-\tau(r_t)}^t x^\top(s)Qx(s)ds$$

$$V_2(\mathbf{x}(t), r_t) = \int_{-\bar{\tau}}^{-\underline{\tau}} \int_{t+\theta}^t x^\top(s)\bar{\lambda}Qx(s)dsd\theta.$$

Let $\mathcal{A}$ be the generator of the Markov process $\{(\mathbf{x}_t, r_t), t \geq 0\}$. Then, by the same argument as in Section 9.4, we get

$$\mathcal{A}V(\mathbf{x}_t, r_t) \leq \xi_t^\top \Theta(r_t, t)\xi_t,$$

where $\xi_t^\top = \begin{pmatrix} x^\top(t) & x^\top(t-\tau) \end{pmatrix}$.

This yields in turn the following inequality:

$$\mathcal{A}V(\mathbf{x}_t, r_t) + x^\top(t)R_1(r_t)x(t) \leq \xi_t^\top \hat{\Theta}(r_t, t)\xi_t.$$

Combining this with (11.35) yields that

$$\int_0^T \mathcal{A}V(\mathbf{x}_t, r_t)dt + \int_0^T x^\top(t)R(r_t)x(t)dt < 0 \qquad (11.36)$$

holds for any $T > 0$.

Using now Dynkin's formula, we have

$$\mathbb{E}\left[\int_0^T \mathcal{A}V(\mathbf{x}_t, r_t)dt|(r_0, \phi(\cdot))\right] = \mathbb{E}[V(\mathbf{x}_T, r_T)] - \mathbb{E}[V(\mathbf{x}_0, r_0)],$$

which combined with (11.36) yields

$$\mathbb{E}\left[\int_0^T x^\top(t)R_1(r_t)x(t)dt\right] \leq -\mathbb{E}\left[\int_0^T \mathcal{A}V(\mathbf{x}_t, r_t)dt|(r_0, \phi(\cdot))\right]$$

$$= \mathbb{E}[V(\mathbf{x}_0, r_0)] - \mathbb{E}[V(\mathbf{x}_T, r_T)]$$

$$\leq \mathbb{E}[V(\mathbf{x}_0, r_0)] = \mathbb{E}[x_0^\top P(r_0)x_0] + \int_{-\bar{\tau}}^0 \phi^\top(s)Q\phi(s)ds$$

$$+ \int_{-\bar{\tau}}^{-\underline{\tau}} \int_\theta^0 \phi^\top(s)\bar{\lambda}Q\phi(s)dsd\theta.$$

This proves (11.34), which completes the proof of Theorem 11.4. $\square$

Corollary 11.3 *Let $\tau(r_t) \equiv \tau$, where $\tau > 0$ is a constant. If there exist symmetric, positive-definite matrices $P = (P(1), \cdots, P(N)) > 0$, $Q > 0$ and a scalar $\varepsilon > 0$ such that for all $i \in \mathcal{S}$*

$$\begin{pmatrix} \# & P(i)A_d(i) & P(i)D_a(i) & P(i)D_d(i) \\ * & -Q + \varepsilon E_d^\top(i)E_d(i) & 0 & 0 \\ * & 0 & -\varepsilon I & 0 \\ * & 0 & 0 & -\varepsilon I \end{pmatrix} < 0 \qquad (11.37)$$

holds, where

$$\# = A^\top(i)P(i) + P(i)A(i) + \sum_{j=1}^N \lambda_{ij}P(j) + Q + \varepsilon E_a^\top(i)E_a(i) + R_1(i),$$

then system (11.30) with $u(t) \equiv 0$ is stochastically stable and the cost defined by (11.31) satisfies the following for all admissible uncertainties:

$$J \leq \mathrm{tr}[P(r_t)x_0 x_0^\top] + \mathrm{tr}\left[Q \int_{-\tau}^0 \phi(s)\phi^\top(s)ds\right].$$

Proof: The proof can be provided by following the same lines as in the proof of Theorem 11.4 and considering a Lyapunov functional candidate as follows:

$$V(\mathbf{x}_t, r_t) = x^\top(t)P(r_t)x(t) + \int_{t-\tau}^t x^\top(s)Qx(s)ds.$$

The details are omitted.   □

11.2.2   State Feedback Controller Design

Now we proceed with the design of a controller of the form of (8.2) that stochastically stabilizes system (11.30) and guarantees the cost is bounded. Substituting (8.2) into (11.30) gives the dynamics of the closed-loop as follows:

$$\dot{x}(t) = \bar{A}(r_t, t)x(t) + A_d(r_t, t)x(t - \tau(r_t)), \qquad (11.38)$$

where $\bar{A}(r_t, t) = \bar{A}(r_t) + D_a(r_t)\Delta(r_t, t)\bar{E}_a(r_t)$ with $\bar{A}(r_t) = A(r_t) + B(r_t)K(r_t)$ and $\bar{E}_a(r_t) = E_a(r_t) + E_b(r_t)K(r_t)$. Then, the cost for the closed-loop system becomes

$$\bar{J}(r_0, \phi(\cdot)) = \mathbb{E}\left[\int_0^\infty x^\top(t)\bar{R}(r_t)x(t)dt|r_0, \phi(\cdot)\right], \qquad (11.39)$$

where $\bar{R}(r_t) = R_1(r_t) + K^\top(r_t)R_2(r_t)K(r_t)$.

In view of Theorem 11.4, we get that if there exist symmetric, positive-definite matrices $P = (P(1), \cdots, P(N)) > 0$, $Q > 0$ and a scalar $\varepsilon > 0$ such that for any $i \in \mathcal{S}$

$$\begin{pmatrix} \mathcal{M}_1(i) & P(i)A_d(i) & P(i)D_a(i) & P(i)D_d(i) \\ * & -Q + \varepsilon E_d^\top(i)E_d(i) & 0 & 0 \\ * & 0 & -\varepsilon I & 0 \\ * & 0 & 0 & -\varepsilon I \end{pmatrix} < 0 \quad (11.40)$$

holds, where

$$\mathcal{M}_1(i) = \bar{A}^\top(i)P(i) + P(i)\bar{A}(i) + \sum_{j=1}^N \lambda_{ij}P(j)$$

$$+ \varrho Q + \varepsilon \bar{E}_a^\top(i)\bar{E}_a(i) + \bar{R}(i),$$

then system (11.30) under the state feedback controller is stochastically stable and the cost defined by (11.39) satisfies

$$\bar{J}(r_0, \phi(\cdot)) \leq \operatorname{tr}[P(r_t)x_0 x_0^\top] + \operatorname{tr}\left[Q \int_{-\tau}^0 \phi(s)\phi^\top(s)ds\right]$$

$$+ \ \bar{\lambda}\operatorname{tr}\left[\int_{-\bar{\tau}}^{-\tau} Q\phi(s)\phi^\top(s)ds\right]. \tag{11.41}$$

Obviously, (11.40) provides a sufficient condition for the closed-loop system to be stochastically stable and guarantees cost (11.41). Using the Schur complement, we obtain that (11.40) is equivalent to

$$\begin{pmatrix} \mathcal{M}_2(i) & P(i)A_d(i) \\ A_d^\top(i)P(i) & -Q + \varepsilon E_d^\top(i)E_d(i) \end{pmatrix} < 0, \tag{11.42}$$

where

$$\mathcal{M}_2(i) = \bar{A}^\top(i)P(i) + P(i)\bar{A}(i) + \sum_{j=1}^N \lambda_{ij}P(j)$$

$$+ \ \varrho Q + \varepsilon \bar{E}_a^\top(i)\bar{E}_a(i) + \bar{R}(i) + \frac{1}{\varepsilon}P(i)D_a(i)D_a^\top(i)P(i)$$

$$+ \ \frac{1}{\varepsilon}P(i)D_d(i)D_d^\top(i)P(i).$$

Moreover, (11.42) holds if and only if

$$\begin{pmatrix} \mathcal{M}_2(i) & P(i)A_d(i) & 0 \\ A_d^\top(i)P(i) & -Q & E_d^\top(i) \\ 0 & E_d(i) & -\frac{1}{\varepsilon}I \end{pmatrix} < 0. \tag{11.43}$$

Letting now $X_i = P^{-1}(i)$, $U = Q^{-1}$ and multiplying both sides of (11.43) by diag$\{X_i, U, I\}$ yields

$$\begin{pmatrix} X_i\mathcal{M}_2(i)X_i & A_d(i)U & 0 \\ UA_d^\top(i) & -U & UE_d^\top(i) \\ 0 & E_d(i)U & -\frac{1}{\varepsilon}I \end{pmatrix} < 0. \tag{11.44}$$

Letting $Y_i = K(i)X_i^{-1}$ in $X_i\mathcal{M}_2(i)X_i$, we obtain

$$X_i\mathcal{M}_2(i)X_i = A(i)X_i + B(i)Y_i + X_iA^\top(i) + Y_i^\top B^\top(i)$$

$$+ \ \lambda_{ii}X_i + X_i\left[\sum_{j\neq i}\lambda_{ij}X_j^{-1}\right]X_i + \varrho X_iU^{-1}X_i$$

$$+ \ \varepsilon[E_a(i)X_i + E_b(i)Y_i]^\top[E_a(i)X_i + E_b(i)Y_i]$$

$$+ \ \frac{1}{\varepsilon}D_a(i)D_a^\top(i) + \frac{1}{\varepsilon}D_d(i)D_d^\top(i)$$

$$+ \ X_iR_1(i)X_i + Y_i^\top R_2(i)Y_i.$$

Using the Schur complement and letting $\eta = 1/\varepsilon$, we obtain that (11.44) holds if and only if

$$\begin{pmatrix} \mathcal{M}_3(i) & A_d(i)U & 0 & \mathcal{M}_{13}(i) & S_i(X) \\ UA_d^\top(i) & -U & UE_d^\top(i) & 0 & 0 \\ 0 & E_d(i)U & -\eta I & 0 & 0 \\ \mathcal{M}_{13}^\top(i) & 0 & 0 & -\mathcal{M}_{33}(i) & 0 \\ S_i^\top(X) & 0 & 0 & 0 & -\mathcal{X}_i \end{pmatrix} < 0, \quad (11.45)$$

where $S_i(X)$, $\mathcal{X}_i$ are defined as in (8.11) and (8.12), respectively, and

$$\mathcal{M}_3(i) = A(i)X_i + B(i)Y_i + X_i A^\top(i) + Y_i^\top B^\top(i)$$
$$+ \lambda_{ii} X_i + \eta D_a(i)D_a^\top(i) + \eta D_d(i)D_d^\top(i)$$
$$\mathcal{M}_{13}(i) = \begin{pmatrix} X_i & X_i E_a^\top(i) + Y_i^\top E_b^\top(i) \end{pmatrix}$$
$$X_i[R_1^{1/2}(i)]^\top \quad Y_i^\top[R_2^{1/2}(i)]^\top \quad)$$

$$\mathcal{M}_{33}(i) = \begin{pmatrix} \frac{U}{\varrho} & 0 & 0 & 0 \\ 0 & \eta I & 0 & 0 \\ 0 & 0 & I & 0 \\ 0 & 0 & 0 & I \end{pmatrix}.$$

Therefore, we obtain the following theorem.

Theorem 11.5 *If there exist symmetric, positive-definite matrices $(X_1, \ldots, X_N) > 0$, matrices $(Y_1, \ldots, Y_N)$, $U > 0$ and a scalar $\eta > 0$ satisfying (11.45), for all $i \in \mathcal{S}$, then controller (8.2) with $K(i) = Y_i X_i^{-1}, i \in \mathcal{S}$, stabilizes system (11.30) and guarantees the following cost bound for all admissible uncertainties:*

$$\bar{J}(r_0, \phi(\cdot)) \leq \text{tr}[X_{r_0}^{-1} x_0 x_0^\top] + \text{tr}\left[U^{-1} \int_{-\tau}^{0} \phi(s)\phi^\top(s)ds\right]$$
$$+ \bar{\lambda}\text{tr}\left[\int_{-\bar{\tau}}^{-\underline{\tau}} U^{-1}\phi(s)\phi^\top(s)ds\right].$$

If the time delay in the system is mode-independent the result is summarized by the following corollary.

Corollary 11.4 *Let the time delay in system (11.30) be a constant, i.e., $\tau(r_t) \equiv \tau$. Then, if there exist symmetric, positive-definite matrices $X = (X_1, \cdots, X_N) > 0$, $U > 0$, matrices $(Y_1, \ldots, Y_n)$ and a scalar $\eta > 0$ satisfying for all $i \in \mathcal{S}$*

$$\begin{pmatrix} \mathcal{M}_3(i) & A_d(i)X_i & 0 & \mathcal{M}_{13}(i) & S_i(X) \\ X_i A_d^\top(i) & -U & UE_d^\top(i) & 0 & 0 \\ 0 & E_d(i)U & -\eta I & 0 & 0 \\ \mathcal{M}_{13}^\top(i) & 0 & 0 & -\hat{\mathcal{M}}_{33}(i) & 0 \\ S_i^\top(X) & 0 & 0 & 0 & -\mathcal{X}_i \end{pmatrix} < 0,$$

where $\mathcal{M}_3(i), \mathcal{M}_{13}(i), S_i(X)$ are the same as in (11.45) and

$$\hat{\mathcal{M}}_{33}(i) = \begin{pmatrix} U & 0 & 0 & 0 \\ 0 & \eta I & 0 & 0 \\ 0 & 0 & I & 0 \\ 0 & 0 & 0 & I \end{pmatrix},$$

then controller (8.2) with $K(i) = Y_i X_i^{-1}, i \in \mathcal{S}$, stabilizes system (11.30) and guarantees the following cost bound for all admissible uncertainties

$$\bar{J}(r_0, \phi(\cdot)) \le \text{tr}[X_{r_0}^{-1} x_0 x_0^\top] + \text{tr} \left[U^{-1} \int_{-\tau}^0 \phi(s) \phi^\top(s) ds \right].$$

Example 11.3 Let us consider a system described by (11.30). The Markov process that represents the system mode-switching takes values in $\mathcal{S} = \{1, 2\}$ and has the generator

$$\Lambda = \begin{pmatrix} -3 & 3 \\ 5 & -5 \end{pmatrix}.$$

The system state contains time delay $\tau_1 = 0.4$ and $\tau_2 = 0.51$ when the system is in mode $r_t = 1$ and $r_t = 2$, respectively. The other parameters are as follows:

$$A(1) = \begin{pmatrix} -1 & 0 \\ 0 & -2 \end{pmatrix} \qquad A(2) = \begin{pmatrix} 2 & 0.8 \\ 0.2 & -2 \end{pmatrix}$$

$$A_d(1) = \begin{pmatrix} -0.3 & 0 \\ 0 & -0.2 \end{pmatrix} \qquad A_d(2) = \begin{pmatrix} 0.1 & -0.1 \\ 0.1 & -0.1 \end{pmatrix}$$

$$B(1) = \begin{pmatrix} 0 \\ -0.1 \end{pmatrix} \qquad B(2) = \begin{pmatrix} 1 \\ 0.1 \end{pmatrix}$$

$$D(1) = \begin{pmatrix} 1 \\ 0.3 \end{pmatrix} \qquad D(2) = \begin{pmatrix} 0 \\ 0.3 \end{pmatrix}$$

$$D_d(1) = \begin{pmatrix} 0.2 \\ 1 \end{pmatrix} \qquad D_d(2) = \begin{pmatrix} 0 \\ 1 \end{pmatrix}$$

$$E_a(1) = \begin{pmatrix} 0 & 0.2 \end{pmatrix} \qquad E_a(2) = \begin{pmatrix} 1 & 0.2 \end{pmatrix}$$

$$E_b(1) = [0.3] \qquad E_b(2) = [0.2]$$

$$E_d(1) = \begin{pmatrix} 0 & 0.1 \end{pmatrix} \qquad E_d(2) = \begin{pmatrix} 0 & -1 \end{pmatrix}.$$

With these data, LMIs (11.45) have the following set of feasible solutions

$$X_1 = \begin{pmatrix} 0.2491 & -0.0473 \\ -0.0473 & 0.3279 \end{pmatrix} \qquad X_2 = \begin{pmatrix} 0.2329 & -0.0353 \\ -0.0353 & 0.3078 \end{pmatrix}$$

$$Y_1 = \begin{pmatrix} 0.0099 & 0.0118 \end{pmatrix} \qquad Y_2 = \begin{pmatrix} -1.1062 & -0.1152 \end{pmatrix}$$

$$U = \begin{pmatrix} 279.8586 & 0.1329 \\ 0.1329 & 0.1905 \end{pmatrix} \qquad \eta = 0.2443.$$

In view of Theorem 11.5, the system under study is robustly stochastically stable under controller (8.2) with gains

$$K(1) = Y_1 X_1^{-1} = \begin{pmatrix} 0.0477 & 0.0430 \end{pmatrix}$$
$$K(2) = Y_2 X_2^{-1} = \begin{pmatrix} -4.8917 & -0.9347 \end{pmatrix}.$$

The results of this section are based on the complete access to the state vector. This assumption is not always valid in practice and an alternative is required. The next section will deal with output feedback guaranteed cost.

11.3   Output Feedback Guaranteed Cost Control

In the previous section, we addressed the design of the state feedback controller under the assumption that the system state $x(t)$ and system mode r_t are perfectly available for feedback. When the system state is not available for feedback we have to recourse to a dynamical output feedback controller, which is the topic of this section.

The objective of this section is to design an output feedback controller of the form of (8.70) that stabilizes system (11.30) in the MSQS sense. For this purpose, let us assume that

$$
\begin{array}{ll}
D_a(i) = D_b(i) = D_1(i) & D_d(i) = D_2(i) \\
D_c(i) = D_f(i) = D_3(i) & D_{cd}(i) = D_4(i) \\
E_a(i) = E_c(i) = E_1(i) & E_b(i) = E_f(i) = E_2(i) \\
E_d(i) = E_3(i) & E_{cd}(i) = E_d(i) = E_3(i).
\end{array}
$$

Let $\hat{x}^\top(t) = \begin{pmatrix} x^\top(t) & \xi^\top(t) \end{pmatrix}$ be the augmented state vector of the closed-loop system. Then $\hat{x}(t)$ satisfies the dynamic

$$\dot{\hat{x}}(t) = \bar{A}(r_t, t)\hat{x}(t) + \bar{A}_d(r_t, t)\hat{x}(t - \tau(r_t)), \tag{11.46}$$

where

$$\bar{A}(r_t, t) = \begin{pmatrix} A(r_t, t) & B(r_t, t)K_C(r_t) \\ K_B(r_t)C(r_t, t) & K_A(r_t) + K_B(r_t)F(r_t, t)K_C(r_t) \end{pmatrix}$$

$$\bar{A}_d(r_t, t) = \begin{pmatrix} A_d(r_t, t) \\ K_B(r_t)C_d(r_t, t) \end{pmatrix}.$$

Let us define the matrices $\bar{A}(i)$, $\bar{A}_d(i)$, $\bar{D}_a(i)$, $\bar{E}_a(i)$, $\bar{D}_d(i)$, and $\bar{E}_d(i)$ as follows:

$$\bar{A}(i) = \begin{pmatrix} A(i) & B(i)K_C(i) \\ K_B(i)C(i) & K_A(i) + K_B(i)F(i)K_C(i) \end{pmatrix}$$

$$\bar{A}_d(i) = \begin{pmatrix} A_d(i) \\ K_B(i)C_d(i) \end{pmatrix} \qquad \bar{D}_a(i) = \begin{pmatrix} D_1(i) \\ K_B(i)D_3(i) \end{pmatrix}$$

$$\bar{E}_a(i) = \begin{pmatrix} E_1(i) & E_2(i)K_C(i) \end{pmatrix}$$

$$\bar{D}_d(i) = \begin{pmatrix} D_2(i) \\ K_B(i)D_4(i) \end{pmatrix} \quad \bar{E}_d(i) = E_3(i).$$

Then we obtain the following expressions for $\bar{A}(r_t, t)$ and $\bar{A}_d(r_t, t)$:

$$\bar{A}(r_t, t) = \bar{A}(r_t) + \bar{D}_a(r_t)\Delta(r, t)\bar{E}_a(r_t)$$
$$\bar{A}_d(r_t, t) = \bar{A}_d(r_t) + \bar{D}_d(r_t)\Delta_d(r_t, t)\bar{E}_d(r_t).$$

Let $\hat{R}(i), i \in \mathcal{S}$ be a set of nonnegative matrices and define cost

$$\hat{J}(r_0, \hat{x}(0)) = \mathbb{E}\left[\int_0^\infty \hat{x}^\top(t)\hat{R}(r_t)\hat{x}(t)dt|(r_0, \hat{x}(0))\right]. \tag{11.47}$$

Using Theorem 11.4, we know that if there exist symmetric, positive-definite matrices $P = (P_1, \ldots, P_N)$ with $0 < P(i) \in \mathbb{R}^{2n_1 \times 2n_1}$, $0 < Q \in \mathbb{R}^{n_1 \times n_1}$ and a scalar $\varepsilon > 0$ satisfying for any $i \in \mathcal{S}$,

$$\begin{pmatrix} \mathcal{U}_1(i) & P(i)\bar{A}_d(i) & P(i)\bar{D}_a(i) & P(i)\bar{D}_d(i) \\ * & -Q + \varepsilon E_d^\top(i)E_d(i) & 0 & 0 \\ * & 0 & -\varepsilon I & 0 \\ * & 0 & 0 & -\varepsilon I \end{pmatrix} < 0, \tag{11.48}$$

where

$$\mathcal{U}_1(i) = \bar{A}^\top(i)P(i) + P(i)\bar{A}(i) + \sum_{j=1}^N \lambda_{ij}P(j)$$
$$+ \varrho I_0^\top Q I_0 + \varepsilon \bar{E}_a^\top(i)\bar{E}_a(i) + \hat{R}(i),$$
$$I_0 = \begin{pmatrix} I & 0 \end{pmatrix},$$

then system (11.46) is stochastically stable and for all admissible uncertainties the cost defined by (11.47) satisfies the following

$$\hat{J}(r_0, \hat{x}(0)) \le \text{tr}[P(r_0)\hat{x}_0\hat{x}_0^\top] + \text{tr}\left[Q\int_{-\tau}^0 \phi(s)\phi^\top(s)ds\right]$$
$$+ \bar{\lambda}\text{tr}\left[\int_{-\bar{\tau}}^{-\tau} Q\phi(s)\phi^\top(s)ds\right]. \tag{11.49}$$

Using the Schur complement, we obtain the following proposition.

Proposition 11.3 *If there exist symmetric, positive-definite matrices $P = (P_1, \ldots, P_n)$ with $0 < P(i) \in \mathbb{R}^{2n_1 \times n_1}$, $0 < Q \in \mathbb{R}^{n_1 \times n_1}$ and a scalar $\varepsilon > 0$ satisfying for any $i \in \mathcal{S}$:*

$$\begin{pmatrix} \mathcal{U}_2(i) & P(i)\bar{A}_d(i) & P(i)\bar{D}_a(i) & P(i)\bar{D}_d(i) \\ \bar{A}_d^\top(i)P(i) & -Q + \varepsilon E_d^\top(i)E_d(i) & 0 & 0 \\ \bar{D}_a^\top(i)P(i) & 0 & -\varepsilon I & 0 \\ \bar{D}_d^\top(i)P(i) & 0 & 0 & -\varepsilon I \\ \bar{E}_a(i) & 0 & 0 & 0 \\ I_0 & 0 & 0 & 0 \\ \left[\hat{R}^{1/2}(i)\right]^\top & 0 & 0 & 0 \end{pmatrix}$$

$$\left(\begin{array}{ccc} \bar{E}_a^\top(i) & I_0^\top & [\hat{R}^{1/2}(i)]^\top \\ 0 & 0 & 0 \\ 0 & 0 & 0 \\ 0 & 0 & 0 \\ -\frac{1}{\varepsilon}I & 0 & 0 \\ 0 & -\frac{1}{\varrho}Q^{-1} & 0 \\ 0 & 0 & -I \end{array}\right) \overset{\Delta}{=} \Theta_g(i) < 0, \quad (11.50)$$

where

$$\mathcal{U}_2(i) = \bar{A}^\top(i)P(i) + P(i)\bar{A}(i) + \sum_{j=1}^{N} \lambda_{ij} P(j),$$

then system (11.46) is stochastically stable and the cost defined by (11.47) satisfies (11.49) for all admissible uncertainties.

Let us now go back to the study of the cost (11.31) for system (11.30) under control (8.70). Substituting this controller into the expression of the cost function, (11.31) becomes

$$\hat{J} = \mathbb{E}\left[\int_0^\infty \hat{x}^\top(t)\hat{R}(r_t)\hat{x}(t)dt\right] \tag{11.51}$$

with $\hat{R}(i) = \begin{pmatrix} R_1(i) & 0 \\ 0 & K_C^\top(i)R_2(i)K_C(i) \end{pmatrix}$.

From the above proposition, we obtain the following lemma.

Lemma 11.2 *Suppose that there exist matrices $K_A(i), K_B(i), K_C(i), i \in \mathcal{S}$ satisfying (11.50). Then there exist symmetric, positive matrices $X_i > 0, Y_i > 0$, and matrices $H_i, G_i, M_i, i \in \mathcal{S}$ satisfying the following LMIs for any $i \in \mathcal{S}$:*

$$\left(\begin{array}{cccccccc} \Gamma_{11} & \Gamma_{12} & \Gamma_{13} & \Gamma_{14} & \Gamma_{15} & \Gamma_{16} & \Gamma_{17}(i) & \tilde{S}_i(Y) \\ \Gamma_{12}^\top & -\Gamma_{22} & 0 & 0 & 0 & 0 & 0 & 0 \\ \Gamma_{13}^\top & 0 & -\varepsilon I & 0 & 0 & 0 & 0 & 0 \\ \Gamma_{14}^\top & 0 & 0 & -\varepsilon I & 0 & 0 & 0 & 0 \\ \Gamma_{15}^\top & 0 & 0 & 0 & -\frac{Q^{-1}}{\varrho} & 0 & 0 & 0 \\ \Gamma_{16}^\top & 0 & 0 & 0 & 0 & -\frac{1}{\varepsilon} & 0 & 0 \\ \Gamma_{17}^\top(i) & 0 & 0 & 0 & 0 & -I & 0 \\ \tilde{S}_i^\top(Y) & 0 & 0 & 0 & 0 & 0 & 0 & -\mathcal{Y}_i \end{array}\right) < 0$$

$$\tag{11.52}$$

$$\begin{pmatrix} Y_i & I \\ I & X_i \end{pmatrix} > 0 \tag{11.53}$$

for some given symmetric, positive matrix $Q >$ and a scalar $\varepsilon > 0$, where

$$\Gamma_{11} = \begin{pmatrix} \begin{bmatrix} A(i)Y_i + Y_i A^\top(i) \\ +B(i)G_i + \lambda_{ii} Y_i \\ +G_i^\top B^\top(i) \end{bmatrix} & M_i^\top + A(i) \\ M_i + A^\top(i) & \begin{bmatrix} A^\top(i)X_i + X_i A(i) \\ +H_i C(i) + C^\top(i) H_i^\top \\ + \sum_{j=1}^N \lambda_{ij} Y_j \end{bmatrix} \end{pmatrix}$$

$$\Gamma_{12} = \begin{pmatrix} A_d(i) \\ X_i A_d(i) + H_i C_d(i) \end{pmatrix}$$

$$\Gamma_{22} = Q - \varepsilon E_d^\top(i) E_d(i)$$

$$\Gamma_{13} = \begin{pmatrix} D_1(i) \\ X_i D_1(i) + H_i D_3(i) \end{pmatrix}$$

$$\Gamma_{14} = \begin{pmatrix} D_2(i) \\ X_i D_2(i) + H_i D_4(i) \end{pmatrix}$$

$$\Gamma_{15} = \begin{pmatrix} Y_i \\ I \end{pmatrix}$$

$$\Gamma_{16} = \begin{pmatrix} Y_i E_1^\top(i) + G_i^\top E_2^\top(i) \\ E_1^\top(i) \end{pmatrix}$$

$$\Gamma_{17}(i) = \begin{pmatrix} Y_i [R_1^{1/2}]^\top(i) & -G_i^\top [R_2^{1/2}]^\top(i) \\ [R_1^{1/2}]^\top(i) & 0 \end{pmatrix}$$

$$\tilde{S}_i(Y) = \begin{pmatrix} \sqrt{\lambda_{i1}} Y_i & \cdots & \sqrt{\lambda_{ii-1}} Y_i & \sqrt{\lambda_{ii+1}} Y_i & \cdots & \sqrt{\lambda_{iN}} Y_i \\ 0 & \cdots & 0 & 0 & \cdots & 0 \end{pmatrix}$$

$$\mathcal{Y}_i = \mathrm{diag}\{Y_1, \cdots, Y_{i-1}, Y_{i+1}, \cdots, Y_N\}.$$

Proof: Let $P(i) > 0$, $K_A(i)$, $K_B(i)$ and $K_C(i)$, $i \in \mathcal{S}$ be a set of feasible solutions of (11.50) and partition $P(i)$ as in (8.74). Without loss of generality, we can suppose that P_{2i} is invertible. Define now the following matrices:

$$\begin{cases} X_i = P_{1i} \\ Y_i = [P_{1i} - P_{2i} P_{3i}^{-1} P_{2i}^\top]^{-1} \\ G_i = -K_C(i) P_{3i}^{-1} P_{2i}^\top Y_i \\ H_i = P_{2i} K_B(i) \\ M_i = X_i A(i) Y_i + X_i B(i) G_i + H_i C(i) Y_i \\ \quad - P_{2i}[K_A(i) + K_B(i)F(i)K_C(i)]P_{3i}^{-1} P_{2i}^\top Y_i \\ \quad + \sum_{j=1}^N \lambda_{ij}[P_{1j} - P_{2j} P_{3i}^{-1} P_{2i}^\top] Y_i \end{cases} \tag{11.54}$$

and

$$S_i = \begin{pmatrix} Y_i & I \\ -P_{3i}^{-1} P_{2i}^\top Y_i & 0 \end{pmatrix}.$$

Obviously, S_i are invertible. Therefore, we obtain

$$0 < S_i^\top P(i) S_i = \begin{pmatrix} Y_i & I \\ I & P_{1i} \end{pmatrix} = \begin{pmatrix} Y_i & I \\ I & X_i \end{pmatrix}, \qquad (11.55)$$

yielding (11.53).

Pre- and post-multiplying both sides of (11.50) by $\Xi_i^\top \triangleq \mathrm{diag}\{S_i^\top, I, I, I, I\}$ and Ξ_i yields

$$\Xi_i^\top \Theta(i) \Xi_i < 0, \ \forall i \in S. \qquad (11.56)$$

Direct manipulation gives

$$S_i^\top P(i) \bar{A}(i) S_i = \begin{pmatrix} A(i)Y_i + B(i)G_i & A(i) \\ \# & P_{1i}A(i) + H_i C(i) \end{pmatrix}, \qquad (11.57)$$

where

$$\begin{aligned}
\# &= [P_{1i}A(i) + P_{2i}K_B(i)C(i)]Y_i - [P_{1i}B(i)K_C(i) \\
&\quad + (P_{2i}(K_A(i) + K_B(i)F(i)K_C(i)))]P_{1i}^{-1}P_{2i}^{-1}Y_i \\
&= P_{1i}A(i)Y_i + H_i C(i)Y_i + X_i B(i)G_i \\
&\quad - P_{2i}[K_A(i) + K_B(i)F(i)K_C(i)]P_{3i}^{-1}P_{2i}^\top Y_i.
\end{aligned}$$

$$\begin{aligned}
&S_i^\top P(j) S_i \\
&= \begin{pmatrix} \begin{bmatrix} Y_i[P_{1j} - P_{2j}P_{3j}^{-1}P_{2j}^\top \\ -P_{2j}P_{3i}^{-1}P_{2i}^\top \\ +P_{2i}P_{3i}^{-1}P_{3j}P_{3i}^{-1}P_{2i}^\top]Y_i \\ P_{1j}Y_i - P_{2j}P_{3i}^{-1}P_{2i}^\top Y_i \end{bmatrix} & Y_i P_{1j} - Y_i P_{2i}P_{3i}^{-1}P_{2j}^\top \\ & P_{1j} \end{pmatrix} \\
&= \begin{pmatrix} \begin{array}{c} Y_i \left[Y_j^{-1} + [P_{2i}P_{3i}^{-1}P_{3j} - P_{2j}] \right. \\ \left. \times P_{3j}^{-1}[P_{2i}P_{3i}^{-1}P_{3j} - P_{2j}]^\top \right] Y_i \\ P_{1j}Y_i - P_{2j}P_{3i}^{-1}P_{2i}^\top Y_i \end{array} & Y_i P_{1j} - Y_i P_{2i}P_{3i}^{-1}P_{2j}^\top \\ & P_{1j} \end{pmatrix}.
\end{aligned}$$
$$(11.58)$$

Combining now (11.57) and (11.58) yields

$$\begin{aligned}
S_i^\top \mathcal{U}_2(i) S_i &= \begin{pmatrix} \begin{bmatrix} A(i)Y_i + Y_i A^\top(i) \\ + B(i)G_i + \lambda_{ii}Y_i \\ + G_i^\top B^\top(i) \end{bmatrix} & M_i^\top + A(i) \\ M_i + A^\top(i) & \begin{bmatrix} A^\top(i)P_{1i} + P_{1i}A(i) \\ + H_i C(i) + C^\top(i)H_i^\top \\ + \sum_{j=1}^{N} \lambda_{ij} P_{1j} \end{bmatrix} \end{pmatrix} \\
&\quad + \sum_{j=1}^{N} \lambda_{ij} \begin{pmatrix} \begin{array}{c} Y_i[Y_j^{-1} + [P_{2i}P_{3i}^{-1}P_{3j} - P_{2j}] \\ \times P_{3j}^{-1}[P_{2i}P_{3i}^{-1}P_{3j} - P_{2j}]^\top] \end{array} & 0 \\ 0 & 0 \end{pmatrix} \\
&= \Gamma_{11} + \sum_{j=1,j\neq i}^{N} \lambda_{ij} \begin{pmatrix} Y_i \left[Y_j^{-1} \right] Y_i & 0 \\ 0 & 0 \end{pmatrix}
\end{aligned}$$

$$+ \sum_{j=1,j\neq i}^{N} \lambda_{ij} \begin{pmatrix} Y_i[P_{2i}P_{3i}^{-1}P_{3j} - P_{2j}]P_{3j}^{-1} & \mathbf{0} \\ \times[P_{2i}P_{3i}^{-1}P_{3j} - P_{2j}]^{\top}Y_i & \\ \mathbf{0} & \mathbf{0} \end{pmatrix}. \quad (11.59)$$

Moreover, simple manipulation yields the following

$$S_i^{\top} P(i) = \begin{pmatrix} I & \mathbf{0} \\ P_{1i} & P_{2i} \end{pmatrix}$$

$$S_i^{\top} P(i)\bar{A}_d(i) = \begin{pmatrix} I & \mathbf{0} \\ P_{1i} & P_{2i} \end{pmatrix} \begin{pmatrix} A_d(i) \\ K_B(i)C_d(i) \end{pmatrix}$$
$$= \begin{pmatrix} A_d(i) \\ X_i A_d(i) + P_{2i}K_B(i)C_d(i) \end{pmatrix}$$
$$= \begin{pmatrix} A_d(i) \\ X_i A_d(i) + H_i C_d(i) \end{pmatrix} = \Gamma_{12} \quad (11.60)$$

$$S_i^{\top} P(i)\bar{D}_a(i) = \begin{pmatrix} I & \mathbf{0} \\ P_{1i} & P_{2i} \end{pmatrix} \begin{pmatrix} D_1(i) \\ K_B(i)D_3(i) \end{pmatrix}$$
$$= \begin{pmatrix} D_1(i) \\ X_i D_1(i) + P_{2i}K_B(i)D_3(i) \end{pmatrix}$$
$$= \begin{pmatrix} D_1(i) \\ X_i D_1(i) + H_i D_3(i) \end{pmatrix} = \Gamma_{13}(i) \quad (11.61)$$

$$S_i^{\top} P(i)\bar{D}_d(i) = \begin{pmatrix} I & \mathbf{0} \\ P_{1i} & P_{2i} \end{pmatrix} \begin{pmatrix} D_2(i) \\ K_B(i)D_4(i) \end{pmatrix}$$
$$= \begin{pmatrix} D_2(i) \\ X_i D_2(i) + P_{2i}K_B(i)D_4(i) \end{pmatrix}$$
$$= \begin{pmatrix} D_2(i) \\ X_i D_2(i) + H_i D_4(i) \end{pmatrix} = \Gamma_{14}(i) \quad (11.62)$$

$$S_i^{\top} I_0^{\top} = \begin{pmatrix} Y_i & -Y_i P_{2i}P_{3i}^{-1} \\ I & \mathbf{0} \end{pmatrix} \begin{pmatrix} I \\ \mathbf{0} \end{pmatrix}$$
$$= \begin{pmatrix} Y_i \\ I \end{pmatrix} = \Gamma_{15}(i) \quad (11.63)$$

$$S_i^{\top} \bar{E}_a^{\top}(i) = \begin{pmatrix} Y_i & -Y_i P_{2i}P_{3i}^{-1} \\ I & \mathbf{0} \end{pmatrix} \begin{pmatrix} E_1^{\top}(i) \\ K_C^{\top}(i)E_2^{\top}(i) \end{pmatrix}$$
$$= \begin{pmatrix} Y_i E_1^{\top}(i) - Y_i P_{2i}P_{3i}^{-1} K_C^{\top}(i)E_2^{\top}(i) \\ E_1^{\top}(i) \end{pmatrix}$$
$$= \begin{pmatrix} Y_i E_1^{\top}(i) + G_i^{\top} E_2^{\top}(i) \\ E_1^{\top}(i) \end{pmatrix} = \Gamma_{16}(i) \quad (11.64)$$

$$S_i^\top \left[\hat{R}^{1/2}(i)\right]^\top = \begin{pmatrix} Y_i & -Y_i P_{2i} P_{3i}^{-1} \\ I & 0 \end{pmatrix}$$

$$\times \begin{pmatrix} [R_1^{1/2}(i)]^\top & 0 \\ 0 & K_C^\top(i)[R_2^{1/2}(i)]^\top \end{pmatrix} = \Gamma_{17}(i). \ (11.65)$$

Noting that

$$\sum_{j=1, j\neq i}^{N} \lambda_{ij}[P_{2i}P_{3i}^{-1}P_{3j} - P_{2j}]P_{3j}^{-1}[P_{2i}P_{3i}^{-1}P_{3j} - P_{2j}]^\top \geq 0,$$

it follows from (11.55)-(11.63) that X_i, Y_i, M_i, H_i, G_i are a set of feasible solutions to (11.52). This completes the proof of Lemma 11.2. ☐

Theorem 11.6 (i) *If there exist symmetric, positive-definite matrices* $P = (P(1), \cdots, P(N)) > 0$, *and matrices* $K_A = (K_A(1), \cdots, K_A(N))$, $K_B = (K_B(1), \cdots, K_B(N))$ *and* $K_C = (K_C(1), \cdots, K_C(N))$ *such that* (11.50) *holds for some matrix* $Q > 0$ *and a scalar* $\varepsilon > 0$ *(which means that the controller* (8.70) *stabilizes system* (11.30) *and guarantees cost bound* (11.49)), *then the following LMIs have a set of feasible solutions* $X_i > 0, Y_i > 0, G_i, H_i, i \in S$:

$$\begin{pmatrix} W_1(i) & A_d(i)Q^{-1}E_3^\top(i) & Y_i E_1^\top(i) + G_i^\top E_2^\top(i) \\ E_3(i)Q^{-1}A_d^\top(i) & -\frac{I}{\varepsilon} + E_3(i)Q^{-1}E_3^\top(i) & 0 \\ E_1(i)Y_i + E_2(i)G_i & 0 & -\frac{1}{\varepsilon}I \\ Y_i & 0 & 0 \\ R_1^{1/2}(i)Y_i & 0 & 0 \\ R_2^{1/2}(i)G_i & 0 & 0 \\ S_i^\top(Y) & 0 & 0 \end{pmatrix}$$

$$\begin{pmatrix} * & Y_i[R_1^{1/2}(i)]^\top & G_i^\top[R_2^{1/2}(i)]^\top & S_i(Y) \\ 0 & 0 & 0 & 0 \\ 0 & 0 & 0 & 0 \\ -\frac{Q^{-1}}{\varrho} & 0 & 0 & 0 \\ 0 & -I & 0 & 0 \\ 0 & 0 & -I & 0 \\ 0 & 0 & 0 & -\mathcal{Y}_i \end{pmatrix} < 0, \quad (11.66)$$

$$\begin{pmatrix} W_2(i) & * & * & * \\ A_d^\top(i)X_i + C_d^\top(i)H_i^\top & -Q + \varepsilon E_d^\top(i)E_d(i) & 0 & 0 \\ D_1^\top(i)X_i + D_3^\top(i)H_i^\top & 0 & -\varepsilon I & 0 \\ D_2^\top(i)X_i + D_4^\top(i)H_i^\top & 0 & 0 & -\varepsilon I \end{pmatrix} < 0 \ (11.67)$$

$$\begin{pmatrix} Y_i & I \\ I & X_i \end{pmatrix} > 0, \quad (11.68)$$

where

$$W_1(i) = A(i)Y_i + Y_iA^\top(i) + B(i)G_i + G_i^\top B^\top(i) + \lambda_{ii}Y_i$$
$$+ A_d(i)Q^{-1}A_d^\top(i) + \frac{1}{\varepsilon}D_1(i)D_1^\top(i) + \frac{1}{\varepsilon}D_2(i)D_2^\top(i)$$
$$W_2(i) = A^\top(i)X_i + X_iA(i) + H_iC(i) + C^\top(i)H_i^\top$$
$$+ \sum_{j=1}^{N}\lambda_{ij}X_j + \varrho Q + \varepsilon E_1^\top(i)E_1(i) + R_1(i).$$

(ii) On the other hand, if there exist a symmetric, positive-definite matrix $Q > 0$ and a scalar $\varepsilon > 0$ such that the LMIs (11.66)-(11.68) have a set of solutions $X_i > 0, Y_i > 0, G_i, H_i, M_i, i \in S$, then there must exist symmetric, positive-definite matrices $P(i) > 0$ and matrices $K_A(i), K_B(i), K_C(i)$ satisfying (11.50).

Proof: *(i)* In view of the Schur complement, LMI (11.50) is feasible for some $X_i > 0, Y_i > 0, G_i, H_i, M_i, i \in S$ if and only if the following holds for every $i \in S$:

$$\Gamma_{11}(i) + \Gamma_{12}(i)\Gamma_{22}^{-1}(i)\Gamma_{12}^\top(i) + \frac{1}{\varepsilon}\Gamma_{13}(i)\Gamma_{13}^\top(i) + \frac{1}{\varepsilon}\Gamma_{14}(i)\Gamma_{14}^\top(i)$$
$$+ \varrho\Gamma_{15}(i)Q\Gamma_{15}^\top(i) + \varepsilon\Gamma_{16}(i)\Gamma_{16}(i) + \Gamma_{17}(i)\Gamma_{17}^\top(i) + \tilde{S}_i(Y)\mathcal{Y}_i\tilde{S}_i^\top(Y)$$
$$\triangleq Z_i = \begin{pmatrix} Z_{1i} & Z_{2i}^\top \\ Z_{2i} & Z_{3i} \end{pmatrix} < 0. \tag{11.69}$$

This implies in turn the following:

$$Z_{1i} < 0 \tag{11.70}$$
$$Z_{3i} < 0. \tag{11.71}$$

Direct computation gives the following:

$$Z_{1i} = A(i)Y_i + Y_iA^\top(i) + B(i)G_i + G_i^\top B^\top(i) + \lambda_{ii}Y_i$$
$$+ A_d(i)[Q - \varepsilon E_3^\top(i)E_3(i)]^{-1}A_d^\top(i)$$
$$+ \frac{1}{\varepsilon}D_1(i)D_1^\top(i) + \frac{1}{\varepsilon}D_2(i)D_2^\top(i) + \varrho Y_iQY_i$$
$$+ \varepsilon[Y_iE_1^\top(i) + G_i^\top E_2^\top][E_1(i)Y_i + E_2(i)G_i]$$
$$+ Y_i\left[\sum_{j\neq i}\lambda_{ij}Y_j^{-1}\right]Y_i + Y_iR_1(i)Y_i + G_i^\top R_2(i)G_i$$

$$Z_{3i} = A^\top(i)X_i + X_iA(i) + H_iC(i) + C^\top(i)H_i^\top + \sum_{j=1}^{N}\lambda_{ij}X_j$$
$$+ [X_iA_d(i) + H_iC_d(i)][Q - \varepsilon E_d^\top(i)E_d(i)]^{-1}[X_iA_d(i) + H_iC_d(i)]^\top$$
$$+ \frac{1}{\varepsilon}[X_iD_1(i) + H_iD_3(i)][X_iD_1(i) + H_iD_3(i)]^\top$$

$$+ \frac{1}{\varepsilon}[X_i D_2(i) + H_i D_4(i)][X_i D_2(i) + H_i D_4(i)]^\top$$

$$+ \varrho Q + \varepsilon E_1^\top(i)E_1(i) + R_1(i)$$

$$Z_{2i} = M_i + A^\top(i) + [X_i A_d(i) + H_i C_d(i)][Q - \varepsilon E_d^\top(i)E_d(i)]^{-1}A_d^\top(i)$$

$$+ \frac{1}{\varepsilon}[X_i D_1(i) + H_i D_3(i)]D_1^\top(i) + \frac{1}{\varepsilon}[X_i D_2(i) + H_i D_4(i)]D_2^\top(i)$$

$$+ \varrho Q Y_i + \varepsilon E_1^\top(i)[E_1(i)Y_i + E_2(i)G_i] + R_1(i)Y_i.$$

Using now the matrix inversion formula, we have

$$[Q - \varepsilon E_3^\top(i)E_3(i)]^{-1}$$

$$= Q^{-1} + Q^{-1}E_3^\top(i)\left[\frac{1}{\varepsilon}I - E_3(i)Q^{-1}E_3^\top(i)\right]^{-1}E_3(i)Q^{-1}.$$

From which it follows that

$$Z_{1i} = A(i)Y_i + Y_i A^\top(i) + B(i)G_i + G_i^\top B^\top(i) + \lambda_{ii}Y_i$$

$$+ A_d(i)Q^{-1}E_3^\top(i)[\frac{1}{\varepsilon}I - E_3(i)Q^{-1}E_3^\top(i)]^{-1}E_3(i)Q^{-1}A_d^\top(i)$$

$$+ \frac{1}{\varepsilon}D_1(i)D_1^\top(i) + A_d(i)Q^{-1}A_d^\top(i) + \frac{1}{\varepsilon}D_2(i)D_2^\top(i)$$

$$+ \varrho Y_i Q Y_i + \varepsilon[Y_i E_1^\top(i) + G_i^\top E_2^\top(i)][E_1(i)Y_i + E_2(i)G_i]$$

$$+ Y_i R_1(i)Y_i + G_i^\top R_2(i)G_i + Y_i \left[\sum_{j \neq i} \lambda_{ij}Y_j^{-1}\right]Y_i.$$

Using the Schur complement, we have that $Z_{1i} < 0$ and $Z_{3i} < 0$ are equivalent to (11.66), and (11.67), respectively. This proves (i).

(ii) On the other hand, if LMIs (11.66), (11.67), and (11.68) have a set of feasible solutions X_i, Y_i, G_i, H_i, $i \in \mathcal{S}$, let us define matrices

$$P(i) = \begin{pmatrix} X_i & Y_i^{-1} - X_i \\ Y_i^{-1} - X_i & X_i - Y_i^{-1} \end{pmatrix}, i \in \mathcal{S},$$

and gains

$$\begin{cases} K_A(i) = \left[X_i - Y_i^{-1}\right]^{-1}\left[\mathcal{M}_i Y_i \right. \\ \qquad + X_i A(i) + X_i B(i)G_i Y_i + H_i C(i) + \sum_{j=1}^N \lambda_{ij}Y_j^{-1} \\ \qquad \left. - H_i F(i)G_i Y_i^{-1}\right] \\ K_B(i) = \left[X_i - Y_i^{-1}\right]^{-1} H_i \\ K_C(i) = G_i Y_i^{-1}, \end{cases} \qquad (11.72)$$

where

$$\mathcal{M}_i = A^\top(i) + [X_i A_d(i) + H_i C_d(i)][Q - \varepsilon E_d^\top(i)E_d(i)]^{-1}A_d^\top(i)$$

$$+ \frac{1}{\varepsilon}[X_i D_1(i) + H_i D_3(i)] D_1^\top(i) + \frac{1}{\varepsilon}[X_i D_2(i) + H_i D_4(i)] D_2^\top(i)$$
$$+ \varrho Q Y_i + \varepsilon E_1^\top(i)[E_1(i)Y_i + E_2(i)G_i] + R_1(i)Y_i.$$

From (11.68) and the Schur complement, we obtain that $P(i), i \in \mathcal{S}$ are symmetric, positive-definite. Direct manipulation yields that $P(i)$, $K_A(i)$, $K_B(i)$, $K_C(i)$, $i \in \mathcal{S}$ defined above satisfy (11.50) if

$$\begin{pmatrix} Z_{1i} & \mathbf{0} \\ \mathbf{0} & Z_{3i} \end{pmatrix} < 0, \tag{11.73}$$

which follows from (11.66) and (11.67). $\square$

Let us now work out a numerical example to show the usefulness of the proposed results.

Example 11.4 *Consider a system described by (11.30) with two modes, i.e., $\mathcal{S} = \{1, 2\}$. Let $\tau_1 = 0.2$, $\tau_2 = 1.67$ and assume that the generator is given by*

$$\Lambda = \begin{pmatrix} -2 & 2 \\ 5 & -5 \end{pmatrix}.$$

Let the rest of data be given by the following parameters

$$A(1) = \begin{pmatrix} -10 & 0 \\ 1.4 & -10 \end{pmatrix} \qquad A(2) = \begin{pmatrix} -30 & -0.8 \\ 0.2 & -12 \end{pmatrix}$$

$$A_d(1) = \begin{pmatrix} -0.3 & 0 \\ 0 & -0.2 \end{pmatrix} \qquad A_d(2) = \begin{pmatrix} 0.1 & -0.1 \\ 0.1 & -0.1 \end{pmatrix}$$

$$B(1) = \begin{pmatrix} 1 & 0 \\ 0 & -0.1 \end{pmatrix} \qquad B(2) = \begin{pmatrix} 0 & 1 \\ 0 & 0.1 \end{pmatrix}$$

$$D_a(1) = \begin{pmatrix} 1 & 0 \\ 0 & 0.3 \end{pmatrix} \qquad D_a(2) = \begin{pmatrix} 0 & 1 \\ 0 & -0.3 \end{pmatrix}$$

$$D_d(1) = \begin{pmatrix} 0 & 0.2 \\ 0 & 1 \end{pmatrix} \qquad D_d(2) = \begin{pmatrix} -1 & 0 \\ 0 & 1 \end{pmatrix}$$

$$D_c(1) = \begin{pmatrix} 0 & 1 \\ 1 & 0 \end{pmatrix} \qquad D_c(2) = \begin{pmatrix} 1 & 0 \\ 0 1 \end{pmatrix}$$

$$E_a(1) = \begin{pmatrix} 1 & 0 \\ 0 & 0 \end{pmatrix} \qquad E_a(2) = \begin{pmatrix} -1 & 0 \\ 0 & 0.2 \end{pmatrix}$$

$$E_d(1) = \begin{pmatrix} 0 & 0.3 \\ 0 & -1 \end{pmatrix} \qquad E_d(2) = \begin{pmatrix} -0.2 & 0 \\ 1 & 0 \end{pmatrix}$$

$$E_b(1) = \begin{pmatrix} 0.2 & 0 \\ 0.2 & 0.1 \end{pmatrix} \qquad E_b(2) = \begin{pmatrix} -0.1 & 0 \\ 0 & -0.2 \end{pmatrix}$$

$$C(1) = \begin{pmatrix} 1 & 0 \\ 0 & 1 \end{pmatrix} \qquad C(2) = \begin{pmatrix} 0 & 1 \\ 1 & 0 \end{pmatrix}$$

$$R_1(1) = R_1(2) = \begin{pmatrix} 1 & 0 \\ 0 & 1 \end{pmatrix} \qquad R_2(1) = R_2(2) = \begin{pmatrix} 0.9 & 0 \\ 0 & 0.9 \end{pmatrix}.$$

Let $\varepsilon_1 = 0.81$ and $Q = \begin{pmatrix} 1.5 & 0 \\ 0 & 1.5 \end{pmatrix}$. With the above data, solving (11.66), (11.67) and (11.68) gives the following set of feasible solutions:

$$X_1 = \begin{pmatrix} 4.0183 & -0.2817 \\ -0.2817 & 3.8436 \end{pmatrix} \qquad X_2 = \begin{pmatrix} 3.1238 & 0.2695 \\ 0.2695 & 3.9513 \end{pmatrix}$$

$$Y_1 = \begin{pmatrix} 1.1257 & -0.0673 \\ -0.0673 & 1.2712 \end{pmatrix} \qquad Y_2 = \begin{pmatrix} 1.1797 & -0.0190 \\ -0.0190 & 1.3327 \end{pmatrix}$$

$$G_1 = \begin{pmatrix} -0.7199 & -0.0637 \\ 0.0230 & 0.0818 \end{pmatrix} \qquad G_2 = \begin{pmatrix} 0 & 0 \\ 1.4373 & -0.0145 \end{pmatrix}$$

$$H_1 = \begin{pmatrix} 166.2262 & -13.1206 \\ 4.2195 & 14.0828 \end{pmatrix}$$

$$H_2 = \begin{pmatrix} 16.6775 & 655.9544 \\ 23.7233 & -4.1990 \end{pmatrix}.$$

Thus, in view of Theorem 11.6 system (11.30) is robustly stabilizable in the MSQS sense. Moreover, substituting $X_i, Y_i, G_i, H_i, i \in S$ into (11.72) yields a set of stabilizing gains as follows:

$$K_A(1) = \begin{pmatrix} -7.0055 & 8.6522 \\ 6.6124 & -3.7192 \end{pmatrix}$$

$$K_A(2) = \begin{pmatrix} 233.4965 & -57.3311 \\ -22.2084 & -0.3287 \end{pmatrix}$$

$$K_B(1) = \begin{pmatrix} 53.9126 & -3.7534 \\ 7.1867 & 4.2064 \end{pmatrix}$$

$$K_B(2) = \begin{pmatrix} 6.5490 & 291.0045 \\ 6.8851 & -24.7127 \end{pmatrix}$$

$$K_C(1) = \begin{pmatrix} -0.6446 & -0.0842 \\ 0.0244 & 0.0656 \end{pmatrix}$$

$$K_C(2) = \begin{pmatrix} 0 & 0 \\ 1.2185 & 0.0065 \end{pmatrix}.$$

11.4 Notes

The first part of this chapter is adapted from [24]. The guaranteed cost control of jump linear systems with norm-bounded uncertainties and time delay is studied in [22], from which the second part of this chapter was adapted to include mode-dependent time delays.

12
Nonlinear Stochastic Control Problem

In the previous chapters we dealt with linear deterministic and stochastic time delay systems, and we solved many problems like stability, stabilizability (using different types of controllers), $\mathcal{H}_\infty$ control, filtering, and robustness of the techniques therein. The tools presented in these chapters will be efficient for dynamical time delay linear systems, but will fail for nonlinear ones. Therefore an alternative for solving stability and stabilizability problems for the class of nonlinear time delay dynamical systems is needed.

This chapter deals with the class of nonlinear time delay systems with Markov jumps and Gaussian noise. Contrary to all that we did previously, the dynamic is nonlinear and Gaussian exogenous disturbance is acting on the system. We consider delayed state only. Our goal here is to study stability and stabilizability problems. Keeping in mind the tractability of the two problems we are dealing with, some LMIs techniques are developed. To the best of our knowledge, the problems we are considering here have never been considered by researchers before. The linear case of the class of systems we are considering without delay has been considered by Morozan [151], in which $\{r_t, t \geq 0\}$ is allowed to be a continuous-time Markov process with finite state space $\mathcal{S} = \{1, 2, \cdots, N\}$ and generator matrix $\Lambda = (\lambda_{ij}), i, j \in \mathcal{S}$, where $\lambda_{ij} > 0, i \neq j$ and $\lambda_{ii} = -\sum_{j \neq i} \lambda_{ij}$.

Let us consider a system with N modes described by $\{r_t, t \geq 0\}$ and a constant time delay $\tau > 0$ and suppose that the system is described by the following dynamics:

$$\begin{cases} dx(t) = f(\mathbf{x}_t, r_t, t)dt + g(\mathbf{x}_t, r_t, t)dB_t \\ x(s) = \phi(s), \ s \in [-\tau, 0], \end{cases} \tag{12.1}$$

where $x(t) \in \mathbb{R}^n, y_t \in \mathbb{R}^k$ are the system state and output, respectively, $\mathbf{x}_t \in \mathbb{C}[-\tau, 0]$ defined by $x_t(s) = x(t + s), s \in [-\tau, 0]$. $f(\mathbf{x}_t, r_t, t) \in \mathbb{R}^n$ and $g(\mathbf{x}_t, r_t, t) \in \mathbb{R}^{n \times m}, h(\mathbf{x}_t, r_t, t) \in \mathbb{R}^k$ are given functionals, $B_t \in \mathbb{R}^m$ is a standard Wiener process, representing the noise in the system, which is assumed to be independent of the system mode $\{r_t, t \geq 0\}$. The initial condition $\phi(\cdot)$ is a stochastic process with continuous trajectories, which is independent of the σ-algebra generated by $\{(r_t, B_t), t \geq 0\}$.

Assumption 12.1 *There exist nondecreasing functions $J_f(t)$ and $J_g(t)$ such that for any $r_t \in \mathcal{S}$, $\psi_1(\cdot), \psi_2(\cdot) \in \mathbb{C}[-\tau, 0]$, the following hold:*

$$\|f(\psi_1, r_t, t) - f(\psi_2, r_t, t)\|^2 \leq \int_{-\tau}^{0} \|\psi_1(t + s) - \psi_2(t + s)\|^2 dJ_f(s) \tag{12.2}$$

$$\|g(\psi_1, r_t, t) - g(\psi_2, r_t, t)\| \leq \int_{-\tau}^{0} \|\psi_1(t + s) - \psi_2(t + s)\|^2 dJ_g(s) \tag{12.3}$$

Let us assume that all the required assumptions for the existence of the solution $x(t, r_0, \phi(\cdot))$ corresponding to the initial conditions $r_0, \phi(\cdot)$ are satisfied.

Definition 12.1 *System (12.1) is said to be*
1) mean square stable if for each $\varepsilon > 0$, there exists a $\delta > 0$ such that the solution $x(t, \phi)$ exists for $0 \leq t < \infty$ and $\mathbb{E}\|x(t, r_0, \phi)\|^2 < \varepsilon(0 \leq t < \infty)$, if only

$$\begin{cases} \sup_{-\tau \leq t \leq 0} \mathbb{E}\|\phi(t)\|^2 < \delta(\varepsilon), \\ \sup_{-\tau \leq t \leq 0} \mathbb{E}\|\phi(t)\|^2 < \infty; \end{cases} \tag{12.4}$$

2) asymptotically square stable if it is mean square stable and there is an $\varepsilon > 0$ such that

$$\lim_{t \to \infty} \mathbb{E}\|x(t, \phi)\|^2 = 0. \tag{12.5}$$

This chapter deals with the class of nonlinear time delay systems and provides some results on stability and stabilizability. For the stability problem two results are established. The first one is applied to the general dynamics (12.1). The second one can be applied only to the special class of nonlinear systems with jumps and time delay as described by dynamics (12.20). For the stabilization problem we develop design algorithms for synthesizing linear and nonlinear state feedback controllers that stabilize the dynamics (12.20).

The rest of this chapter is organized as follows: Section 12.1 establishes sufficient conditions for system (12.1) to be asymptotically stable. Section

12.2 addresses the stabilization problem for a special class of nonlinear stochastic systems.

12.1   Stability

Let us now consider the class of systems with the dynamics described by (12.1) and restrict our study to stability. The following theorem establishes a sufficient condition for this class of systems to be asymptotically stable.

Theorem 12.1 *If there exists a continuous functional*

$$V(t, r_t, \mathbf{x}_t) : [0, \infty) \times \mathcal{S} \times \mathbb{C}[-\tau, 0] \to \mathbb{R}$$

such that for any solution of (12.1) the inequalities

$$V(t, r_t, \mathbf{x}_t) \geq C_1 \|x(t)\|^2, \tag{12.6}$$

$$\mathbb{E}[V(t, r_t, \mathbf{x}_t)] \leq C_2 \sup_{-\tau \leq \theta \leq 0} \mathbb{E}\|x(t+\theta)\|^2 \tag{12.7}$$

hold for every $r_t \in \mathcal{S}$ and for any $t \geq s \geq 0$,

$$\mathbb{E}[V(t, r_t, \mathbf{x}_t)] - E[V(s, r_s, \mathbf{x}_s)] \leq -C_3 \int_s^t \mathbb{E}\|x(\theta)\|^2 d\theta, \tag{12.8}$$

where $C_1, C_2\ C_3$ are positive constants. Then the trivial solution of (12.1) is asymptotically mean square stable.

Proof: From (12.6) and (12.7) it follows that

$$C_1 \mathbb{E}\|x(t)\|^2 \leq C_2 \sup_{-\tau \leq \theta \leq 0} \mathbb{E}\|\phi(\theta)\|^2, \tag{12.9}$$

which proves mean square stability.

Next, we proceed to prove (12.5). Using (12.8), we obtain

$$\mathbb{E}[V(t, r_t, \mathbf{x}_t)] \leq \mathbb{E}[V(0, r_0, \phi)] - C_3 \int_0^t \mathbb{E}\|x(s)\|^2 ds, \tag{12.10}$$

which combined with (12.9) yields

$$\begin{cases} \int_0^\infty \mathbb{E}\|x(t)\|^2 dt < \infty \\ \sup_{t \geq 0} E\|x(t)\|^2 < \infty. \end{cases} \tag{12.11}$$

Therefore, in order to prove asymptotic stability, it suffices to show that $\mathbb{E}[\|x(t)\|^2]$ is Lipschitz in t. In fact, let $\mathcal{A}$ be the infinitesimal generator of the Markov process $\{(\mathbf{x}_t, r_t), t \geq 0\}$. Then applying Ito's formula to $\|x(t)\|^2$, we obtain

$$\mathcal{A}[x^\top(t)x(t)] = 2x^\top(t)dx(t) + \text{tr}[g(\mathbf{x}_t, r_t, t)g^\top(\mathbf{x}_t, r_t, t)]dt. \tag{12.12}$$

Based on Dynkin's formula, we obtain that for any $t_2 \geq t_1 \geq 0$

$$\mathbb{E}[\|x(t_2)\|^2 - \|x(t_1)\|^2] = \int_{t_1}^{t_2} 2x^\top(t) f(\mathbf{x}_t, r_t, t) dt$$
$$+ \int_{t_1}^{t_2} tr[g(\mathbf{x}_t, r_t, t) g^\top(\mathbf{x}_t, r_t, t)] dt. \quad (12.13)$$

Using (12.1) and (12.11), we obtain that

$$2\mathbb{E}[x^\top(t) f(\mathbf{x}_t, r_t, t) \leq \mathbb{E}\|x(t)\|^2 + \mathbb{E}[f(\mathbf{x}_t, r_t, t) f^\top(\mathbf{x}_t, r_t, t)]$$
$$\leq \mathbb{E}\|x(t)\|^2 + \mathbb{E}\left[\int_{-\tau}^0 \|x(t+\theta)\|^2 dJ_f(\theta)\right] \leq C_4 \quad (12.14)$$

and

$$tr[g(t, r_t, \mathbf{x}_t) g^\top(t, r_t, \mathbf{x}_t)] \leq \mathbb{E}\int_{-\tau}^0 \|x(t+\theta)\|^2 dJ_g(\theta) \leq C_4 \quad (12.15)$$

hold for some constant C_4.

Therefore, we get

$$\mathbb{E}[\|x(t_2)\|^2 - \|x(t_1)\|^2] \leq C_4\|t_2 - t_1\|. \quad (12.16)$$

Combining this with (12.11) proves the asymptotic stability of the system under study. $\qquad \square$

From the above theorem, we obtain the following corollary:

Corollary 12.1 *Let $\mathcal{A}$ be the infinitesimal generator of the Markov process $\{r_t, \mathbf{x}_t\}, t \geq 0\}$. If there exists a continuous functional*

$$V(t, r_t, \mathbf{x}_t) : [0, \infty) \times \mathcal{S} \times \mathbb{C}[-\tau, 0] \to \mathbb{R}$$

such that for any solution of (12.1) the inequalities

$$V(t, r_t, \mathbf{x}_t) \geq C_1\|x(t)\|^2, \quad (12.17)$$
$$\mathbb{E}V(t, r_t, \mathbf{x}_t) \leq C_2 \sup_{-\tau \leq \theta \leq 0} \mathbb{E}\|x(t+\theta)\|^2 \quad (12.18)$$

hold for some positive constants C_1, C_2, and there exists a constant C_3, such that

$$\mathcal{A}[V(t, r_t, \mathbf{x}_t)] \leq -C_3\|x(t)\|^2, \quad (12.19)$$

then the trivial solution of (12.1) is asymptotically mean square stable.

Proof: From (12.19), it follows that for any $0 \leq s \leq t$

$$\mathbb{E}\left[\int_s^t \mathcal{A}V(\theta, r_\theta, \mathbf{x}_\theta) d\theta\right] \leq -C_3 \int_s^t \mathbb{E}\|x(\theta)\| d\theta,$$

which combined with Dynkin's formula yields (12.8). Thus, the proof of Corollary 12.1 follows from Theorem 12.1. $\qquad \square$

Let us now consider a special class of nonlinear systems described by the following dynamics:

$$\begin{cases} dx(t) = A(r_t)x(t) + f(x(t), r_t, t) + A_d(r_t)x(t - \tau) \\ \qquad\quad + f_d(x(t - \tau), r_t, t) + C(r_t)x(t)dB_t \\ x(s) = \phi(s), \quad -\tau \le s \le 0, \end{cases} \tag{12.20}$$

where $f(x(t), r_t, t)$ and $f_d(x(t - \tau), r_t, t)$ denote the nonlinear parts in the system dynamics.

In the sequel of this chapter, we will assume that the following hold for every $r_t \in \mathcal{S}$:

$$\begin{cases} \|f(x(t), r_t, t)\|^2 \le \alpha(r_t)\|x(t)\|^2 \\ \|f_d(x(t - \tau), r_t, t)\|^2 \le \beta(r_t)\|x(t - \tau)\|^2. \end{cases} \tag{12.21}$$

The stability test for this class of systems is given by the following theorem.

Theorem 12.2 *If there exist symmetric, positive-definite matrices* $P = (P(1), \cdots, P(N)) > 0, Q > 0$, *and scalars* $\varepsilon_1 > 0$, $\varepsilon_2 > 0$ *such that*

$$\Theta(i) = \begin{pmatrix} J(i) & P(i) & P(i) \\ P(i) & -\varepsilon_1 I & 0 \\ P(i) & 0 & -\varepsilon_2 I \end{pmatrix} < 0 \tag{12.22}$$

$$Q - \varepsilon_2 \beta(i) I - A_d^\top(i) P(i) A_d(i) \ge 0 \tag{12.23}$$

hold for all $i \in \mathcal{S}$, *where*

$$J(i) = A^\top(i)P(i) + P(i)A(i) + \varepsilon_1 \alpha(i)I$$
$$+ C^\top(i)P(i)C(i) + Q + \sum_{j=1}^{N} \lambda_{ij} P(j) + P(i),$$

then system (12.20) is asymptotically stable.

Proof: Let $P = (P(1), \cdots, P(N)) > 0$ and $Q > 0$ be a set of solutions of (12.22) and (12.23). Let us consider a Lyapunov functional candidate as follows:

$$V(\mathbf{x}_t, r_t, t) = x^\top(t)P(r_t)x(t) + \int_{t-\tau}^{t} x^\top(s)Qx(s). \tag{12.24}$$

Obviously, we can conclude that

$$\min_{i \in \mathcal{S}}\{\lambda_{\min}(P(i))\}\|x(t)\|^2 \le V(\mathbf{x}_t, r_t, t) \tag{12.25}$$

$$V(\mathbf{x}_t, r_t, t) \le C_0 \sup_{\theta \in [-\tau, 0]} \|x(t + \theta)\|^2 \tag{12.26}$$

where $C_0 = \tau \max_{i \in \mathcal{S}}\{\max\{\lambda_{\max}(P(i)), \lambda_{\max}(Q)\}\}$.

Let $\mathcal{A}$ be the infinitesimal generator of the Markov process $\{(\mathbf{x}_t, r_t), t \geq 0\}$. Then, using Ito's formula, we obtain

$$\mathcal{A}V(\mathbf{x}_t, r_t, t) = x^\top(t) \left[\sum_{j=1}^N \lambda_{rtj} P(j) \right] x(t)$$

$$+ 2x^\top(t)P(r_t) \Big[A(t)x(t) + f(x(t), r_t, t)$$

$$+ A_d(r_t)x_{t-\tau} + f_d(x(t-\tau), r_t, t) \Big]$$

$$+ x^\top(t)C^\top(r_t)P(r_t)C(r_t)x(t)$$

$$+ x^\top(t)Qx(t) - x^\top(t-\tau)Qx(t-\tau). \qquad (12.27)$$

Using Lemma A.5 and Assumption 12.21, we obtain that for any $\varepsilon_1 > 0, \varepsilon_2 > 0$,

$$2x^\top(t)P(r_t)f(x(t), r_t, t) \leq \frac{1}{\varepsilon_1}x^\top(t)P^2(r_t)x(t)$$

$$+ \varepsilon_1 f^\top(x(t), r_t, t)f(x(t), r_t, t)$$

$$\leq \frac{1}{\varepsilon_1}x^\top(t)P^2(r_t)x(t) + \varepsilon_1\alpha(r_t)x^\top(t)x(t)$$

$$(12.28)$$

$$2x^\top(t)P(r_t)f_d(x(t), r_t, t) \leq \frac{1}{\varepsilon_2}x^\top(t)P^2(r_t)x(t)$$

$$+ \varepsilon_2 f_d^\top(x(t-\tau), r_t, t)f_d(x(t-\tau), r_t, t)$$

$$\leq \frac{1}{\varepsilon_2}x^\top(t)P^2(r_t)x(t) + \varepsilon_2\beta(r_t)x^\top(t-\tau)x(t-\tau)$$

$$(12.29)$$

and

$$2x^\top(t)P(r_t)A_d(r_t)x(t-\tau)$$
$$= x^\top(t)P(r_t)x(t) + x^\top(t-\tau)A_d^\top(r_t)P(r_t)A_d(r_t)x(t-\tau)$$
$$- [x(t) - A_d(r_t)x(t-\tau)]^\top P(r_t)[x(t) - A_d(r_t)x(t-\tau)]$$
$$\leq x^\top(t)P(r_t)x(t) + x^\top(t-\tau)A_d^\top(r_t)P(r_t)A_d(r_t)x(t-\tau). \quad (12.30)$$

Combining (12.28)-(12.30) yields

$$\mathcal{A}V(\mathbf{x}_t, r_t, t) \leq x^\top(t)\Xi(r_t)x(t) - x^\top(t-\tau)\Big[Q - \varepsilon_2\beta(r_t)I$$

$$- A_d(r_t)P(r_t)A_d(r_t)\Big]x(t-\tau), \qquad (12.31)$$

where

$$\Xi(r_t) = P(r_t)A(r_t) + A^\top(r_t)P(r_t) + \sum_{j=1}^{N} \lambda_{r_t j} P(j) + \frac{1}{\varepsilon_1} P^2(r_t)$$

$$+ \varepsilon_1 \alpha(i)I + C^\top(r_t)P(r_t)C(r_t) + \frac{1}{\varepsilon_2} P^2(r_t) + Q + P(r_t). \quad (12.32)$$

Using the Schur complement, it follows from (12.23) that

$$Q - \varepsilon_2 \beta(r_t)I - A_d(r_t)P(r_t)A_d(r_t) \geq 0, \; \forall r_t \in \mathcal{S}.$$

Therefore, we obtain that

$$\mathcal{A}V(\mathbf{x}_t, r_t, t) \leq x^\top(t)\Xi(r_t)x(t). \quad (12.33)$$

Furthermore, using the Schur complement, we conclude that $\Xi(r_t) < 0$ is equivalent to $\Theta(r_t) < 0$. Therefore, we get

$$\mathcal{A}V(\mathbf{x}_t, r_t, t) \leq -\min_{i \in \mathcal{S}}\{\lambda_{\min}(-\Xi(i))\}\|x(t)\|^2. \quad (12.34)$$

According to Corollary 12.1, combining (12.25), (12.26), and (12.34) proves the stability of the class of systems under study, which completes the proof of Theorem 12.2. $\qquad \square$

In the above theorem, the upper bounds $\alpha(i), \beta(i)$ are assumed to be constants. In fact, they can be optimized by solving a quasiconvex optimization problem. Let $\alpha(i) = \beta(i) \triangleq \frac{1}{\gamma}, \; i \in \mathcal{S}$. Then, the LMI (12.22) can be rewritten as

$$\begin{pmatrix} \varepsilon_1 I & \mathbf{0} & \mathbf{0} \\ \mathbf{0} & \mathbf{0} & \mathbf{0} \\ \mathbf{0} & \mathbf{0} & \mathbf{0} \end{pmatrix} < \gamma \begin{pmatrix} -J_0(i) & -P(i) & -P(i) \\ -P(i) & \varepsilon_1 I & \mathbf{0} \\ -P(i) & \mathbf{0} & \varepsilon_2 I \end{pmatrix} \quad (12.35)$$

where

$$J_0(i) = A^\top(i)P(i) + P(i)A(i) + C^\top(i)P(i)C(i) + Q$$

$$+ \sum_{j=1}^{N} \lambda_{ij} P(j) + P(i).$$

(12.23) can also be rewritten as

$$\varepsilon_2 I \leq \gamma(A - A_d^\top(i)P(i)A_d(i)). \quad (12.36)$$

Therefore, we get the following theorem.

Theorem 12.3 *Let γ_0 be the solution of the following optimization problem*

$$\min_{\substack{P = (P(1), \cdots, P(N)) > 0, \\ Q > 0, \gamma > 0, \varepsilon_1 > 0, \varepsilon_2 > 0, \\ s.t. \; (12.35), \; (12.36)}} \gamma \quad (12.37)$$

then for any $\alpha(i) \leq 1/\gamma, \beta(i) \leq 1/\gamma$, *the system under study is asymptotically stable.*

To illustrate the usefulness of the above algorithm, let us give an example.

Example 12.1 *Let us consider a system described by the dynamics (12.20) and suppose that the system data are as follows:* $\mathcal{S} = \{1,2\}, \Lambda = \begin{pmatrix} -4 & 4 \\ 1 & -1 \end{pmatrix},$

$$A(1) = \begin{pmatrix} -5 & 1 \\ 0 & -2 \end{pmatrix} \quad A(2) = \begin{pmatrix} -1 & 0 \\ 0 & -2 \end{pmatrix}$$

$$A_d(1) = \begin{pmatrix} -1 & 0 \\ -1 & -1 \end{pmatrix} \quad A_d(2) = \begin{pmatrix} -1 & 0 \\ -0.1 & -1 \end{pmatrix}$$

$$C(1) = \begin{pmatrix} 2 & 1 \\ 0 & 1 \end{pmatrix} \quad C(2) = \begin{pmatrix} 2 & 1 \\ 0 & 1 \end{pmatrix}.$$

With this set of data, solving the optimization problem (12.37) gives the following solution:

$$P(1) = \begin{pmatrix} 67.2512 & 6.1536 \\ 6.1536 & 86.2428 \end{pmatrix}$$

$$P(2) = \begin{pmatrix} 17.8892 & -30.0504 \\ -30.0504 & 94.5869 \end{pmatrix}$$

$$Q = \begin{pmatrix} 61.6193 & 0.3954 \\ 0.3954 & 61.8240 \end{pmatrix}$$

$$\varepsilon_1 = 79.5191 \quad \varepsilon_2 = 79.5619$$

$$\gamma = 1.2998.$$

According to Theorem 12.3, the system under study is asymptotically stable for any $\alpha(i) \leq 0.7694$ *and* $\beta(i) \leq 0.7694$.

In this section we considered the stability problem of the class of nonlinear stochastic time delay systems and gave two results. The first one dealt with the general dynamics (12.1) and the second is applied only to the special nonlinear dynamics (12.20).

12.2   Stabilization

Let us now restrict our study to the class of systems (12.20) and see how we can stabilize them. This section addresses the design problem of memoryless controllers that stabilize system (12.20) in the asymptotically stable sense. Two kinds of controllers (linear state feedback and nonlinear state feedback) are considered and design algorithms are developed to compute the appropriate gains. We will assume complete access to the state space vector for the rest of this section.

12.2.1   Linear State Feedback Controller Design

This subsection considers a linear memoryless state feedback controller of the form

$$u(t) = K(r_t)x(t) \tag{12.38}$$

where $K(r_t)$ is a gain matrix to be determined for each $r_t \in \mathcal{S}$.

Substituting (12.38) into (12.20) yields the following dynamics of the closed-loop system:

$$\begin{aligned} dx(t) = {} & \bar{A}(r_t)x(t) + f(x(t), r_t, t) + A_d(r_t)x(t - \tau) \\ & + f_d(x(t), r_t, t) + C(r_t)x(t)dB_t, \end{aligned} \tag{12.39}$$

where $\bar{A}(r_t) = A(r_t) + B(r_t)K(r_t)$

Using Theorem 12.2, we find that if there exist symmetric, positive-definite matrices $P = (P(1), \cdots, P(N)) > 0, Q > 0$, and scalars $\varepsilon_1 > 0$, $\varepsilon_2 > 0$ such that

$$\begin{pmatrix} \bar{J}(i) & P(i) & P(i) \\ P(i) & -\varepsilon_1 I & 0 \\ P(i) & 0 & -\varepsilon_2 I \end{pmatrix} < 0 \tag{12.40}$$

and

$$\begin{pmatrix} \varepsilon_2 \beta(i)I - Q & A_d^\top(i) \\ A_d(i) & -P^{-1}(i) \end{pmatrix} < 0 \tag{12.41}$$

hold for all $i \in \mathcal{S}$, where

$$\begin{aligned} \bar{J}(i) = {} & \bar{A}^\top(i)P(i) + P(i)\bar{A}(i) + \varepsilon_1 \alpha(i)I \\ & + C^\top(i)P(i)C(i) + Q + \sum_{j=1}^{N} \lambda_{ij}P(j) + P(i), \end{aligned}$$

then controller (12.38) asymptotically stabilizes system (12.20).

Using the Schur complement, we obtain that (12.40) is equivalent to

$$\begin{aligned} & \bar{A}^\top(i)P(i) + P(i)\bar{A}(i) + \varepsilon_1 \alpha(i)I + C^\top(i)P(i)C(i) + Q \\ & + P(i) + \sum_{j \in \mathcal{S}} \lambda_{ij}P(j) + \frac{1}{\varepsilon_2}P^2(i) + \frac{1}{\varepsilon_1}P^2(i) < 0. \end{aligned} \tag{12.42}$$

Let $X_i = P^{-1}(i), i \in \mathcal{S}$. Pre- and post-multiplying both sides of the above inequality by X_i and letting $Y_i = K(i)X_i$, $\eta_i = 1/\varepsilon_i, i = 1, 2$ yields

$$\begin{aligned} & J_1(i) + \frac{1}{\eta_1}\alpha(i)X_iX_i + X_iC^\top(i)X_i^{-1}C(i)X_i \\ & + X_iQX_i + X_i \left[\sum_{j \neq i} \lambda_{ij}X_j^{-1} \right] X_i < 0, \end{aligned} \tag{12.43}$$

where

$$J_1(i) = A(i)X_i + X_i A^\top(i) + B(i)Y_i + Y_i^\top B^\top(i) + \lambda_{ii} X_i \\ + X_i + \eta_1 I + \eta_2 I.$$

Note that

$$X_i \left[\sum_{j \neq i} \lambda_{ij} X_j^{-1} \right] X_i = S_i(X) \mathcal{X}_i S_i^\top(X), \qquad (12.44)$$

where

$$S_i(X) = \left(\sqrt{\lambda_{i1}} X_i \quad \cdots \quad \sqrt{\lambda_{ii-1}} X_i \quad \sqrt{\lambda_{ii+1}} X_i \quad \cdots \quad \sqrt{\lambda_{iN}} X_i \right) \\ \mathcal{X}_i = \text{diag}\{X_1, \cdots, X_{i-1}, X_{i+1}, \cdots, X_N\}.$$

Using the Schur complement, we get that (12.43) is equivalent to

$$\begin{pmatrix} J_1(i) & X_i & X_i C^\top(i) & X_i & S_i(X) \\ X_i & -\frac{\eta_1}{\alpha(i)} I & & & \\ C(i)X_i & & -X_i & & \\ X_i & & & -Q^{-1} & \\ S_i^\top(X) & & & & -\mathcal{X}_i \end{pmatrix} < 0. \qquad (12.45)$$

Moreover, using the Schur complement, (12.41) is equivalent to

$$\begin{pmatrix} \frac{1}{\eta_2} \beta(i)I - Q & A_d^\top(i) \\ A_d(i) & -X_i \end{pmatrix} < 0, \qquad (12.46)$$

which is equivalent to

$$\begin{pmatrix} -Q & A_d^\top(i)X_i & I \\ X_i A_d(i) & -X_i & 0 \\ -I & 0 & -\frac{\eta_2}{\beta(i)} I \end{pmatrix} < 0. \qquad (12.47)$$

Therefore, we get the following theorem.

Theorem 12.4 *If there exist symmetric, positive-definite matrices $X = (X_1, \cdots, X_N) > 0, Y = (Y_1, \cdots, Y_N)$, satisfying (12.45) and (12.47), then controller (12.38) with $K(i) = Y_i X_i^{-1}, i \in \mathcal{S}$ asymptotically stabilizes system (12.20).*

In this theorem, the upper bounds for the nonlinear terms of the system are assumed to be constants. In fact, if we let $\alpha(i) = \beta(i) = \frac{1}{\gamma}$, we can optimize γ by solving an appropriate optimization problem. In fact, note that (12.45) can be rewritten as

$$\begin{pmatrix} J_1(i) & X_i & X_i C^\top(i) & X_i & S_i(X) \\ X_i & -\Gamma_1 & & & \\ C(i)X_i & & -X_i & & \\ X_i & & & -Q^{-1} & \\ S_i^\top(X) & & & & -\mathcal{X}_i \end{pmatrix} < 0, \qquad (12.48)$$

$$\Gamma_1 \leq \gamma \eta_1 I, \tag{12.49}$$

and (12.47) is equivalent to

$$\begin{pmatrix} -Q & A_d^\top(i) & I \\ A_d(i) & -X_i & \mathbf{0} \\ -I & \mathbf{0} & -\Gamma_2 \end{pmatrix} < 0, \tag{12.50}$$

and

$$\Gamma_2 \leq \eta_2 \gamma I, \tag{12.51}$$

where Γ_1, Γ_2 are two auxiliary symmetric, positive-definite matrices.

Theorem 12.5 *Let the matrices $X_i > 0, \Gamma_1 > 0, \Gamma_2 > 0, Y_i$, the scalars η_1, η_2, and γ_0 be a set of optimal solutions to the following optimization problem:*

$$\min_{\substack{\gamma > 0, X = (X_1, \cdots, X_N) > 0, \\ Y = (Y_1, \cdots, Y_N) \\ \Gamma_1 > 0, \Gamma_2 > 0, \eta_1 > 0, \eta_2 > 0, \\ s.t.(12.48), (12.51)}} \gamma \tag{12.52}$$

Then, for any $0 \leq \alpha(i), \beta(i) \leq \frac{1}{\gamma}$, controller (12.38) with $K(i) = Y_i X_i^{-1}, i \in \mathcal{S}$, asymptotically stabilizes system (12.20).

This theorem gives a design algorithm for the linear memoryless controller that asymptotically stabilizes the class of systems (12.20). In the next section we consider a more general controller with a special nonlinear form.

12.2.2 Nonlinear State Feedback Controller Design

This subsection consider the problem of designing a nonlinear state feedback control of the form

$$u(t) = K_1(r_t)x(t) + K_2(r_t)f(x(t), r_t, t), \tag{12.53}$$

which stochastically stabilizes system (12.20).

Substituting (12.53) into (12.20) yields the dynamics of the closed-loop system as follow:

$$dx(t) = [\bar{A}(r_t)x(t) + \bar{f}(x(t), r_t, t) + A_d(r_t)x(t - \tau) \\ + f_d(x(t), r_t, t)]dt + C(r_t)x(t)dB_t, \tag{12.54}$$

where

$$\bar{A}(r_t) = A(r_t) + B(r_t)K_1(r_t)$$
$$\bar{f}(x(t), r_t, t) = f(x(t), r_t, t) + B(r_t)K_2(r_t)f(x(t), r_t, t).$$

Lemma 12.1 *If there exist symmetric, positive-definite matrices* $P =$ $(P(1), \cdots, P(N)) > 0, Q > 0,$ *scalars* $\varepsilon_1 > 0, \varepsilon_2 > 0$ *and* $\gamma > 0$ *such that*

$$J_2(i) < 0 \tag{12.55}$$

$$-Q + \frac{1}{\varepsilon_2}\beta(i) + A_d^\top(i)P(i)A_d(i) < 0 \tag{12.56}$$

$$\begin{pmatrix} -\rho I & [I + B(i)K_2(i)]^\top \\ [I + B(i)K_2(i)] & -\varepsilon_1 I \end{pmatrix} < 0 \tag{12.57}$$

hold for all $i \in \mathcal{S}$, *where*

$$J_2(i) = \bar{A}^\top(i)P(i) + P(i)\bar{A}(i) + \sum_{j=1}^N \lambda_{ij}P(j) + C^\top(i)P(i)C(i)$$
$$+ Q + P(i) + \varepsilon_1 P^2(i) + \varepsilon_2 P^2(i) + \rho\alpha(i)I,$$

then the closed-loop system (12.54) is asymptotically stable.

Proof: Consider the Lyapunov functional candidate as in (12.24). Then, we have

$$\mathcal{A}V(\mathbf{x}_t, r_t, t) = x^\top(t)\left[\sum_{j=1}^N \lambda_{r_t j}P(j)\right]x(t)$$
$$+ 2x^\top(t)P(r_t)\Big[\bar{A}(r_t)x(t) + \bar{f}(x(t), r_t, t)$$
$$+ A_d(r_t)x(t - \tau) + f_d(x(t - \tau), r_t, t)\Big]$$
$$+ x^\top(t)C^\top(r_t)P(r_t)C(r_t)x(t)$$
$$+ x^\top(t)Qx(t) - x^\top(t - \tau)Qx(t - \tau). \tag{12.58}$$

Using Lemma A.5, we obtain

$$2x^\top(t)P(r_t)\bar{f}(x(t), r_t, t)$$
$$\leq \varepsilon_1 x^\top(t)P^2(r_t)x(t) + \frac{1}{\varepsilon_1}\bar{f}^\top(x(t), r_t, t)\bar{f}(x(t), r_t, t)$$
$$\leq \varepsilon_1 x^\top(t)P^2(r_t)x(t) + \frac{1}{\varepsilon_1}f^\top(x(t), r_t, t)$$
$$\times [I + B(r_t)K_2(r_t)]^\top[I + B(r_t)K_2(r_t)]f(x(t), r_t, t). \tag{12.59}$$

Using the Schur complement, (12.57) implies that

$$\frac{1}{\varepsilon_1}[I + B(i)K_2(i)]^\top[I + B(i)K_2(i)] \leq \rho I, \tag{12.60}$$

which combined with (12.59) and the assumption (12.21) yields

$$2x^\top(t)P(r_t)\bar{f}(x(t), r_t, t) \leq \varepsilon_1 x^\top(t)P^2(r_t)x(t) + \rho\alpha(r_t)x^\top(t)x(t).$$

Similarly, we have

$$2x^\top(t)P(r_t)f_d(x(t-\tau),r_t,t)$$
$$\leq \varepsilon_2 x^\top(t)P^2(r_t)x(t) + \frac{1}{\varepsilon_2}\beta(r_t)x^\top(t-\tau)x(t-\tau) \qquad (12.61)$$

and

$$2x^\top(t)P(r_t)A_d(r_t)x(t-\tau)$$
$$= x^\top(t)P(r_t)x(t) + x^\top(t-\tau)A_d^\top(r_t)P(r_t)A_d(r_t)x(t-\tau)$$
$$- [x(t) - A_d(r_t)x(t-\tau)]^\top P(r_t)[x(t) - A_d(r_t)x(t-\tau)]$$
$$\leq x^\top(t)P(r_t)x(t) + x^\top(t-\tau)A_d^\top(r_t)P(r_t)A_d(r_t)x(t-\tau). \quad (12.62)$$

Therefore, we have

$$\mathcal{A}V(\mathbf{x}_t,r_t,t) \leq x^\top(t)J_2(r_t)x(t) + x^\top(t-\tau)[-Q + \frac{1}{\varepsilon_2}\beta(r_t)I$$
$$+ A_d^\top(r_t)P(r_t)A_d(r_t)]x(t-\tau)$$
$$\leq x^\top(t)J_2(r_t)x(t),$$

which is negative-definite according to (12.55). The last inequality is due to (12.56). This completes the proof of Lemma 12.1. □

With the above lemma, we are ready to proceed with the design of a stabilizing controller of the form of (12.53).

Let $X_i = P^{-1}$. Pre- and post-multiplying both sides of (12.55) by X_i yields

$$X_i\bar{A}^\top(i) + \bar{A}(i)X_i + X_iC^\top(i)X_i^{-1}C(i)X_i + X_iQX_i$$
$$+ X_i\left[\sum_{j=1}^N \lambda_{ij}X_j^{-1}\right]X_i + X_i + \varepsilon_1 I + \varepsilon_2 I + \rho\alpha(i)X_iX_i < 0.$$

Using the Schur complement and letting $Y_i = K_1(i)X_i$ and $\varrho = 1/\rho$ leads to

$$\begin{pmatrix} J_3(i) & X_iC^\top(i) & X_i & X_i & S_i(X) \\ C(i)X_i & -X_i & & & \\ X_i & & -Q^{-1} & & \\ X_i & & & -\frac{\varrho}{\alpha(i)}I & \\ S_i^\top(X) & & & & -\mathcal{X}_i \end{pmatrix} < 0, \qquad (12.63)$$

where

$$J_3(i) = A(i)X_i + B(i)Y_i + X_iA^\top(i) + Y_i^\top B^\top(i) + X_i$$
$$+ \varepsilon_1 I + \varepsilon_2 I + \lambda_{ii}X_i.$$

Using the Schur complement, (12.56) is equivalent to

$$\begin{pmatrix} -Q + \frac{1}{\varepsilon_2}\beta(i)I & A_d^\top(i) \\ A_d(i) & -X_i \end{pmatrix} < 0,$$

which is equivalent to

$$\begin{pmatrix} -Q & A_d^\top(i)X_i & I \\ X_i A_d(i) & -X_i & 0 \\ I & 0 & -\frac{\varepsilon_2}{\beta(i)}I \end{pmatrix} < 0. \tag{12.64}$$

Pre- and post-multiplying both sides of (12.57) by $\text{diag}\{\varrho I, I\}$ and letting $\hat{K}_2(i) = \varrho K_2(i)$ yields

$$\begin{pmatrix} -\varrho I & [\varrho I + B(i)\hat{K}_2(i)]^\top \\ \varrho I + B(i)\hat{K}_2(i) & -\varepsilon_1 I \end{pmatrix} < 0. \tag{12.65}$$

From above derivation, we obtain the following theorem.

Theorem 12.6 *If there exist symmetric, positive-definite matrices* $X = (X(1),\ldots,X(N)) > 0$, *matrices* $Y = (Y(1),\ldots,Y(N))$, $\hat{K} = (K(1),\ldots, K(N))$ *and scalars* $\varepsilon_1 > 0$, $\varepsilon_2 > 0$, $\varrho > 0$ *satisfying (12.63)-(12.65), then controller (12.53) with*

$$\begin{cases} K_1(i) = Y_i X_i^{-1} \\ K_2(i) = \frac{1}{\varrho}\hat{K}_2(i) \end{cases} \tag{12.66}$$

stabilizes system (12.20).

Likewise, we can design a stabilizing controller that optimizes the upper bound of the nonlinear terms in the system. Let $\alpha(i) = \beta(i) = \frac{1}{\gamma}$. Note that (12.63) can be rewritten as

$$\begin{pmatrix} J_3(i) & X_i C^\top(i) & X_i & X_i & S_i(X) \\ C(i)X_i & -X_i & & & \\ X_i & & -Q^{-1} & & \\ X_i & & & -\Gamma_1 & \\ S_i^\top(X) & & & & -\mathcal{X}_i \end{pmatrix} < 0 \tag{12.67}$$

$$\Gamma_1 \leq \varrho\gamma I \tag{12.68}$$

and (12.64) can be rewritten as

$$\begin{pmatrix} -Q & A_d^\top(i) & I \\ A_d(i) & -X_i & 0 \\ I & 0 & -\Gamma_2 \end{pmatrix} < 0 \tag{12.69}$$

$$\Gamma_2 \leq \gamma\varepsilon_2 I. \tag{12.70}$$

Using Theorem 12.6, we get the following theorem.

Theorem 12.7 *Let $\gamma_0, \varrho > 0, X_i, Y_i, \hat{K}_2(i)$ be a set of optimal solutions of the optimization problem*

$$
\begin{array}{c}
\min \qquad \qquad \gamma \qquad \qquad \text{(12.71)} \\
\gamma > 0, X = (X_1, \cdots, X_N) > 0, \\
Y = (Y_1, \cdots, Y_N), \hat{K}_2(i), \\
\varepsilon_1 > 0, \varepsilon_2 > 0, , \varepsilon > 0, \\
s.t. \ (12.67)\text{-}(12.70), \ and \ (12.65)
\end{array}
$$

then for all $0 \le \alpha(i), \beta(i) \le \frac{1}{\gamma_0}$, (12.53) with

$$
\left\{
\begin{array}{l}
K_1(i) = Y_i X_i^{-1} \\
K_2(i) = \frac{1}{\varrho} \hat{K}_2(i)
\end{array}
\right.
$$

asymptotically stabilizes system (12.20).

Theorem 12.7 gives an algorithm to design a state feedback memoryless controller that stabilizes system (12.20). To show the usefulness of the above design technique, let us give a numerical example.

Example 12.2 *Consider a system with the dynamics described by (12.20) and suppose that it has two modes, i.e., $\mathcal{S} = \{1, 2\}$ and the system parameters are as follows: $\Lambda = \begin{pmatrix} -14 & 14 \\ 8 & -8 \end{pmatrix}$,*

$$
A(1) = \begin{pmatrix} 5 & 1 \\ 0 & -2 \end{pmatrix} \qquad A(2) = \begin{pmatrix} 1 & 0 \\ 0 & -2 \end{pmatrix}
$$

$$
A_d(1) = \begin{pmatrix} 0.1 & 0 \\ -0.1 & -0.1 \end{pmatrix} \qquad A_d(2) = \begin{pmatrix} 0.1 & 0 \\ 0.1 & 0.1 \end{pmatrix}
$$

$$
C(1) = \begin{pmatrix} 0.2 & 1 \\ 0 & 1 \end{pmatrix} \qquad C(2) = \begin{pmatrix} 0.2 & 0.1 \\ 0 & 0.1 \end{pmatrix}
$$

$$
B(1) = \begin{pmatrix} 1, \\ -1 \end{pmatrix} \qquad B(2) = \begin{pmatrix} 1 \\ 0 \end{pmatrix}.
$$

With this set of data, solving the optimization problem (12.71), we get the following solution

$$
X_1 = \begin{pmatrix} 6.6771 & -5.8368 \\ -5.8368 & 6.3261 \end{pmatrix} \qquad X_2 = \begin{pmatrix} 30.9426 & -0.7710 \\ -0.7710 & 1.2057 \end{pmatrix}
$$

$$
\Gamma_1 = \begin{pmatrix} 0.6732 & 0.0063 \\ 0.0063 & 0.6606 \end{pmatrix} \qquad \Gamma_2 = \begin{pmatrix} 1.5049 & 0.0227 \\ 0.0227 & 1.4186 \end{pmatrix}
$$

$$
Y_1 = 10^6 \left(-9.7272 \quad 9.7272 \right) \qquad Y_2 = 10^7 \left(-1.8761 \quad -0.0000 \right)
$$

$$
\hat{K}(1) = \left(-0.1648 \quad 0.1648 \right) \qquad \hat{K}(2) = \left(-0.3297 \quad 0 \right)
$$

$$
\varepsilon_1 = 0.4945 \ \varepsilon_2 = 1.0553 \qquad \varrho = 0.3297 \ \gamma_0 = 1.4620.
$$

Therefore, according to Theorem 12.7, we conclude that for any $0 \leq \alpha(i), \beta(i) \leq 1/\gamma_0 = 0.6840$, controller (12.53) with

$$K_1(i) = Y_i X_i^{-1} = 10^6 \left(\begin{array}{cc} -0.5824 & 1.0002 \end{array} \right)$$

$$K_2(i) = \frac{1}{\varrho} \hat{K}_2(i) = 10^5 \left(\begin{array}{cc} -6.1613 & -3.9399 \end{array} \right)$$

asymptotically stabilizes system (12.20).

12.3 Notes

Stochastic nonlinear systems with time delay and with disturbance represented by a standard Wiener process is studied by Kalmanovskii and Myshkis [117]. Nonlinear systems with time delays represented by a continuous Markov process is addressed by Kolmanovsky and Maizenberg [119].

Appendix A
Linear Matrix Inequality and Preliminary Lemmas

The goal of this appendix is to recall some important notions on linear matrix inequality (LMI) in order to render our book self-contained. Mainly, we define the LMI problem and the related problems like the feasibility problem (FEAS), minimization of a linear objective under LMI constraints (MINCX), and the generalized eigenvalue minimization problem (GEVP).

A.1 LMI Functions

A linear matrix inequality (LMI) has the form

$$F(z) \triangleq F_0 + \sum_{i=1}^{m} z_i F_i < 0, \tag{A.1}$$

where $z = (z_1, \ldots, z_m) \in \mathbb{R}^m$ is the variable to be determined and the symmetric matrices $F_i \in \mathbb{R}^{n \times n}, 0 \leq i \leq m$ are given. The inequality symbol in (A.1) means that $F(x)$ is negative-definite, i.e., $v^\top F(x)v < 0$ for all nonzero $v \in \mathbb{R}^n$.

For example, a linear system with the following dynamic

$$\dot{x}(t) = Ax(t) \tag{A.2}$$

where $x(t) \in \mathbb{R}^2, A \in \mathbb{R}^{2 \times 2}$, is stable if and only if there exists a symmetric and positive-definite matrix $P > 0$ such that

$$A^\top P + PA < 0. \tag{A.3}$$

This problem, in fact, can be solved using the LMI toolbox. Let us now see how we can put this problem in the form of (A.1). For this purpose, let $A = \begin{pmatrix} a_1 & a_2 \\ a_2 & a_3 \end{pmatrix}$ and suppose that $P = \begin{pmatrix} z_1 & z_2 \\ z_2 & z_3 \end{pmatrix}$, where z_1, z_2, and z_3 are design parameters. Then,

$$A^\top P + PA = z_1 \begin{pmatrix} 2a_1 & a_2 \\ a_2 & 0 \end{pmatrix} + z_2 \begin{pmatrix} 2a_2 & a_1 + a_3 \\ a_1 + a_3 & 2a_2 \end{pmatrix}$$
$$+ z_3 \begin{pmatrix} 0 & a_2 \\ a_2 & 2a_3 \end{pmatrix}.$$

Therefore, (A.3) is a standard LMI feasibility problem that can be solved using the LMI toolbox.

There are three kinds of generic LMI problems often referred to in this book. Related to these generic LMI problems, several functions are intensively used: *lmivar, lmiterm, setlmis, getlmis, feasp, mincx, gevp, mat2dec*. Before giving numerical examples, let us give some information on these functions. The material below is based on the LMI Matlab toolbox help, from which the descriptions of these functions are reproduced.

```
X = lmivar(type,struct)
[X,ndec,Xdec] = lmivar(type,struct)
```

Adds a new matrix variable X to the LMI system currently described. A label X can be optionally attached to this new variable to facilitate future reference to it.

```
Input:
 TYPE      structure of X:
           1 -> symmetric block diagonal
           2 -> full rectangular
           3 -> other
 STRUCT    additional data on the structure of X
           TYPE=1: the i-th row of STRUCT describes the
           i-th diagonal block of X
             STRUCT(i,1) -> block size
             STRUCT(i,2) -> block type, i.e.,
                             0 for scalar blocks t*I
                 1 for full block
                             -1 for zero block
      TYPE=2: STRUCT = [M,N]  if X is a MxN matrix
      TYPE=3: STRUCT is a matrix of same dimension as X
           where STRUCT(i,j) is either
             0  if X(i,j) = 0
            +n  if X(i,j) = n-th decision variable
```

                    -n if X(i,j) = (-1)* n-th decision var
Output:
  X Optional: identifier for the new matrix variable.
     Its value is k if k-1 matrix variables have already
     been declared. This identifier is not affected by
              subsequent modifications of the LMI system
  NDEC total number of decision variables so far
  XDEC entry-wise dependence of X on the decision variables
              (Xdec = struct for Type 3)

lmiterm(termID,A,B,flag)

  Adds one term to some LMI in the LMI system currently
  described. A term is either an outer factor, a constant
  matrix, or a variable term A*X*B or A*X'*B where X is
  a matrix variable.

  IMPORTANT: Because the OFF-DIAGONAL blocks (i,j) and (j,i)
             are transposed of one another, specify the term
             content of ONLY ONE of these two blocks.

  Input:
   TERMID 4-entry vector specifying the term location and
          nature
          Which LMI?
          TERMID(1) = +n -> left-hand side of the n-th LMI
          TERMID(1) = -n -> right-hand side of the n-th LMI
          Which block?
          For outer factors, set TERMID(2:3) = [0 0].
          Otherwise, set TERMID(2:3) = [i j] if the term
          belongs to the (i,j) block of the LMI
          What type of term?
          TERMID(4) = 0 -> constant term
          TERMID(4) = X -> variable term A*X*B
          TERMID(4) = -X -> variable term A*X'*B
          where X is the variable identifier returned by
          LMIVAR
   A   value of the outer factor, constant term, or
          left coefficient in variable terms A*X*B or
          A*X'*B
   B   right coefficient in variable terms A*X*B or
          A*X'*B
   FLAG   quick way of specifying the expression
          A*X*B+B'*X'*A' in a DIAGONAL block. Set
          FLAG='s' to specify it with only one LMITERM

command

setlmis(lmisys0)

   Initializes the description of a new LMI system.

   To start from scratch, type

                   setlmis([])

   To add on to an existing LMI system LMISYS0, type

                   setlmis(lmisys0)

   See also LMIVAR, LMITERM, GETLMIS.

   lmisys = getlmis

   Returns the internal representation LMISYS of an LMI
   system once this LMI system has been fully described
   with LMIVAR and LMITERM.

   The internal representation LMISYS can be passed
   directly to the LMI solvers or any other LMI-Lab
   function.

   See also SETLMIS, LMIVAR, LMITERM.

   [tmin,xfeas] = feasp(lmis,options,target)

   Solves the feasibility problem defined by the system LMIS
   of LMI constraints. When the problem is feasible, the
   output XFEAS is a feasible value of the vector of (scalar)
   decision variables.

   Given a feasibility problem of the form   L(x) < R(x),
   FEASP solves the auxiliary convex program:

           Minimize   t   subject to   L(x) < R(x) + t*I

   The system of LMIs is feasible iff. the global minimum TMIN
   is negative. The current best value of t is displayed by
   FEASP at each iteration.

   Input:

LMIS   array describing the system of LMI constraints
OPTIONS optional: five-entry vector of control parameters.
         Default values are selected by setting OPTIONS(i)=0.
         OPTIONS(1): not used
         OPTIONS(2): max. number of iterations (Default=100)
         OPTIONS(3): feasibility radius R. R>0 constrains
                     x to x'*x < R^2   (Default=1e9).
                     R<0 means "no bound"
         OPTIONS(4): when set to an integer value L > 1,
                     forces termination when t has not
                     decreased by more than 1%over the last
                     L iterations (Default = 10).
         OPTIONS(5): when nonzero, the trace of execution is
                     turned off.
TARGET   optional: target for TMIN. The code terminates as
                    soon as
                    t < TARGET   (Default=0)
  Output:
  TMIN   value of t upon termination. The LMI system is
            feasible   iff. TMIN <= 0
  XFEAS   corresponding minimizer. If TMIN <= 0, XFEAS is
            a feasible vector for the set of LMI constraints.
            Use DEC2MAT to get the matrix variable values
            from XFEAS.

See also MINCX, GEVP, DEC2MAT.

[copt,xopt] = mincx(lmis,c,options,xinit,target)

Solves the LMI problem:

       Minimize   c'*x   subject to   L (x) < R(x)

where x is the vector of DECISION VARIABLES.

  Input:
  LMIS   description of the system of LMI constraints
  C   vector of the same size as x (see DECNBR).
            Use DEFCX to specify the objective c'*x
            directly in terms of matrix variables
  OPTIONS   optional: five-entry vector of control
            parameters. The default value is used when
            OPTIONS(i)=0 .
       OPTIONS(1): relative accuracy required on the
                   optimal value of the objective

                          (Default = 0.01)
        OPTIONS(2): max. number of iterations (Default=100)
        OPTIONS(3): feasibility radius R.   R>0 constrains x
                    to x'*x < R^2.   R<0 means "no bound"
                                              (Default=1e9)
        OPTIONS(4): integer value L.   The code
                    terminates when the objective value
                    has decreased by less than
                    OPTIONS(1) during the last
                    L iterations   (Default=10)
        OPTIONS(5): when nonzero, the trace of execution is
                    turned off   (Default=0)
    XINIT   optional: initial guess for X   ([] if none,
                    ignored when infeasible)
    TARGET   optional: target for the objective value.
                    The code terminates as soon as a feasible x is
                    found such that
                    c'*x < TARGET   (Default=-1e20)
    Output:
    COPT   global minimum of the objective c'*x
    XOPT   minimizing value of the vector x of decision
              variables. Use DEC2MAT to get the corresponding
              matrix variable values.

    See also FEASP, GEVP, DEFCX, DEC2MAT.

    [tmin,xopt] = gevp(lmis,nlfc,options,t0,x0,target)

    Solves the generalized eigenvalue minimization problem

                    Minimize   t

    subject to the LMI constraints:

            $C(x) < 0$

             $0 < Bj(x)$ (j=1,..,NLFC)

            $Aj(x) < t * Bj(x)$ (j=1,..,NLFC)

    Here x denotes the vector of (scalar) decision variables.
    The positivity constraints $Bj(x) > 0$ must be specified
    for well-posedness, and the LMIs involving t should be
    specified last.

Input:

LMIS description of the system of LMI constraints

NLFC number of linear fractional constraints (LMIs involving t)

OPTIONS optional: five-entry vector of control parameters. The default value is used when OPTIONS(i)=0

    OPTIONS(1): relative accuracy required on TMIN (Default = 1.0e-2)

    OPTIONS(2): max. number of iterations (Default=100)

    OPTIONS(3): feasibility radius R. R>0 constrains x to x'*x < R^2 (Default=1e8). R<0 means "no bound"

    OPTIONS(4): integer value L. The code terminates when t has decreased by less than OPTIONS(1) during the last L iterations (Default = 5)

    OPTIONS(5): when nonzero, the trace of execution is turned off.

TO,XO optional: initial guesses for t,x (ignored when unfeasible)

TARGET optional: target for TMIN. The code terminates as soon as t falls below this value (DEFAULT = -1e5)

Output:

TMIN minimal value of t

XOPT minimizing value of the vector x of decision variables Use DEC2MAT to get the corresponding matrix variable values.

See also FEASP, MINCX, DEC2MAT.

decvars = mat2dec(lmisys,X1,X2,X3,...)

For given values X1, X2, X3, ... of the matrix variables involved in the LMI system LMISYS, MAT2DEC computes the value DECVARS of the vector of decision variables. This operation is the converse of that performed by DEC2MAT

Input:

  LMISYS array describing the LMI system

  X1, X2, X3,...

      values of the matrix variables. MAT2DEC accepts up to 20 values. An error is issued

```
                    if some matrix variable remains unassigned.
   Output:
      DECVARS    vector of the decision variable values.
```

See also DEC2MAT, DECINFO, DEFCX.

A.2   LMI Problems

As we mentioned earlier, there exist three main problems that we have used extensively in this book. These problems are:

- the feasibility problem

- the linear optimization problem

- the generalized eigenvalue minimization problem.

A.2.1   Feasibility Problem

The LMI Feasibility Problem (FEASP) consists of determining the variable $x \in \mathbb{R}^m$ such that $F(x) < 0$ holds. This problem can be solved using the function "feasp" of the LMI toolbox. In the rest of this subsection we will see how to state the problem and how to solve using "feasp" function.

A typical situation for the feasibility problem is the stability test for dynamical systems. In fact, based on control theory a system with the dynamic (A.2) is stable if and only if there exists a symmetric and positive-definite matrix $P > 0$ such that (A.3) is satisfied.

The goal in this case is to find a matrix $P > 0$ such that the inequality (A.3) is satisfied. The "feasp" can be used to solve our problem. The next example demonstrates the use of such a function.

Example A.1 *Consider a system described by (A.2) and suppose that the matrix A is given by $A = \begin{pmatrix} -2 & 1 \\ 1 & -1 \end{pmatrix}$. Let us now see how we can solve the stability problem. Based on what we described previously, we can write a code in LMI Matlab toolbox. To solve our stability problem, type the following code and save it as, for example, app1.m.*

```
clear all

A = [-2 1;1 -1];

setlmis([])
```

```
P = lmivar(1, [2, 1]);
lmiterm([1 1 1 P],A,1, 's')
lmiterm([-2 1 1 P],1,1)

LMI = getlmis;
[copt xopt] = feasp(LMI);
P = dec2mat(LMI, xopt, P)
```

Then in Matlab environment, running app1.m yields

```
P =
    0.6587      0.5646
    0.5646      1.2233
```

Therefore, the system under study is stable.

Next, let us consider the stabilization problem of the system

$$\dot{x}(t) = Ax(t) + Bu(t) \tag{A.4}$$

where $u(t)$ is the control input. Using the stability result, a state feedback memoryless controller $u(t) = Kx(t)$ stabilizes system (A.4) if and only if there exists a symmetric and positive matrix P such that

$$P(A + BK) + (A + BK)^\top P < 0. \tag{A.5}$$

Pre- and post-multiplying both sides of (A.5) by $X = P^{-1}$ and letting $Y = KX$ yields that (A.5) is equivalent to

$$AX + BY + XA^\top + Y^\top B^\top < 0. \tag{A.6}$$

This problem represents another example of a feasible LMI problem. To see how this result can be used, let us consider the following numerical example.

Example A.2 *Suppose that the system parameters are as follows: $A = \begin{pmatrix} 2 & 1 \\ 1 & 1 \end{pmatrix}$, $B = \begin{pmatrix} 1 \\ 2 \end{pmatrix}$. Obviously, this free system (with $u(t) \equiv 0$) is unstable. To compute a stabilizing gain, let us type the following code:*

```
clear all
A = [2 1;1 1];
B = [1; 2];

setlmis([])
X = lmivar(1, [2, 1]);
Y = lmivar(2, [1,2]);

lmiterm([1 1 1 X],A,1,'s')
lmiterm([1 1 1 Y],B,1,'s')
```

```
lmiterm([-2 1 1 X],1,1)

LMI = getlmis;
[copt xopt] = feasp(LMI);
X = dec2mat(LMI, xopt, X)
Y = dec2mat(LMI, xopt, Y)
K = Y*inv(X)
```

Running the above code yields

```
X =
    86.3203    53.3647
    53.3647   833.4266

Y =
  -320.2027 -521.1294

K =
     -3.4599    -0.4037
```

Therefore, a stabilizing controller is obtained.

A.2.2 Minimization of a Linear Objective under LMI Constraints

Minimization of a linear objective under LMI constraints (MINCX) is another interesting problem that we have used extensively. The MINCX problem is stated as follows:

$$\min_{x \in \mathbb{R}^m} \quad C^\top x$$
$$\text{s.t. } F(x) < 0$$

where $C \in \mathbb{R}^m$ is a given vector. This problem can be solved using "mincx" of LMI Toolbox.

To provide an example using "mincx", let us consider an $\mathcal{H}_2$ control problem. For this purpose let us assume the system dynamics are given by

$$\begin{cases} \dot{x}(t) = Ax(t) + Bw(t) \\ y(t) = Cx(t) \end{cases} \tag{A.7}$$

where $w(t)$ is a white noise disturbance with unit covariance. Suppose that the $\mathcal{H}_2$ performance is defined by

$$\|\mathcal{H}\|_2^2 = \lim_{t \to \infty} \mathbb{E}\left(\frac{1}{t} \int_0^t y^\top(s)y(s)ds \right).$$

Then, it can be shown that the solution to this problem is given by (see for example [38]):

$$\|\mathcal{H}\|_2^2 = \min\{tr(CPC^\top) : AP + PA^\top + BB^\top < 0\}.$$

Obviously, this optimization problem is equivalent to minimizing $\mathrm{tr}(Q)$ subject to

$$AP + PA^\top + BB^\top < 0 \tag{A.8}$$
$$CPC^\top \leq Q. \tag{A.9}$$

Using the Schur complement, (A.9) is equivalent to

$$\begin{pmatrix} -Q & CP \\ PC^\top & -P \end{pmatrix} < 0.$$

Example A.3 *Let us now suppose that the system parameters are*

$$A = \begin{pmatrix} -2 & 1 \\ 1 & -1 \end{pmatrix} \quad B = \begin{pmatrix} 1 \\ 2 \end{pmatrix} \quad C = \begin{pmatrix} 2 & 1 \\ -1 & 0 \end{pmatrix}$$

and see how we can use the mincx function to minimize our $\mathcal{H}_2$ *performance. Let us type the following Matlab code:*

```
%% apped3
clear all

A = [-2 1;1 -1];
B=[1;2];
C=[2 1;-1 0];

setlmis([])

Q = lmivar(1, [2, 1]);
P = lmivar(1, [2, 1]);
lmiterm([1 1 1 P],A, 1, 's');
lmiterm([1 1 1  0],B*B');

lmiterm([-2 1 1 Q],1,1)
lmiterm([-2 1 2 P],C,1)
lmiterm([-2 2 2 P],1,1)

lmiterm([-3 1 1 P],1,1);
lmiterm([-4 1 1 Q],1,1);

c=[1 0 1 0 0 0];

LMI = getlmis;
[copt xopt] = mincx(LMI,c);

 Q = dec2mat(LMI, xopt, Q)
 P = dec2mat(LMI, xopt,P)
 optrace = copt
```

Running the above code yields the following solution:

```
Q =
    22.8395    -6.1685
    -6.1685     1.6728
P =
     1.6684     2.8322
     2.8322     4.8348

optrace =
    24.5122
```

A.2.3 Generalized Eigenvalue Minimization Problem

The generalized eigenvalue minimization problem (GEVP) is the third interesting problem used extensively in this book. This problem is stated as follows:

$$\min_{x \in \mathbb{R}^m} \lambda$$
$$\text{s.t. } F_1(x) < \lambda F_2(x)$$

where $F_1(x), F_2(x)$ are two matrices of form (A.1). The GEVP is quasi-convex with respect to the design parameters x and λ, which can be solved by using "gevp" of the LMI toolbox.

Remark A.1 *When using 'gevp', $F_2(x)$ should be stated to be positive-definite.*

The decay rate of system (A.2) is defined as the largest γ such that $\lim_{t \to \infty} e^{\gamma t} \|x(t)\| = 0$. Let us consider a Lyapunov function candidate $V(x(t) = x^\top(t)Px(t)$. If we can establish that

$$\frac{dV(x(t))}{dt} \leq -2\gamma V(x(t)) \tag{A.10}$$

holds for all trajectories, then the decay rate of system (A.2) is at least γ.
Noting that (A.10) holds if and only if

$$A^\top P + PA + 2\gamma P \leq 0, \tag{A.11}$$

we conclude that the largest lower bound on the decay rate can be found by solving the GEVP in P and γ

$$\max_{P>0} \gamma$$
$$\text{s.t. (A.11)}.$$

To solve this optimization problem, let us rewrite (A.11) as

$$2P \leq \frac{1}{\gamma}(-PA - A^\top P).$$

Example A.4 *Let us now consider a dynamical system with A given by* $A = \begin{pmatrix} -2 & 0.2 \\ 2 & -5 \end{pmatrix}$ *and see how we can compute the decay rate. For this purpose, type the following code and save it.*

```
%% apped4
clear all
A = [-2 .2;2 -5];

setlmis([])
P = lmivar(1, [2, 1]);

lmiterm([-1 1 1 P],1,1)
lmiterm([-2 1 1 P],-1, A, 's')

lmiterm([3 1 1 P],2,1)
lmiterm([-3 1 1 P],-1,A,'s')

LMIs = getlmis;
[copt xopt] = gevp(LMIs, 1);

P = dec2mat(LMIs, xopt, P)
alpha = 1/copt
```

Running this code yields

```
P =
  1.0e-005 *
    0.1254   -0.1937
   -0.1937    0.3031

alpha =
    1.8720
```

Therefore, the optimal decay rate of system (A.2) with $A = \begin{pmatrix} -2 & 0.2 \\ 2 & -5 \end{pmatrix}$ *is 1.8720, which is, actually, the smallest eigenvalue of A.*

Since this book deals with stochastic systems, let us give a numerical example for checking the stability of a jump linear system using the feasibility problem.

Consider a jump linear system with the following dynamics:

$$\begin{cases} \dot{x}(t) = A(r_t)x(t), & x(0) = x_0, \\ y(t) = C(r_t)x(t) \end{cases} \tag{A.12}$$

where $x(t) \in \mathbb{R}^n$ is the system state, $y(t) \in \mathbb{R}^m$ is the output of the system, and $r_t \in \mathcal{S} = \{1, 2, \cdots, N\}$ is a Markov process with infinitesimal generator $\Lambda = (\lambda_{ij}), i, j \in \mathcal{S}, \lambda_{ii} = -\sum_{j \neq i} \lambda_{ij}, \lambda_{ij} \geq 0, j \neq i$. Based on control theory the system (A.12) is said to be stochastically stable if for any initial condition (x_0, r_0), the following holds

$$\int_0^\infty \mathbb{E}\{\|x(t)\|^2|(x_0, r_0)\} < \infty.$$

For system (A.12), we have the following result (see [62]): System (A.12) is stochastically stable if and only there exist symmetric and positive-definite matrices $P = (P(1), \cdots, P(N)) > 0$ such that for all $i \in \mathcal{S}$, the following hold:

$$A^\top(i)P(i) + P(i)A(i) + \sum_{j=1}^N \lambda_{ij}P(j) < 0. \tag{A.13}$$

Example A.5 *Let us consider a system with* $\mathcal{S} = \{1, 2\}$ *and* $\Lambda = \begin{pmatrix} -4 & 4 \\ 7 & -7 \end{pmatrix}$ *and assume that* $A(1) = \begin{pmatrix} -2 & 0.2 \\ 2 & -5 \end{pmatrix}$, $A(2) = \begin{pmatrix} 1 & 0.2 \\ 0 & -5 \end{pmatrix}$.
Type the following code and save it.

```
%% apped5
clear all

A1 = [-2 0.2;2 -5];
A2 = [1 .2;0 -5];
Lambda = [-4 4;7 -7];

setlmis([])
P1 = lmivar(1, [2, 1]);
P2 = lmivar(1, [2, 1]);

lmiterm([1 1 1 P1],1,A1,'s');
lmiterm([1 1 1 P1],1,Lambda(1,1));
lmiterm([1 1 1 P2],1,Lambda(1,2));
lmiterm([2 1 1 P2],1,A2,'s');
lmiterm([2 1 1 P2],1,Lambda(2,2));
lmiterm([2 1 1 P1],1,Lambda(2,1));
lmiterm([-3 1 1 P1],1,1)
lmiterm([-4 1 1 P2],1,1)

LMIs = getlmis;

[copt xopt] = feasp(LMIs);

P1 = dec2mat(LMIs, xopt, P1)
P2 = dec2mat(LMIs, xopt,P2)
```

Running the above code yields the following solution:

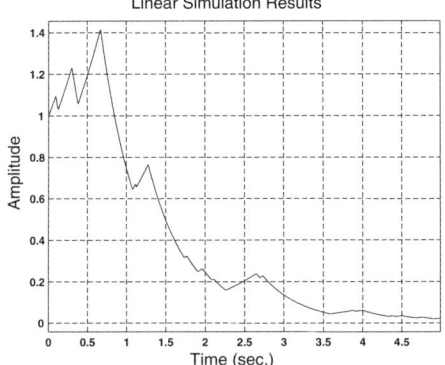

Figure A.1. Behavior of the system trajectory.

P1 =

0.9673	0.0653
0.0653	0.1389

P2 =

1.5775	0.0689
0.0689	0.1378

Therefore, the system under study is stochastically stable. In the case of $C(1) = C(2) = \begin{pmatrix} 1 & 0 \end{pmatrix}$, *a trajectory of the system is plotted in Figure A.1.*

A.3   Lemmas

This section presents a certain number of results that are used extensively in this book in various proofs of the proposed theorems.

The following lemma is critical for casting a nonlinear problem into the framework of LMI.

Lemma A.1 *Let* X, Y *be real constant matrices of compatible dimensions. Then*

$$X^\top Y + Y^\top X \leq \varepsilon X^\top X + \frac{1}{\varepsilon} Y^\top Y$$

holds for any $\varepsilon > 0$.

Proof: The proof follows from the inequality

$$0 \leq \left(\sqrt{\varepsilon} X^\top - \frac{1}{\sqrt{\varepsilon}} Y^\top \right) \left(\sqrt{\varepsilon} X - \frac{1}{\sqrt{\varepsilon}} Y \right).$$

$\square$

Lemma A.2 *(Schur Complement) Let the symmetric matrix M be partitioned as*

$$M = \begin{pmatrix} X & Y \\ Y^\top & Z \end{pmatrix}$$

with X, Z being symmetric matrices. We have

(i) M is nonnegative-definite if and only if either

$$\begin{cases} Z \geq 0 \\ Y = L_1 Z \\ X - L_1 Z L_1^\top \geq 0 \end{cases} \tag{A.14}$$

or

$$\begin{cases} X \geq 0 \\ Y = X L_2 \\ Z - L_2^\top X L_2 \geq 0 \end{cases} \tag{A.15}$$

hold, where L_1, L_2 are some (nonunique) matrices of compatible dimensions.

(ii) M is positive-definite if and only if either

$$\begin{cases} Z > 0 \\ X - Y Z^{-1} Y^\top > 0 \end{cases} \tag{A.16}$$

or

$$\begin{cases} X > 0 \\ Z - Y^\top X^{-1} Y > 0. \end{cases} \tag{A.17}$$

Matrix $X - Y Z^{-1} Y^\top$ is called the Schur complement $X(Z)$ in M.

Proof: We begin with proving (A.14). *Necessity.* Clearly, $Z \geq 0$ is necessary. We must prove the necessity of $X - Y Z^{-1} Y^\top$. Letting x be a vector and partition it as $x = \begin{pmatrix} x_1 \\ x_2 \end{pmatrix}$ according to the partitioning of M, we have

$$x^\top M x = x_1^\top X x_1 + 2 x_1^\top Y x_2 + x_2^\top Z x_2. \tag{A.18}$$

Let x_2 be such that $Z x_2 = 0$. If $Y x_2 \neq 0$, let $x_1 = -\alpha Y x_2, \alpha > 0$. Then

$$x^\top M x = \alpha^2 x_2^\top Y^\top X Y x_2 - 2 \alpha x_2^\top Y^\top Y x_2,$$

which is negative for a sufficiently small $\alpha > 0$. Therefore

$$X x_2 = 0 \Longrightarrow Y x_2 = 0, \ \forall x_2,$$

which implies

$$Y = L_1 Z \tag{A.19}$$

for some (nonunique) L_1.

Since $M \geq 0$, the quadratic form (A.18) has, for any x_1, a minimum over x_2. Thus differentiating (A.18) with respect to $x_2^\top$ we have

$$0 = \frac{\partial(x^\top M x)}{\partial x_2^\top} = 2Y^\top x_1 + 2Z x_2 = 2Z L_1^\top x_1 + 2Z x_2$$

whence

$$ZL^\top x_1 = -Z x_2. \tag{A.20}$$

Using (A.19) and (A.20) in (A.18), we find that the minimum of $x^\top M x$ over x_2 and for any x_1 is

$$\min_{x_2} x^\top M x = x_1^\top (X - L_1 Z L_1^\top) x_1,$$

which proves the necessity of $X - L_1 Z L_1^\top \geq 0$.

Sufficiency. The conditions (A.14) are therefore necessary for $M \geq 0$, and since together they imply that the minimum of $x^\top Q x$ over x_2 for any x_1 is nonnegative, they are also sufficient.

By the same argument, conditions (A.15) can be derived as those of (A.14) are, by starting with X.

(ii) is a direct corollary of (i). This completes the proof of Lemma A.2. $\square$

Let $U \in \mathbb{R}^{n \times k}$. $U_\perp$ is said to be the complement of U if $U^\top U_\perp = 0$ and $[UU_\perp]$ is of maximum rank (which means that $[UU_\perp]$ is nonsingular).

Lemma A.3 *Let G, U, V be given matrices with G being symmetric.*

(i) Then there exists matrix X such that

$$G + U X V^\top + V X^\top U^\top > 0 \tag{A.21}$$

if and only if

$$U_\perp^\top G U_\perp > 0, \quad V_\perp^\top G V_\perp > 0 \tag{A.22}$$

hold, where $U_\perp, V_\perp$ are orthogonal complements of U and V, respectively.

(ii) $U_\perp^\top G U_\perp > 0$ holds if and only if there exists a scalar σ such that

$$G - \sigma U U^\top > 0.$$

Proof: See [38] pp. 32-33.

Using Lemma A.3, one can eliminate some matrix variables in a matrix inequality, and therefore a nonlinear problem can be cast into the LMI framework. This reduces the computation burden significantly for the problem under consideration.

Lemma A.4 *Let $X \in \mathbb{R}^{n \times n}$ and $Y \in \mathbb{R}^{n \times n}$ be symmetric, positive-definite matrices. Then, there exists a symmetric, positive-definite matrix $P > 0$ satisfying $P = \begin{pmatrix} Y & \# \\ \# & \# \end{pmatrix}, P^{-1} = \begin{pmatrix} X & \# \\ \# & \# \end{pmatrix}$ if and only if $X - Y^{-1} \geq 0$.*

Lemma A.5 *(See [214]) Let Y be a symmetric matrix, and H, E be given matrices of appropriate dimensions, and F satisfy $F^{\top}F \leq I$. Then, we have*

(i) *For any $\varepsilon > 0$, $HFE + E^{\top}F^{\top}H^{\top} \leq \varepsilon HH^{\top} + \frac{1}{\varepsilon}E^{\top}E$.*

(ii) *$Y + HFE + E^{\top}F^{\top}H^{\top} < 0$ holds if and only if there exists a scalar $\varepsilon > 0$ such that $Y + \varepsilon HH^{\top} + \varepsilon^{-1}E^{\top}E < 0$.*

Lemma A.6 *(See [209]) Let A, D, Δ, E be real matrices of appropriate dimensions with $\|\Delta\| \leq 1$. Then, we have*

(i) *for any matrix $P > 0$ and scalar $\varepsilon > 0$ satisfying $\varepsilon I - EPE^{\top} > 0$,*

$$(A + D\Delta E)P(A + D\Delta E)^{\top}$$
$$\leq APA^{\top} + APE^{\top}(\varepsilon I - EPE^{\top})^{-1}EPA^{\top} + \varepsilon DD^{\top} \quad \text{(A.23)}$$

(ii) *for any matrix $P > 0$ and scalar $\varepsilon > 0$ satisfying $P - \varepsilon DD^{\top} > 0$,*

$$(A + D\Delta E)^{\top}P^{-1}(A + D\Delta E) \leq A^{\top}(P - \varepsilon DD^{\top})^{-1}A + \frac{1}{\varepsilon}E^{\top}E.$$

Notes

The proof of the Schur complement lemma is borrowed from [122], of which another derivation using pseudo inverses can be found in [3]. Elimination of matrix variables is related to a matrix dilation problem considered in [61], which was used in control problems in [85, 107, 161, 162, 163]. The proof of Lemma A.3 is from [38].

Appendix B
Matrix Inversion Formulas

Let A be a square matrix partitioned as follows:

$$A = \left(\begin{array}{cc} A_{11} & A_{12} \\ A_{21} & A_{22} \end{array} \right)$$

with A_{11}, A_{22} being square matrices. If $A_{11}(A_{22})$ is invertible, then matrix $\Delta = A_{22} - A_{21}A_{11}^{-1}A_{12}$ ($\tilde{\Delta} = A_{11} - A_{12}A_{22}^{-1}A_{21}$) is called the Schur complement of A_{11} (A_{22}). Matrix A is invertible if and only if A_{11} and Δ are nonsingular, or A_{22} and $\tilde{\Delta}$ are nonsingular.

Furthermore, if A is invertible, then

$$\left(\begin{array}{cc} A_{11} & A_{12} \\ A_{21} & A_{22} \end{array} \right)^{-1} = \left(\begin{array}{cc} \Gamma_1 & -A_{11}^{-1}A_{12}\Delta^{-1} \\ -\Delta^{-1}A_{21}A_{11}^{-1} & \Delta^{-1} \end{array} \right) \qquad (B.1)$$

and

$$\left(\begin{array}{cc} A_{11} & A_{12} \\ A_{21} & A_{22} \end{array} \right)^{-1} = \left(\begin{array}{cc} \tilde{\Delta}^{-1} & -\tilde{\Delta}^{-1}A_{12}A_{22}^{-1} \\ -A_{22}^{-1}A_{21}\tilde{\Delta}^{-1} & \Gamma_2 \end{array} \right) \qquad (B.2)$$

where

$$\Gamma_1 = A_{11}^{-1} + A_{11}^{-1}A_{12}\Delta^{-1}A_{21}A_{11}^{-1},$$
$$\Gamma_2 = A_{22}^{-1} + A_{22}^{-1}A_{21}\tilde{\Delta}^{-1}A_{12}A_{22}^{-1}.$$

Suppose that the matrices A_{11} and A_{22} are nonsingular, then we have

$$[A_{11} - A_{12}A_{22}^{-1}A_{21}]^{-1} = A_{11}^{-1} + A_{11}^{-1}A_{12}[A_{22} - A_{21}A_{11}^{-1}A_{12}]^{-1}A_{21}A_{11}^{-1} \quad (B.3)$$

Appendix C
Kronecker Product

Let $A = (a_{ij}) \in \mathbb{R}^{p \times q}$ and $B \in \mathbb{R}^{m \times n}$ be two matrices. The Kronecker product of A and B denoted by $A \otimes B$ is defined by

$$A \otimes B = \begin{pmatrix} a_{11}B & a_{12}B & \cdots & a_{1q}B \\ a_{21}B & a_{22}B & \cdots & a_{2q}B \\ \vdots & \vdots & \ddots & \vdots \\ a_{p1}B & a_{p2}B & \cdots & a_{pq}B \end{pmatrix}.$$

The following properties of the Kronecker product are easily established:

$$1 \otimes A = A$$
$$(A + B) \otimes C = A \otimes C + B \otimes C$$
$$(A \otimes B)(C \otimes D) = AC \otimes BD$$
$$(A \otimes B)^\top = A^\top \otimes B^\top$$
$$(A \otimes B)^{-1} = A^{-1} \otimes B^{-1}.$$

Let I_n denote the $n \times n$ unit matrix. The q-dimensional vector which has value one in the kth element and zero elsewhere is called the unit vector and is denoted by

$$\underset{(q)}{e_k} \triangleq \delta_{ik}.$$

The parenthetical underscore is omitted if the dimension can be inferred from the context. The *elementary matrix*

$$E_{ik}^{(p \times q)} \triangleq \underset{(p)(q)}{e_i e_k^\top}$$

has dimensions $p \times q$, and has value one in the $i - k$th element and zero elsewhere. Define the *permutation matrix* by

$$\mathbf{U}_{p \times q} \triangleq \sum_{i=1}^{p} \sum_{j=1}^{q} E_{ik}^{(p \times q)} \otimes E_{ki}^{(q \times p)}.$$

This matrix is square $(pq \times pq)$ and has precisely a single one value in each row and in each column. Let $X \in \mathbb{R}^{p \times q}$ and let $\text{vec}(X)$ denote the vector formed by stacking the columns of X into one long vector, that is,

$$\text{vec}(X) = \begin{pmatrix} x_{11} \\ x_{21} \\ \vdots \\ x_{p1} \\ x_{12} \\ x_{22} \\ \vdots \\ x_{1q} \\ x_{2q} \\ \vdots \\ x_{pq} \end{pmatrix}.$$

Then we have the following properties (see [39] for the proof).

Property 1: $\text{vec}(X^\top) = \mathbf{U}_{p \times q} \text{vec}(X)$.

Property 2: Let A, B, X be matrices of appropriate dimensions. Then
$\text{vec}(AXB) = (B^\top \otimes A)\text{vec}(X)$.

Appendix D
Markov Process

In this appendix we will give some results on stochastic processes to help the reader understand the material covered in this book. The Markovian property is recalled and some definitions and equalities are given.

D.1 Continuous-Time Finite State Markov Process

Definition D.1 *Let $(\Omega, \mathcal{F}, P)$ be a probability space and $\{r_t, t \geq 0\}$ be a stochastic process taking values in $\mathcal{S} = \{1, 2, \cdots, N\}$. Then, $\{r_t, t \geq 0\}$ is said to be a Markov process with state space $\mathcal{S}$ if*

$$P(r(t) = i | r(w) : w \leq s) = P(r(t) = i | r(s))$$

holds for all $0 \leq s \leq t$ and $i \in \mathcal{S}$.

For any $i, j \in \mathcal{S}$, let $P_{ij}(s, t)$ denote the transition probability, i.e., $P_{ij}(s, t) = P(r(t) = j | r(s) = i)$. Matrix $P(s, t) = (P_{ij}(s, t)), i, j \in \mathcal{S}$ is said to be the transition matrix. Under the continuity condition $\lim_{t \to 0+} P(t) = I$, it follows that for $0 \leq s \leq u \leq t$,

$$P_{ij}(s, t) \geq 0, \ i, j \in \mathcal{S}$$

$$\sum_{j \in \mathcal{S}} P_{ij}(s, t) = 1, \ i \in \mathcal{S}$$

$$P_{ij}(s, t) = \sum_{k \in \mathcal{S}} P_{ik}(s, u) P_{kj}(u, t), \ i, j \in \mathcal{S}.$$

The last identity is usually referred to as the Chapman–Kolmogorov equation.

If the transition probability $P_{ij}(s,t)$ depends only on $t - s$, $\{r(t), t \geq 0\}$ is said to be stationary. Otherwise, the process is nonstationary. In this book, the Markov processes are assumed to be stationary.

For a stationary Markov process, it can be shown [84] that the following hold

$$\lim_{t \to 0} \frac{1 - P_{ii}(t)}{t} = \lambda_i < \infty$$

$$\lim_{t \to s} \frac{P_{ij}(t)}{t} = \lambda_{ij} < \infty.$$

Obviously, $\lambda_i = \sum_{j \neq i} \lambda_{ij}$. The parameter matrix $\Lambda = (\lambda_{ij}), i, j \in \mathcal{S}$ with, by convention, $\lambda_{ii} = -\lambda_i$ gives the infinitesimal generator of the Markov process $\{r(t), t \geq 0\}$ and $\lambda_{ij}s$ are called the infinitesimal characteristics. The transition probability and the generator matrices of the process satisfy the following forward Chapman–Kolmogorov equation:

$$\frac{dP_{ij}(t)}{dt} = -P_{ij}(t)\lambda_j + \sum_{k \neq j} P_{ik}(t)\lambda_{kj}, \ i, j \in \mathcal{S}.$$

D.2 Strong Markov Property and Infinitesimal Generator of Stochastic Hybrid Systems

This section addresses the Markov property of a hybrid stochastic system with time delay. Let $\mathbb{C}[-\tau, 0]$ be the space of continuous functions on the real interval $[-\tau, 0], \tau > 0$ (the time delay in the system), and let $x(t)$ be a vector-valued stochastic process. Define the process $\mathbf{x}_t$ with values in $\mathbb{C}[-\tau, 0]$ by

$$\mathbf{x}_t(s) = x(t + s), s \in [-\tau, 0].$$

Let us define the norm of $\mathbf{x}_t$ by $\|\mathbf{x}_t\| = \sup_{s \in [-\tau, 0]} \|x(t + s)\|$. Let us consider a hybrid system with N modes and mode-switching governed by a Markov process $\{r_t, t \geq 0\}$ taking values in $\mathcal{S} = \{1, 2, \cdots, N\}$ and having infinitesimal generator $\Lambda = (\lambda_{ij}), i, j \in \mathcal{S}$, where $\lambda_{ij} \geq 0, j \neq i$, and $\lambda_{ii} = -\sum_{j \neq i} \lambda_{ij}$.

Suppose that $x(t)$ is described by the following dynamic:

$$dx(t) = f(r_t, \mathbf{x}_t)dt + g(r_t, \mathbf{x}_t)dB(t), \ \mathbf{x}_0 = \phi(\cdot), \tag{D.1}$$

where $f(\cdot, \cdot) : \mathcal{S} \times \mathbb{C}[-\tau, 0] \to R^n$, $g(\cdot, \cdot) : \mathcal{S} \times \mathbb{C}[-\tau, 0] \to \mathbb{R}^{n \times m}$; $B(t) \in \mathbb{R}^m$ is a standard Brownian process and is assumed to be independent of $\{r_t, t \geq 0\}$; and $\phi(\cdot)$ is the initial condition.

This model contains all the models used in the book as special cases. This section addresses the Markov property of this process. For each given mode $i \in \mathcal{S}$, let $f_j(i, \cdot), 1 \leq j \leq n$ denote the jth component of $f(i, \cdot)$ and let $g_{jk}(i, \cdot)$ denote the (i, j)th component of $g(i, \cdot)$. Define the vector and matrix norms as $\|f(i, \cdot)\|^2 = \sum_{j=1}^{n} f_j^2(i, \cdot), \|g(i, \cdot)\|^2 = \sum_{j=1}^{n} \sum_{k=1}^{m} g_{jk}^2(i, \cdot)$, respectively. Assume that the following assumptions hold for every given $i \in \mathcal{S}$.

Assumption D.1 $f_j(i, \cdot), g_{jk}(i, \cdot)$ *are continuous real-valued functions on* $\mathbb{C}[-\tau, 0]$.

Assumption D.2 *In the interval* $[-\tau, 0]$, $x(t)$ *is continuous with probability one, is independent of* $B(s) - B(0), s \geq 0$, *and satisfies* $\mathbb{E}\|\phi(s)\|^4 < \infty$.

Assumption D.3 *There is a constant* $M < \infty$ *and a bounded measure* μ *on* $[-\tau, 0]$ *so that for any* $\psi_1, \psi_2 \in \mathbb{C}[-\tau, 0]$,

$$\begin{cases} \|f(i, \psi_1) - f(i, \psi_2)\| + \|g(i, \psi_1) - g(i, \psi_2)\| \\ \quad \leq \int_{-\tau}^{0} |\psi_1(s) - \psi_2(s)| d\mu(s) \\ \|f(i, 0)\| + |g(i, 0)| \leq M. \end{cases} \qquad (D.2)$$

Assumption D.4 *For each positive real number* ρ *there is a bounded measure* μ_ρ *on* $[-\tau, 0]$ *so that for* $\|\psi\| \leq \rho$ *and* $\|\phi\| \leq \rho$, *(D.2) is valid with* μ_ρ *replacing* μ.

Let $\mathcal{C}$ be the collection of open sets in $\mathcal{S} \times \mathbb{C}[-\tau, 0]$ and let $\mathcal{B}$ be the Borel field over $\mathcal{C}$. We suppose that the probability measure space introduced in the sequel is complete with respect to whatever measures are imposed on them. Let Ω denote the probability sample space, and ω denote the generic element of Ω. Define $\tilde{M}_t^{(i,\mathbf{x})}$ and $\tilde{N}_t^{(i,\mathbf{x})}$ as the least σ-fields on Ω over which $x(s), -\tau \leq s \leq t$, and $x(s), t - \tau \leq s \leq t$, are measurable, respectively, for fixed initial condition $r(0) = i, \mathbf{x}_0 = \mathbf{x}$. Let $P_{(i,\mathbf{x})}$ be the probability measure on

$$\tilde{M}^{(i,\mathbf{x})} = \cup_{t \geq 0} \tilde{M}_t^{(i,\mathbf{x})}.$$

With this notation defined, we have the following theorem.

Theorem D.1 *Assume that Assumptions D.1-D.3 hold and let* $\mathbf{x}_0 = \phi(\cdot) \in \mathbb{C}[-\tau, 0]$. *Then* $\{(r(t), \mathbf{x}_t), t \geq 0\}$ *is a strong Markov process on the topological state space* $\{(\mathcal{S} \times \mathbb{C}[-\tau, 0], \mathcal{C}, \mathcal{B})\}$.

Proof: To prove the Markov property, we need to check the conditions of Dynkin [71], pp. 77-80. That is, we need to show that

1) the function p defined by $p(t, (i, \mathbf{x}), \Gamma) = P_{(i,\mathbf{x})}((r(t), \mathbf{x}_t) \in \Gamma)$ for arbitrary $\Gamma \in \in \mathcal{S} \times \mathbb{C}[-\tau, 0]$ is measurable, and

2) $P_{(i,\mathbf{x})}((r(t+\Delta), \mathbf{x}_{t+\Delta}) \in \Gamma | \tilde{M}_t^{(i,\mathbf{x})}) = p(\Delta, (r(t), \mathbf{x}_t), \Gamma)$ with probability one.

To prove the strong Markov property, it suffices to prove that if $w(\cdot, \cdot)$ is bounded and continuous on $(\mathcal{S} \times \mathbb{C}[-\tau, 0]$, then $\mathbb{E}_{(i,\mathbf{x})} w(r(t), \mathbf{x}_t) \overset{\Delta}{=} \gamma(i, \mathbf{x})$ is continuous in $(i, \mathbf{x})$, where $\mathbb{E}_{(i,x)}$ is the expectation operator corresponding to $P_{(i,\mathbf{x})}$. The details of the proof are omitted here. $\qquad\qquad\square$

A real-valued function $V(\cdot, \cdot)$ on $\mathbb{C}[-\tau, 0]$ is said to be in the domain of $\mathcal{A}$, the infinitesimal operator, if the limits

$$\lim_{t \to 0} \frac{\mathbb{E}[V(r(t), \mathbf{x}_t)|r(0) = i, \mathbf{x}_0 = \mathbf{x}] - V(i, \mathbf{x})}{t} = q(i, \mathbf{x})$$

$$\lim_{t \to 0} \mathbb{E}[q(r(t), \mathbf{x}_t)] = q(i, \mathbf{x})$$

exist pointwise in $\mathbb{C}[-\tau, 0]$ and $i \in \mathcal{S}$ and the limit is uniformly bounded in $\mathbf{x}$. Then, we write $q(i, \mathbf{x}) = \mathcal{A}V(i, \mathbf{x})$.

For example, let us consider the following Lyapunov function for system (D.1):

$$V(r_t, x(t)) = x^\top(t) P(r_t) x(t).$$

Then,

$$\begin{aligned}
\mathcal{A}V(r_t, x(t)) &= \mathbb{E}_{(r_t, \mathbf{x}_t)} \left[2x^\top(t) P(r_t)(f(r_t, \mathbf{x}_t) + g(r_t, \mathbf{x}_t) dB(t)) \right. \\
&\qquad \left. + \frac{1}{2}\mathrm{tr}\left(g(r_t, \mathbf{x}_t) g^\top(r_t, \mathbf{x}_t) \frac{\partial V(r_t, x(t))}{\partial x^2} \right) \right] \\
&= 2x^\top(t) P(r_t)(f(r_t, \mathbf{x}_t) dt + \sum_{j=1}^N \lambda_{r_t j} V(r_t, \mathbf{x}_t) \\
&\qquad + \mathrm{tr}\left(g(r_t, \mathbf{x}_t) g^\top(r_t, \mathbf{x}_t) P(r_t) \right).
\end{aligned}$$

In particular, if $m = 1, g(r_t, \mathbf{x}_t) = g(r_t) x(t)$ with $g(r_t) \in \mathbb{R}^{n \times n}$, then

$$\begin{aligned}
\mathcal{A}V(r_t, x(t)) &= 2x^\top(t) P(r_t)(f(r_t, \mathbf{x}_t) dt + \sum_{j=1}^N \lambda_{r_t j} V(r_t, \mathbf{x}_t) \\
&\qquad + x^\top(t) g^\top(r_t) P(r_t) g(r_t) x(t).
\end{aligned}$$

Let β be a random time with $\mathbb{E}[\beta|(r(0), \mathbf{x}_0)] < \infty$ and let $V(r(t), \mathbf{x}_t)$ be in the domain of $\mathcal{A}$. Then

$$\mathbb{E}[V(r(\beta), \mathbf{x}_\beta)|(r(0), \mathbf{x}_0)] - V(r(0), \mathbf{x}_0)$$

$$= \mathbb{E}\left[\int_0^\beta \mathcal{A}V(r(s), \mathbf{x}_s) ds | (r(0), \mathbf{x}_0) \right]. \qquad (\text{D.3})$$

The above equation is called *Dynkin's formula*. The deterministic counterpart is the basic formula of the calculus: $f(x(t)) - f(x(0)) = \int_0^t \dot{f}(x(s)) ds$.

References

[1] Abou-Kandil, H., Freiling, G., and Jank, G. (1994). Solution and asymptotic behavior of coupled Riccati equations in jump linear systems. *IEEE Transactions on Automatic Control*, **39**(8), 1631–1636.

[2] Abou-Kandil, H., Freiling, G., and Jank, G. (1995). On the solution of discrete-time Markovian jump linear quadratic control problems. *Automatica*, **31**(5), 765–768.

[3] Albert, A. (1969). Conditions for positive and nonnegative definiteness in terms of pseudo inverses. *SIAM Journal on Applied Mathematics*, **17**, 434–440.

[4] Aliyu, M.D.S. and Boukas, E.K. (2000). Finite and infinite horizon H_∞ control for stochastic nonlinear systems. *IMA, Journal of Mathematical Control and Information*, **17**(3), 265–279.

[5] Aliyu, M.D.S. and Boukas, E.K. (2000). Robust H_∞ control for Markovian jump nonlinear systems. *IMA, Journal of Mathematical Control and Information*, **17**(3), 295–308.

[6] Anderson, B.D.O. and Moore, J.B. (1971). *Linear Optimal Control.* Prentice–Hall, Englewood Cliffs, New Jersey.

[7] Basar, T. and Bernhard, P. (1990). H_∞-*Optimal Control and Related Minimax Problems.* Birkhäuser, Berlin.

[8] Benjelloun, K. and Boukas, E.K. (1998). Mean square stochastic stability of linear time-delay system with Markov jumping parameters. *IEEE Transactions on Automatic Control*, **43**(10), 1456–1459.

[9] Benjelloun, K. Boukas, E.K., and Costa, O.L.V. (2000). H_∞-control for linear time-delay systems with Markovian jumping parameters. *Journal of Optimization Theory and Applications*, **105**(1), 73–95.

[10] Benjelloun, K., Boukas, E.K., and Yang, H. (1996). Robust stabilizability of uncertain linear time-delay systems with Markovian jumping parameters. *Journal of Dynamic Systems, Measurement, and Control*, **118**, 776–783.

[11] Blair, W.P. and Sworder, D.D. (1975). Feedback control of a class of linear discrete systems with jump parameters and quadratic cost criteria. *International Journal of Control*, **21**(5), 833–841.

[12] Bolzern, P., Colaneri, P., and de Nicolao, G. (1994). Optimal robust filtering with time-varying parameter uncertainty. *IEEE Transactions on Automatic Control*, **39**(3), 623–626.

[13] Bolzern, P., Colaneri, P., and de Nicolao, G. (1996). Optimal robust filtering with time-varying parameter uncertainty. *International Journal of Control*, **63**(3), 557–576.

[14] Boukas, E.K. (1987). *Commande Optimale Stochastique Appliquée aux Systèmes de Production*. École Polytechnique de Montréal.

[15] Boukas, E.K. (1993). Control of systems with controlled jump Markov disturbances. *Control Theory and Advanced Technology*, **9**(2), 577–595.

[16] Boukas, E.K. (1995). *Systèmes Asservis*. Éditions de l'École Polytechnique de Montréal.

[17] Boukas, E.K. and Haurie, A. (1990). Manufacturing flow control and preventive maintenance, a stochastic control approach. *IEEE Transactions on Automatic Control*, **35**(7), 1024–1031.

[18] Boukas, E.K. and Liu, Z.K. (1999). Robust stability and H_∞ control of discrete-time jump linear systems with time-delay, An LMI approach. *Submitted for publication*.

[19] Boukas, E.K. and Liu, Z.K. (2000). Delay-dependent output feedback stabilization of Markov jump systems with time-delay. *Submitted for publication*.

[20] Boukas, E.K. and Liu, Z.K. (2000). Delay-dependent robust stability and stabilization of uncertain linear systems with discrete and distributed time-delays. *Submitted for publication*.

[21] Boukas, E.K. and Liu, Z.K. (2000). Jump linear quadratic regulator with controlled jump rate. *IEEE Transactions on Automatic Control*, **45**(2).

[22] Boukas, E.K. and Liu, Z.K. (2000). Output feedback guaranteed cost control for uncertain time-delay systems with Markov jump. *Proceedings of ACC*.

[23] Boukas, E.K. and Liu, Z.K. (2000). Robust H_∞ filtering for polytopic uncertain time-delay systems with Markov jumps. *Submitted for publication*.

[24] Boukas, E.K. and Liu, Z.K. (2000). *Systems with Time-Delay, Stability, Stabilization, H_∞ and their Robustness*. A chapter.

[25] Boukas, E.K. and Liu, Z.K. (2001). *Deterministic and Stochastic Time-Delay Systems*. Birkhäuser, Boston.

[26] Boukas, E.K., Liu, Z.K., and Liu, G.X. (2001). Delay-dependent robust stability and H_∞ control of jump linear systems with time-delay. *International Journal of Control*.

[27] Boukas, E.K. and Shi, P. (1998). Stochastic stability and guaranteed cost control of discrete-time uncertain systems with Markovian jumping parameters. *International Journal of Robust and Nonlinear Control*, **8**(13), 1155–1167.

[28] Boukas, E.K. and Shi, P. (1999). H_∞ control for discrete-time linear systems with Frobenius norm-bounded uncertainties. *Automatica*, **35**(9), 1625–1631.

[29] Boukas, E.K., Shi, P., and Andijani, A. (1999). Optimal inventory-production control problem with stochastic demand. *Optimal Control Applications and Methods*, **20**(1), 1–20.

[30] Boukas, E.K., Shi, P., and Benjelloun K. On stabilization of uncertain linear systems with Markovian jumping parameters. *International Journal of Control*, **72**, 842–850, 1999.

[31] Boukas, E.K. and Yang, H. (1995). Stability of discrete-time linear systems with Markovian jumping parameters. *Mathematics Control, Signals and Systems*, **8**, 390–402.

[32] Boukas, E.K. and Yang, H. (1996). Optimal control of manufacturing flow and preventive maintenance. *IEEE Transactions on Automatic Control*, **41**(6), 881–885.

[33] Boukas, E.K. and Yang, H. (1999). Exponential stability of stochastic Markovian jumping parameters. *Automatica*, **35**(8), 1437–1441.

[34] Boukas, E.K., Yang H., and Zhang, Q. (1995). Minmax production planning in failure-prone manufacturing systems. *Journal of Optimization Theory and Applications*, **87**(2), 269–286.

[35] Boukas, E.K., Swierniak, A., and Yang, H. (1999). Suboptimality of a decentralized feedback controller. *Journal of Dynamic Systems, Measurement, and Control*, **121**(2), 305–308.

[36] Boukas, E.K., Zhang, Q., and Yin, G. (1995). Robust production and maintenance planning in stochastic manufacturing systems. *IEEE Transactions on Automatic Control*, **40**(6), 1098–1102.

[37] Boukas, E.K., Zhu, Q., and Zhang, Q. (1994). Piecewise deterministic Markov process model for flexible manufacturing systems with preventive maintenance. *Journal of Optimization Theory and Applications*, **81**(2), 259–275.

[38] Boyd, S. Ghaoui, L.E., Feron, E., and Balakrishnan, V. (1994). *Linear Matrix Inequalities in System and Control Theory*. SIAM, Philadelphia.

[39] Brewef, J.W. (1978). Kronecker products and matrix calculus in system theory. *IEEE Transactions on Circuits and Systems*, **CAS-25**(9).

[40] Cao, Y.Y. and Lam, J. (1999). Stochastic stabilizability and H_∞ control for discrete-time jump linear systems with time-delay. In *Proceedings of the 14th IFAC World Congress*, Beijing, China.

[41] Cao, Y.Y. and Lam, J. (2000). Robust H_∞ control of uncertain Markovian jump systems with time-delay. *IEEE Transactions on Automatic Control*, **45**(1).

[42] Chao, G.L., Pering, J.W., and Han, K.W. (1998). Robust stability analysis of time-delay systems using parameter-plane and parameter-space methods. *Journal of Franklin Institute*, **335B**(7), 1249–1262.

[43] Cao, Y.Y., Sun Y.X., and Lam, J. (1998). Delay-dependent robust H_∞ control for uncertain systems with time-varying delays. *IEE Proceedings D, Control Theory and Applications*, **145**(3).

[44] Chang, Y.H. (1999). Robust regional stability analysis of continuous time-delay systems. *IEE Proceedings D, Control Theory and Applications*, **146**(4), 311–317.

[45] Chen, J.D., Lien, C.H., Fan, K.K., and Cheng, J.S. (2000). Delay-dependent stability criterion for neutral time-delay systems. *Electronics Letters*, **36**(22), 1897–1898.

[46] Chen, Y.H., Wang, W.J., and Mau, L.G. (1996). Robust stabilization of large-scale time-delay systems with estimated state feedback. *Journal of Optimization Theory and Applications*, **89**(3), 543–559.

[47] Cheres, E., Palmor, Z.J., and Gutman, S. (1989). Quantitative measure of robustness for systems including delayed perturbations. *IEEE Transactions on Automatic Control*, **34**, 1203–1204.

[48] Chilali, M. and Gahinet, P. (1996). H_∞ design with pole placement constraints, An LMI approach. *IEEE Transactions on Automatic Control*, **41**(3), 358–367.

[49] Choi, H.H. and Chung, M.J. (1995). Memoryless H_∞ controller design for linear systems with delayed state and control. *Automatica*, **31**(6), 917–919.

[50] Costa, O.L.V. (1994). Linear minimum mean square error estimation for discrete-time Markovian jump linear systems. *IEEE Transactions on Automatic Control*, **39**(8), 1685–1689.

[51] Costa, O.L.V. (1996). Mean-square stabilizing solutions for discrete-time coupled algebraic Riccati equations. *IEEE Transactions on Automatic Control*, **41**(4), 593–598.

[52] Costa, O.L.V. and Boukas, E.K. (1998). Necessary and sufficient condition for robust stability and stabilizability of continuous-time linear systems with Markovian jumps. *Journal of Optimization Theory and Applications*.

[53] Costa, O.L.V., do Val, J.B.R., and Geromel, J.C. (1997). A convex programming approach to H_2 control of discrete-time Markovian jump linear systems. *International Journal of Control*, **66**(4), 557–579.

[54] Costa, O.L.V., do Val, J.B.R., and Geromel, J.C. (1999). Continuous-time state-feedback H_2 control of Markovian jump linear systems via convex analysis. *Automatica*, **35**(2), 259–268.

[55] Costa, O.L.V., Filho, E.O.A., Boukas, E.K., and Marquez, R.P. (1999). Constrained quadratic control of discrete-time Markovian jump linear systems. *Automatica*, **35**(4), 617–627.

[56] Costa, O.L.V. and Fragoso, M.D. (1991). Necessary and sufficient conditions for mean square stability of discrete-time linear systems subject to Markovian jumps. In *Proceedings of the 9th International Symposium on Mathematics Theory of Networks and Systems*, pages 85–86, Kobe, Japan.

[57] Costa, O.L.V. and Fragoso, M.D. (1993). Stability results for discrete-time linear systems with Markovian jumping parameters. *Journal of Mathematical Analysis and Applications*, **179**(2), 154–178.

[58] Costa, O.L.V. and Fragoso, M.D. (1995). Discrete-time LQ-optimal control problems for infinite Markov jump parameter systems. *IEEE Transactions on Automatic Control*, **40**(12), 2076–2088.

[59] Costa, O.L.V. and Marques, R.P. (1998). Mixed H_2/H_∞ control of discrete-time Markovian jump linear systems. *IEEE Transactions on Automatic Control*, **43**, 95–100.

[60] Datko, R. (1974). Neutral autonomous functional equations with quadratic cost. *SIAM Journal Control*, **12**, 70–82.

[61] Davis, C., Kahan, W.M., and Weingerger, H.F. (1982). Norm-preserving dilations and their applications to optimal error bounds. *SIAM Journal of Numerical Analysis*, **19**(3), 445–479.

[62] de Farias, D.P., Geromel, J.C., do Val, J.B.R., and Costa, O.L.V. (2000). Output feedback control of Markov jump linear systems in continuous-time. *IEEE Transactions on Automatic Control*, **45**(5).

[63] de Souza, C.E. and Fragoso, M.D. (1993). H_∞ control for linear systems with Markovian jumping parameters. *Control-Theory and Advanced Technology*, **9**(2), 457–466.

[64] Doyle, J.C., Glover, K., Khargonekar, P.P., and Francis, B.A. (1989). State space solutions to the standard H^2 and H^∞ control problems. *IEEE Transactions on Automatic Control*, **34**(8), 831–847.

[65] Dragan, V. and Morozan, T. (1997). Global solutions to a game-theoretic Riccati equations of stochastic control. *Journal of Differential Equations*, **138**(2), 328–350.

[66] Dragan, V. and Morozan, T. (1999). Stability and robust stabilization of linear stochastic systems described by differential equations with Markovian jumping and multiplicative white noise. Technical report, 17/1999, Institute of Mathematics of Romanian Academy Science.

[67] Dragan, V., Shi, P., and Boukas, E.K. (1999). Control of singularly perturbed systems with Markovian jump parameters, an H_∞ approach. *Automatica*, **35**(8), 1369–1378.

[68] Dufour, F. and Bertrand, P. (1994). The filtering problem for continuous-time linear systems with Markovian switching coefficients. *Systems and Control Letters*, **23**(5), 453–461.

[69] Dufour, F. and Bertrand P. (1996). An image-based filter for discrete-time Markovian jump linear systems. *Automatica*, **32**(2), 241–247.

[70] Dugard, L. and Verriest, E.I. (Eds). (1997). *Stability and control of time-delay systems*. Number 228. Lecture Notes on Control and Information Science, New York, Springer.

[71] Dynkin, E.B. (1965). *Markov Process*. Springer, Berlin.

[72] Esfahani, S.H., Moheimani, S.O.R., and Petersen, I.R. (1998). LMI approach to suboptimal guaranteed cost control for uncertain time-delay systems. *IEE Proceedings D, Control Theory and Applications*, **145**(6), 491–498.

[73] Esfahani, S.H. and Petersen, I.R. (1999). An LMI approach to the output feedback guaranteed cost control for uncertain time-delay systems. *Preprint*.

[74] Ezzine, J. and Haddad, A.H. (1989). On the controllability and observability of hybrid systems. *International Journal of Control*, **49**(6), 2045–2055.

[75] Ezzine, J. and Karvanoglu, D. (1997). On almost-sure stabilization of discrete-time parameter systems, An LMI approach. *International Journal of Control*, **68**(5), 1129–1146.

[76] Fattouh, A., Sename, O., and Dion, J.M. (1998). H_∞ observer design for time-delay systems. In *Proceedings of 37th IEEE Conference on Decision and Control*, pages 4545–4546, Tampa, Florida.

[77] Feliachi, A. and Thowsen, A. (1981). Memoryless stabilization of linear delay-differential systems. *IEEE Transactions on Automatic Control*, **26**, 586–587.

[78] Feng, X. and Loparo, K.A. (1990). Almost sure instability of the random harmonic oscillator. *SIAM Journal on Applied Mathematics*, **50**(3), 744–759.

[79] Feng, X., Loparo, K.A., Ji, Y., and Chizeck, H.J. (1992). Stochastic stability properties of jump linear systems. *IEEE Transactions on Automatic Control*, **37**(1), 38–53.

[80] Fragoso, M.D. (1988). On a partially observable LQG problem for systems with Markovian jumping parameters and quadratic cost. *Systems and Control Letters*, **10**(5), 349–356.

[81] Fragoso, M.D. (1989). Discrete-time jump LQG problem. *International Journal of Systems Science*, **20**(12), 2539–2545.

[82] Fragoso, M.D., do Val, J.B.R., and Pinto Jr., D.L. (1995). Jump linear H_∞ control, The discrete-time case. *Control Theory and Advanced Technology*.

[83] Francis, B.A. (1987). *A Course in H_∞ Control Theory*. Springer-Verlag, New York.

[84] Freedman, D. (1983). *Markov Chain*. Springer-Verlag, New York.

[85] Gahinet, P. (1992). A convex parametrization of H_∞ suboptimal controllers. *Proceedings of IEEE Conference on Decision and Control*, pages 1742–1729.

[86] Gajic, Z. and Borno, I. (1995). Lyapunov iterations for optimal control of jump linear systems at steady state. *IEEE Transactions on Automatic Control*, **40**, 1971–1975.

[87] Gallegos, J.A. (1994). Disturbance rejection of time-delay systems. *International Journal of Systems Science*, **25**(6), 1081–1091.

[88] Ge, J.H., Frank, P.M., and Lin, C.F. (1997). H_∞ control via output feedback for state delayed systems. *International Journal of Control*.

[89] Geromel, J.C., Oliveira, M.C., and Bernussou, J. (1999). Robust filtering of discrete-time linear systems with parameters dependent Lyapunov functions. In *Proceedings of the 38th IEEE Conference on Decision and Control*, Phoenix.

[90] Geromel, J.C., Peres, P.L.D., and Bernussou, J. (1991). On a convex parameter space method for linear control design of uncertain systems. *SIAM Journal on Control and Optimization*, **29**, 381–402.

[91] Gorecki, H., Fuksa, S., Grabowski, P., and Korytowski, A. (1989). *Analysis and Synthesis of Time Delay Systems*. John Wiley and Sons, Warszawa.

[92] Griffiths, B.E. and Loparo, K.A. (1985). Optimal control of jump-linear gaussian systems. *International Journal of Control*, **42**(4), 791–819.

[93] Gu, K. (1997). Discretized LMI set in the stability problem of linear uncertain time-delay systems. *International Journal of Control*, **68**, 923–934.

[94] Gu, K. (1999). A generalized discretization scheme of Lyapunov functional in the stability problem of linear uncertain time-delay systems. *International Journal of Robust and Nonlinear Control*, **9**, 1–14.

[95] Gu, K. and Niculescu, S.L. (2000). Additional dynamics in transformed time-delay systems. *IEEE Transactions on Automatic Control*, **45**(3), 572–575.

[96] Guojun, J. and Wenzhong, S. (2001). Stability of bilinear time-delay systems. *IMA Journal of Mathematical Control and Information*, **18**, 55–60.

[97] Hale, J.K. (1977). *Theory of Functional Differential Equations*. New York, Springer-Verlag.

[98] Hale, J.K. and Lunel, S.M.V. (1993). *Introduction to Functional Differential Equations*. Springer-Verlag, New York.

[99] Hmamed, A. (1991). Further results on the stability of uncertain time-delay systems. *International Journal of System Science*, **22**(3), 605–614.

[100] Hmamed, A., Benzaouia, A., and Bensalah, H. (1995). Regulator problem for linear continuous-time delay systems with nonsymmetrical constrained control. *IEEE Transactions on Automatic Control*, **40**(9), 1615–1619.

[101] Hsiao, F.H. and Hwang, H.D. (1996). Criteria for asymptotic stability of uncertain multiple time-delay systems. *Electronics Letters*, **32**(4), 410–412.

[102] Hsiao, F.H., Pan, S.T., and Teng, C.C. (1997). Robust controller design for uncertain multiple time-delay systems. *Transactions of ASME Journal of Dynamic Systems, Measurement, and Control*, **119**, 122–127.

[103] Huang, Y.P. and Zhou, K. (2000). On the robustness of uncertain time-delay systems with structured uncertainties. *Systems and Control Letters*, **41**, 367–376.

[104] Huang, Y.P. and Zhou, K. (2000). Robust stability of uncertain time-delay systems. *IEEE Transactions on Automatic Control*, **45**(11), 2169–2173.

[105] Ikeda, M. and Ashida, T. (1979). Stabilization of linear systems with time-varying delay. *IEEE Transactions on Automatic Control*, **24**, 369–370.

[106] Ivanescu, D., Dion, J.M., Dugard, L., and Niculescu, S.I. (2000). Dynamical compensation for time-delay systems, an LMI approach. *International Journal of Robust and Nonlinear Control*, **10**, 611–628.

[107] Iwasaki, T. and Skelton, R E. (1993). All controllers for the general H_∞ control problem, LMI existence conditions and state space formulas. *Automatica*, **30**(8), 1307–1317.

[108] Jeung, E.T., Kim, J.H., and Park, H.B. (1996). Robust control for parameter uncertain delay systems in state and control input. *Automatica*, **32**(9), 1337–1339.

[109] Jeung, E.T., Kim, J.H., and Park, H.B. (1998). H_∞-output feedback controller design for linear systems with time-varying delayed state. *IEEE Transactions on Automatic Control*, **43**(7).

[110] Jeung, E.T., Oh, D.C., Kim, J.H., and Park, H.B. (1996). Robust controller design for uncertain systems with time delays, LMI approach. *Automatica*, **32**(8), 1229–1231.

[111] Ji, Y. and Chizeck, H.J. (1990). Controllability, stabilizability, and continuous-time Markovian jump linear quadratic control. *IEEE Transactions on Automatic Control*, **35**(7), 777–788.

[112] Ji, Y. and Chizeck, H.J. (1992). Jump linear quadratic Gaussian control in continuous time. *IEEE Transactions on Automatic Control*, **37**(12), 1884–1892.

[113] Jiang, X., Fei, S., and Feng, C.B. (2000). Comments on "bounded real criteria for linear time-delay systems". *IEEE Transactions on Automatic Control*, **45**(7), 1409–1410.

[114] Kharitonov, V.L. (1999). Robust stability analysis of time-delay systems, A survey. *Annual Reviews in Control*, **23**, 185–196.

[115] Kharitonov, V.L. and Zhabko, A.P. (1994). Robust stability of time-delay systems. *IEEE Transactions on Automatic Control*, **39**(12), 2388–2397.

[116] Kojima, A. and Ishijima, S. (1995). Explicit formulas for operator Riccati equations arising in H_∞ control with delays. *Proceedings of the IEEE 34th Conference on Decision and Control*, New Orleans, LA.

[117] Kolmanovskii, V.B. and Myshkis, A. (1999). *Introduction to the Theory and Applications of Functional Differential Equations*. Dordrecht, Kluwer Academic Publishers.

[118] Kolmanovskii, V.B. and Richard, J.P. (1999). Stability of some linear systems with delay. *IEEE Transactions on Automatic Control*, **44**(5), 984–989.

[119] Kolmanovsky, I. and Maizenberg, T.L. (2000). Stochastic stability of a class of nonlinear systems with randomly varying time-delay. In *Proceedings of the American Control Conference*, pages 4304–4308, Chicago, Illinois.

[120] Krasovskii, N.N. and Lidskii, E.A. (1961). Analysis design of controller in systems with random attributes–Part 1. *Automatic Remote Control*, **22**, 1021–1025.

[121] Krasovskii, N.N. and Lidskii, E.A. (1961). Analysis design of controller in systems with random attributes–Part 2. *Automatic Remote Control*, **22**, 1141–1146.

[122] Kreindler, E. and Jameson, A. (1972). Conditions for nonnegativeness of partitioned matrices. *IEEE Transactions on Automatic Control*, **17**(2), 147–148.

[123] Kung, F.C. and Lee, C.H. (1996). Decentralized robust control design for large-scale time-delay systems with time-varying uncertainties. *JSME International Journal, Series C*, **39**(3), 528–533.

[124] Lee, C.H. (1998). Simple stabilization criteria and memoryless state feedback control for time-delay systems with time-varying perturbations. *International Journal of Control*, **45**(11), 1211–1215.

[125] Lee, C.H., Li, T.H., and Kung, F.C. (1995). On the robustness of stability for uncertain time-delay systems. *International Journal of Systems Science*, **26**(2), 457–465.

[126] Li, H., Niculescu, S.I., Dugard, L., and Dion, J.M. (1998). Robust guaranteed cost control of uncertain linear time-delay systems using dynamic output feedback. *Mathematics and Computers in Simulation*, **45**, 349–358.

[127] Li, X. and de Souza, C.E. (1995). LMI approach to delay-dependent robust stability of uncertain linear systems. *Proceedings of the 34th Conference on Decision and Control*, New Orleans, LA, pages 3614–3648.

[128] Li, X. and de Souza, C.E. (1997). Delay-dependent robust stability and stabilization of uncertain linear delay systems: a linear matrix inequality approach. *IEEE Transactions on Automatic Control*, **42**(8), 1144–1148.

[129] Lien, C.H. and Hsieh, J.G. (2000). New results on global exponential stability of interval time-delay systems. *JSME International Journal Series C.*, **43**(2), 306–320.

[130] Lisong, Y. (1996). Robust analysis and synthesis of linear time-delay systems with norm-bounded time-varying uncertainty. *Systems and Control Letters*, **42**, 281–289.

[131] Liu, P.L. and Su, T.J. (1998). Robust stability of interval time-delay systems with delay-dependence. *Systems and Control Letters*, **33**, 231–239.

[132] Loiseau, J.J. (2000). Algebraic tools for the control and stabilization of time-delay systems. *Annual Reviews in Control*, **24**, 135–149.

[133] Luc, C.G. and Habets, J.M. (1996). On the genericity of stabilizability for time-delay systems. *SIAM Journal on Control and Optimization*, **34**(3), 833–854.

[134] Luo, M., de La Sen, M., and Rodellar, J. (1997). Robust stabilization of a class of uncertain time-delay systems in sliding mode. *International Journal of Robust and Nonlinear Control*, **7**, 59–74.

[135] Mahmoud, M.S. (2000). *Robust Control and Filtering for Time-Delay Systems*. Marcel Dekker, New York.

[136] Mahmoud, M.S. (2001). Control of uncertain state-delay systems: guaranteed cost approach. *IMA Journal of Mathematical Control and Information*, **18**, 109–128.

[137] Mahmoud, M.S. and Al-Muthairi, N.F. (1994). Design of robust controllers for time-delay systems. *IEEE Transactions on Automatic Control*, **39**, 995–999.

[138] Mahmoud, M.S. and Bingulac, S. (1998). Robust design of stabilizing controllers for interconnected time-delay systems. *Automatica*, **34**(6), 795–800.

[139] Mahmoud, M.S., Terro, M.J., and Abdel-Rohman, M. (1998). An LMI approach to H_∞-control of time-delay systems for the benchmark problem. *Earthquake Engineering and Structural Dynamics*, **27**, 957–976.

[140] Mahmoud, M.S. and Xia, L. (1999). Stability and positive realness of time-delay systems. *Journal of Mathematical Analysis and Applications*, **239**, 7–19.

[141] Malek-Zavarei, M. and Jamshidi, M. (1987). *Time-delay systems, analysis, optimization and applications*. North-Holland Systems and Control Series.

[142] Mariton, M. (1990). *Jump Linear Systems in Automatic Control*. Marcel Dekker, New York.

[143] Mariton, M. and Bertrand, P. (1985). Output feedback for a class of linear systems with stochastic jump parameters. *IEEE Transactions on Automatic Control*, **30**(9), 898–900.

[144] Moheimani, S.O.R. and Petersen, I.R. (1996). Quadratic guaranteed cost control with robust pole placement in a disk. *IEE Proceedings D, Control Theory and Applications*, **143**(1), 37–43.

[145] Moheimani, S.O.R. and Petersen, I.R. (1997). Optimal quadratic guaranteed cost control of a class of uncertain time-delay systems. *IEE Proceedings D, Control Theory and Applications*, **144**(2), 183–188.

[146] Moheimani, S.O.R., Savkin, A., and Petersen, I.R. (2000). Synthesis of minimax optimal controllers for uncertain time-delay systems with structured uncertainty. *International Journal of Systems Science*, **31**(2), 137–147.

[147] Mori, T. et al. (1982). On an estimate of the decay rate for stable linear delay systems. *International Journal of Control*, **36**(1), 95–97.

[148] Mori, T. and Kokame, H. (1989). Stability of $\dot{x}(t) = Ax(t) + Bx(t - \tau)$. *IEEE Transactions on Automatic Control*, **34**(4).

[149] Mori, T., Noldus, E., and Kuwahara, M. (1983). A way to stabilize linear systems with delayed state. *Automatica*, **19**(5), 571–573.

[150] Morozan, T. (1983). Optimal stationary control for dynamic systems with Markov perturbations. *Stochastic Analysis and Applications*, **1**(3), 299–325.

[151] Morozan, T. (1995). Stability and control for linear systems with jump Markov perturbations. *Stochastic Analysis and Applications*, **13**(1), 91–110.

[152] Morozan, T. (1999). Parametrized Riccati equations and input-output operators for time-varying stochastic differential equations with state dependent noise. *Studii si Cercetari Matematice*, (1).

[153] Naimark, L., Kogan, J., and Zeheb, E. (2000). Stabilizability consideration and design of rational controllers for a class of time-delay systems. *Automatica*, **36**, 475–480.

[154] Niculescu, S.I. (1998). H_∞ memoryless control with an α-stability constraint for time-delay systems: an LMI approach. *IEEE Transactions on Automatic Control*, **43**(5), 739–743.

[155] Niculescu, S.I., de Souza C.E., Dion, J.M., and Dugard, L. (1994). Robust stability and stabilization of uncertain linear systems with state delay: single delay case. *Proceedings of the IFAC Symposium on Robust Control Design*, Rio de Janeiro.

[156] Niculescu, S.I., de Souza, C.E., Dugard, L., and Dugard, J.M. (1994). Robust stability and stabilization of uncertain linear systems with state delay:

single delay case (1). *Proceedings of the IFAC Symposium on Robust Control Design*, Rio De Janeiro, pages 469–474.

[157] Niculescu, S.I., Dion, J.M., and Dugard, L. (1996). Robust stabilization for uncertain time-delay systems containing saturating actuators. *IEEE Transactions on Automatic Control*, **41**(5), 742–747.

[158] Niculescu, S.I., Verriest, E.I., Dugard, L., and Dugard, J.M. (1997). Stability of linear systems with delayed state: a guide tour. In *Stability and Control of Time-Delay Systems*, L. Dugard and E. I Verriest, Eds, Lecture Notes in Control and Information Sciences, **228**, 1–71, Springer-Verlag, London,

[159] O'Connor, D.A. and Tarn, T.J. (1983). On stabilization by state feedback for neutral differential difference equations. *IEEE Transactions on Automatic Control*, **28**, 615–618.

[160] Oucheriah, S. (2000). Exponential stabilization of a class of uncertain time-delay systems with bounded controllers. *IEEE Transactions on Automatic Control*, **47**(4), 606–609.

[161] Packard, A. (1994). Gain scheduling via linear fractional transformations. *Systems and Control Letters*, **22**, 79–92.

[162] Packard, A., Zhou, K., Pandey, P., and Becker, G. (1991). A collection of robust control problems leading to LMIs. In *Proceedings of IEEE Conference on Decision and Control*, pages 1245–1250.

[163] Packard, A., Zhou, K., Pandey, P., and Leonhardson, J. (1992). Optimal constant input/output similarity scaling for full-information and state-feedback control problem. *Systems and Control Letters*, pages 271–280.

[164] Palhares, R.M., de Souza, C.E., and Peres, P.L.D. (1999). Robust H_∞ filter design for uncertain discrete-time state-delayed systems: an LMI approach. *Proceedings of the 38th IEEE Conference on Decision and Control*, Phoenix.

[165] Palhares, R.M., Peres, P.L.D., and de Souza, C.E. (1999). Delay-dependent robust H_∞-filtering for convex polytopic uncertain systems via LMI. In *Proceedings of the 38th IEEE Conference on Decision and Control*, Phoenix.

[166] Pila, A.W., Shaked, U., and de Souza, C.E. (1999). H_∞ filtering for continuous-time linear systems with delay. *IEEE Transactions on Automatic Control*, **44**(7).

[167] Pila, A.W., Shaked, U., and de Souza, C.E. (1996). H_∞ filtering for continuous-time linear systems with delay. In *Proceedings of 13th IFAC World Congress*, San Francisco, USA, pages 49–54.

[168] Rami, M.A. and El Ghaoui, L. (1996). LMI optimization for nonstandard Riccati equations arising in stochastic control. *IEEE Transactions on Automatic Control*, **41**(11), 1666–1671.

[169] Rami, M.A. and El Ghaoui, L. (1996). Robust state-feedback stabilization of jump linear systems via LMIs. *International Journal of Robust and Nonlinear Control*, (6), 1015–1022.

[170] Ramos, J.L. and Pearson, A.E. (2000). Output feedback stabilizing controller for time-delay systems. *Automatica*, **36**, 613–617.

[171] Razimukhin, B.S. (1960). Application of Lyapunov method to problems in stability of systems with dealay. *Automatica i Telemechanica*, **21**, 740–749.

[172] Richard, J.P. (1998). Some trends and tools for the study of time-delay systems. In *Proceedings of the IMACS-IEEE Multiconference, CESA'98*, pages 27–43, Hammamet, Tunisia.

[173] Rosenblueth, J.F. (2001). Certain classes of properly relaxed delayed controls. *IMA Journal of Mathematical Control and Information*, **18**, 31–52.

[174] Shaked, U., Yaesh, I., and de Souza, C.E. (1998). Bounded real criteria for linear time-delay systems. *IEEE Transactions on Automatic Control*, **43**(7), 1016–1022.

[175] Shen, J., Chen, B., and Kung, F. (1991). Memoryless stabilization of uncertain dynamic delay systems, Riccati equation approach. *IEEE Transactions on Automatic Control*, **36**, 638–640.

[176] Shi, P. (1996). Robust filtering for uncertain systems with sampled measurements. *International Journal of Systems Science*, **27**(12), 1403–1415.

[177] Shi, P. (1998). Filtering on sampled-data systems with parametric uncertainty. *IEEE Transactions on Automatic Control*, **43**(7), 1022–1027.

[178] Shi, P. and Agarwal, R.K. (2001). Robust disturbance attenuation of fluid dynamics systems with norm-bounded uncertainties. *IMA Journal of Mathematical Control and Information*, **18**, 73–82.

[179] Shi, P., Agarwal, R.K., Boukas, E.K., and Shue, S.P. (2000). Robust H_∞ state feedback control of discrete time-delay linear systems with norm bounded uncertainty. *International Journal of Systems Science*, **31**(4), 409–415.

[180] Shi, P. and Boukas, E.K. (1997). H_∞ control for Markovian jumping linear systems with parametric uncertainty. *Journal of Optimization Theory and Applications*, **95**(1), 75–99.

[181] Shi, P., Boukas, E.K., and Agarwal, R.K., (1999). Control of Markovian jump discrete-time systems with norm bounded uncertainty and unknown delays. *IEEE Transactions on Automatic Control*, **44**(11), 2139–2144.

[182] Shi, P., Boukas, E.K., and Agarwal, R.K. (1999). Kalman filtering for continuous-time uncertain systems with Markovian jumping parameters. *IEEE Transactions on Automatic Control*, **44**(8), 1592–1597.

[183] Shi, P., Boukas, E.K., and Agarwal, R.K. (1999). Robust control for Markovian jumping discrete-time systems. *International Journal of Systems Science*, **30**(8), 787–797.

[184] Shyu, K.K. and Chien, Y.C. (1995). Robust tracking and model following for uncertain time-delay systems. *International Journal of Systems Science*, **62**(3), 589–600.

[185] Shyu, K.K. and Yan, J.J. (1993). Robust stability of uncertain time-delay systems and its stabilization by variable structured control. *International Journal of Control*, **57**(1), 237–246.

[186] Slemrod, M. and Infante, E.F. (1972). Asymptotic stability criteria for linear systems of differential equations of neutral type and their discrete analogies. *Journal of Mathematical Analysis and Applications*, **38**, 399–415.

[187] Su, H. and Chu, J. (1999). Robust H_∞ control for linear time-varying uncertain time-delay systems via dynamic output feedback. *International Journal of Systems Science*, **30**(10), 1093–1107.

[188] Su, H. and Chu, J. (1999). Stabilization of a class of uncertain time-delay systems containing saturating actuators. *International Journal of Systems Science*, **30**(11), 1193–1203.

[189] Su, H. (1994). Further results on the robust stability of linear systems with a single time delay. *Systems and Control Letters*, **23**, 375–379.

[190] Su, T.J. and Huang, C.G. (1992). Robust stability of delay dependence for linear uncertain systems. *IEEE Transactions on Automatic Control*, **37**, 1656–1659.

[191] Sworder, D.D. (1968). On the stochastic maximum principle. *Journal of Mathematical Analysis and Applications*, **24**, 627–640.

[192] Sworder, D.D. (1969). Feedback control of a class of linear systems with jump parameters. *IEEE Transactions on Automatic Control*, **14**(1), 9–14.

[193] Sworder, D.D. (1972). Control of jump parameter systems with discontinuous state trajectories. *IEEE Transactions on Automatic Control*, **20**(10), 740–741.

[194] Sworder, D.D. and Chou, D.S. (1985). A survey of some design methods for random parameter systems. In *Proceedings of the 24th IEEE Conference on Decision and Control*, pages 894–898, Ft. Lauderdale, Florida.

[195] Sworder, D.D. and Rogers, R.O. (1983). An LQ-solution to a control problem associated with a solar thermal central receiver. *IEEE Transactions on Automatic Control*, **28**(8), 971–978.

[196] Szita, G. and Sanathan, S. (1997). Robust design for disturbance rejection in time-delay systems. *Journal of Franklin Institute*, **334B**(4), 611–629.

[197] Tadmor, G. (1995). H_∞ control in systems with a single input delay. In *Proceedings of ACC, Seattle*, Washington, pages 321–325.

[198] Tarbouriech, S. and da Silva, J.M.G. (2000). Synthesis of controllers for continuous-time delay systems with saturating controls via LMIs. *IEEE Transactions on Automatic Control*, **45**(1), 105–111.

[199] Trinh, H. (1999). An observer design procedure for time-delay systems. *Control and Intelligent Systems*, **27**(1), 6–10.

[200] Trinh, H. (1999). Linear functional state observer for time-delay systems. *International Journal of Control*, **72**(18), 1642–1658.

[201] Trinh, H. and Aldeen, M. (1994). On the stability of linear systems with delayed perturbations. *IEEE Transactions on Automatic Control*, **39**, 1948–1951.

[202] Trinh, H. and Aldeen, M. (1996). Output tracking for linear uncertain time-delay systems. In *IEE Proceedings D, Control Theory and Applications*, **143**(6).

[203] Trinh, H., Aldeen, M., and Fernando, T. (1999). An observer for interconnected discrete time-delay systems. *Control and Intelligent Systems*, **27**(2), 70–78.

[204] Tsay, J.T., Liu, P.L., and Su, T.J. (1996). Robust stability for perturbed large-scale time-delay systems. In *IEE Proceedings D, Control Theory and Applications*, **143**(3), 233–236.

[205] Verriest, E.I. (1996). Riccati type conditions for robust stability of delay systems. In *Proceedings of MTNS*, St. Louis, Missouri.

[206] Verriest, E.I., Fan, K.H., and Kullstam, J. (1993). Frequency domain robust stability criteria for linear delay systems. In *Proceedings of the 32nd IEEE Conference on Decision and Control*, San Antonio, Texas.

[207] Wang, S.S., Chen, B.S., and Lin, T.P. (1987). Robust stability of uncertain time-delay system. *International Journal of Control*, **46**, 963–976.

[208] Wang, W.J. and Wang, R.J. (1995). New stability criteria for linear time-delay systems. *Control Theory and Advanced Technology*, **10**(2), 1213–1222.

[209] Wang, Y., Xie, L., and de Souza, C.E. (1992). Robust control of a class of uncertain systems. *Systems and Control Letters*, **19**, 139–149.

[210] Won, S. and Park, J.H. (1999). Design of observer-based controller for perturbed time-delay systems. *JSME International Journal*, **42**(1), 129–132.

[211] Wonham, W.M. (1971). Random differential equations in control theory. In *Probabilistic Methods in Applied Mathematics*, volume 2, pages 131–213, A.T. Bharucha-Reid, Ed., Academic Press, New York.

[212] Wu, H. and Mizukami, K. (1995). Robust stability criteria for dynamical systems including delayed perturbations. *IEEE Transactions on Automatic Control*, **40**, 487–490.

[213] Xiaohong, N. (2000). Robust stability for a type of uncertain time-delay systems. *Applied Mathematics and Mechanics*, **21**(4), 479–484.

[214] Xie, L. (1996). Output feedback H_∞ control of systems with parameter uncertainty. *International Journal of Control*, **63**(4), 741–750.

[215] Xie, L. and de Souza, C.E. (1993). Robust stabilization and disturbance attenuation for uncertain delay systems. In *Proceedings of the 193 European Control Conference*, Groningen, The Netherlands.

[216] Xu, B. and Lam, J. (1999). Decentralized stabilization of large scale interconnected time-delay systems. *Journal of Optimization Theory and Applications*, **103**(1), 231–240.

[217] Xue, X. and Qiu, D. (2000). Robust H_∞ compensation design for time-delay systems with norm-bounded time-varying uncertainties. *IEEE Transactions on Automatic Control*, **45**(5), 606–609.

[218] Yao, Y.X., Zhang, Y.M., and Kovacevic, R. (1996). Parameterization of observers for time-delay systems and its application in observer design. In *IEE Proceedings D, Control Theory and Applications*, **143**(3), 225–232, 1996.

[219] Yao, Y.X., Zhang, Y.M., and Kovacevic, R. (1997). Functional observer and state feedback for input time-delay systems. *International Journal of Control*, **66**(4), 603–617.

[220] Yu, L. and Chu, J. (1999). An LMI approach to guaranteed cost control of linear uncertain time-delay systems. *Automatica*, **35**, 1139–1159.

[221] Yu, L., Chu, J., and Su, H. (1996). Robust memoryless H_∞ controller design for linear time-delay systems with norm-bounded time-varying uncertainty. *Automatica*, **32**(12), 1759–1762.

Index